# About Island Press

Island Press is the only nonprofit organization in the United States whose principal purpose is the publication of books on environmental issues and natural resource management. We provide solutions-oriented information to professionals, public officials, business and community leaders, and concerned citizens who are shaping responses to environmental problems.

In 2004, Island Press celebrates its twentieth anniversary as the leading provider of timely and practical books that take a multidisciplinary approach to critical environmental concerns. Our growing list of titles reflects our commitment to bringing the best of an expanding body of literature to the environmental community throughout North America and the world.

Support for Island Press is provided by the Agua Fund, Brainerd Foundation, Geraldine R. Dodge Foundation, Doris Duke Charitable Foundation, Educational Foundation of America, the Ford Foundation, the George Gund Foundation, the William and Flora Hewlett Foundation, Henry Luce Foundation, the John D. and Catherine T. MacArthur Foundation, the Andrew W. Mellon Foundation, the Curtis and Edith Munson Foundation, National Environmental Trust, National Fish and Wildlife Foundation, the New-Land Foundation, Oak Foundation, the Overbrook Foundation, the David and Lucile Packard Foundation, the Pew Charitable Trusts, the Rockefeller Foundation, the Winslow Foundation, and other generous donors.

The opinions expressed in this book are those of the author(s) and do not necessarily reflect the views of these foundations.

# About SCOPE

The Scientific Committee on Problems of the Environment (SCOPE) was established by the International Council for Science (ICSU) in 1969. It brings together natural and social scientists to identify emerging or potential environmental issues and to address jointly the nature and solution of environmental problems on a global basis. Operating at an interface between the science and decision-making sectors, SCOPE's interdisciplinary and critical focus on available knowledge provides analytical and practical tools to promote further research and more sustainable management of the Earth's resources. SCOPE's members, forty national science academies and research councils and twenty-two international scientific unions, committees, and societies, guide and develop its scientific program.

# The Global Carbon Cycle

THE SCIENTIFIC COMMITTEE ON PROBLEMS OF THE ENVIRONMENT (SCOPE)

## SCOPE Series

SCOPE 1–59 in the series were published by John Wiley & Sons, Ltd., U.K. Island Press is the publisher for SCOPE 60 as well as subsequent titles in the series.

SCOPE 60: *Resilience and the Behavior of Large-Scale Systems,* edited by Lance H. Gunderson and Lowell Pritchard Jr.

SCOPE 61: *Interactions of the Major Biogeochemical Cycles: Global Change and Human Impacts,* edited by Jerry M. Melillo, Christopher B. Field, and Bedrich Moldan

SCOPE 62: *The Global Carbon Cycle: Integrating Humans, Climate, and the Natural World,* edited by Christopher B. Field and Michael R. Raupach

 SCOPE 62

# The Global Carbon Cycle

*Integrating Humans, Climate,*
*and the Natural World*

### Edited by
### Christopher B. Field
### and Michael R. Raupach

A project of SCOPE, the Scientific Committee on
Problems of the Environment, of the
International Council for Science

## ISLAND PRESS

Washington • Covelo • London

Copyright © 2004 Scientific Committee on Problems of the Environment (SCOPE)

All rights reserved under International and Pan-American Copyright Conventions. No part of this book may be reproduced in any form or by any means without permission in writing from the publisher: Island Press, 1718 Connecticut Ave., NW, Suite 300, Washington, DC 20009

Permission requests to reproduce portions of this book should be addressed to SCOPE (Scientific Committee on Problems of the Environment, 51 Boulevard de Montmorency, 75016 Paris, France).

Inquiries regarding licensing publication rights to this book as a whole should be addressed to Island Press (1718 Connecticut Avenue, NW, Suite 300, Washington, DC 20009, USA).

ISLAND PRESS is a trademark of The Center for Resource Economics.

*Library of Congress Cataloging-in-Publication Data*

The global carbon cycle : integrating humans, climate, and the natural world / edited by Christopher B. Field and Michael R. Raupach.

    p. cm. — (SCOPE series)

Includes bibliographical references and index.

ISBN 1-55963-526-6 (cloth : alk. paper) -- ISBN 1-55963-527-4 (pbk. : alk. paper)   1. Atmospheric carbon dioxide.   2. Carbon cycle (Biogeochemistry)   3. Greenhouse gas mitigation.   4. Nature — Effect of human beings on.   I. Field, Christopher B.   II. Raupach, M. R. III.  SCOPE report ; 62

QC879.8.G55 2004

363.738'746 — dc22                                                        2003024792

*British Cataloguing-in-Publication data available.*

Printed on recycled, acid-free paper ♻

Manufactured in the United States of America
10 9 8 7 6 5 4 3 2 1

# Contents

## Part I: Crosscutting Issues

# Part II: Overview of the Carbon Cycle

# Part III: The Carbon Cycle of the Oceans

# Part IV: The Carbon Cycle of the Land

# Part V: The Carbon Cycle of Land-Ocean Margins

# Part VI: Humans and the Carbon Cycle

# Part VII: Purposeful Carbon Management

# List of Colorplates, Figures, Tables, Boxes, and Appendixes

## Colorplates

## Figures

## Tables

## Boxes

## Appendixes

# Foreword

The Scientific Committee on Problems of the Environment (SCOPE) publishes this book as the second in a series of rapid assessments of the important biogeochemical cycles that are essential to life on this planet. SCOPE's aim is to make sure that experts meet on a regular basis to discuss and summarize recent advances within disciplines and evaluate their possible significance in understanding environmental problems and potential solutions. The SCOPE rapid assessment series attempts to ensure that the information so generated is published and made available within a year from the date of the synthesis. These assessments provide timely, definitive syntheses of important issues for scientists, students, and policy makers.

The present volume is intended to be a successor to SCOPE carbon books of the 1970s and 1980s and to complement recent Intergovernmental Panel on Climate Change reports on the scientific basis of climate change, the impacts of climate change, and the potential for mitigation of climate change. This volume's main concept is that the carbon cycle, climate, and humans work together as a single system. This type of system-level approach focuses the science on a number of issues that are almost certain to be important in the future. It should provide a timely examination of the practical consequences of this knowledge and facilitate its application in enhancing the sustainability of ecosystems affected by humans.

This synthesis volume is a joint project of two bodies sponsored by the International Council of Science (ICSU): SCOPE and the Global Carbon Project (GCP). SCOPE is one of twenty-six interdisciplinary bodies established by the ICSU to address cross-disciplinary issues. In response to emerging environmental concerns, the ICSU established SCOPE in 1969 in recognition that many of these concerns required scientific input spanning several disciplines represented within its membership. Representatives of forty member countries and twenty-two international, disciplinary-specific unions, scientific committees, and associates currently participate in the work of SCOPE, which directs particular attention to developing countries. The mandate of SCOPE is to assemble, review, and synthesize the information available on environmental changes attributable to human activity and the effects of these changes on humans; to assess and evaluate methodologies for measuring environmental parameters; to provide an intelligence service on current research; and to provide informed advice to agencies engaged in studies of the environment.

The recently formed Global Carbon Project is a shared partnership between the International Geosphere-Biosphere Programme (IGBP), the International Human Dimensions Programme on Global Environmental Change (IHDP), and the World Climate Research Programme (WCRP). The attention of the scientific community, policy makers, and the general public increasingly focuses on the rising concentration of greenhouse gases, especially carbon dioxide ($CO_2$), in the atmosphere and on the carbon cycle in general. Initial attempts, through the United Nations Framework Convention on Climate Change and its Kyoto Protocol, are underway to slow the rate of increase of greenhouse gases in the atmosphere. These societal actions require a scientific understanding of the carbon cycle and are placing increasing demands on the international science community to establish a common, mutually agreed knowledge base to support policy debate and action. The Global Carbon Project aims to meet this challenge by developing a complete picture of the global carbon cycle, including both its biophysical and human dimensions together with the interactions and feedbacks between them.

John W. B. Stewart, Editor-in-Chief

SCOPE Secretariat
51 Boulevard de Montmorency, 75016 Paris, France
Executive Director: Véronique Plocq-Fichelet

# Acknowledgments

The very able Scientific Steering Committee for the rapid assessment project (RAP) on the carbon cycle included Niki Gruber, Jingyun Fan, Inez Fung, Jerry Melillo, Rich Richels, Chris Sabine, Riccardo Valentini, and Reynaldo Victoria. Reynaldo was a superb local host, spreading the resources of Brazil before the meeting participants. The laboratory Reynaldo directs, the Centro de Energia Nuclear na Agricultura (CENA), provided a wealth of institutional support, including able assistance from Daniel Victoria, Vicki Ballaster, and Alex Krusche. Technical support from the Inter-American Institute (IAI) was invaluable, with Luis Marcelo Achite helping to keep the meeting on track. Without Luiz Martinelli's contributions as beach patrol agent, the meeting's sessions would have never reached a quorum. Jan Brown was a dedicated and efficient volunteer editor. The SCOPE editorial team, John Stewart (editor in chief) and Susan Greenwood Etienne, crafted a smooth transition from meeting mode to book mode. The Island Press editorial team, including Todd Baldwin, James Nuzum, David Peattie, and Heidi Fritschel, was efficient and professional. Véronique Plocq-Fichelet, the executive director of SCOPE, and Pep Canadell, the project officer for the Global Carbon Project, both opened many doors by making this RAP a priority.

The RAP for the carbon cycle was supported by funds from the A. W. Mellon Foundation, the National Science Foundation (USA), the National Aeronautics and Space Administration (USA), the National Oceanic and Atmospheric Administration (USA), the National Institute for Environmental Studies (Japan), and the European Union. Additional travel funds came from the Electric Power Research Institute. Thanks to Yoshiki Yamagata for help with funding from the National Institute for Environmental Studies and to Riccardo Valentini for help with funding from the European Union.

Critical support for the RAP concept came from the International Council for Science (ICSU), the United Nations Educational, Scientific, and Cultural Organization (UNESCO), and the U.S. National Science Foundation (NSF). Jerry Melillo developed the RAP concept and assembled the resources to get it started.

The city of Ubatuba, Brazil, provided a stimulating and enjoyable venue for developing the ideas discussed in this volume. The Carnegie Institution of Washington and the Commonwealth Scientific and Industrial Research Organization (CSIRO) generously provided the time to let us steer this project.

Finally, it is a privilege to work with and share the excitement of scientific discovery with the community of carbon cycle researchers. Their dedication and insight help lay the foundations for a sustainable future.

Chris Field                         Mike Raupach
Stanford, California, USA            Canberra, Australia

October 2003

# 1

# The Global Carbon Cycle: Integrating Humans, Climate, and the Natural World

Christopher B. Field, Michael R. Raupach, and Reynaldo Victoria

## The Carbon-Climate-Human System

It has been more than a century since Arrhenius (1896) first concluded that continued emissions of carbon dioxide from the combustion of fossil fuels could lead to a warmer climate. In the succeeding decades, Arrhenius's calculations have proved both eerily prescient and woefully incomplete. His fundamental conclusion, linking fossil-fuel combustion, the radiation balance of the Earth system, and global climate, has been solidly confirmed. Both sophisticated climate models (Cubasch et al. 2001) and studies of past climates (Joos and Prentice, Chapter 7, this volume) document the link between atmospheric $CO_2$ and global climate. The basic understanding of this link has led to a massive investment in detailed knowledge, as well as to political action. The 1992 United Nations Framework Convention on Climate Change is a remarkable accomplishment, signifying international recognition of the vulnerability of global climate to human actions (Sanz et al., Chapter 24, this volume).

Since Arrhenius's early discussion of climate change, scientific understanding of the topic has advanced on many fronts. The workings of the climate system, while still uncertain in many respects, are well enough known that general circulation models accurately reproduce many aspects of past and present climate (McAvaney et al. 2001). Greenhouse gas (GHG) emissions by humans are known with reasonable accuracy (Andres et al. 1996), including human contributions to emissions of greenhouse gases other than $CO_2$ (Prinn, Chapter 9, this volume). In addition, a large body of literature characterizes land and ocean processes that release or sequester greenhouse gases in the

1

context of changing climate, atmospheric composition, and human activities. Much of the pioneering work on land and ocean aspects of the carbon cycle was collected in or inspired by three volumes edited by Bert Bolin and colleagues and published by SCOPE (Scientific Committee on Problems of the Environment) in 1979 (Bolin et al. 1979), 1981 (Bolin 1981), and 1989 (Bolin et al. 1989).

The Intergovernmental Panel on Climate Change (IPCC), established by the United Nations as a vehicle for synthesizing scientific information on climate change, has released a number of comprehensive assessments, including recent reports on the scientific basis of climate change (Houghton et al. 2001), impacts of climate change (McCarthy et al. 2001), and potential for mitigating climate change (Metz et al. 2001). These assessments, which reflect input from more than 1,000 scientists, summarize the scientific literature with balance and precision. The disciplinary sweep and broad participation of the IPCC efforts are great strengths.

This volume is intended as a complement to the IPCC reports and as a successor to the SCOPE carbon-cycle books of the 1970s and 1980s. It extends the work of the IPCC in three main ways. First, it provides an update on key scientific discoveries in the past few years. Second, it takes a comprehensive approach to the carbon cycle, treating background and interactions with substantial detail. Managed aspects of the carbon cycle (and aspects subject to potential future management) are discussed within the same framework as the historical and current carbon cycle on the land, in the oceans, and in the atmosphere. Third, this volume makes a real effort at synthesis, not only summarizing disciplinary perspectives, but also characterizing key interactions and uncertainties between and at the frontiers of traditional disciplines.

This volume's centerpiece is the concept that the carbon cycle, climate, and humans work together as a single system (Figure 1.1). This systems-level approach focuses the science on a number of issues that are almost certain to be important in the future and that, in many cases, have not been studied in detail. Some of these issues concern the driving forces of climate change and the ways that carbon-climate-human interactions modulate the sensitivity of climate to greenhouse gas emissions. Others concern opportunities for and constraints on managing greenhouse gas emissions and the carbon cycle.

The volume is a result of a rapid assessment project (RAP) orchestrated by SCOPE (http://www.icsu-scope.org) and the Global Carbon Project (GCP, http://www.global carbonproject.org). Both are projects of the International Council for Science (ICSU, http://www.icsu.org), the umbrella organization for the world's professional scientific societies. The GCP has additional sponsorship from the World Meteorological Organization (http://www.wmo.ch) and the Intergovernmental Oceanographic Commission (http://ioc.unesco.org/iocweb/). The RAP process assembles a group of leading scientists and challenges them to extend the frontiers of knowledge. The process includes mutual education through a series of background papers and an intensive effort to develop cross-disciplinary perspectives in a series of collectively written synthesis papers. To provide timely synthesis on rapidly changing issues, the timeline is aggres-

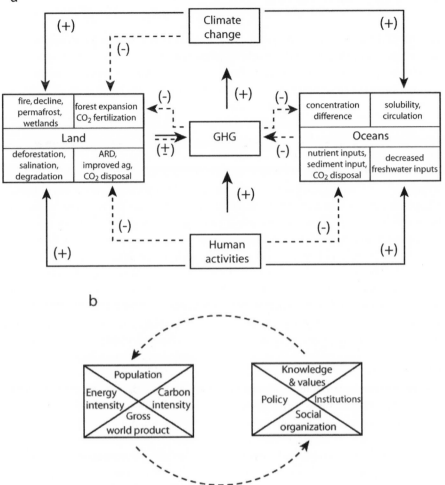

**Figure 1.1.** (a) Schematic representation of the components of the coupled carbon-climate-human system and the links among them. Solid lines and (+) indicate positive feedbacks, feedbacks that tend to release carbon to the atmosphere and amplify climate change. Dashed lines and (−) indicate negative feedbacks, feedbacks that tend to sequester carbon and suppress climate change. GHG, in the center box, is greenhouse gases. ARD, in the lower right of the land box, is afforestation, reforestation, deforestation, the suite of forestry activities identified as relevant to carbon credits in the Kyoto Protocol. Over the next century, the oceans will continue to operate as a net carbon sink, but the land (in the absence of fossil emissions) may be either a source or sink. (b) Two complementary perspectives on human drivers of carbon emissions. In the Kaya identity widely used for economic analysis (left), emissions are seen as a product of four factors: population, per capita gross world product, the energy intensity of the gross world product, and the carbon intensity of energy production. From a political science perspective (right), the drivers emerge from interactions among policy, institutions, social organization, and knowledge and values.

sive. All of the authors worked with the editors and the publisher to produce a finished book within nine months of the synthesis meeting.

The book is organized into seven parts. Part I contains the crosscutting chapters, which address the current status of the carbon cycle (Sabine et al., Chapter 2), the future carbon cycle of the oceans and land (Gruber et al., Chapter 3), possible trajectories of carbon emissions from human actions (Edmonds et al., Chapter 4), approaches to reducing emissions or sequestering additional carbon (Caldeira et al., Chapter 5), and the integration of carbon management in the broader framework of human and Earth-system activities (Raupach et al., Chapter 6). Part II surveys the carbon cycle, including historical patterns (Joos and Prentice, Chapter 7), recent spatial and temporal patterns (Heimann et al., Chapter 8), greenhouse gases other than $CO_2$ (Prinn, Chapter 9), two-way interactions between the climate and the carbon cycle (Friedlingstein, Chapter 10), and the socioeconomic trends that drive carbon emissions (Nakicenovic, Chapter 11). Parts III through VII provide background and a summary of recent findings on the carbon cycle of the oceans (Le Quéré and Metzl, Chapter 12; Greenblatt and Sarmiento, Chapter 13), the land (Foley and Ramankutty, Chapter 14; Baldocchi and Valentini, Chapter 15; Nabuurs, Chapter 16), land-ocean margins (Richey, Chapter 17; Chen, Chapter 18), humans and the carbon cycle (Romero Lankao, Chapter 19; Lebel, Chapter 20; Tschirley and Servin, Chapter 21), and purposeful carbon management (Sathaye, Chapter 22; Edmonds, Chapter 23; Sanz et al., Chapter 24; Manne and Richels, Chapter 25; Bakker, Chapter 26; Brewer, Chapter 27; Smith, Chapter 28; and Robertson, Chapter 29).

The key messages from this assessment focus on five main themes that cut across all aspects of the carbon-climate-human system. The overarching theme of the book is that all parts of the carbon cycle are interrelated. Understanding will not be complete, and management will not be successful, without a framework that considers the full set of feedbacks, a set that almost always transcends both human actions and unmanaged systems. This systems perspective presents many challenges, because the interactions among very different components of the carbon cycle tend to be poorly recognized and understood. Still, the field must address these challenges. To do that, we must start with four specific themes that link the ideas discussed throughout the book. These four themes are (1) inertia and the consequence of entrained processes in the carbon, climate, and human systems, (2) unaccounted-for vulnerabilities, especially the prospects for large releases of carbon in a warming climate, (3) gaps between reasonable expectations for future approaches to managing carbon and the requirements for stabilizing atmospheric $CO_2$, and (4) the need for a common framework for assessing natural and managed aspects of the carbon cycle. Each of these themes is previewed here and discussed extensively in the following chapters.

## Inertia

Many aspects of the carbon-climate-human system change slowly, with a strong tendency to remain on established trajectories. As a consequence, serious problems may be

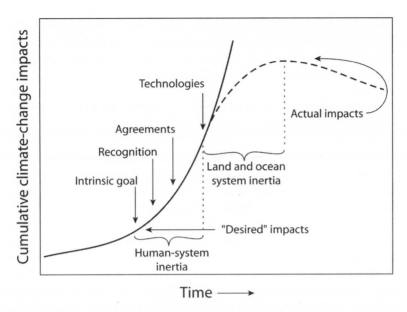

**Figure 1.2.** Effects of inertia in the coupled carbon-climate-human system. If there are delays associated with (1) assembling the evidence that climate has moved outside an acceptable envelope, (2) negotiating agreements on strategy and participation, and (3) developing new technologies to accomplish the strategies, then there will be additional delays associated with internal dynamics of the land and ocean system. As a consequence, the actual climate change may be far greater than that originally identified as acceptable.

effectively entrained before they are generally recognized (Figure 1.2). Effective management may depend on early and consistent action, including actions with financial costs. The political will to support these costs will require the strongest possible evidence on the nature of the problems and the efficiency of the solutions.

The carbon-climate-human system includes processes that operate on a wide range of timescales, including many that extend over decades to centuries. The slow components have added tremendously to the challenge of quantifying human impacts on ocean carbon (Sabine et al., Chapter 2) and ocean heat content (Levitus et al. 2000). They also prevent the ocean from quickly absorbing large amounts of anthropogenic carbon (Sabine et al., Chapter 2) and underlie the very long lifetime of atmospheric $CO_2$.

Several new results highlight the critical role of inertia for the carbon cycle on land. It is increasingly clear that a substantial fraction of the current terrestrial sink, perhaps the majority, is a consequence of ecosystem recovery following past disturbances. Across much of the temperate Northern Hemisphere, changes in forestry practices, agriculture, and fire management have allowed forests to increase in biomass or area (Nabuurs, Chapter 16). Evidence that much of the recent sink on land is a result of land management has important implications for the future trajectory of the carbon cycle. Beginning with Bacastow and Keeling (1973), most estimates of future carbon sinks have assumed that recent sinks were a consequence of $CO_2$ fertilization of plant growth and that past responses could be projected into the future with a $CO_2$-sensitivity coefficient or beta factor (Friedlingstein et al. 1995). To the extent that recent sinks are caused by management rather than $CO_2$ fertilization, past estimates of future sinks from $CO_2$ fertilization are likely to be too optimistic (Gruber et al., Chapter 3). Eventual saturation in sinks from management (Schimel et al. 2001) gives them a very different trajectory from that of sinks from $CO_2$ fertilization, especially those calculated by models without nutrient limitation (Prentice 2001).

In the human system, inertia plays a number of critical roles. The dynamics of development tend to concentrate future growth in carbon emissions in countries with developing economies (Romero Lankao, Chapter 19). This historical inertia, combined with potentially limited resources for carbon-efficient energy systems (Sathaye, Chapter 22), creates pressure for massive future emissions growth. Slowly changing institutions and incentive mechanisms in all countries (Lebel, Chapter 20) tend to entrain emissions trajectories further.

Inertia is profoundly important in the energy system, especially in the slow pace for introducing new technologies. The slow pace reflects not only the long time horizon for research and development, but also the long period required to retire existing capital stocks (Caldeira et al., Chapter 5). The long time horizon for bringing technologies to maturity and retiring capital stocks is only part of the timeline for the non-emitting energy system of the future, which also depends on the development of fundamentally new technologies (Hoffert et al. 2002). The search for fundamentally new energy sources cannot, however, constitute the entire strategy for action, because the entrained damage may be unacceptably large before new technologies are ready (Figure 1.2). A diverse portfolio of energy efficiency, new technologies, and carbon sequestration offers the strongest prospects for stabilizing atmospheric $CO_2$ (Caldeira et al., Chapter 5).

## Vulnerability

A fundamental goal of the science of the carbon-climate-human system is to understand and eventually reduce the Earth's vulnerability to dangerous changes in climate. This agenda requires that we understand the mechanisms that drive climate change, develop strategies for minimizing the magnitude of the climate change that does occur, and create approaches for coping with the climate change that cannot be avoided. Successful pursuit of this agenda is simpler when the carbon-climate-human system generates negative feedbacks (that tend to suppress further climate change), and it is more complicated when the system generates positive feedbacks (Figure 1.1). Positive feedbacks are especially challenging if they occur suddenly, as threshold phenomena, or if they involve coupled responses of the atmosphere, land, oceans, and human activities.

We are entering an era when we need not—and in fact must not—view the question of vulnerability from any single perspective. The carbon-climate-human system generates climate change as an integrated system. Attempts to understand the integrated system must take an integrated perspective. Mechanistic process models, the principal tools for exploring the behavior of climate and the carbon cycle on land and in the oceans, are increasingly competent to address questions about interactions among major components of the system (Gruber et al., Chapter 3). Still, many of the key interactions are only beginning to appear in models or are not yet represented. For these interactions, we need a combination of dedicated research and other tools for taking advantage of the available knowledge. In assessing the vulnerability of the carbon cycle to the possibility of large releases in the future, we combine results from mechanistic simulations with a broad range of other kinds of information.

Several new lines of information suggest that past assessments have underestimated the vulnerability of key aspects of the carbon-climate-human system. Several of these concern climate-carbon feedbacks. Simulations with coupled climate-carbon models demonstrate a previously undocumented positive feedback between warming and the terrestrial carbon cycle, in which $CO_2$ releases that are stimulated by warming accelerate warming and further $CO_2$ releases (Friedlingstein, Chapter 10, this volume). The experiments to date are too limited to support an accurate quantification of this positive feedback, but the range of results highlights the importance of further research. The behavior of two models of comparable sophistication is so different that, with similar forcing, they differ in atmospheric $CO_2$ in 2100 by more than 200 parts per million (ppm).

The models that simulate the future carbon balance of land are still incomplete. At least three mechanisms either not yet represented or represented in the models in a rudimentary way have the potential to amplify positive feedbacks to climate warming (Gruber et al., Chapter 3). The first of these is the respiration of carbon currently locked in permanently frozen soils. General Circulation Model (GCM) simulations indicate that much of the permafrost in the Northern Hemisphere may disappear over the next century. Because these soils contain large quantities of carbon (Michaelson et al. 1996), and because much of this carbon is relatively labile once thawed, potential releases over a century could be in the range of 100 PgC (Gruber et al., Chapter 3). Wetland soils are similar, containing vast quantities of carbon, which is subject to rapid decomposition when dry and aerated. Drying can allow wildfires, such as those that released an estimated additional 0.8 to 3.7 PgC from tropical fires during the 1997–1998 El Niño (Langenfelds et al. 2002). Drying wetland soils might result in a decrease in methane emissions, along with an increase in $CO_2$ emissions, requiring a careful analysis of overall greenhouse forcing (Manne and Richels, Chapter 25). A third aspect of the terrestrial biosphere with the potential for massive carbon releases in the future is large-scale wildfire, especially in tropical and boreal forest ecosystems (Gruber et al., Chapter 3). Climate changes in both kinds of ecosystems could push large areas past a threshold where they are dry enough to support large wildfires (Nepstad et al. 2001), and a fundamental change in the fire regime could effectively eliminate large areas of forest. None of these three mechanisms is thoroughly addressed in current ecosystem or carbon-cycle models. As a consequence, it is not yet feasible to estimate either the probability of the changes or the likely carbon emissions. Still, ignoring the potential for these large releases is not responsible, and the vulnerability of the climate system to them should be explored.

Vulnerability of ecosystems used for carbon management highlights other aspects of the need for an integrated perspective on the carbon-climate-human system. Ocean fertilization and deep disposal both create altered conditions for ocean ecosystems (Bakker, Chapter 26; Brewer, Chapter 27). To date, the consequences of these alterations are poorly known. Ecosystem alteration is also an issue for terrestrial sequestration through afforestation. Especially where afforestation involves plantations of a single tree species or non-native species, it is important to assess how any extra vulnerability to loss of ecosystem services alters the overall balance of costs and benefits (Raupach et al., Chapter 6).

## The Energy Gap

Humans interact with nearly every aspect of the carbon cycle. In the past, trajectories of emissions and land use change unfolded with little or no reference to their impacts on climate. Now much of the world is ready to make carbon management a priority. The United Nations Framework Convention on Climate Change and its Kyoto Protocol establish initial steps toward stabilizing the climate (Sanz et al., Chapter 24, this volume). In the future, however, much more will need to be done, especially if $CO_2$ concentrations are to be stabilized at a concentration of 750 ppm or lower. The basic problem is that world energy demand continues to grow rapidly. With a business-as-usual strategy, global carbon emissions could exceed 20 PgC per year ($y^{-1}$) (about three times current levels) by 2050 (Nakicenovic, Chapter 11).

Many technologies present options for decreasing emissions or sequestering carbon. Unfortunately, no single technology appears to have the potential to solve the energy problem comprehensively within the next few decades (Caldeira et al., Chapter 5, this volume). Indeed, meeting world energy demands without carbon emissions may require fundamental breakthroughs in energy technology (Hoffert et al. 2002). Even with future breakthroughs, the best options for managing the future energy system are very likely to involve a portfolio of approaches, including strategies for extracting extra energy from carbon-based fuels, technologies for generating energy without carbon emissions, and approaches to increasing sequestration on the land and in the oceans (Caldeira et al., Chapter 5).

Increases in energy efficiency (measured as energy per unit of carbon emissions) typically accompany economic development, and it is reasonable to assume that efficiency increases will continue in the future (Sathaye, Chapter 22). Even with aggressive assumptions about increases in efficiency, reasonable scenarios for the future may result

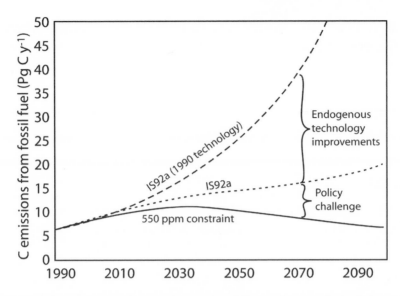

**Figure 1.3.** The energy gap, showing the growing difference between the emissions projected in a widely used scenario (IS92a) and the emissions required to stabilize atmospheric $CO_2$ at 550 ppm (with the WRE 550 scenario [Edmonds et al., Chapter 4, this volume]). This energy gap is the target for climate policy. Also shown is the emissions trajectory for IS92a in the absence of endogenous technology improvements. The very large improvements can be expected based on past experience, but they may involve many of the options that are also candidates for closing the energy gap between the emissions scenario (IS92a) and the stabilization scenario (550 ppm constraint). Redrawn from Edmonds et al., Chapter 4.

in $CO_2$ levels well above widely discussed stabilization targets (i.e., 450, 550, and 750 ppm $CO_2$). This is the case for many of the scenarios explored in the IPCC Special Report on Emission Scenarios (Nakicenovic, Chapter 11), leading to a gap between emissions consistent with reasonable advances in energy technology and those required to reach a particular stabilization target. This gap needs to be filled through active policies and could include incentives for new technologies, sequestration, or decreased energy consumption (Edmonds et al., Chapter 4).

The juxtaposition of the portfolio of future options for energy and carbon management with the gap between many economic scenarios and $CO_2$ stabilization creates a problem. A priori, it is not possible to identify a set of options available for filling the energy gap because most or even all of the available options may have already been used in the increased energy efficiency that occurs as a natural part of technological advance (Figure 1.3). Because there is no way to predict the mechanisms that will appear endogenously, there is no simple way to identify an additional set that should be the targets for policy intervention. From a carbon management perspective, the efficiency increases that occur spontaneously make some aspects of the carbon problem simpler, and they make some aspects more difficult to solve. On the one hand, if economic pressures consistently lead to efficiency increases, additional policy tools may not be necessary, at least for some of the efficiency increases. On the other hand, if the efficiency increases in the economic scenarios consume most of the options for carbon management, the costs of developing options for closing the gap may be very high (Edmonds et al., Chapter 4) or they may entail unacceptable trade-offs with other sectors (Raupach et al., Chapter 6).

## Toward a Common Framework

Some of the greatest challenges in managing the carbon-climate-human system for a sustainable future involve establishing appropriate criteria for comparing options. Ultimately, we need a framework where any option can be explored in terms of its implications for the climate system, its implications for energy, and its other impacts on ecosystems and humans (Raupach et al., Chapter 6). Many of the challenges involve processes that operate on different timescales. The sensitivity to time frame of the relative value of mitigating $CO_2$ and $CH_4$ emissions illustrates the problem. On a timescale of a few years, decreasing $CH_4$ emissions has a large impact on climate, but this impact decreases over decades as a consequence of the relatively short atmospheric life of $CH_4$ (Manne and Richels, Chapter 25). Carbon management through reforestation and afforestation potentially yields benefits over many decades, but these benefits disappear or reverse when forests stop growing, are harvested, or are disturbed. A decision about using a plot for a forest plantation versus a photovoltaic array needs to be based on a common framework for assessing the options, a framework that includes not only time frames, but also ancillary costs and benefits (Edmonds, Chapter 23).

All of the decisions that underlie the transition to a sustainable energy future require placing the decision in a larger context (Raupach et al., Chapter 6). Institutions, culture, economic resources, and perspectives on intergenerational equity all shape opportunities for and constraints on managing the carbon cycle.

## Meeting Future Challenges

Each of the themes that emerged from the RAP on the carbon cycle tends to make the climate problem more difficult to solve. The role of land management in current sinks suggests that future sinks from $CO_2$ fertilization will be smaller than past estimates. Inertia in the human system extends the timeline for developing and implementing solutions. Land ecosystems appear to be vulnerable to large releases of carbon, including releases from several mechanisms that have been absent from or incomplete in the models used for past assessments. Strategies for increased energy efficiency, carbon sequestration, and carbon-free energy are abundant, but no single technology is likely to solve the climate problem completely in the next few decades. A portfolio approach is the best option, but many of the elements of the portfolio are implicitly present in economic scenarios that fail to meet stabilization targets. Finally, each of the strategies for increased energy efficiency, carbon sequestration, or carbon-free energy involves a series of ancillary costs and benefits. In the broad context of societal issues, the ancillary effects may dominate the discussion of implementation.

How should an appreciation of the new dimensions of the climate problem change strategies for finding and implementing solutions? The most obvious conclusion is that the problem of climate change warrants more attention and higher priority. It also warrants a broader discussion of strategies, a discussion that should move beyond land, atmosphere, oceans, technology, and economics to include serious consideration of equity, consumption, and population.

## Acknowledgments

The very able Scientific Steering Committee for the RAP on the carbon cycle included Niki Gruber, Jingyun Fan, Inez Fung, Jerry Melillo, Rich Richels, Chris Sabine, and Riccardo Valentini. Assistance from Susan Greenwood, John Stewart, Véronique Ploq-Fichelet, and Daniel Victoria made the process run smoothly. Jan Brown was a dedicated and efficient volunteer editor. The RAP for the carbon cycle was supported by funds from the A. W. Mellon Foundation, the National Science Foundation (USA), the National Aeronautics and Space Administration (USA), the National Oceanic and Atmospheric Administration (USA), the National Institute for Environmental Studies (Japan), and the European Union. The city of Ubatuba, Brazil, provided a stimulating and enjoyable venue for developing the ideas discussed in this volume.

## Literature Cited

Andres, R. J., G. Marland, I. Fung, and E. Matthews. 1996. A 1° × 1° distribution of carbon dioxide emissions from fossil fuel consumption and cement manufacture. *Global Biogeochemical Cycles* 10:419–430.

Arrhenius, S. 1896. On the influence of carbonic acid in the air upon the temperature of the ground. *The London, Edinburgh and Dublin Philosophical Magazine and Journal of Sciences* 41:237–276.

Bacastow, R., and C. D. Keeling. 1973. Atmospheric carbon dioxide and radiocarbon in the natural carbon cycle. II. Changes from A.D. 1700 to 2070 as deduced from a geochemical reservoir. Pp. 86–135 in *Carbon and the biosphere*, edited by G. M. Woodwell and E. V. Pecan. Springfield, VA: U.S. Department of Commerce.

Bolin, B., ed. 1981. *Carbon cycle modelling.* SCOPE 16. Chichester, U.K.: John Wiley and Sons for the Scientific Committee on Problems of the Environment.

Bolin, B., E. T. Degens, S. Kempe, and P. Ketner, eds. 1979. *The global carbon cycle.* SCOPE 13. Chichester, U.K.: John Wiley and Sons for the Scientific Committee on Problems of the Environment.

Bolin, B., B. R. Döös, J. Jäger, and R. A. Warrick, eds. 1989. *The greenhouse effect: Climatic change and ecosystems.* SCOPE 29. Chichester, U.K.: John Wiley and Sons for the Scientific Committee on Problems of the Environment.

Cubasch, U., G. A. Meehl, G. J. Boer, R. J. Stouffer, M. Dix, A. Noda, C. A. Senior, S. Raper, and K. S. Yap. 2001. Projections of future climate change. Pp. 525–582 in *Climate change 2001: The scientific basis (Contribution of Working Group I to the Third Assessment Report of the Intergovernmental Panel on Climate Change)*, edited by J. T. Houghton, Y. Ding, D. J. Griggs, M. Noguer, P. J. van der Linden, X. Dai, K. Maskell, and C. A. Johnson. Cambridge: Cambridge University Press.

Friedlingstein, P., K. C. Prentice, I. Y. Fung, J. G. John, and G. P. Brasseur. 1995. Carbon-biosphere-climate interactions in the last glacial maximum climate. *Journal of Geophysical Research (Atmospheres)* 100:7203–7221.

Hoffert, M. I., K. Caldeira, G. Benford, D. R. Criswell, C. Green, H. Herzog, A. K. Jain, H. S. Kheshgi, K. S. Lackner, J. S. Lewis, H. D. Lightfoot, W. Manheimer, J. C. Mankins, M. E. Mauel, L. J. Perkins, M. E. Schlesinger, T. Volk, and T. M. L. Wigley. 2002. Advanced technology paths to global climate stability: Energy for a greenhouse planet. *Science* 298:981–987.

Houghton, J. T., Y. Ding, D. J. Griggs, M. Noguer, P. J. van der Linden, X. Dai, K. Maskell, and C. A. Johnson, eds. 2001. *Climate change 2001: The scientific basis (Contribution of Working Group I to the Third Assessment Report of the Intergovernmental Panel on Climate Change).* Cambridge: Cambridge University Press.

Langenfelds, R., R. Francey, B. Pak, L. Steele, J. Lloyd, C. Trudinger, and C. Allison. 2002. Interannual growth rate variations of atmospheric $CO_2$ and its $\delta13C$, $H_2$, $CH_4$, and CO between 1992 and 1999 linked to biomass burning. Art. no. 1048. *Global Biogeochemical Cycles* 16:1048.

Levitus, S., J. I. Antonov, T. P. Boyer, and C. Stephens. 2000. Warming of the world ocean. *Science* 287:2225–2229.

McAvaney, B. J., C. Covey, S. Joussaume, V. Kattsov, A. Kitoh, W. Ogana, A. J. Pitman, A. J. Weaver, R. A. Wood, and Z.-C. Zhao. 2001. Model evaluation. Pp. 471–523 in *Climate change 2001: The scientific basis (Contribution of Working Group I to the Third Assessment Report of the Intergovernmental Panel on Climate Change)*, edited by J. T. Houghton, Y. Ding, D. J. Griggs, M. Noguer, P. J. van der Linden, X. Dai, K. Maskell, and C. A. Johnson. Cambridge: Cambridge University Press.

McCarthy, J. J., O. F. Canziani, N. A. Leary, D. J. Dokken, and K. S. White, eds. 2001. *Climate change 2001: Impacts, adaptation, and vulnerability (Contribution of Working Group II to the Third Asessment Report of the Intergovernmental Panel on Climate Change).* Cambridge: Cambridge University Press.

Metz, B., O. Davidson, R. Swart, and J. Pan, eds. 2001. *Climate change 2001: Mitigation (Contribution of Working Group II to the Third Assessment Report of the Intergovernmental Panel on Climate Change).* Cambridge: Cambridge University Press.

Michaelson, G., C. Ping, and J. Kimble. 1996. Carbon storage and distribution in tundra soils of Arctic Alaska, USA. *Arctic and Alpine Research* 28:414–424.

Nepstad, D., G. Carvalho, A. Barros, A. Alencar, J. Capobianco, J. Bishop, P. Moutinho, P. Lefebvre, U. Silva, and E. Prins. 2001. Road paving, fire regime feedbacks, and the future of Amazon forests. *Forest Ecology and Management* 154:395–407.

Prentice, I. C. 2001. The carbon cycle and atmospheric carbon dioxide. Pp. 183–237 in *Climate change 2001: The scientific basis (Contribution of Working Group I to the Third*

*Assessment Report of the Intergovernmental Panel on Climate Change),* edited by J. T. Houghton, Y. Ding, D. J. Griggs, M. Noguer, P. J. van der Linden, X. Dai, K. Maskell, and C. A. Johnson. Cambridge: Cambridge University Press.

Schimel, D. S., J. I. House, K. A. Hibbard, P. Bousquet, P. Ciais, P. Peylin, B. H. Braswell, M. J. Apps, D. Baker, A. Bondeau, J. Canadell, G. Churkina, W. Cramer, A. S. Denning, C. B. Field, P. Friedlingstein, C. Goodale, M. Heimann, R. A. Houghton, J. M. Melillo, B. M. Moore III, D. Murdiyarso, I. Noble, S. W. Pacala, I. C. Prentice, M. R. Raupach, P. J. Rayner, R. J. Scholes, W. L. Steffen, and C. Wirth. 2001. Recent patterns and mechanisms of carbon exchange by terrestrial ecosystems. *Nature* 414:169–172.

# PART I
## Crosscutting Issues

# PART I

# Crosscutting Issues

# 2

# Current Status and Past Trends of the Global Carbon Cycle

Christopher L. Sabine, Martin Heimann, Paulo Artaxo,
Dorothee C. E. Bakker, Chen-Tung Arthur Chen,
Christopher B. Field, Nicolas Gruber,
Corinne Le Quéré, Ronald G. Prinn, Jeffrey E. Richey,
Patricia Romero Lankao, Jayant A. Sathaye, and
Riccardo Valentini

In a global, long-term perspective, the record of atmospheric $CO_2$ content documents the magnitude and speed of climate-driven variations, such as the glacial-interglacial cycles (which drove $CO_2$ variations of ~100 parts per million [ppm] over 420,000 years). These observations also, however, document a remarkable stability, with variations in atmospheric $CO_2$ of <20 ppm during at least the last 11,000 years before the Industrial Era (Joos and Prentice, Chapter 7, this volume). In this longer-term context, the anthropogenic increase of ~100 ppm during the past 200 years is a dramatic alteration of the global carbon cycle. This atmospheric increase is also a graphic documentation of profound changes in human activity. The atmospheric record documents the Earth system's response to fossil-fuel releases that increased by more than 1,200 percent between 1900 and 1999 (Nakicenovic, Chapter 11, this volume).

To understand and predict future changes in the global carbon cycle, we must first understand how the system is operating today. In many cases the current fluxes of carbon are a direct result of past processes affecting these fluxes (Nabuurs, Chapter 16, this volume). Thus, it is important to understand the current carbon cycle in the context of how the system has evolved over time. The Third Assessment Report (TAR) of the Intergovernmental Panel on Climate Change (IPCC) recently compiled a global carbon budget (Prentice et al. 2001). While that budget reflected the state of the art at that time, this chapter presents a revised budget based on new information from model studies and oceanographic observations. The IPCC-TAR budget focused on the overall carbon balance between the major active reservoirs of land, atmosphere, and ocean. In this chapter we present a somewhat more comprehensive representation of the connections between the reservoirs, together with our current understanding of the biogeochemical processes and human driving forces controlling these exchanges. We also introduce the key processes involved in controlling atmospheric concentrations of $CO_2$ and relevant non-$CO_2$ gases (e.g., $CH_4$, $N_2O$) that may be susceptible to changes in the future, either through deliberate management or as direct and indirect consequences of global change.

## The Global Carbon Budget

Over the past 200 years humans have introduced ~ 400 petagrams of carbon (PgC) to the atmosphere through deforestation and the burning of fossil fuels. Part of this carbon was absorbed by the oceans and terrestrial biosphere. Table 2.1, section 1, shows the global budget recently compiled by the IPCC Third Assessment Report (Prentice et al. 2001). The global carbon budget quantifies the relative importance of these two reservoirs today, and the budget uncertainty reflects our understanding of the exchanges between these reservoirs. The IPCC assessment partitions the uptake into net terrestrial and oceanic components based primarily on observations of the concurrent global trends of atmospheric $CO_2$ and oxygen. Since the compilation of the IPCC report, new evidence from model studies and oceanographic observations shows that this budget should be slightly revised to account for a previously ignored oceanic oxygen flux. This flux is the result of enhanced oceanic mixing, as inferred from observed changes in oceanic heat content (Le Quéré et al. 2003). The revised values are presented in section 2 of Table 2.1.

Section 3 of Table 2.1 shows the breakdown of the net land-atmosphere flux through the 1990s into emissions from changes in land use and a residual terrestrial sink, based on the updated land use change emissions of Houghton (2003). This breakdown has recently been challenged based on new estimates of land use change determined from remote sensing data (Table 2.1, sections 4 and 5). These new estimates lie at the lower end of the range of estimates based on data reported by individual countries (Houghton 2003). If correct, they imply a residual terrestrial sink in the 1990s that is about 40 percent smaller than previous estimates.

This budget, of course, does not attempt to represent the richness of the global carbon cycle. The land-atmosphere-ocean system is connected by a multitude of exchange fluxes. The dynamical behavior of this system is determined by the relative sizes of the different reservoirs and fluxes, together with the biogeochemical processes and human driving forces controlling these exchanges. Colorplate 1 shows the globally aggregated layout of the carbon cycle, together with the pools and exchange fluxes that are relevant on timescales of up to a few millennia. Panel a in Colorplate 1 presents a basic picture of the global carbon cycle, including the preindustrial (thin) and anthropogenic (bold) ocean-atmosphere and land-atmosphere exchange fluxes. The anthropogenic fluxes are average values for the 1980s and 1990s. Panel a also shows components of the long-term

**Table 2.1.** The global carbon budget (PgC $y^{-1}$)

|  | *1980s* | *1990s* |
|---|---|---|
| *1. Prentice et al. 2001* | | |
| Atmospheric increase | +3.3 ± 0.1 | +3.2 ± 0.1 |
| Emissions (fossil fuel, cement) | +5.4 ± 0.3 | +6.3 ± 0.4 |
| Ocean-atmosphere flux | −1.9 ± 0.6 | −1.7 ± 0.5 |
| Net land-atmosphere flux | −0.2 ± 0.7 | −1.4 ± 0.7 |
| Land use change | +1.7 (+0.6 to +2.5) | - |
| Residual terrestrial sink | −1.9 (−3.8 to +0.3) | - |
| *2. Le Quéré et al. 2003* | | |
| Ocean corrected | −1.8 ± 0.8 | −1.9 ± 0.7 |
| Net land-atmosphere flux | −0.3 ± 0.9 | −1.2 ± 0.8 |
| *3. Houghton 2003* | | |
| Land use change | +2.0 (+0.9 to +2.8) | +2.2 (+1.4 to +3.0) |
| Residual terrestrial sink | −2.3 (−4.0 to −0.3) | −3.4 (−5.0 to −1.8) |
| *4. DeFries et al. 2002* | | |
| Land use change | +0.6 (+0.3 to +0.8) | +0.9 (+0.5 to +1.4) |
| Residual terrestrial sink | −0.9 (−3.0 to 0) | −2.1 (−3.4 to −0.9) |
| *5. Achard et al. 2002* | | |
| Land use change | | +1.0 ± 0.2 |
| Residual terrestrial sink | | −2.2 (−3.2 to −1.2) |

*Note:* Positive values represent atmospheric increase (or ocean/land sources); negative numbers represent atmospheric decrease (sinks). Residual terrestrial sink determined by difference (net land/atmosphere flux minus land use change).

geological cycle and the composite estimates of $CO_2$ emissions from geological reservoirs (i.e., fossil fuels and the production of lime for cement). Panels *b* and *c* provide more detailed pictures of the ocean and terrestrial fluxes, respectively. Individual component pools and fluxes, including key climatic and anthropogenic drivers, are discussed later in this chapter and in subsequent chapters.

Although Colorplate 1 focuses primarily on fluxes directly related to $CO_2$, a number of non-$CO_2$ trace gases also play significant roles in the global carbon cycle (e.g., CO, $CH_4$, non-methane hydrocarbons) and/or climate forcing (e.g., $CH_4$, $N_2O$, chlorofluorocarbons). The global cycles of these trace gases, which share many of the processes driving the $CO_2$ cycle, are briefly discussed later in this chapter.

## Reservoir Connections

The background chapters in this volume discuss the various carbon reservoirs and the processes relevant to controlling atmospheric $CO_2$ and related trace gas concentrations. To appreciate the Earth's carbon cycle and its evolution, it is necessary to examine the connections among the various carbon pools. This analysis must be done within a framework that provides an integrated perspective across both disciplinary and geographic boundaries, with particular emphasis on the carbon cycle as an integral part of the human-environment system.

## Fossil Fuel–Atmosphere Connections

The world energy system delivered approximately 380 exajoules (EJ [$10^{18}$ J]) of primary energy in 2002 (BP 2003). Of this, 81 percent was derived from fossil fuels, with the remainder derived from nuclear, hydroelectric, biomass, wind, solar, and geothermal energy sources (Figure 2.1). The fossil-fuel component released 5.2 PgC in 1980 and 6.3 PgC in 2002 (CDIAC 2003). Cement production is the other major industrial source of carbon, and its release increased to 0.22 PgC in 1999. The combined release of 5.9 PgC shown in Colorplate 1a represents an average emission for the 1980s and 1990s. Underground coal fires, which are poorly known and only partly industrial, may be an additional as yet unaccounted for source of carbon to the atmosphere as large as cement manufacturing (Zhang et al. 1998). In terms of energy released, the current mix of fossil fuels is approximately 44 percent oil, 28 percent coal, and 27 percent natural gas (Figure 2.1). At current rates of consumption, conventional reserves of coal, oil, and gas (those that can be economically produced with current technology; see Colorplate 1c) are sufficient to last 216, 40, and 62 years, respectively (BP 2003). If estimates of undiscovered oil and gas fields are included with the conventional reserves, oil and gas lifetimes increase to 101 and 142 years, respectively (Ahlbrandt et al. 2000).

Conventional reserves represent only a fraction of the total fossil carbon in the Earth's crust. A much larger quantity of fossil reserves is in tar and heavy oil. These reserves cannot be economically produced with existing technology but are likely to become accessible in the future. The best estimates for total fossil resources that might ultimately be recovered are in the range of 6,000 Pg (Nakicenovic, Chapter 11, this volume), or about five times the conventional reserves. In addition, vast quantities of methane, exceeding all known fossil-fuel reserves, exist in the form of methane hydrates under continental shelf sediments around the world, in the Arctic permafrost, and in various marginal seas. With current technology, however, these reserves do not appear viable as a future energy source.

Although not included in Table 2.1 or Colorplate 1, combustion of fossil fuels also releases a number of non-$CO_2$ carbon gases. In particular, carbon monoxide (CO) can be used as an effective tracer of fossil-fuel combustion in atmospheric gas measurements. The relative impact of these gases is discussed in a later section.

## Land-Atmosphere Connections

The exchange of carbon between the terrestrial biosphere and the atmosphere is a key driver of the current carbon cycle. Global net primary production (NPP) by land plants

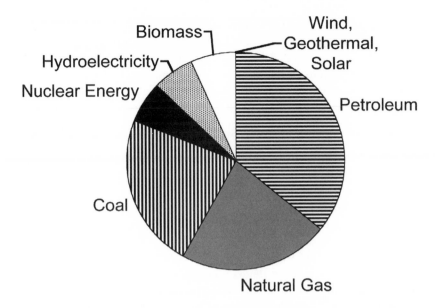

**Figure 2.1.** The distribution of sources for the world energy system in 2000. The values for all sources except biomass are from BP (2003). The value for biomass is from the U.S. Department of Energy (http://www.eere.energy.gov).

is about 57 PgC per year ($y^{-1}$) (Colorplate 1). Of this, about 4 PgC $y^{-1}$ is in crops. Humans co-opt a much larger fraction of terrestrial NPP, probably about 40 percent, where co-opting is defined as consuming, removing some products from, or altering natural states of the terrestrial biosphere through ecosystem changes (Vitousek et al. 1986). Total NPP is approximately 40 percent of gross primary production (GPP), with the remainder returned to the atmosphere through plant respiration. For many purposes, NPP is the most useful summary of terrestrial plant activity. NPP can be assessed with inventories and harvest techniques, and it represents the organic matter passed to other trophic levels (Lindemann 1942). For other purposes, including isotope studies and scaling from eddy flux, GPP is a more useful index.

Most of the annual flux in NPP is returned to the atmosphere through the respiration of heterotrophs, including microorganisms, saprophytes, and animals. A smaller fraction, 5–10 percent, is released to the atmosphere through combustion (Colorplate 1). Approximately 5 percent of NPP leaves land ecosystems in organic form, as $CH_4$ or volatile organic carbon (Prinn, Chapter 9, this volume). In recent decades the land was close to carbon neutral (in the 1980s) or was a net sink (in the 1990s) for atmospheric $CO_2$ (Table 2.1). This net flux represents a balance between substantial emissions from biomass clearing or fires and enhanced uptake as a physiological response to rising $CO_2$ or the regrowth of previously cleared areas.

Land plants contain slightly less carbon than the atmosphere; soils contain substantially more (Colorplate 1). The estimates for soil carbon in Colorplate 1 are higher than shown in previous budgets for two reasons. First, previous budgets estimated soil carbon to a depth of 1 meter (m). Jobaggy and Jackson (2000) extended these to 3 m, adding about 55 percent to the known stock. Second, previous budgets ignored carbon in wetland and permanently frozen soils. The exact magnitudes of these stocks are very uncertain. The potential for substantial carbon losses from these stocks in coming decades, however, is a strong incentive for careful quantification and further analysis.

Tropical forests contain the largest carbon pool of terrestrial biota (see Table 2.2) and also the largest NPP (Saugier et al. 2001). Averaged over several years, tropical forests traditionally have been believed to be close to carbon neutral, with uptake from NPP balanced by releases from decomposition and fire. The difference between NPP and the sum of all of the processes that release carbon from the land is net ecosystem production (NEP). Recent studies based on in situ flux measurements and forest inventory techniques indicate net carbon uptake or positive NEP of 100–700 gC per square meter $(m^{-2})$ $y^{-1}$, corresponding to ~1 PgC $y^{-1}$ across the tropical evergreen forest biome. The duration and spatial scale of these studies is not yet extensive enough for solid extrapolation. Still, the estimated uptake is in the proper range to compensate for emissions from tropical land use (Achard et al. 2002; DeFries et al. 2002; Foley and Ramankutty, Chapter 14, this volume). The tropical NEP estimates, however, are also in the range of recent estimates of $CO_2$ releases from tropical rivers, caused mainly by the decomposition of material transported from the land (Richey, Chapter 17, this volume). A full carbon accounting has yet to be accomplished, but these individual flux estimates need to be reconciled with top-down estimates from atmospheric inverse modeling studies, which show that the overall net carbon balance of the tropical land areas (30°S–30°N) must be close to zero, albeit with large uncertainty ranges (Gurney et al. 2002; Heimann et al., Chapter 8, this volume).

Temperate forests cover about 60 percent of the area of tropical forests and contain the second largest pool of plant carbon. Together, tropical and boreal forests account for approximately 75 percent of the world's plant carbon and for 40 percent of the non-wetland, nonfrozen soil carbon (Table 2.2).

Carbon emissions from land use and land management have increased dramatically over the past two centuries because of the expansion of cropland and pasture, infrastructure extension, and other effects driven by market growth, pro-deforestation policies, and demographic pressures (Geist and Lambin 2001). Before about 1950, carbon emissions from land use change were mainly from temperate regions. In recent decades, however, carbon releases from land use change have been concentrated in the tropics (Achard et al. 2002; DeFries et al. 2002; Houghton 2003). Cumulative emissions from land use, estimated at approximately 185 PgC, entail substantial fluxes from every continent except Antarctica (DeFries et al. 1999). Before about 1970, cumulative emissions from land use and land management were larger than cumulative emissions from fossil-fuel combustion.

**Table 2.2.** Plant carbon, soil carbon, and net primary production in the world's major biomes

| Ecosystem | Area $(10^6 \ km^2)$ | NPP $(PgC \ y^{-1})$ | Plant C $(PgC)$ | Soil C $(Pg \ C)$ |
|---|---|---|---|---|
| Tropical forests | 17.5 | 20.1 | 340 | 692 |
| Temperate forests | 10.4 | 7.4 | 139 | 262 |
| Boreal forests | 13.7 | 2.4 | 57 | 150 |
| Arctic tundra | 5.6 | 0.5 | 2 | 144 |
| Mediterranean shrublands | 2.8 | 1.3 | 17 | 124 |
| Crops | 13.5 | 3.8 | 4 | 248 |
| Tropical savannas and grasslands | 27.6 | 13.7 | 79 | 345 |
| Temperate grasslands | 15 | 5.1 | 6 | 172 |
| Deserts | 27.7 | 3.2 | 10 | 208 |
| *Subtotal* | *149.3* | *57.5* | *652* | *2,344* |
| Wetlands | | | | 450 |
| Frozen soils | 25.5 | | | 400 |
| *Total* | *174.8* | *57.5* | *652* | *3,194* |

*Note:* Plant carbon is from Saugier et al. (2001). NPP is from Saugier et al. (2001), scaled to give the terrestrial total as the satellite study of Behrenfeld et al. (2001). Soil carbon is from Jobbagy and Jackson (2000), to a depth of 3 m. Wetland soil carbon is from (Gorham 1991). Carbon in frozen (nonwetland) soils is from (Zimov et al. 1997).

As with fossil-fuel combustion, carbon is lost from terrestrial vegetation in a variety of non-$CO_2$ gases, including biogenic volatile organic compounds (VOCs). Globally VOC emissions are $0.2-1.4$ PgC y$^{-1}$ (Prinn, Chapter 9, this volume). For grassland and cultivated land, large emissions of other greenhouse gases such as $N_2O$ and $CH_4$ can also occur. These emissions are strongly affected by land management and land use change (Smith, Chapter 28, and Robertson, Chapter 29, both this volume).

## Ocean-Atmosphere Connections

The oceans contain about fifty times more $CO_2$ than the atmosphere and ten times more than the latest estimates of the plant and soil carbon stores (Colorplate 1). $CO_2$ moves between the atmosphere and the ocean by molecular diffusion, when there is a difference between the $CO_2$ gas pressure (p$CO_2$) in the oceans and the atmosphere. For example, when the atmospheric p$CO_2$ is higher than the surface ocean p$CO_2$, $CO_2$ diffuses across the air-sea boundary into the seawater. Based on about 940,000 measurements of surface water p$CO_2$ obtained since the 1960s, the climatological, monthly distribution of p$CO_2$ in the global surface waters has been calculated with a spatial resolution of $4° \times 5°$ (Takahashi et al. 2002). Although the published value for the net

air-sea flux indicates a net ocean uptake of 2.1 PgC for the reference year 1995, this estimate was based on inappropriate wind speed estimates (T. Takahashi, personal communication). The winds used for these estimates correspond to approximately 40 m
above mean sea level, but the gas exchange coefficient formulas are typically related to
winds at 10 m above sea level. The near-surface gradient in wind speed is such that the
10 m winds are about 1 m per second ($s^{-1}$) slower than the 40 m winds. Thus, Takahashi's corrected net global ocean uptake is 1.5 (−19 percent to +22 percent) PgC for
the reference year 1995 (Colorplate 2). The asymmetrical error estimates reflect the
paucity of data on the spatial and temporal variability in sea surface $pCO_2$ concentrations, plus limitations in the non-linear wind speed parameterizations for the gas
exchange coefficient. Ocean models and observations suggest that the interannual variability in the global ocean $CO_2$ flux is around ±0.5 PgC $y^{-1}$ (Greenblatt and Sarmiento,
Chapter 13, this volume).

The gross exchanges of $CO_2$ across the air-sea interface, as shown in Colorplate 1,
are much larger than the net flux. The global budget presented in this figure shows the
preindustrial oceans as a net source of ~0.6 PgC $y^{-1}$ to the atmosphere, partially offsetting the addition of carbon to the oceans from rivers. The total net modern flux of Takahashi et al. (2002), after the wind speed correction discussed previously (ocean uptake
= 1.5 PgC $y^{-1}$), is consistent with the modern balance implied by Colorplate 1 (21.9 −
20 + 70 − 70.6 = 1.3 PgC $y^{-1}$), despite the fact that the coastal zones are not well represented in the Takahashi et al. analysis. The coastal zone fluxes represent the largest
unknown in the $CO_2$ balance of the oceans and are a topic of active research (Chen,
Chapter 18, this volume).

During the past decade, significant advances have been made in separating the
anthropogenic component from the large background of ocean dissolved inorganic
carbon (DIC). Data-based approaches estimate a global inventory of anthropogenic
$CO_2$ in the oceans to be ~112 ± 17 PgC for a nominal year, 1994 (Colorplate 3) (Lee
et al. 2003). Inventories are generally high in the mid-latitudes and lowest in the high-
latitude Southern Ocean and near the Equator.

The high-inventory regions are convergence zones, where waters with relatively high
anthropogenic concentrations are moving into the ocean's interior. Roughly 25 percent
of the total inventory of anthropogenic carbon is in the North Atlantic, one of the main
regions of deepwater formation. The low-inventory waters are generally regions of
upwelling, where waters with low anthropogenic concentrations are brought near the
surface. The high-latitude Southern Ocean generally has very low anthropogenic $CO_2$
inventories and very shallow penetration. The Southern Hemisphere mode and intermediate waters at around 40–50°S, on the other hand, contain some of the largest
inventories of anthropogenic $CO_2$. More than 56 percent of the total anthropogenic
$CO_2$ inventory is stored in the Southern Hemisphere.

Over the long term (millennial timescales), the ocean has the potential to take up
approximately 85 percent of the anthropogenic $CO_2$ that is released to the atmosphere.

The reason for the long time constant is the relatively slow ventilation of the deep ocean. Most of the deep and intermediate waters have yet to be exposed to anthropogenic $CO_2$. As long as atmospheric $CO_2$ concentrations continue to rise, the oceans will continue to take up $CO_2$. This reaction, however, is reversible. If atmospheric $CO_2$ were to decrease in the future, the recently ventilated waters would start releasing part of the accumulated anthropogenic $CO_2$ back to the atmosphere, until a new equilibrium is reached.

## *Land-River-Ocean Connections*

Carbon is transported from the land to the oceans via rivers and groundwater. The transfer of organic matter from the land to the oceans via fluvial systems is a key link in the global carbon cycle. Rivers also provide a key link in the geological-scale carbon cycle by moving weathering products to the ocean (Colorplate 1). The conventional perspective is that rivers are simply a conduit for transporting carbon to the ocean. Reevaluation of this model suggests that the overall transfer of terrestrial organic matter through fluvial systems may be more complex (Richey, Chapter 17, this volume). A robust evaluation of the role of rivers, however, is complicated by both the diverse dynamics and multiple time constants involved and by the fact that data are scarce, particularly in many of the most affected systems.

Humans have had a significant impact on the concentrations of carbon and nutrients in river systems. Intensifying agriculture has led to extensive erosion, mobilizing perhaps 10–100 times more sediment, and its associated organic carbon, than undisturbed systems. Sewage, fertilizers, and organic waste from domestic animals also contribute to the carbon and nutrient loads of rivers. Not all of these materials make it to the ocean. Much can be deposited near and along river channels. The retention of particulate material in aquatic systems has increased since preindustrial times because of the proliferation of dams (primarily in the 30–50°N regions), which has increased the average residence time of waters in rivers. If the carbon mobilized via erosion were subsequently replaced by newly fixed carbon in agriculture, then a sink on the order of 0.5–1.0 PgC y$^{-1}$ would be created. But organic carbon does not move passively through river systems; even very old and presumably recalcitrant soil carbon may be at least partially remineralized in aquatic systems. Remineralization of organic carbon during transport leads to elevated levels of dissolved $CO_2$ in rivers, lakes, and estuaries worldwide. These high concentrations subsequently lead to outgassing to the atmosphere on the order of ~1 PgC y$^{-1}$, with the majority in the humid tropics (Richey, Chapter 17, this volume).

Despite the increased particulate retention in rivers, a significant amount of carbon escapes to the ocean. DIC, particulate inorganic carbon (PIC), dissolved organic carbon (DOC), and particulate organic carbon (POC) exported from rivers to the coastal ocean are 0.4, 0.2, 0.3, and 0.2 PgC y$^{-1}$, respectively (Chen, Chapter 18, this volume). These fluxes include a pronounced, but difficult to quantify, anthropogenic component. These values are poorly constrained by direct measurements and may represent minimum estimates of inputs to the coastal ocean (Richey, Chapter 17, this volume). Groundwater discharges, which make up about 10 percent of the surface flow to the ocean, also contribute poorly known contributions of carbon and nutrients to the coastal oceans.

The final step in the land-to-ocean pathway is the marine fate of fluvial carbon. Previous estimates have suggested that the marginal seas are net heterotrophic. Chen (Chapter 18, this volume) has suggested that continental shelves are in fact net autotrophic, mostly because of production from coastal upwelling of nutrient-rich waters. If this is correct, biological production in the coastal zone may decrease the thermodynamic drive to outgas the terrestrial carbon delivered by rivers, resulting in greater preservation in the marine environment.

## Trade and Commerce Connections

The lateral fluxes of carbon are poorly constrained by measurements and are frequently excluded from models. The river fluxes outlined in the previous section are only beginning to be quantified. The lateral transport of carbon through trade has not been constrained or quantified in global carbon budgets based on inventories of terrestrial carbon stocks and fluxes. Approximately one-third of the 4 PgC fixed annually through agriculture represents harvested products that are directed toward some form of human or domestic animal consumption (Tschirley and Servin, Chapter 21, this volume). In 2001, US$547 million in agricultural and forest-related products entered into international trade, amounting to 9 percent of total trade (excluding services). Imports and exports of cereals, wood, and paper products accounted for about 0.72 PgC of "embodied" carbon traded in 2000.

As we move from the basic carbon budget presented in Table 2.1 to a more detailed review of the fluxes shown in Colorplate 1 and to regional assessments of carbon budgets, it becomes increasingly important to recognize and quantify the lateral transports of carbon. The transport of carbon through trade and commerce represents a flux that is nearly as large as the net terrestrial and oceanic sinks. This carbon is assumed to be respired back to the atmosphere, but if the $CO_2$ is taken out of the atmosphere on one continent, then transported to another continent before being respired, it can have a significant impact on regional budget assessments.

Embodied carbon represents the carbon that resides directly within the product itself (for cereals, wood, and paper, about half of the final weight is carbon). In addition to embodied carbon, at least two other components are usually excluded from estimates of carbon content in international trade. One is the "production" carbon, which comprises the inputs required to produce a final product, primarily energy-related consumption to provide inputs (e.g., pesticides, fertilizers) and processing (e.g., machinery for harvesting, milling, sawing). A second element is the "transport" carbon, which represents the energy costs of transporting the processed products to a final destination. Countries that use intensive production systems and engage in significant exports (primarily in the developed countries) would have higher ratios of carbon content to product mass exported, possibly up to 20 percent, and would therefore have higher carbon emissions to the atmosphere than currently stated, under a system of full carbon accounting. Other countries would have lower carbon release to the atmosphere under full carbon accounting. Trade and commerce carbon accounting has a political background and differs from budget studies by atmospheric inversion and inventory checks. For example, $CO_2$ emissions during production and transport of products across regional boundaries would fully and partly, respectively, be accounted for by the country of origin using inventory studies. Inverse models based on atmospheric measurements, however, would reflect the emissions as they were actually distributed over the regions. As we begin to move from the global carbon budget to regional carbon budgets, fluxes like those in trade and commerce will need to be better quantified.

## Non-$CO_2$ Trace Gas Connections

Rising concentrations of a large number of potent anthropogenic greenhouse gases other than carbon dioxide have collectively contributed an amount of radiative forcing comparable to that of $CO_2$ since preindustrial times. Many of these non-$CO_2$ gases (e.g., $CH_4$, $N_2O$, $CF_2Cl_2$, $SF_6$) are emitted at the Earth's surface and contribute directly to this forcing (Prinn, Chapter 9, this volume). They are characterized by atmospheric lifetimes of decades to millennia (lifetime as used here is the amount of the gas in the global atmosphere divided by its global rate of removal). Other non-$CO_2$ gases (e.g., isoprene, terpenes, nitric oxide [NO], carbon monoxide [CO], sulphur dioxide [$SO_2$], dimethyl sulphide [$CH_3$]$_2$S), most of which are also emitted at the Earth's surface, contribute indirectly to this forcing through production of either tropospheric ozone (which is a powerful greenhouse gas) or tropospheric aerosols (which directly reflect sunlight back to space, absorb it, or indirectly change the reflection properties of clouds). This second group of climatically important non-$CO_2$ gases is characterized by much shorter lifetimes (hours to months).

The role of non-$CO_2$ carbon gases is not typically included in global carbon budgets, because the sources and sinks for these gases are not well understood. Figure 2.2 summarizes, albeit very simply, the basic cycles and fluxes of the major non-$CO_2$ trace gases relevant to climate. Note that the emissions of the carbon-containing gases alone contribute about 2.3 PgC y$^{-1}$ to the carbon cycle (Prinn, Chapter 9, this volume). To aid the handling of the non-$CO_2$ gases in the policy processes under the United Nations Framework Convention on Climate Change (UNFCCC), scientists have calculated so-called global warming potentials (GWPs). These dimensionless GWPs, which range from 20 to 20,000 for the major non-$CO_2$ gases, are intended to relate the time-integrated radiative forcing of climate by an emitted unit mass of a non-$CO_2$ trace gas to the forcing caused by emission of a unit mass of $CO_2$. The GWP concept has difficulties, because the removal mechanisms for many

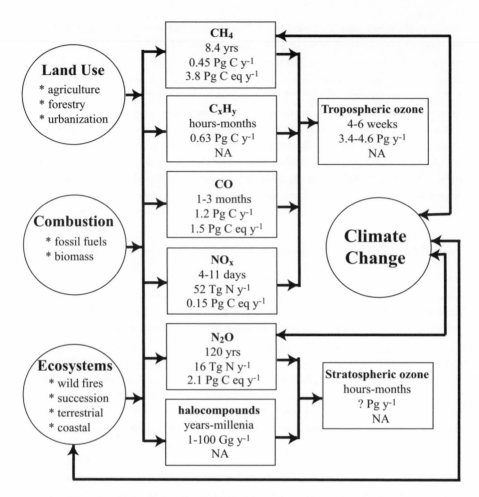

**Figure 2.2.** A summary of the sources, quantities, lifetimes, and consequences of carbon-containing non-CO$_2$ gases in the atmosphere. The data are from sources summarized in Prinn (Chapter 9, this volume). NA = not available.

gases (including CO$_2$ itself) involve complex chemical and/or biological processes and because the time period (e.g., decade, century) over which one integrates the instantaneous radiative forcing of a gas to compute its GWP is somewhat arbitrary (Manne and Richels, Chapter 25, this volume). Nevertheless, by multiplying the emissions of each major non-CO$_2$ greenhouse gas by its GWP, we obtain equivalent amounts of CO$_2$ emissions, which are comparable to the total actual emissions of CO$_2$ (e.g., current total emissions of CH$_4$ and N$_2$O are equivalent to 3.8 and 2.1 PgC y$^{-1}$, respectively). For this reason, a "CO$_2$ emissions only" approach for global warming policy may lead to significant biases in the estimation of global warming abatement costs (Manne and Richels, Chapter 25, this volume). Thus, a multi-gas approach for studying the carbon cycle and how it relates to climate change needs to be implemented.

## Carbon Cycle Processes

To fully understand the significance of the reservoir connections, as well as the possibility that these fluxes might change in the future, one must consider the processes responsible for controlling the fluxes within and between reservoirs. In some cases human activities have already significantly affected these processes. In other cases, the processes may be vulnerable to climate change in the future.

### Terrestrial Processes

On seasonal timescales the gains and losses from the terrestrial biosphere are reflected in the cyclic variations in atmospheric $CO_2$ concentrations. Although there is significant interannual variability in the seasonal balance between the gains and losses, it is the trends in this net balance between NPP and carbon losses through decomposition, fires, and disturbances that can affect climate on timescales of decades to centuries. The scientific community currently believes that the terrestrial biosphere has been acting as a net sink for atmospheric carbon for the past few decades (Foley and Ramankutty, Chapter 14, this volume). For example, Pacala et al. (2001) recently estimated the carbon sink of the coterminous Unites States using a combination of inventory and atmospheric concentration inversions. They determined a consistent land- and atmosphere-based carbon sink for the United States in the range of $0.37-0.71$ PgC $y^{-1}$. Similar studies have also shown a net sink in Europe (Janssens et al. 2003). The mechanisms underlying the sinks have been the subject of much recent research. The leading hypothesis in the 1970s and 1980s was that the sinks were mainly a result of more rapid plant growth from elevated $CO_2$ and climate change. This hypothesis has gradually been replaced with a multi-mechanism explanation, including contributions from changes in forest management, agriculture, long-lived products (e.g., wood), aquatic systems, and nitrogen deposition, in addition to $CO_2$ fertilization and changes in plant growth and soil carbon pools resulting from climate change (Schimel et al. 2001).

NPP is sensitive to a broad range of factors, including climate, soil fertility, atmospheric $CO_2$, atmospheric pollutants, and human management. An increase in NPP can lead to a carbon sink, but only if it is not matched by a corresponding increase in carbon losses. In general, processes that promote gradual increases in NPP can lead to carbon sinks, because increases in respiration tend to follow changes in biomass and soil carbon and not NPP directly (Field 1999). For the past 30 years, carbon cycle researchers have hypothesized that gradual increases in NPP in response to the 30 percent rise in atmospheric $CO_2$ since preindustrial times explain much or all of the terrestrial sink inferred from atmospheric studies (Bacastow and Keeling 1973). Experimental studies at the ecosystem level often demonstrate accelerated plant growth in response to elevated atmospheric $CO_2$ (Mooney et al. 1999), but these growth rates only explain a small fraction of the sink required to balance the global atmospheric $CO_2$ budget (Friedlingstein et al. 1995). Other factors that can increase NPP, including warming, greater precipitation and humidity, nitrogen deposition, and changes in plant species composition, may also contribute to terrestrial carbon sinks.

Processes that retard carbon releases can also lead to terrestrial sinks. Evidence indicates that a substantial fraction of the forest sink in temperate forests is a result of changes in land use and land management. In essence, forests, cut in the past, are regrowing, with a growth rate that is faster than the rate of harvesting (Goodale et al. 2002). Some of this change is a result of shifts in land use, especially the abandonment of agriculture over large regions of North America and parts of Europe (Foley and Ramankutty, Chapter 14, this volume). Fire suppression and the thickening (more trees per unit of area) of marginal forests also contribute to increased forest biomass in some areas (Pacala et al. 2001).

Changes in agriculture can also promote terrestrial carbon sinks. Increases in yield, incorporation of crop residue, and areas in perennial crops, as well as a reduction in tillage, can all contribute to carbon sinks (Lal et al. 1998).

Recent terrestrial carbon sinks reflect a number of mechanisms (Pacala et al. 2001). The potential for sinks to persist varies from mechanism to mechanism. Sinks due to $CO_2$ fertilization are likely to persist until NPP is limited by another factor. In some settings, this limitation might occur very soon, and in others it might be far in the future. Sinks caused by forest regrowth saturate as the forests mature (Nabuurs, Chapter 16, this volume). In general, sinks saturate when increases in NPP are outpaced by increases in the sum of decomposition and combustion. The persistence of terrestrial sinks in the future is by no means assured. Many lines of evidence suggest that the prospect of increasing carbon sources in the terrestrial biosphere is a real possibility (Gruber et al., Chapter 3, this volume).

## Ocean Processes

Air-sea gas exchange is a physico-chemical process, primarily controlled by the air-sea difference in gas concentrations and the exchange coefficient, which determines how quickly a molecule of gas can move across the ocean-atmosphere boundary (see Le Quéré and Metzl, Chapter 12, this volume). It takes about one year to equilibrate $CO_2$ in the surface ocean with atmospheric $CO_2$, so it is not unusual to observe large air-sea differences in $CO_2$ concentrations (Colorplate 2). Most of the differences are caused by variability in the oceans due to biology and ocean circulation. The oceans contain a large reservoir of carbon that can be exchanged with the atmosphere (see Colorplate 1). Since air-sea exchange can only occur at the surface, however, the rate at which carbon exchanges between the surface and the ocean interior ultimately regulates how well the atmosphere equilibrates with the ocean as a whole.

Two basic mechanisms control the natural distribution of carbon in the ocean interior: the solubility pump and the biological pump. The solubility pump is driven by two principle factors. First, more $CO_2$ can dissolve into cold polar waters than in the warm equatorial waters. As major ocean currents (e.g., the Gulf Stream) move waters from the tropics to the poles, they are cooled and can take up more $CO_2$ from the atmosphere. Second, the high-latitude zones are places where deepwater is formed. As the water cools, it becomes denser and sinks into the ocean's interior, taking with it the $CO_2$ accumulated at the surface.

The biological pump also transports $CO_2$ from the surface to the deep ocean. Growth of phytoplankton uses $CO_2$ and other chemicals from the seawater to form plant tissue. Roughly 70 percent of the $CO_2$ taken up by phytoplankton is recycled near the surface, and the remaining 30 percent sinks into the deeper waters before being converted back into $CO_2$ by marine bacteria (Falkowski et al. 1998). Only about 0.1 percent of the organic carbon fixed at the surface reaches the seafloor to be buried in the sediments. The carbon that is recycled at depth is transported large distances by currents to areas where the waters return to the surface (upwelling regions). When the waters regain contact with the atmosphere, the $CO_2$ originally taken up by the phytoplankton is returned to the atmosphere. This exchange helps to control atmospheric $CO_2$ concentrations over decadal and longer time scales.

The amount of organic carbon that is formed and sinks out of the surface ocean is limited by the availability of light and nutrients (mainly nitrate, phosphate, silicate, and iron) and by temperature. The plankton types present in the water also play a role. Plankton that bloom create favorable conditions for the formation of fast-sinking particles, particularly when they have shells of calcium carbonate or silicate (Klaas and Archer 2002). The formation of calcium carbonate shells affects carbon chemistry in such a way that it works to counteract the drawdown of $CO_2$ by soft tissue production. The biological pump also removes inorganic nutrients from surface waters and releases them at depth. Since productivity is limited by the availability of these nutrients, the large-scale thermohaline circulation (THC) of the oceans has a strong impact on global ocean productivity by regulating the rate at which nutrients are returned to the surface.

Up to now, humans have had a relatively small direct impact on the global-scale ocean carbon cycle. This is primarily because humans generally only transit across the ocean and because the ocean naturally contains orders of magnitude more carbon than the atmosphere and the terrestrial biosphere. The ocean does, however, act as a significant sink for $CO_2$ ultimately derived from anthropogenic activities (Table 2.1). Because biology is not limited by carbon in the oceans, it is thought that increasing $CO_2$ levels have not significantly affected ocean biology. The current distribution of anthropogenic $CO_2$ is assumed to result from physico-chemical equilibration of the surface ocean with rising atmospheric $CO_2$ and slow mixing of the anthropogenic $CO_2$ into the ocean's interior. The long residence time for the deep oceans means that most of the deep ocean waters have not been exposed to the rising atmospheric $CO_2$ concentrations observed over the past couple of centuries. Although the oceans have the potential to absorb 85 percent of the anthropogenic $CO_2$ released to the atmosphere, today's oceans are only at about 15 percent capacity (Le Quéré and Metzl, Chapter 12, this volume). Average penetration depth for anthropogenic $CO_2$ in the global ocean is only about 800 m (Sabine et al. 2002). There is growing evidence, however, that changes in ocean mixing and biology may be occurring as a result of climate change.

Since the solubility of $CO_2$ is a function of temperature, warming of the ocean will decrease its ability to absorb $CO_2$. Furthermore, changes in temperature and precipitation may lead to significant alterations of ocean circulation and the transport of carbon and nutrients to and from the surface. Because of its effect on ocean carbon distributions and biological productivity, changes in the THC have been used to help explain past excursions in climate and atmospheric $CO_2$, including glacial-interglacial and Dansgard-Oeschger events (Joos and Prentice, Chapter 7, this volume).

Ocean productivity can also be affected by atmospheric inputs that may change as a result of human activity. Iron in oceanic surface waters originates from terrestrial dust deposited over the ocean, deep ocean waters, continental shelves, and to a lesser extent river inflow. Because of the spatial distribution of dust deposition and other iron sources, large regions of the ocean show a deficit in iron (and to a smaller extent in silicate), although other nutrients are plentiful. These are called high-nutrient low-chlorophyll (HNLC) regions. There is a potential for enhanced biological productivity in these regions if the ocean can be "fertilized" by iron. The potential for $CO_2$ reduction in surface waters and in the atmosphere through this artificial sequestration, however, depends on factors like the composition of plankton types and oceanic circulation, and can lead to undesirable side effects (Bakker, Chapter 26, this volume).

There is also increasing evidence that rising $CO_2$ levels may directly affect ocean productivity and ecosystem structure. For example, Riebesell et al. (2000) showed a significant reduction in the ability of two different species of coccoliths to secrete calcium carbonate shells under elevated $CO_2$ conditions. Similar reductions in calcification have been observed in corals and coralline algae. As atmospheric $CO_2$ concentrations continue to rise, the potential for significantly altering the current balance between the amount of carbon moved into the ocean's interior by the biological pump versus the solubility pump in the ocean increases. The net effect on the ability of the ocean to act as a sink for anthropogenic $CO_2$ is not clear.

## Coastal Ocean

Although the coastal zones, consisting of the continental shelves with depths less than 200 m including bays and estuaries, occupy only 10 percent of the total ocean area, they play a crucial role in the global carbon cycle. Carbon is transported to the coastal zone by riverine inputs and transport of inorganic carbon from the open ocean. Estuaries and proximal coastal seas are believed to be sources of $CO_2$ because of the decay of terrestrial organic carbon. Since the riverine flux of nutrients has risen continually over the past few decades, however, these areas may now have enhanced biological productivity and hence may be releasing less $CO_2$. At present a large fraction (~80 percent) of the land-derived organic and inorganic materials that are transported to the ocean is trapped on the proximal continental shelves (Mackenzie and Ver 2001). The much wider open shelves, on the other hand, probably serve as sinks for atmospheric carbon. A recent overview suggests that the global coastal waters and marginal seas (extending to a water depth of 200 m) are now absorbing about 0.36 PgC y$^{-1}$ from the atmosphere (Chen, Chapter 18, this volume).

Across most of the coastal seas and continental margins, surface waters are transported offshore because of fresh water inputs from land. This surface transport draws nutrient-rich subsurface waters from the open ocean onto the shelves. Such external sources of nutrients support high primary productivity. Most of the organic material produced is respired and recycled on the shelves. The organic matter that is not recycled either accumulates in the sediments or is exported to the slopes and open oceans. The coastal zone may account for 30–50 percent of the total calcium carbonate accumulation and up to 80 percent of the organic carbon accumulation in ocean sediments (Mackenzie and Ver 2001). Globally, the shelf seas are estimated to transport 0.6 PgC y$^{-1}$ of DOC, 0.5 PgC y$^{-1}$ of POC, and 0.2 PgC y$^{-1}$ of PIC to the open oceans (Chen, Chapter 18, this volume). Although these transports have large uncertainties because of high variability and inadequate data coverage, they represent an important and often neglected link in the global carbon cycle. Shelves and estuaries are also important sources of other greenhouse or reactive gases, such as methane and dimethyl sulfide.

Finally, although humans appear to have had only a small direct impact on the open ocean, they have had a profound and poorly understood impact on the coastal oceans. Furthermore, direct and indirect human perturbations vis-à-vis the continental margins (e.g., pollution, eutrophication) are likely to have large and dire consequences on marine ecosystems in the future. As much as 40–60 percent of the world population lives in coastal areas, depending on the definitions and methodologies applied. The coastal regions have the most rapidly growing populations, because of migration from rural areas. Most megacities in this century will develop in coastal zones, approximately half of them in Asia (IHDP 2000). Such patterns of urbanization have affected and will continue to affect coastal and marine systems, through processes such as land use change, pressure on infrastructure (water, sewage, and transportation), coastal resource depletion and degradation, eutrophication, and other carbon-relevant impacts. Coastal zones support more than 60 percent of the global commercial fish production (World Resources Institute 1996). It is not clear how these fisheries will be affected by the increasing human pressures on the coastal zone. Additional studies of the coastal zone and the interaction between humans and the coastal biogeochemical systems are needed.

## Human Systems

Humans have had a profound impact on carbon cycling in the atmosphere, the terrestrial biosphere, and, to a lesser extent, the oceans. To understand how humans will continue to interact with these reservoirs in the future, one must understand the drivers responsible for how humans interact with the environment.

Energy use has historically been viewed as an essential commodity for economic growth. The paradigm of "grow or die" historically meant increasing demands for additional resources and a commensurate increase in the amount of pollutants released. Over the past 30 years, however, a complex debate has ensued about how to decouple economic growth and resource consumption. Since this debate started, many scholars, policy makers, and nongovernmental organizations have asked themselves how to reshape this paradigm of development. Answers have ranged from doing nothing (business-as-usual approach) to slowing and even stalling economic growth, so as to not exceed the Earth's carrying capacity. Carrying capacity here refers to the ability of natural resources and ecosystems to cope with anthropogenic pressures, such as use of renewable natural resources, emission of pollutants, and modification of ecosystem structures without crossing "critical thresholds of damage beyond which [these resources] lose their ability for self-renewal and slide inexorably into deeper degradation" (Board on Sustainable Development Policy Division 1999).

Another more tenuous response is based on the idea that one could keep economic growth as a development goal and at the same time find mechanisms aimed at reducing the amount of material input and emissions or aimed at finding replacements for nonrenewable or dangerous resources (e.g., fossil fuels). This proposal has resulted in environmental policies and instruments promoting technological innovations that seek to increase the efficiency of economic activities and to decarbonize economies.

Energy intensity (the ratio of total domestic energy primary consumption to gross domestic product or physical output) in the United States, Japan, and other developed countries has increased far slower than economic growth. China, too, has made significant gains, and the growth rate of its energy consumption has been half that of its gross domestic product (GDP) since the early 1980s. While the energy use per unit of GDP has fallen or stabilized since 1970, energy use per capita has increased in most of the developed countries (Sathaye, Chapter 22, this volume). This statistic may be misleading however, because there has also been a change in consumption and production patterns away from materially intensive commodities toward less-intensive services, and production of many of the materially intensive commodities has moved out of the developed countries. This does not necessarily mean that the developed countries are using fewer resources, but rather that the regional distribution of this consumption has changed. Countries belonging to the former Soviet Union (FSU) have also experienced decreases in energy intensity but for very different reasons. Because of sociopolitical changes, economic activity declined substantially in these countries, resulting in a dramatic decline in energy use and associated carbon emissions.

Over the past century and a half, the use of biomass has successively given way to expanded use of coal, oil, and natural gas as the primary fuels to supply energy (Figure 2.3). Over the past two decades, the use and share of nuclear and other forms of renewable energy have increased. The carbon intensity (PgC/gigajoule [GJ]) of each successive fuel, beginning with coal, is lower, and this decline has led to the decarbonization of the global fuel mix over the past century. Critical inventions, such as the steam and internal combustion engines, vacuum tubes, and airplanes, have accompanied and fostered the use of successive fuels and electricity.

The speed of energy consumption and land use changes has increased during the past two centuries. Three spatial and historical variations of carbon-relevant social tendencies need to be considered in developing an understanding of where we are today.

- Historically, industrialized countries have been the main releasers of carbon from combustion of fossil fuels. Although developing countries are expected to increase their share of emissions, the relocation strategies of corporations based in industrialized regions may actually significantly contribute to increased emissions in these developing countries. In 1925, for example, Australia, Japan, the United States, and Western Europe were responsible for about 88 percent of the world's fossil-fuel carbon dioxide emissions (Houghton and Skole 1990). Data from 1950–1995 indicate that 26 countries fell above the average cumulative emissions figure. These countries include almost all developed, or Annex I countries, plus Brazil, China, India, and several other larger developing countries that have seen significant growth in foreign investments (Claussen and McNeilly 1998).
- Until the 20th century, most of the conversion from forest area to cultivated land occurred in the developed countries. During the past few decades, however, most of the deforestation has occurred in tropical forests.
- Urbanization, a key driver of energy and land use, was primarily a feature of industrialized countries until the middle of the 20th century. The largest and demographically most dynamic urban agglomerations, like London, New York, and Tokyo, were in developed countries. As of the beginning of the 21st century, however, most urban agglomerations are situated in developing countries. These developing-country urban agglomerations have different production systems and living standards than the urban centers in developed countries. These differences affect the carbon emissions of various urban areas (Romero Lankao, Chapter 19, this volume).

Although technology is perceived as the answer to decarbonizing economies, other societal factors work as constraints and windows of opportunity for that purpose (e.g., institutional settings and economic dynamics; see Raupach et al., Chapter 6, this volume). In addition, new technological paradigms of production and consumption patterns emerge only over decades. The main energy and production components of the *engineering* epoch, for instance (1850–1940), took at least 20 years to develop. Hence, it will likely take decades to set up alternative energy sources and materials aimed at decarbonizing industrial and agricultural activities. Despite efforts aimed at decoupling

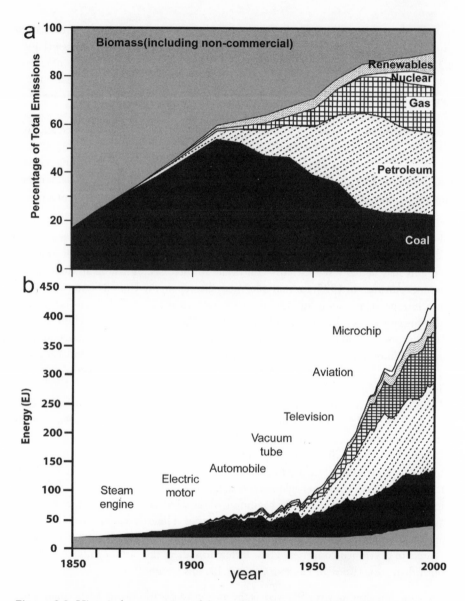

**Figure 2.3.** Historical composition of the world energy system, in (a) percent contributions and (b) energy contributions in EJ (1 EJ = $10^{18}$ J = 1.05 Quad = $1.05*10^{15}$ BTU). In 2000 the global consumption of primary energy was about 400 EJ.

economic growth from its carbon impacts, $CO_2$ emissions from the combustion of fossil fuels and from land use change have been increasing. Many features of the development trends need to be studied and better understood before opportunities for modifications in life styles, technologies, institutions, and other drivers of carbon emissions can be fully addressed.

## Geological Processes

The discussion thus far has focused on reservoirs and processes relevant to human timescales. With a residence time of about 300 million years, the huge reserves of carbon stored in the sedimentary rocks are not expected to play a large role in the short-term carbon budget. On timescales greater than 500,000 years, about 80 percent of the $CO_2$ exchange between the solid earth and the atmosphere is controlled by the carbonate-silicate cycle (Kasting et al. 1988). In this cycle, atmospheric $CO_2$ is used to weather calcium-silicate rock minerals on land, which are then transported to the ocean via rivers as calcium and bicarbonate ions. In the oceans, plankton and other organisms incorporate the ions into calcium carbonate shells. A portion of the calcium carbonate is deposited onto the ocean floor, and eventually $CO_2$ is returned to the atmosphere through volcanic and diagenetic processes.

Vast quantities of carbon stored are in ocean sediments as methane hydrates and as calcium carbonate. The methane hydrates are relatively stable but could be released if ocean temperatures increase sufficiently through global warming (Harvey and Huang 1995). The carbonate sediments are likely to be a significant sink for fossil fuel $CO_2$ on millennial timescales (Archer et al. 1999). As the oceans continue to take up anthropogenic $CO_2$, the $CO_2$ will penetrate deeper into the water column, lowering the pH and making the waters more corrosive to calcium carbonate. Dissolution of sedimentary carbonates binds the carbon in a dissolved form that is not easily converted back into atmospheric $CO_2$. Carbonate dissolution is typically thought to occur in the deep ocean, well removed from the anthropogenic $CO_2$ taken up in the surface waters. In portions of the North Atlantic and North Pacific Oceans, however, anthropogenic $CO_2$ may have already penetrated deep enough to influence the dissolution of calcium carbonate in the water column and shallow sediments (Feely et al. 2002).

Although the processes of $CO_2$ uptake through weathering and $CO_2$ release from volcanism and diagenesis appear to have a small net effect on the global carbon cycle on millennial timescales, short-term variability in one of these fluxes can affect the carbon cycle on timescales relevant to humans. For example, explosive volcanic eruptions result in the emission of $CO_2$, dust, ash, and sulfur components. These sulfur particles may rain out and promote acid rain. Atmospheric sulfur particles block sunlight and cool the regional climate. Particles that reach the stratosphere reduce global temperatures for several years. The incidental occurrence of cataclysmic volcanic eruptions or the coincidence of several large eruptions (e.g., the eruptions of 1783 or the 1991

Mount Pinatubo eruption) affect the global carbon cycle directly by their $CO_2$ emissions and indirectly by their impact on marine and terrestrial primary production (Hamblyn 2001; Sarmiento and Gruber 2002).

## Understanding Today's Carbon Cycle

Although our understanding of the contemporary global carbon cycle has increased dramatically over the past few decades, many aspects are still not well understood. Several specific areas where additional studies are needed have been mentioned. There are two general areas of research, however, where our understanding is exceptionally weak and focused research is necessary. One area that needs improvement is understanding regional variability. A second, somewhat related topic is how changes in the carbon cycle may be linked to different modes of climate variability.

### Regional Budgets

Although atmospheric $CO_2$ concentrations are changing on a global scale, the spatial scales of natural processes, as well as the scales of human interventions and the associated societal mechanisms, have a profound regional character. Industrialized countries, for instance, dominate fossil fuel $CO_2$ emissions by direct release and through trade, whereas developing countries have become the primary $CO_2$ emitters through land use changes (Romero Lankao, Chapter 19, this volume). Insufficient observations limit our ability to conduct regional-scale assessments globally. However, the potential benefits of such approaches in identifying the biogeochemical and human processes responsible for controlling fluxes makes these studies very important. A regional-scale carbon budget assessment also provides a unique opportunity to verify and bridge independent methods and observations made over a range of spatial scales (e.g., top-down atmospheric inversion estimates versus land-based or ocean-based bottom-up observations).

In the land-based bottom-up method, carbon sinks and sources from various ecosystems (forests, croplands, grasslands, and organic soil wetlands) are aggregated over all regions to provide a large-scale perspective. The land-based approach can therefore provide information about which ecosystems and regions are accumulating carbon and which are losing carbon to the atmosphere. The diversity of land mosaics, the complexity of human activities, and the lateral transports of carbon in different components make it difficult, however, to provide a comprehensive carbon budget.

Ocean-based bottom-up methods can be used to assess the complex and often competitive controls of heat flux, mixing, and biology on air-sea gas exchange, as well as to develop proxies for extrapolating limited observations to larger time and space scales. Monitoring changes in ocean interior properties can also provide valuable information on surface processes and fluxes.

Atmosphere-based approaches, in contrast, give no information about which ecosystems or processes are contributing to a sink or source but produce a more consistent large-scale assessment of the net carbon flux to the atmosphere. No one approach holds the key to understanding regional variability in the carbon cycle. A suite of approaches much be used. In addition to providing different, but complementary, information, a combination of the bottom-up and top-down methods puts independent constraints on the integrated carbon balance of regional budgets.

The use of multiple approaches can also be used to identify key fluxes that the inventory approach may have missed. For example, Janssens et al. (2003) used inventories to estimate that the European continent sink is on the order of 0.11 PgC y$^{-1}$, compared with a mean atmospheric inversion estimate of 0.29 PgC y$^{-1}$. The discrepancy between these estimates is attributed to the intercontinental displacement of organic matter via trade, emissions of non-$CO_2$ gases, and a slight overestimation of fossil-fuel emissions. This example illustrates not only the need for comprehensive assessments of the major vertical and lateral fluxes, but also the power of multiple approaches for highlighting areas where additional studies are needed.

## *Variability*

Observing and quantifying variability in natural processes and the impact of human interventions is of primary importance for three reasons:

- It provides a key diagnostic of how climatic factors and societal dynamics affect exchange fluxes and provides information that is needed to develop and validate comprehensive process-based carbon cycle models that include human dynamics.
- Several terrestrial and oceanic carbon cycle components effectively contain a longer-term "memory"—that is, their present state is partly a result of past natural and human perturbations. For example, enhanced fire frequencies or management strategies during a drier past period will be reflected in the age structure of a present forest, or the carbon content in particular oceanic deepwater masses will reflect climate-driven variations in deepwater formation in the past.
- Climate variability often tends to mask the slow, longer-term signals in the carbon cycle that are of primary interest; for example, many studies have shown that the uptake rates of the terrestrial biosphere or the ocean are highly variable. In this context, climate variability constitutes "noise" from which the signals must be distinguished. Indeed, a significant fraction of the uncertainty in the global budget in Table 2.1 can be traced to incomplete quantification of possible climate perturbations in the pertinent observations.

On a global scale, climate-driven variability can be inferred from atmospheric time series of $CO_2$ and associated variables, such as $^{13}C/^{12}C$ and $O_2/N_2$ ratios. Continental or ocean basin–scale variations can be detected on timescales of up to several years by means of top-down inversions of atmospheric $CO_2$ concentration measurements (Heimann et al., Chapter 8, this volume). It is difficult, however, to separate the terrestrial signals from the ocean signals using these large-scale approaches (Greenblatt and Sarmiento, Chapter 13, this volume). On land and in the oceans, local-scale direct observations of variability exist from in situ flux measurements at a few time-series stations during the past few years. On a regional scale, however, an observational gap exists. The present atmospheric network is not dense enough to resolve carbon sources and sinks on regional scales by atmospheric inversion, while the upscaling of measured in situ oceanic and terrestrial carbon fluxes is extremely difficult. On land, the large heterogeneity of terrestrial ecosystems and complex atmospheric transport patterns resulting from topography make it difficult to scale up local measurements. In the oceans, the complex interplay between physical and biological controls on sea surface $pCO_2$ and the dynamics of air-sea exchange complicate the extrapolation of the in situ observations.

Interannual variability in climate leads to large changes in atmospheric temperature and rainfall patterns, as well as changes in ocean surface temperatures and circulation. All of these changes can have dramatic effects on biological productivity and complex effects on $CO_2$ exchanges with the atmosphere on land and in the oceans. For example, during El Niño events, the warming of ocean surface waters and the reduction in biological productivity in the Equatorial Pacific Ocean should lead to enhanced outgassing of $CO_2$. Since the upwelling of carbon-rich deepwaters, which release $CO_2$ to the atmosphere, is reduced during El Niños, however, the net effect is a significant reduction in the outgassing of $CO_2$ in this region. On the other hand, warmer temperatures and anomalous rainfall patterns during El Niños can lead to increased terrestrial biosphere respiration, forest fires, and droughts. The timing of these effects and the teleconnections between them have a direct impact on atmospheric $CO_2$ concentrations that is still not completely understood.

Overall, the general consensus is that the interannual variability in air-sea $CO_2$ fluxes is smaller than that of terrestrial $CO_2$ fluxes, but the exact amplitude and spatial distributions remain uncertain (Greenblatt and Sarmiento, Chapter 13, this volume). To better determine these signals, an expanded network of time-series $CO_2$ measurements must be maintained for the atmosphere, oceans, and land systems. These measurements, together with intensive process studies, a better use of satellite data, atmospheric observations, and rigorously validated models, will help us better understand the current global carbon cycle and how it is evolving over time.

## Conclusions

The current global carbon cycle is in a state of transition. Human activities over the past few centuries have had a profound impact on many aspects of the system. As we begin to assess ways to monitor and potentially manage the global carbon cycle, it is imperative that we better understand how the system, including human dynamics and the biogeochemical processes controlling $CO_2$ and other carbon gases in the atmosphere, has operated in the past and how it is operating today. How will the land-ocean-atmosphere processes respond to human activities, and what are the societal dynamics that will determine how humans will respond to changes in the land, ocean, and atmosphere systems? We have made tremendous progress over the past few decades at reducing the uncertainties in Table 2.1 and Colorplate 1, but until we can confidently explain and model the contemporary carbon cycle, our ability to predict future changes in atmospheric $CO_2$ concentrations will be limited.

## Literature Cited

Achard, F., H. D. Eva, H.-J. Stibig, P. Mayaux, J. Gallego, T. Richards, and J.-P. Malingreau. 2002. Determination of deforestation rates of the world's humid tropical forests. *Science* 297:999–1002.

Ahlbrandt, T. S., R. R. Charpentier, T. R. Klett, J. W. Schmoker, C. J. Schenk, and G. F. Ulmishek. 2000. Future oil and gas supplies of the world. *Geotimes* 45:24–25.

Archer, D., H. Kheshgi, and E. Maier-Reimer. 1999. Dynamics of fossil fuel $CO_2$ neutralization by marine $CaCO_3$. *Global Biogeochemical Cycles* 12:259–276.

Bacastow, R., and C. D. Keeling. 1973. Atmospheric carbon dioxide and radiocarbon in the natural carbon cycle. II. Changes from A.D. 1700 to 2070 as deduced from a geochemical reservoir. Pp. 86–135 in *Carbon and the biosphere,* edited by G. M. Woodwell and E. V. Pecan. Springfield, VA: U.S. Department of Commerce.

Behrenfeld, M. J., J. T. Randerson, C. R. McClain, G. C. Feldman, S. O. Los, C. J. Tucker, P. G. Falkowski, C. B. Field, R. Frouin, W. E. Esaias, D. D. Kolber, and N. H. Pollack. 2001. Biospheric primary production during an ENSO transition. *Science* 291 (5513): 2594–2597.

Board on Sustainable Development Policy Division and National Research Council. 1999. *Our common journey: A transition toward sustainability.* Washington, DC: National Academies Press.

BP. 2003. *BP statistical review of world energy.* http://www.bp.com/centres/energy/index.asp.

CDIAC (Carbon Dioxide Information Analysis Center). 2003. *Trends online: A compendium of data on global change.* Oak Ridge, TN: Oak Ridge National Laboratory, U.S. Department of Energy. http://cdiac.esd.ornl.gov/trends/trends.htm.

Claussen, E., and L. McNeilly. 1998. *Equity and global climate change: The complex elements of global fairness.* Arlington, VA: Pew Center on Global Climate Change. http://www.pewclimate.org/report2.html.

DeFries, R. S., C. B. Field, I. Fung, J. Collatz, and L. Bounoua. 1999. Combining satellite data and biogeochemical models to estimate global effects of human-induced land cover change on carbon emissions and primary productivity. *Global Biogeochemical Cycles* 13:803–815.

DeFries, R. S., R. A. Houghton, M. C. Hansen, C. B. Field, D. Skole, and J. Townshend. 2002. Carbon emissions from tropical deforestation and regrowth based on satellite observations for the 1980s and 1990s. *Proceedings of the National Academy of Sciences* 99:14256–14261.

Falkowski, P. G., R. T. Barber, and V. Smetacek. 1998. Biogeochemical controls and feed-backs on ocean primary production. *Science* 281:200–206.

Feely, R. A., C. L. Sabine, K. Lee, F. J. Millero, M. F. Lamb, D. Greeley, J. L. Bullister, R. M. Key, T.-H. Peng, A. Kozyr, T. Ono, and C. S. Wong. 2002. In situ calcium carbonate dissolution in the Pacific Ocean. *Global Biogeochemical Cycles* 16 (4): 1144, doi:10.1029/2002GB001866.

Field, C. B. 1999. Diverse controls on carbon storage under elevated $CO_2$: Toward a synthesis. Pp. 373–391 in *Carbon dioxide and environmental stress*, edited by Y. Luo. San Diego: Academic Press.

Friedlingstein, P., I. Fung, E. A. Holland, J. John, G. Brasseur, D. Erikson, and D. Schimel. 1995. On the contribution of the biospheric $CO_2$ fertilization to the missing sink. *Global Biogeochemical Cycles* 9:541–556.

Geist, H. J., and E. F. Lambin. 2001. What drives tropical deforestation? LUCC Report Series 4. Louvain-la-Neuve, Belgium: Land Use and Land Cover Change (LUCC). http://www.geo.ucl.ac.be/LUCC/lucc.html.

Goodale, C. L., M. J. Apps, R. A. Birdsey, C. B. Field, L. S. Heath, R. A. Houghton, J. C. Jenkins, G. H. Kohlmaier, W. Kurz, S. Liu, G.-J. Nabuurs, S. Nilsson, and A. Z. Shvidenko. 2002. Forest carbon sinks in the northern hemisphere. *Ecological Applications* 12:891–899.

Gorham, E. 1991. Northern peatlands: Role in the carbon-cycle and probable responses to climatic warming. *Ecological Applications* 1:182–195.

Gurney, K. R., R. M. Law, A. S. Denning, P. J. Rayner, D. Baker, P. Bousquet, L. Bruhwiler, Y. H. Chen, P. Ciais, S. Fan, I. Y. Fung, M. Gloor, M. Heimann, K. Higuchi, J. John, T. Maki, S. Maksyutov, K. Masarie, P. Peylin, M. Prather, B. C. Pak, J. Randerson, J. Sarmiento, S. Taguchi, T. Takahashi, and C. W. Yuen. 2002. Towards robust regional estimates of $CO_2$ sources and sinks using atmospheric transport models. *Nature* 415:626–630.

Hamblyn, R. 2001. *The invention of clouds: How an amateur meteorologist forged the language of the skies.* New York: Picador.

Harvey, L. D. D., and Z. Huang. 1995. Evaluation of the potential impact of methane clathrate destabilization on future global warming. *Journal of Geophysical Research* 100:2905–2926.

Houghton, R. 2003. Why are estimates of the terrestrial carbon balance so different? *Global Change Biology* 9:500–509.

Houghton, R.A., and D. L. Skole. 1990. Carbon. Pp. 393–408 in *The Earth as transformed by human action*, edited by B. L. Turner II, W. C. Clark, R. W. Kates, J. F. Richards, J. T. Mathews, and W. B. Meyer. Cambridge: Cambridge University Press.

IHDP (International Human Dimensions Programme on Global Environmental Change). 2000. *IHDP/START International Human Dimensions Workshop: "Human Dimensions in the Coastal Zones."* IHDP Proceeding No. 03. Bonn, Germany. http://www.ihdp.uni-bonn.de/html/publications/publications.html.

Janssens, I., A. Freibauer, P. Ciais, P. Smith, G. Nabuurs, G. Folberth, B. Schlamadinger, R. Hutjes, R. Ceulemans, E. Schulze, R. Valentini, and A. Dolman. 2003. Europe's terrestrial biosphere absorbs 7 to 12% of European anthropogenic $CO_2$ emissions. *Science* 300:1538–1542.

Jobbagy, E., and R. Jackson. 2000. The vertical distribution of soil organic carbon and its relation to climate and vegetation. *Ecological Applications* 10:423–436.

Kasting, J. F., O. B. Toon, and J. B. Pollack. 1988. How climate evolved on the terrestrial planets. *Scientific American* 2:90–97.

Klaas, C., and D. E. Archer. 2002. Association of sinking organic matter with various types of mineral ballast in the deep sea: Implications for the rain ratio. *Global Biogeochemical Cycles* 16 (4): 1116, doi:10.1029/2001GB001765.

Lal, R., J. M. Kimble, R. F. Follett, and C. V. Cole. 1998. *The potential of U.S. cropland to sequester carbon and mitigate the greenhouse effect.* Ann Arbor, MI: Chelsea Press.

Lee, K., S.-D. Choi, G.-H. Park, R. Wanninkhof, T.-H. Peng, R. M. Key, C. L. Sabine, R. A. Feely, J. L. Bullister, and F. J. Millero. 2003. An updated anthropogenic $CO_2$ inventory in the Atlantic Ocean. *Global Biogeochemical Cycles* (in press).

Le Quéré, C., O. Aumont, L. Bopp, P. Bousquet, P. Ciais, R. Francey, M. Heimann, C. D. Keeling, R. F. Keeling, H. Kheshgi, P. Peylin, S. C. Piper, I. C. Prentice, and P. J. Rayner. 2003. Two decades of ocean $CO_2$ sink and variability. *Tellus* 55B:649–656.

Lindemann, R. L. 1942. The trophic-dynamic aspect of ecology. *Ecology* 23:399–418.

Mackenzie, F. T., and L. M. Ver. 2001. Land-sea global transfers. Pp. 1443–1453 in *Encyclopedia of ocean sciences,* edited by J. H. Steele, K. K. Turekian, and S. A. Thorpe. St. Louis, MO: Academic Press. doi:10.1006/rwos.2001.0073.

Mooney, H. A., J. Canadell, F. S. Chapin III, J. Ehleringer, C. Körner, R. McMurtrie, W. J. Parton, L. Pitelka, and E.-D. Schulze. 1999. Ecosystem physiology responses to global change. Pp. 141–189 in *The terrestrial biosphere and global change: Implications for natural and managed ecosystems,* edited by B. H. Walker, W. L. Steffen, J. Canadell, and J. S. I. Ingram. Cambridge: Cambridge University Press.

Pacala, S. W., G. C. Hurtt, R. A. Houghton, R. A. Birdsey, L. Heath, E. T. Sundquist, R. F. Stallard, D. Baker, P. Peylin, P. Ciais, P. Moorcroft, J. Caspersen, E. Shevliakova, B. Moore, G. Kohlmaier, E. Holland, M. Gloor, M. E. Harmon, S.-M. Fan, J. L. Sarmiento, C. Goodale, D. Schimel, and C. B. Field. 2001. Convergence of land- and atmosphere-based U.S. carbon sink estimates. *Science* 292:2316–2320.

Prentice, C., G. D. Farquhar, M. J. R. Fasham, M. L. Goulden, M. Heimann, V. J. Jaramillo, H. S. Kheshgi, C. Le Quéré, R. J. Scholes, and D. W. R. Wallace. 2001. The carbon cycle and atmospheric carbon dioxide. Pp. 183–237 in *Climate change 2001: The scientific basis (Contribution of Working Group I to the Third Assessment Report of the Intergovernmental Panel on Climate Change),* edited by J. Houghton, Y. Ding, D. J. Griggs, M. Noguer, P. J. van der Linden, X. Dai, K. Maskell, and C. A. Johnson. Cambridge: Cambridge University Press.

Riebesell, U., I. Zondervan, B. Rost, P. D. Tortell, R. E. Zeebe, and F. M. M. Morel. 2000. Reduced calcification of marine plankton in response to increased atmospheric $CO_2$. *Nature* 407 (6802): 364–367

Sabine, C. L., R. M. Key, K. M. Johnson, F. J. Millero, J. L. Sarmiento, D. W. R. Wallace, and C. D. Winn. 1999. Anthropogenic $CO_2$ inventory of the Indian ocean. *Global Biogeochemical Cycles* 13:179–198.

Sabine, C. L., R. A. Feely, R. M. Key, J. L. Bullister, F. J. Millero, K. Lee, T.-H. Peng, B. Tilbrook, T. Ono, and C. S. Wong. 2002. Distribution of anthropogenic $CO_2$ in the Pacific Ocean. *Global Biogeochemical Cycles* 16 (4): 1083, doi: 10.1029/2001GB001639.

Sarmiento, J. L., and N. Gruber. 2002. Sinks for anthropogenic carbon. *Physics Today* (August): 30–36.

Saugier, B., J. Roy, and H. A. Mooney. 2001. Estimations of global terrestrial productivity: Converging toward a single number? Pp. 543–557 in *Terrestrial global productivity*, edited by J. Roy, B. Saugier, and H. A. Mooney. San Diego: Academic Press.

Schimel, D. S., J. I. House, K. A. Hibbard, P. Bousquet, P. Ciais, P. Peylin, B. H. Braswell, M. J. Apps, D. Baker, A. Bondeau, J. Canadell, G. Churkina, W. Cramer, A. S. Denning, C. B. Field, P. Friedlingstein, C. Goodale, M. Heimann, R. A. Houghton, J. M. Melillo, B. Moore III, D. Murdiyarso, I. Noble, S. W. Pacala, I. C. Prentice, M. R. Raupach, P. J. Rayner, R. J. Scholes, W. L. Steffen, and C. Wirth. 2001. Recent patterns and mechanisms of carbon exchange by terrestrial ecosystems. *Nature* 414:169–172.

Takahashi, T., S. C. Sutherland, C. Sweeney, A. Poisson, N. Metzl, B. Tilbrook, N. Bates, R. Wanninkhof, R. A. Feely, C. Sabine, J. Olafsson, and Y. Nojiri. 2002. Global sea-air $CO_2$ flux based on climatological surface ocean $pCO_2$, and seasonal biological and temperature effects. *Deep-Sea Research II* 49:1601–1623.

Vitousek, P. M., P. R. Ehrlich, A. H. Ehrlich, and P. A. Matson. 1986. Human appropriation of the products of photosynthesis. *BioScience* 36:368–373.

World Resources Institute. 1996. *World resources: A guide to the global environment.* New York: Oxford University Press.

Zhang, X. M., C. J. S. Cassells, and J. L. van Genderen. 1998. Multi-sensor data fusion for the detection of underground coal fires. *Geologie en Mijnbouw* 77:117.

Zimov, S. A., Y. V. Voropaev, I. P. Semiletov, S. P. Davidov, S. F. Prosiannikov, F. S. Chapin III, M. C. Chapin, S. Trumbore, and S. Tyler. 1997. North Siberian lakes: A methane source fueled by pleistocene carbon. *Science* 277:800–802.

# 3

# The Vulnerability of the Carbon Cycle in the 21st Century: An Assessment of Carbon-Climate-Human Interactions

Nicolas Gruber, Pierre Friedlingstein, Christopher B. Field, Riccardo Valentini, Martin Heimann, Jeffrey E. Richey, Patricia Romero Lankao, E.-Detlef Schulze, and Chen-Tung Arthur Chen

In most scenario calculations to date, emissions from fossil-fuel burning are prescribed, and a carbon cycle model computes the time evolution of atmospheric $CO_2$ as the residual between emissions and uptake by land and ocean, typically without considering feedbacks of climate on the carbon cycle (see, e.g., Schimel et al. 1996). The global carbon cycle is, however, intimately embedded in the physical climate system and tightly interconnected with human activities. As a consequence, climate, the carbon cycle, and humans are linked in a network of feedbacks, of which only those between the physical climate system and the carbon cycle have been explored so far (Friedlingstein, Chapter 10, this volume). One example of a carbon-climate feedback begins with the modification of climate through increasing atmospheric $CO_2$ concentration. This modification affects ocean circulation and consequently ocean $CO_2$ uptake (e.g., Sarmiento et al. 1998; Joos et al. 1999; Matear and Hirst 1999). Similar feedbacks occur on land. For example, rising temperatures lead to higher soil respiration rates, which lead to greater releases of carbon to the atmosphere (e.g., Cox et al. 2000; Friedlingstein et al. 2003). Human actions can also lead to feedbacks on climate. If climate change intensifies pressure to convert forests into pastures and cropland, then the climate change may be amplified by the human response (Raupach et al., Chapter 6, this volume). These positive feedbacks increase the fraction of the emitted $CO_2$ that stays in the atmosphere, increasing the growth rate of atmospheric $CO_2$ and accelerating climate change. Negative feedbacks are also possible. For example, a northward extension of forest or increased rates of plant growth in a warmer climate could increase rates of carbon storage, constraining further climate change.

A detailed quantitative assessment of a broad range of positive and negative feedbacks will require the use of fully coupled carbon-climate-human models. Such models do not exist yet. For some kinds of interactions (e.g., changes in institutional incentives and constraints on land use), predictive frameworks may be far in the future. Nevertheless, some experience has been gained already using coupled climate-carbon cycle models.

Two such coupled models that simulate the anthropogenic perturbation during the 21st century based on emission scenarios of the Intergovernmental Panel on Climate Change (IPCC) (Cox et al. 2000; Dufresne et al. 2002) show dramatically different magnitudes of climate-carbon cycle interactions (Figure 3.1). Both models simulate an accelerated increase of atmospheric $CO_2$ as a result of impacts of climate change on the carbon cycle. The magnitude of this feedback varies, however, by a factor of 4 between the two simulations. Without the carbon-climate interaction, both models reach an atmospheric concentration of 700 parts per million (ppm) by 2100. When the carbon-climate feedback is operating, the Hadley Centre model (Cox et al. 2000) reaches 980 ppm, leading to an average near-surface warming of +5K, while the IPSL model (Dufresne et al. 2002) attains only 780 ppm and a warming of +3K. This different behavior can be traced to the higher sensitivity of the land carbon cycle to warming in the Hadley Centre model, and to the larger ocean uptake in the IPSL model (Friedlingstein et al. 2003).

Although these pioneering model simulations represent a large step forward in scientists' ability to elucidate the interactions of the physical climate system with the global carbon cycle, they are also subject to important limitations. In both models, key processes are highly parameterized and poorly constrained. For example, emissions from land use changes are prescribed as an external input, and the associated changes in land cover are not explicitly modeled. In addition, a substantial number of processes and carbon pools are not included in such models. Several of these pools and processes could be vulnerable—that is, they are at risk of losing large amounts of carbon to the atmosphere as a result of a changing climate and/or human drivers (e.g., markets, policies, and demographic dynamics). Incorporating of all these pools and processes into coupled models is a difficult task, particularly because some processes operate through low-probability, high-consequence stochastic events. As a result, coupled climate-carbon models currently have limited capability for assessing the full magnitude and all risks associated with carbon-climate-human feedbacks, and their capability will continue to be limited for the foreseeable future.

An alternative and in many respects complementary approach is that of risk assessment, a method often used to identify and assess the risks associated with the operation of complex systems, such as nuclear power plants or chemical factories (Kammen and Hassenzahl 2001). In a risk analysis, the magnitude and likelihood of certain processes that may lead to catastrophic failures are determined and assessed in order to arrive at

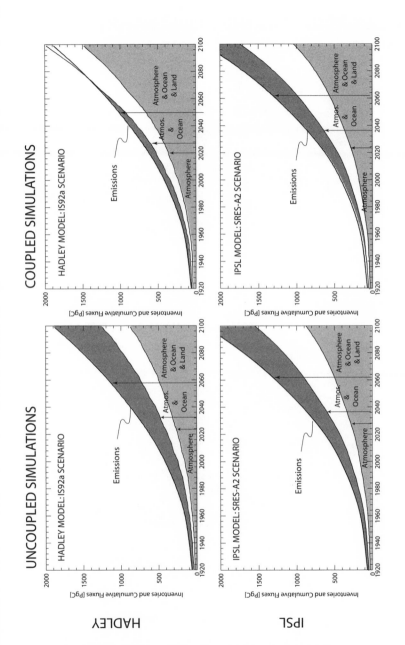

**Figure 3.1.** Cumulative fossil-fuel emissions and their redistribution among the three major reservoirs of the global carbon cycle over the period from 1920 to 2100 as modeled by two Earth system models (Cox et al. 2000; Dufresne et al. 2002). The left column represents the result of uncoupled simulations, in which changes in the physical climate system did not affect the carbon cycle, whereas results of the right column include such effects. The Hadley model (upper row; Cox et al. 2000) responds very sensitively to such feedbacks, whereas the carbon cycle within IPSL model (lower row; Dufresne et al. 2002) shows a much lower sensitivity. See also Friedlingstein, Chapter 10, this volume.

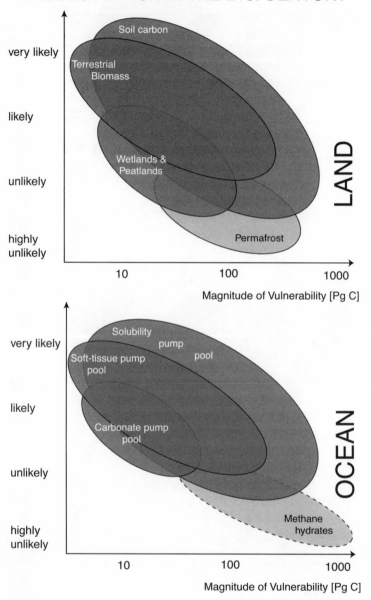

**Figure 3.2.** Vulnerability of the carbon pools in the 21st century. Shown are the estimated magnitudes and likelihood for these various pools to release carbon into the atmosphere. The exact size of the circles is not known and the boundaries therefore have to be viewed as being approximate only. See Table 3.1 for more details.

the outcomes and likelihood of a number of scenarios. Such approaches are most often used when dealing with systems whose dynamics cannot be captured with differential equations, and hence not encapsulated into numerical or analytical models.

In this chapter, we adopt an approach that is similar to a risk analysis and attempt to identify the magnitude and likelihood of key feedbacks in the carbon-climate-human system. We focus on the major carbon pools and quantify the maximum fraction of these pools that are potentially at risk over the next two decades and over the next century. We contrast these results with the maximum sink strengths of these pools over the same periods. Although we take advantage of model results, we mostly rely on data-based analyses. This approach to quantifying vulnerability is inherently subjective and uncertain. It offers, however, the possibility of going beyond the practical and sometimes fundamental restrictions associated with models and of including all pools and processes that are known to have the potential to influence atmospheric $CO_2$. We would like to emphasize that we restrict our analysis here to those changes that affect atmospheric $CO_2$. Other effects of human actions and climate change (e.g., loss of biodiversity and of economic activities currently and potentially related to it) may have impacts that are as large as or even larger than those on the carbon cycle, but they are beyond the scope of this discussion.

Figure 3.2 shows a summary of the vulnerable carbon pools, their magnitude, and the likelihood of the release of the carbon stored in these pools into the atmosphere over the 21st century. This figure is based on the detailed information in Table 3.1. Several key carbon pools may be vulnerable over the timescale of this century. The total amount of carbon stored in these pools is very large, two to five times as large as the total carbon in the current atmosphere. If a substantial fraction of the carbon in these pools were released to the atmosphere, the consequence would be a major increase in atmospheric $CO_2$ and hence a severe reduction of the anthropogenic $CO_2$ emissions permitted, if the atmosphere is to not exceed a certain target $CO_2$ concentration. As discussed in detail later in this chapter, we expect the magnitude of potential releases to increase with increasing stabilization targets, as the higher targets tend to be associated with a larger climate change. The potential release of carbon from these pools has to be viewed in the context of the sinks for atmospheric $CO_2$, which will probably continue to operate through the 21st century, although likely with gradually falling efficiency (Schimel et al. 2001). Because the potentially vulnerable pools are large, and because they can be rapidly mobilized for release, we believe that the majority of the feedbacks within the carbon-climate-human system are positive—that is, that they lead to an acceleration of the greenhouse gas–induced climate change.

This conclusion does not take into account the possibility of surprises. A small number of very large carbon pools may become vulnerable during the 21st century, especially the carbon stored in the very deep permafrost soils in Siberia and Alaska and the carbon stored in methane hydrates (Figure 3.2). Although we view the release of large quantities of carbon from these two pools as very unlikely, scientists' present understanding of the dynamics and controls of these two pools is insufficient.

**Table 3.1.** Summary of pools and their vulnerability over the next 20 years and over the course of the 21st century

| Pools | Stock size (PgC) | Sink capacity 20 yrs | Sink capacity 100 yrs | Vulnerability in 20 yrs[a] PgC | % | Confidence[b] | Vulnerability in 100 yrs[a] PgC | % | Confidence[b] | Processes | Direct human drivers |
|---|---|---|---|---|---|---|---|---|---|---|---|
| *Land* | | | | | | | | | | | |
| Total soils | 3,150 | | 300 | | | | | | | | Deforestation, cattle, management practices |
| Soils | 2,300 | | | 10 | 0.4% | Medium | 400 | 17% | Medium | Rh/erosion/fire | |
| Permafrost | 400 | | | 5 | 1.3% | Low | 100 | 25% | Low | Rh (T, moisture) | |
| Wetlands | 450 | | | 10 | 2.2% | Low | 100 | 22% | Low | Rh vs. methanogenesis/ T, water level | Rice paddies, cattle |
| Terrestrial living biomass | 650 | 40 | 150 | 50 | 7.7% | Medium | 200 | 31% | Medium | | |
| Tropical | 420 | | | | | | | | | Logging/fire/NPP vs Ra/Rh | Logging, fire, land use change, reforestation |
| Boreal | 57 | | | | | | | | | Logging/fire/NPP vs Ra/Rh | Logging, fire, land use change, reforestation |
| Temperate | 145 | | | | | | | | | Pest/diseases/fire/ logging/NPP vs. R | Land use change, reforestation, aforestation |
| TOTAL LAND (summed) | 3,800 | 40 | 450 | | | | | | | | |

*Ocean*

| | | | | | | | | | | | |
|---|---|---|---|---|---|---|---|---|---|---|---|
| Inorganic ocean carbon pool | 38,000 | | | | | | | | | | |
| "Inert" | 32,200 | 60–80 | 200–700 | 10 | 0.0% | Medium | 400 | 1% | Medium | Circulation | None |
| Solubility pump associated | 2,700 | | | 15 | 0.6% | Medium | 150 | 6% | | Circulation/salinity/T | None |
| Soft-tissue pump associated[c] | 2,500 | | | 10 | 0.4% | Low | 150 | 6% | Low | Circulation/nutrients/internal dynamics | Coastal eutrophication/deliberate fertilization/fisheries |
| Carbonate pump associated[c] | 600 | | | 2 | 0.3% | Low | 20 | 3.3% | Low | Circulation/nutrients/internal dynamics | Coastal eutrophication/deliberate fertilization/fisheries |
| DOC/POC | 725 | | | | | | | | | | |
| TOTAL OCEAN (summed) | 38,725 | 60–80 | 200–700 | | | | | | | | |
| Methane hydrates | >10,000 | | | ? | | | ? | | Extremely low | Ocean temperature | |

*Note*: T = temperature; R = respiration; Ra = autotrophic respiration; Rh = heterotrophic respiration; NPP = net primary production

[a]We define vulnerability as the maximum part of the total carbon stock that could be released into the atmosphere over the time scale indicated.

[b]Subjective assessment of our confidence

[c]Only the fraction that causes a change in atmospheric $CO_2$ is given here. The true changes in the pool are about five times larger

While the past 10 years have seen a large increase in the understanding of the present-day global carbon cycle and its anthropogenic $CO_2$ perturbation (Sabine et al., Chapter 2, this volume), scientists' understanding of how the carbon cycle will evolve into the 21st century and how it will interact with human actions and the physical climate system is overall still poor. This chapter is intended as one step on a long road to accurately incorporating the full suite of carbon-climate-human feedbacks in numerical models.

In assessing the future dynamics of the global carbon cycle, the natural starting point is the current status of the carbon cycle. This aspect is extensively reviewed by Sabine et al. (Chapter 2). We, therefore, concentrate only on those aspects that may cause a change in the carbon cycle over the next 100 years. In the next sections, we first briefly review the expected future trends in the climate system under typical IPCC scenarios. From these changes, we attempt to determine the strengths and vulnerability of the land and ocean carbon sinks in the 21st century.

## Setting the Stage: The Earth System in the 21st Century

The vulnerability of many carbon cycle processes and pools depends on the magnitude of future climate change. The magnitude of future climate change, in turn, depends on the vulnerability of the carbon cycle. Therefore there is no unique answer, but it is nevertheless instructive for our ensuing analysis to put some bounds on the expected climate change.

Cubasch et al. (2001) provide a comprehensive review of the projected climate change for the 21st century on the basis of coupled atmosphere-ocean-land climate models and for a large range of radiative forcing projections. For the IS92a emission scenario, they project a mean warming of 1.3°C for the mid-21st century (2021–2050) relative to the 1961–1990 average, with a range from +0.8°C to +1.7°C. This estimate includes the effect of sulphate aerosols, whereas the effect from greenhouse gases alone would be about 0.3°C higher. For the Special Report on Emissions Scenarios (SRES) A2 and B2 emission scenarios,[1] the mean and range are very similar for the same time period. For the end of the 21st century (2071–2100), the mean global surface temperature is projected to increase by 3.0°C, with a range from 1.3°C to 4.5°C, while for the B2 scenario the mean temperature change is +2.2°C and the range is +0.9°C to +3.4°C. These individual scenarios can be compared with the full set of scenarios, for which Cubasch et al. (2001) give a range of +1.4°C to +5.8°C. Unfortunately, no comprehensive coupled climate models have simulated climate change using WRE (Wigley, Richels, Edmonds 1996) stabilization trajectories. Instead, to estimate the temperature changes associated with the WRE scenarios, Cubasch et al. (2001) used a simple climate model (Raper and Cubasch 1996) that has been tuned to the results of a number of comprehensive climate models. They find for the WRE 450 scenario (assuming stabilization at an atmospheric $CO_2$ concentration of 450 ppm), a temperature change between +1.2°C and +2.3°C for the latter part of this century, increasing to +1.6°C to +2.9°C for WRE550, and +1.8°C to +3.2°C for WRE650. The general pattern of warming is relatively uniform across a large range of models and generally consists of maximum warming in the high latitudes of the Northern Hemisphere and a minimum in the Southern Ocean. There is also a general tendency for the land to warm more than the ocean.

In some cases, even more important for the global carbon cycle is the projected change in the hydrological cycle. While uncertainties and intermodel differences are substantially larger for the hydrological cycle than for the projected temperature changes, the following trends appear to be quite robust: a general increase in the intensity of the hydrological cycle—that is, an increase in mean water vapor, evaporation, and precipitation; and a tendency for an increase in precipitation in the tropics and in the high latitudes and a reduction in precipitation in the subtropical latitudes (Cubasch et al. 2001).

As the coupling between the carbon cycle and human systems is tight, particularly with regard to the terrestrial carbon cycle, common features and regional differences in the development of population, the economy, and social systems in the 21st century are crucial in determining whether certain pools become at risk (Romero Lankao, Chapter 19, this volume). The SRES story lines of Nakicenovic et al. (2000) provide a basis for evaluating these interactions. In the A1 and B1 scenarios, global population is expected to increase to 9 billion people in 2050 and to decrease thereafter to about 7 billion people in 2100. In the A2 and B2 scenarios, global population is expected to grow continuously in the 21st century. Raupach et al. (Chapter 6, this volume) assess some impacts of these scenarios on land use change. We expect that the A2 scenario will lead to highest pressures on local resources, because of the underlying assumption of regionally different impacts on land, and "low incomes in developing countries lessening their ability to meet food demand by improved management or technology" (Raupach et al., Chapter 6).

In summary, the expected stresses on the global carbon cycle might range from relatively small to very substantial, depending on greenhouse gas emission pathways, the magnitude of climate sensitivity, and the development of human systems with regard to their pressure on terrestrial carbon resources. Given this non-negligible risk, it is of great importance to assess the magnitude of the carbon system feedbacks in greater detail. We do not adopt any specific emission scenario or climate sensitivity but tend to use the upper-bound scenarios, consistent with quantifying the maximum expected vulnerability (note that these are nonintervention scenarios, i.e., scenarios that do not contain purposeful human action to curb emissions or increase sinks).

## Land Sinks and Their Vulnerability

How will future anthropogenic emission pathways affect the terrestrial carbon cycle? As summarized in Sabine et al. (Chapter 2, this volume), the land biosphere has acted as a substantial net sink for carbon over the past 20 years, removing on average about 0.6

petagrams of carbon (PgC) per year ($y^{-1}$) from the atmosphere. This net sink is the residual of a land use change flux of the order of 1.2 PgC $y^{-1}$ (DeFries et al. 2002) and an inferred uptake of the order of 1.8 PgC $y^{-1}$ (Table 2.1 and Colorplate 1 in Sabine et al., Chapter 2). Assessing the responses of the terrestrial carbon cycle to the projected climate and human factors in the 21st century will require separating the processes that lead to sources from those that lead to sinks. This attribution task is complicated, because the carbon balance on land is, in general, a small difference between large fluxes in and out. As a consequence, modest proportional changes in fluxes in or out of a pool can have a large impact on the magnitude of the residual sink.

## Terrestrial Living Biomass

Living biomass on land contains about 650 PgC, which is slightly less than the amount of carbon in the atmosphere. With an average residence time of about 10 years, the turnover of this pool is relatively fast, making a substantial fraction of this pool potentially vulnerable on timescales of decades to centuries. It is also likely that most of the carbon gained by the terrestrial biosphere over the past few decades has entered this pool and has not progressed in any substantial manner into the longer-lived soil pools (Schlesinger and Lichter 2001). The biomass pool can expand only slowly as a result of increased plant growth or decreased disturbance, but this pool can be degraded quickly, especially with natural and human-induced disturbances.

### Land Sinks in the 21st Century

As discussed in Sabine et al. (Chapter 2), scientists' perspective on the current terrestrial carbon balance has changed dramatically over the past decade. A decade ago, it was assumed that the entire land sink is a result of $CO_2$ fertilization, and perhaps warming (Houghton et al. 1990). This view has been supplanted by an appreciation of the fact that the current and future land sinks are most likely the result of many factors, including land use history and land cover management, and that $CO_2$ fertilization is likely smaller than previously assumed (e.g., Pacala et al. 2001).

These changes in knowledge are only beginning to be incorporated into terrestrial carbon cycle models (McGuire et al. 2001). In both the Hadley and IPSL models discussed earlier (Cox et al. 2000; Dufresne et al. 2002), $CO_2$ fertilization is assumed to be the main driver of the present and future terrestrial carbon sink. In the absence of climate change, the projection from these two models is that $CO_2$ fertilization would be responsible for a cumulative land uptake of 600 PgC for a doubling of atmospheric $CO_2$ (Friedlingstein et al. 2003). In the Hadley model, a substantial fraction of the carbon that is being released back into the atmosphere as a result of surface warming comes from soil pools that have grown significantly during the 20th and 21st centuries as a result of $CO_2$ fertilization. Projections of large sinks from $CO_2$ fertilization are difficult to reconcile with the present understanding of the current land carbon cycle, where a number of constraints (Field et al. 1992) are likely to limit potential land sinks from $CO_2$ fertilization to the order of 100 PgC over the next 100 years.

The contribution of land use history to the future land sink is rather uncertain. In general, forest sinks due to regrowth of previously harvested forests tend to saturate as a forest matures or are rapidly converted to large sources in forests disturbed by fire, storms, or disease. Key controllers of the future trajectory of sinks in Northern hemisphere forests include the forest age structure, changes in the disturbance regime, and changes in management (Nabuurs, Chapter 16, this volume). The land sink in the 21st century as a result of historical land use change is, at maximum, on the order of 170 PgC (House et al. 2002). This amount assumes that all carbon lost as a result of the past land use change could be stored back on land. A realistic estimate should be less than half this number, since humans still use most of the land deforested in the past and since the land not used for food production is often of poor quality.

From the perspective of the carbon budget of the 21st century, a slowing of deforestation has the same effect as a decrease in fossil emissions or an increase in a land or ocean sink. In addition, reforestation and afforestation may be increasingly important aspects of future carbon management. The extent to which new forest sinks can be realized must be consistent with other societal constraints on available land (Raupach et al., Chapter 6, this volume). The need for expanded areas for agriculture and for replacement of degraded agricultural lands, plus constraints of climate and water availability, are likely to be profoundly important. Given these constraints, reforestation and afforestation are likely to continue to generate carbon sinks throughout the 21st century, but their magnitudes are unlikely to be much more than a few Pg for the next 20 years and a few tens of Pg for the entire 21st century.

Woody encroachment, including the establishment of trees and shrubs on grassland, as well as the thickening of open forests, has been an important aspect of land dynamics on several continents over the last century (Sabine et al., Chapter 2). The driving factors for woody encroachment are too poorly known to support a quantitative estimate of future changes, particularly since its net contribution to atmospheric $CO_2$ can be positive or negative. We expect this process, however, to be a relatively small contributor, and one whose future carbon trajectory will be primarily determined by human actions.

Fires are a major factor in the terrestrial carbon cycle, annually returning more than 3 PgC to the atmosphere (see Colorplate 1 of Sabine et al., Chapter 2). Because fire has always been a part of the land biosphere and recovery from fires is slow, its role as a carbon source or sink depends on fire frequency and intensity trends over decades to centuries rather than on present burning. Over the past few decades, some regions have probably experienced carbon sinks as a consequence of the success of fire suppression (Sabine et al., Chapter 2). It is unclear whether this trend will continue, as fuels accumulate. We believe that the carbon sink that arises from fire suppression is going to be very small in this coming century but that an increase in fire frequency and intensity is a significant concern (see below).

In summary, the sink potential in living terrestrial biomass over the next 20 years is modest, on the order of a few tens of Pg. Terrestrial sinks may grow somewhat from current levels, but societal and natural limits from land use, fire, and nutrient availability will operate as constraints. Over the next 100 years, some mechanisms and societal drivers contributing to land sinks will continue to operate, but not all (see Raupach et al., Chapter 6, this volume) giving rise to a maximum potential sink of the order of 150 PgC.

## POTENTIAL VULNERABILITY OF THE LAND POOLS

The carbon pools on land also have the potential to release substantial amounts of carbon into the atmosphere. Many of the mechanisms that are responsible for generating current sinks are closely related to mechanisms that can lead to future sources of carbon to the atmosphere.

Probably the largest future vulnerability of land pools arises from land use change and deforestation, particularly if the world develops along lines similar to those described in the A2 story line—that is, a world in which countries develop in a very heterogeneous pattern. The exact magnitude of the pools at risk is difficult to assess, but we expect that the carbon stored in the living biomass in tropical and subtropical ecosystems is the most vulnerable. Based on past trends, we estimate that on the order of 40 PgC carbon stored in the living biomass might be at risk over the next 20 years, and up to 100 PgC over the course of the 21st century. More subtle changes to land ecosystems, such as human-induced degradation and increases in mortality due to pests might increase this estimate, but probably not by a significant amount.

Deforestation has a number of other implications for climate and the carbon cycle. One set of effects concerns albedo. Boreal deforestation leads to large increases in albedo, with a cooling effect that may be larger than the warming related to the $CO_2$ release (Bonan et al. 1992; Betts et al. 1997). In the tropics, the albedo effect is much smaller, and effects of rooting depth on transpirable soil moisture may feed back to increase local temperature and decrease precipitation (Shukla et al. 1990). Finally, deforestation also removes the potential that growth of an established forest can serve as a carbon sink.

A second important mechanism that could lead to substantial losses of carbon from the terrestrial biosphere is the redistribution of species and biomes as a result of climate change. It is estimated that the entire terrestrial biosphere gained about 500 PgC between the last glacial maximum (about 20,000 years ago) and the ensuing Holocene (Sigman and Boyle 2000). Whereas the postglacial warming of about 5K occurred over 10,000 years, however, global warming over the next century might result in a similar temperature change, but over a timescale of 100 years. This climate perturbation is likely to occur at a rate exceeding the capacity of plant species to adapt, resulting in a total rearrangement of species and communities. This change will result in transient carbon losses that can eventually be recovered, but only over a much longer timescale (Smith and Shugart 1993).

In the Northern Hemisphere, a warming may lead to a forest dieback at the southern boundaries of presently forested regions, although increase in spring or fall temperature may extend the growing season length of temperate and boreal ecosystems. Invasive species could spread in these regions. Direct human action on species composition, however, including establishment of plantations of non-native "exotic" species, is also likely to continue. The net effect on carbon reservoirs is difficult to estimate but is more likely to be a source than a sink, and the total forest area is more likely to shrink than to expand. In the tropics, warming and increased aridity may lead to a size reduction of the tropical forests (Cox et al. 2000). Given that tropical forests contain the largest carbon pool of terrestrial biota and also the largest net primary production (see Table 2.2 in Sabine et al., Chapter 2), this reduction will likely lead to a release of carbon to the atmosphere.

In summary, we estimate that on the order of 40 PgC carbon stored in the living biomass might be at risk over the next 20 years, and up to 100 PgC over the course of the 21st century. This risk arises mostly from deforestation and other anthropogenic land use changes, An additional 10 PgC (20 years) and 100 PgC (100 years) is probably at risk as a result of climate change—that is, the inability of communities to adapt to a changed climate.

## Soil Carbon

Soil carbon is by far the largest carbon pool in terrestrial ecosystems. As summarized in Colorplate 1 of Sabine et al. (Chapter 2), about 2300 PgC is stored in the upper 3 meters (m) of tropical, temperate, and boreal soils. Carbon stocks in wetland and permanently frozen soils are much less certain and are usually not included in carbon cycle models. These stocks are in the range of 200–800 PgC in wetland soils and 200–800 PgC in (nonwetland) permafrost soils. There is therefore more than five times more carbon stored in soils than in the living biomass on land. This large organic carbon reservoir is the result of a delicate balance between biomass accumulation by photosynthesis and oxidation by respiration. Most of this soil carbon is in long-lived pools and therefore relatively inert. Over the course of this century, however, a substantial fraction of this pool might be mobilized. Given the very different dynamics of the various soil pools, we review separately the wetland, the frozen, and the "other" soil pools.

### TROPICAL, TEMPERATE, AND BOREAL SOILS: POTENTIAL SINKS AND VULNERABILITY

Carbon sequestration in soil can be achieved through changes in agricultural practices (i.e., no tillage) or with land use changes, such as reforestation of degraded lands. The global potential for carbon sequestration in soils is estimated to be about $0.9 \pm 0.3$ PgC $y^{-1}$, with different patterns and rates in European cropland (Smith et al. 2000), U.S. cropland (Lal et al. 1998), and degraded lands (Lal 2003). These estimates of carbon sequestration potential do not account for limitations due to land use, erosion, climate change or economic, social, and political factors (Raupach et al., Chapter 6, this volume). Increased net primary production (NPP) resulting from fertilization of plant growth by elevated $CO_2$ and nitrogen deposition will tend to increase soil carbon. If these increases are not offset by increased decomposition in a warmer climate, the equilibrium increases in soil carbon should reach the same proportional size as the NPP stimulation. Assuming a fertilization effect of the order of 20 percent of NPP and equilibrium conditions for organic matter decomposition, the maximum possible increase in soil C from $CO_2$ and N effects is on the order of 300 PgC over the 21st century. This is considerably smaller than the more than 600 PgC accumulation simulated in the Hadley model.

Losses of soil carbon can be caused by decreased inputs from NPP, accelerated decomposition, and increased losses from erosion and combustion. Each of these mechanisms has human and natural drivers (e.g., deforestation, increasing energy consumption, diversion of water bodies). Boreal soils contain huge stocks of carbon. These stocks could increase if tree growth increases in a warmer climate (Dufresne et al. 2002), but they could also decrease, especially if increases in decomposition are larger than increases in NPP or if an increase in fire frequency and magnitude reduces the input of carbon to the soil. Increasing temperature and more frequent droughts (due to increase of climate variability) can enhance desertification in subtropical and tropical regions, stimulating losses of soil carbon. Soil erosion due to changes in the hydrological cycle is also a major disturbance to soil carbon in the semi-arid regions. Climate regulates carbon release by soils through effects of temperature and soil moisture changes. The way in which these two factors interact, however, is highly nonlinear and often opposite (i.e., a temperature increase will increase soil carbon losses while a soil moisture decrease will reduce decomposition but increase soil temperature). In areas where increased temperature and associated drought are expected, the predictions of increased losses of soil carbon losses are debatable. The two coupled climate-carbon simulations from Hadley and IPSL simulate a total loss of soil carbon as high as 600 PgC by the end of the 21st century. As stated earlier, however, these estimates should be seen as unlikely upper estimates as the carbon release is fueled by a large accumulation of soil carbon, resulting from $CO_2$ fertilization. We estimate from FLUXNET data (Sanderman et al. 2003) that about 8 kilograms (Kg) C per square meter ($m^{-2}$) in 100 years could be potentially released by soils as a result of changes in climate, assuming a temperature change of 4°C. This amount reflects a potential loss of about 280 PgC in the next century.

Land use change, however, is the major driver of carbon losses in the soil. Since 1850 between 45 and 90 PgC (Houghton et al. 1999; Lal 1999), have been lost through cultivation and disturbances. One of the main characteristics of land use effects is that carbon fluxes are pulsed in response to disturbance frequency and intensity. The effects of such pulses can persist for several years after disturbances and are usually one of the major uncertainties in current models.

Land use change pulses of carbon are also directly influenced by climate-human interactions (i.e., desertification, which is the result of human pressure and aggravated temperature and water impacts). Thus the interaction of direct human effects with climate impacts can amplify the carbon losses, particularly in developing regions. An analysis of recent studies on the impact of land use conversion on soil carbon stocks suggests that up to 50 percent of the total pool of soil carbon can be lost in 50–100 years after conversion (Schulze et al. 2003). Of course not all the land use changes occur in organic rich soils. We therefore estimate that about 25 percent of the total soil carbon pool is vulnerable over the course of this century, equivalent to about 400 PgC.

## WETLANDS AND PEATLANDS: POTENTIAL SINKS AND VULNERABILITY

Wetland carbon fluxes are controlled by climate and land use change. Climate drivers include temperature effects on carbon and other greenhouse gas (mainly $CH_4$) fluxes. In addition, changes in hydrology can have pronounced effects on both carbon and other greenhouse gas fluxes. Temperature increases stimulate both $CO_2$ and methane emissions, while changes in the water table have contrasting effects on those two fluxes. Drainage stimulates oxidation of organic material and release of $CO_2$; in contrast, increasing water levels reduces $CO_2$ emission but stimulates anaerobic decomposition and methane fluxes to the atmosphere.

To estimate the vulnerability of wetland pools of carbon to climate effects requires a detailed analysis of the interaction between $CO_2$ and other trace gases. The global warming potential (GWP) of methane is about 20 times larger than that of $CO_2$. Therefore, a change in the ratio of methane to $CO_2$ respiration will have a climate impact. Christensen et al. (1996) showed that large northern wetlands, for example, are leading to a net increase in radiative forcing because of their large methane emissions even though they are a net carbon sink. On a global scale, this net source of radiative forcing is equivalent to about 0.5 PgC y$^{-1}$. Assessing the future evolution of the wetland and peatland emissions of $CO_2$ and $CH_4$ is very difficult, as this depends critically on temperature and water levels. If we accept the climate projections that regions that are already receiving substantial rainfall will receive even more in a future warmer climate, then we would expect an increase in the water level, stimulating the production of methane. In addition, higher temperatures would lead to an increase in respiration rates. Both factors conspire to give a relatively high sensitivity of the carbon pools in wetlands and peatlands. A rough estimate is that a $CO_2$ equivalent of about 100 PgC will become vulnerable in the next 100 years.

An interesting aspect of the interaction between $CO_2$ and $CH_4$ emissions in both natural and human-managed wetlands and peatlands is that drainage control provides the possibility of generating a sink potential, as a lowered water table reduces $CH_4$ emissions at the expense of an increase in the $CO_2$ emissions. Other environmental concerns (loss of biodiversity, habitats, etc.) should also be considered, however.

PERMAFROST: POTENTIAL VULNERABILITY

Permanently frozen soils contain a very large pool of carbon, the result of a gradual accumulation over thousands of years. As long as these soils are frozen, the carbon they contain is essentially locked away from the active part of the carbon cycle. When they thaw, however, the carbon becomes vulnerable to rapid loss. Using results from several climate models, all scaled to a global mean warming of 2°C, Anisimov et al. (1999) estimate that the area of permafrost could shrink by about 25 percent. They also estimate an increased summer thaw depth in the areas that maintain permafrost. Current knowledge is too incomplete to convert this loss of area into a loss of carbon, but the pool of carbon vulnerable to this relatively modest warming could total more than 100 PgC. If a substantial fraction of the C loss occurs as methane emissions, as is typical of wetlands, the impact on radiative forcing will be amplified. Assuming that total permafrost carbon is 400 Pg and 20 percent is lost in the next century, the potential release is 80 PgC.

# The Ocean Sink and Its Vulnerability

The ocean represents the largest reservoir of carbon in the global carbon cycle (see Colorplate 1 of Sabine et al., Chapter 2) and has taken up approximately 30 percent of the total anthropogenic emissions since the beginning of the industrial period, constituting the second-largest sink of anthropogenic carbon after the atmosphere itself (Le Quéré and Metzl, Chapter 12; Sabine et al., Chapter 2). Therefore, relatively small changes in the oceanic reservoir and its current sink activity may have a large impact on the future trajectory of atmospheric $CO_2$.

## The Ocean Sink Potential in the 21st Century

On millennial timescales, more than 95 percent of the anthropogenic $CO_2$ will be taken up by the ocean (Archer et al. 1997). On timescales of decades to centuries, however, this large sink potential can be realized only to a limited degree, with the rate-limiting step being the transport of the anthropogenic carbon from the surface into the interior of the ocean (Sarmiento et al. 1992). Because the primary driving force for the oceanic uptake of anthropogenic $CO_2$ is the atmospheric $CO_2$ concentration, the magnitude of the oceanic sink in the 21st century depends on the magnitude of the atmospheric perturbation. In the absence of major feedbacks, paradoxically, the higher the anthropogenic $CO_2$ burden in the atmosphere, the higher the oceanic sink is going to be.

Throughout the anthropocene, the oceanic uptake of anthropogenic $CO_2$ has scaled nearly linearly with the anthropogenic perturbation of atmospheric $CO_2$ (Gloor et al. 2003) ($0.027 \pm 0.008$ PgC $y^{-1}$ ppm$^{-1}$). This is because the atmospheric $CO_2$ so far has grown quasi exponentially, and the rate-limiting process is linear. This scaling cannot be extrapolated far into the future, as the atmospheric growth rate may change substantially. More important, this scaling is expected to decrease over time as the continued uptake of $CO_2$ by the ocean decreases the oceanic uptake capacity for a given change in atmospheric $CO_2$ (Greenblatt and Sarmiento, Chapter 13, this volume). Using this linear scaling and projected atmospheric $CO_2$, we estimate the maximum oceanic sink potential for the next 20 years in the absence of climate change to be on the order of 60–80 PgC. This estimate compares well with estimates from ocean models that were forced using the IS92a scenario (Watson and Orr 2003). The maximum oceanic sink for the next 100 years is more difficult to estimate as it depends strongly on the atmospheric $CO_2$ level attained. If we adopt 1,000 ppm as the upper envelope, the cumulative ocean uptake over the next century, in the absence of climate change, might approach 600–700 PgC. In the case of the stabilization scenarios, the oceanic sink is going to be substantially smaller, as the slower growth rate and the smaller atmospheric $CO_2$ burden both lead to a reduction in the uptake.

## Oceanic Carbon Pools and Processes at Risk

How sensitive is this oceanic sink to natural and human-induced changes over the next 20 to 100 years? Greenblatt and Sarmiento (Chapter 13, this volume) review and discuss several detailed studies on the basis of coupled atmosphere-ocean models. Plattner et al. (2001) provide a very detailed analysis of the nature and magnitude of the various feedbacks using a model of intermediate complexity. The goal here is to put these results in perspective by analyzing and assessing the pools and processes that might become at risk in the 21st century from a systems perspective. It is thereby instructive to differentiate between the factors that affect the physical/chemical uptake of anthropogenic $CO_2$ and those that change the natural carbon cycle in the ocean. The former feedbacks are associated with the anthropogenic $CO_2$ perturbation in the atmosphere, whereas the latter feedbacks are independent of the anthropogenic $CO_2$ changes in the atmosphere and arise because of changes in climate and other factors. Table 3.2 gives a summary of the six feedbacks that can either accelerate or decelerate the flux of carbon from the atmosphere into the ocean.

### CHEMISTRY FEEDBACK

The feedback that is understood and quantified best is the reduction of the oceanic uptake capacity as a consequence of the uptake of $CO_2$ from the atmosphere (see Figure 13.3 in Greenblatt and Sarmiento, Chapter 13, this volume). In addition to lowering the uptake capacity for anthropogenic $CO_2$, this reaction also lowers the pH of the ocean, which could have potentially important consequences for ocean biology and coral reefs (discussed later in this chapter). Sarmiento et al. (1995) demonstrated that the magnitude of the chemistry feedback for the next 20 years remains small but could lead to reduction of more than 30 percent in the cumulative uptake over the next 100 years, with the exact magnitude depending on the size of the anthropogenic perturba-

**Table 3.2.** Summary of marine carbon cycle feedbacks

| Process | Feedback | 20 years (PgC) | 100 years (PgC) | Uncertainty/ understanding |
|---|---|---|---|---|
| *Anthropogenic $CO_2$ uptake feedbacks* | | | | |
| Chemical feedback | Positive | <5 | 300 | Low/high |
| Circulation feedback (anthropogenic $CO_2$) | Positive | 6–8 | 400 | Medium/medium |
| *Natural carbon cycle feedbacks* | | | | |
| Temperature/salinity feedback | Positive | 15 | 150 | Low/medium to high |
| Ocean biota feedback | Positive/ negative | 10–15 | 150 | High/low |
| Circulation feedback (natural) | Negative | –20 | –400 | Medium/medium |
| Methane feedback | Positive | 0 | ? | Extremely high/low |

tion (see also Yi et al. 2001). This positive feedback is fully implemented in ocean carbon cycle models and therefore seldom explicitly discussed.

OCEAN CIRCULATION FEEDBACK ON ANTHROPOGENIC $CO_2$

Still relatively well understood but more difficult to quantify is the potential impact of ocean circulation changes on the uptake of anthropogenic $CO_2$. Current climate models tend to show that the warming of the surface ocean, together with a decrease in high latitude salinity as a result of increased precipitation, will lead to a reduction in the surface ocean density relative to that of the underlying waters, thereby increasing vertical stratification. Such an increase in stratification will lead to a reduction of the exchange of surface waters with deeper layers, reducing the downward transport of anthropogenic $CO_2$ and hence reduce the oceanic uptake of anthropogenic $CO_2$ from the atmosphere (Sarmiento et al. 1998). Less well established is the impact of climate change on deepwater formation rates and the meridional overturning circulation (Greenblatt and Sarmiento, Chapter 13, this volume).

To estimate the magnitude of the ocean circulation feedback on the oceanic uptake of anthropogenic $CO_2$, we assume that over the next 20 years the surface to mid-thermocline density gradient changes by about 10 percent (equivalent to a temperature change of 1.5°C). If we further assume that the reduction of the anthropogenic $CO_2$ uptake scales inversely with the increase of the vertical density gradient, we estimate that these changes in upper ocean stratification lead to a reduction in the cumulative uptake over the next 20 years on the order of 10 percent, or about 6–8 PgC. Over the next 100 years, the ocean circulation feedback could probably be up to three to four times as large, reducing the oceanic uptake of anthropogenic $CO_2$ by up to 40 percent, with the absolute magnitude depending on the size of oceanic sink. Our estimate here is quite a bit larger than the model-based estimates of 3 percent to 21 percent (Sarmiento and Le Quéré 1996; Sarmiento et al. 1998, Joos et al. 1999; Matear and Hirst 1999; Plattner et al. 2001; see summary by Greenblatt and Sarmiento, Chapter 13) mainly because we assumed a larger change in stratification than was simulated by these models.

The temperature, salinity, and circulation changes already discussed not only influence the uptake of anthropogenic $CO_2$, but also affect the natural carbon cycle within the ocean, as well as the vast quantities of methane hydrates that are stored along the continental shelves. We look at these changes in turn, focusing first on the individual effects and thereafter at their interactions.

## TEMPERATURE AND SALINITY FEEDBACKS

The amount of dissolved inorganic carbon (DIC) in seawater for a given atmospheric partial pressure and ocean total alkalinity is highly temperature dependent, with a reduction of about 8 millimoles (mmol) per cubic meter ($m^{-3}$) in DIC for each degree of warming. Therefore, assuming a maximum surface ocean warming of 2°C and a 0.5°C warming over the upper 300 m in the next 20 years would lead to an equilibrium loss of about 15 PgC. Similarly, we estimate that a maximum surface warming of 5°C and a 2°C warming over the upper 1,000 m over the next 100 years would lead to an equilibrium loss of carbon into the atmosphere of more than 150 PgC. These warming-induced losses represent about a 20 percent reduction in the net oceanic uptake for atmospheric $CO_2$. Models generally find a smaller magnitude of the temperature feedback with a reduction of the order of 50–70 PgC (10–14 percent) over the 21st century (Greenblatt and Sarmiento, Chapter 13, this volume), mostly because their warming is smaller than the magnitude we assumed. Changes in surface salinity also have the potential to alter the ocean atmosphere distribution of inorganic carbon, on the one hand through changes in the solubility of $CO_2$ and on the other hand through changes in surface ocean alkalinity (dilution effect). We consider the salinity feedback to be a minor factor on the global scale for the 21st century, as we expect relatively small changes in global mean surface salinity, an assessment shared by the model simulations of Plattner et al. (2001).

## CIRCULATION FEEDBACKS (OCEAN BIOTA CONSTANT)

The increase in vertical stratification and the other circulation changes not only reduce the uptake of anthropogenic $CO_2$ from the atmosphere, but also affect the natural carbon cycling substantially, even in the absence of any changes in ocean productivity. An important distinction between this circulation feedback and that associated with the anthropogenic $CO_2$ uptake is that the latter feedback affects only the rate at which the atmospheric $CO_2$ perturbation is equilibrated with the ocean and does not change the long-term equilibrium, whereas the former feedback changes the equilibrium distribution of carbon between the ocean and atmosphere.

As it turns out, in the presence of the biological pumps, these circulation changes appear to lead to a negative feedback—that is, an increase in the uptake of atmospheric $CO_2$ from the atmosphere. The main reason for this somewhat surprising result is that a slowdown of the surface to deep mixing also reduces the upward transport of re-mineralized DIC from the thermocline into the upper ocean, while the downward transport of biologically produced organic carbon remains nearly unchanged.

To estimate the magnitude of this feedback, we assume that global export production by particles remains at about 10 PgC $y^{-1}$ and that the upward supply of DIC to compensate in the steady state for this downward organic carbon transport changes inversely proportionally to the increase in vertical stratification. Adopting our previously estimated changes in surface stratification (10 percent over the next 20 years and 40 percent over the next 100 years), we estimate the cumulative reduction in the upward supply of DIC to be 20 PgC over the next 20 years and 400 PgC over the next 100 years. Not all of this reduced supply of DIC will be compensated for by an additional uptake of $CO_2$ from the atmosphere. Model simulations suggest that this ratio is about half, leading to an estimate of this negative feedback of about 10 PgC over the next 20 years and about 200 PgC over the next 100 years. The model simulations summarized by Greenblatt and Sarmiento (Chapter 13) suggest a range from 33 to more than 110 PgC over the anthropocene. They also show that the models that have a larger positive circulation feedback associated with the anthropogenic $CO_2$ uptake tend to have a large negative circulation feedback associated with the natural carbon cycle. This result is not unexpected, as the two feedbacks are tightly linked with each other.

## MARINE BIOTIC FEEDBACKS (AT CONSTANT CIRCULATION)

The marine biota affects the cycling of carbon in the ocean in two fundamentally different ways. One is through the formation of organic matter in the surface ocean by photosynthesis and the subsequent export of this material into the ocean interior (soft-tissue pump), and the other is through the biogenic formation of $CaCO_3$ shells in the upper ocean, which also tend to sink and dissolve at depth (carbonate pump) (see also Sabine et al., Chapter 2). We will first focus on the soft-tissue pump, because it is quantitatively more important, and discuss the carbonate pump thereafter.

The pool of organic matter in the ocean is small, but most of it turns over on timescales shorter than a year. As a consequence, when considering the impact of climate change on the natural carbon cycle and in particular ocean biota, it is insufficient to look at the pool size of organic matter in the ocean. Instead we have to focus on the time-integrated effect of ocean biology on the oceanic DIC distribution, in particular, the fraction of the surface-to-deep gradient in DIC in the ocean that is induced by the soft-tissue pump. Gruber and Sarmiento (2002) estimated that about 50 percent of the surface-to-deep gradient in oceanic DIC is due to the soft-tissue pump (about 150 micromoles [μmol] kg$^{-1}$). Integrated over the global ocean, this inorganic carbon pool of biological origin amounts to about 2,500 PgC. Because the ocean $CO_2$ system is strongly buffered, we estimate that only about 10–20 percent of this "biological" carbon comes from the atmosphere, with the remainder coming from the oceanic DIC pool. As a consequence, only a fraction of any loss from this biological DIC pool ends up changing atmospheric $CO_2$. Model simulations suggest, for example, that a complete die-off of ocean life would lead to an atmospheric $CO_2$ increase of about 150–200 ppm only (300–400 PgC) (Gruber and Sarmiento 2002).

The biological pump in the ocean does not operate at full strength, however, as there are many regions where surface nutrients are not entirely used. Only about half of the global nitrate pool is currently associated with the biological carbon pool, offering the possibility of almost doubling the biological DIC pool to about 5,000 PgC. If we assume again that about 10 percent of this carbon comes out of the atmosphere, we find that the absolute upper and lower bounds of how changes in organic matter export can influence atmospheric $CO_2$ are on the order of ± 250 PgC, an estimate consistent with the recent model simulations by Archer et al. (2000). The fraction of the change in the biological DIC pool that shows up in the atmospheric $CO_2$ pool appears to be highly model dependent, for reasons not fully understood (Archer et al. 2000).

As very little is known about the sensitivity of marine export production to climate change, estimating how it might change over the next 20–100 years is by necessity very uncertain. Our estimates, therefore, must be viewed with caution. It is nevertheless instructive to determine possible upper bounds and to put them into perspective with the other pools and processes that might change atmospheric $CO_2$ in the 21st century.

Over the next 20 years, we estimate that global export production will not change by more than about about 3 PgC y$^{-1}$ (25 percent). This estimate would lead to a maximum change in the biological carbon pool of about 60 PgC (4 percent), which would lead to a change in the atmospheric $CO_2$ pool of only about 6–12 PgC. There are many possible reasons for changes in global marine export production, including changes in surface ocean physical properties, changes in the delivery of nutrients from land by rivers and atmosphere, and internal dynamics of the ocean biota, including complex predator-prey interactions and fisheries-induced pressures on marine predators (see Boyd and Doney 2003 for a comprehensive review). Over the next 100 years, we estimate that the maximum change of the biological carbon in the ocean is likely less than about half of the total pool or maximally about 800 PgC. Taking into account the century timescales for this perturbation in the biological pool to equilibrate with the atmosphere, we believe that the maximum change in the atmospheric $CO_2$ pool over the next 100 years due to this effect is about 100 to 150 PgC, or of a magnitude similar to the glacial-interglacial $CO_2$ changes.

There exist possibilities for surprises, though, as our understanding of the mechanisms controlling marine productivity and the subsequent cycling of organic matter in the ocean is limited. For example, while marine export production is generally believed to be controlled primarily by the supply of nutrients and the availability of light (bottom-up control), grazing pressure (top-down control) also plays a major role. With the enormous fishing pressure on the top predators in the ocean, marine foodwebs have already been altered fundamentally in many regions (Jackson et al. 2001) and are expected to change further. On the basis of our current understanding, this human-induced change in foodweb structure should have a relatively small impact on global marine export production, but not enough is known today to exclude the possibility for a bigger role.

We next turn to the impact of climate change on the carbonate pump. Gruber and Sarmiento (2002) estimated that the surface-to-deep gradient generated by this pump amounts to about 60 $\mu mol/kg$, giving rise to a calcium carbonate ($CaCO_3$) pump-induced DIC pool in the ocean of about 600 Pg. The effect of changes in this pool on atmospheric $CO_2$ depends on the timescale considered. On timescales less than a couple of hundred years, an increase in this pool size tends to increase atmospheric $CO_2$. This increase is because the formation of $CaCO_3$ lowers the alkalinity of the surface ocean more than it lowers DIC, thereby reducing the buffer capacity of the ocean. On very long timescales, this effect is counteracted by interactions with the ocean sediments (Archer et al. 1997).

Therefore, if changes in the export of $CaCO_3$ from the surface ocean are proportional to the changes in the export of organic matter (a constant rain ratio) as expected—for example, if $CaCO_3$ plays an important role as a mineral ballast (Armstrong et al. 2002; Klaas and Archer 2002)—this process would have a counteracting effect on atmospheric $CO_2$ changes. Conversely, if the formation and export of $CaCO_3$ was a process that is very independent of the export of organic matter, then the two processes could reinforce each other, leading to a larger ocean biota feedback.

One example of how the production and export of $CaCO_3$ could be affected is through changes in pH. There is increasing evidence that the calcification rate by coccolithophorids (the dominant class of phytoplankton that produces $CaCO_3$ shells) might be significantly reduced in response to a lowering of the surface ocean pH (Riebesell et al. 2000). As a reduction of calcification leads to an increase in the ocean uptake capacity for atmospheric $CO_2$, this represents a negative feedback for climate change. Zondervan et al. (2001), however, estimated the magnitude of this effect to be on the order of 10–20 PgC only for the 21st century. A lower pH also affects the calcification rates of corals (Gattuso et al. 1998; Kleypas et al. 1999). This effect, together with the enhanced sea-surface temperatures, poses a significant threat to coral reefs. The impact will likely be relatively small on atmospheric $CO_2$ but could be very large on ecosystem services such as fish recruitment, biodiversity, and tourism.

## METHANE HYDRATES

Vast quantities of methane, exceeding all known fossil-fuel reserves, exist in the form of methane hydrates. These occur primarily under continental shelf sediments around the world and in the Arctic permafrost. Hydrates are a crystalline solid of gas trapped in a frozen cage of six water molecules. Geologic evidence suggests massive releases of methane from the ancient sea floor, associated with episodes of global warming (Dickens et al. 1997). Such events can trigger very large undersea landslides and associated tsunamis. Could such a large-scale release happen over the next 100 years? The mechanisms associated with gas hydrate instabilities are very poorly understood, but calculations of heat penetration into sediments suggest that the probability is very low.

### INTERACTIONS BETWEEN THE VARIOUS FEEDBACKS

Are the various feedbacks described here additive, or is it possible that some of these feedbacks interact with each other to create synergies? The ocean biota feedback and the ocean circulation feedbacks are tightly coupled, because ocean circulation is a prime determinant of marine productivity, mainly because circulation controls the resupply of nutrients from the deep ocean to the light-lit upper ocean. This resupply of nutrients is also coupled with the resupply of the carbon that is associated with the biological pump. Therefore, while a decrease in the supply of nutrients (such as caused by an increase in upper ocean stratification) tends to cause a decrease in marine export production (positive feedback), at the same time it decreases the resupply of the carbon associated with the biological pump (negative feedback). As a consequence, there is a strong tendency for the two processes to counteract each other, with the sign of the combined feedback being very uncertain (see Greenblatt and Sarmiento, Chapter 13, this volume).

In contrast to the nutrient-limited regions, the expected increase in vertical stratification would create a tendency to increase export production in light-limited regions such as the high latitudes. On the basis of nutrient-light-circulation interactions alone, one would expect a relatively small change in global export production but large regional changes, similar to the results of Bopp et al. (2001). Many other interactions, however, might occur (e.g., associated with the delivery of iron and interactions with community structure) that make it nearly impossible to make accurate projections now. Despite all these uncertainties, the circulation feedback and the ocean biota feedbacks appear to be generally additive.

Reviewing the currently available literature, we are not aware of any strong nonlinear interactions between the various feedbacks, so the net oceanic response to the anthropogenic $CO_2$ perturbation in the atmosphere can be understood from the sum of its parts. Table 3.2 summarizes the six feedbacks that we discussed.

## Atmosphere-Land-Ocean Interactions

The carbon cycle on land, in the atmosphere, and in the ocean are tightly linked with each other. After having specified the vulnerabilities within the various subsystems, it is important to investigate the interactions of these feedbacks from a global carbon cycle perspective.

## *Trade-off between Land and Ocean Uptake through Atmospheric $CO_2$*

Since atmospheric $CO_2$ is one of the main drivers controlling the oceanic uptake of anthropogenic $CO_2$ and probably has some control over the sink strength of the terrestrial biosphere, there exists a trade-off between these two sinks. For example, if a given amount of carbon is lost from one of the land carbon pools into the atmosphere, the resulting higher atmospheric $CO_2$ concentrations would in turn foster increased oceanic uptake, thereby reducing the overall radiative forcing associated with the initial loss. The magnitude of the oceanic compensation can be significant. Cox et al. (2000), for example, find a compensation of approximately 50 percent. The magnitude of this compensation approaches nearly 100 percent over millennial timescales and played an important role in buffering the large loss of carbon from the land biosphere during transitions from interglacial periods to glacial periods. In a similar manner, a decrease in ocean uptake by, for example, increased stratification would also be partly compensated by increased terrestrial uptake. The magnitude of this trade-off effect, however, in this case depends on the sensitivity of the terrestrial biosphere to atmospheric $CO_2$.

## *Potential Vulnerability of the Land-Ocean "Conveyor Belt"*

The vulnerability of the land-river-marine transport and fate system ("conveyor belt") depends on the regional functioning of the hydrological cycle and on the way its dynamics are driven by patterns of water use, changes in land use practices affecting mobilization of carbon and nutrients to and through channels, and ultimately how the seas of the continental margins respond to these changes in forcing. Under scenarios of an increased hydrological cycle and changes in land use, river flow could increase, leading to increased ability of rivers to carry materials. Conversely, those areas subject to drier conditions would incur reduced flow and a reduced capacity of rivers to mobilize materials. Regardless of climate changes, an increased demand for water for agricultural and for urban and industrial practices would reduce water available to be routed down channels. Changes in forcing from upstream would affect coastal receiving waters. Decreases in river flow (from either drier climates or increased irrigation or retention in reservoirs) would reduce the cross-shelf water exchange because of a reduced buoyancy effect, resulting in a diminished onshore nutrient supply (Chen, Chapter 18, this volume). Primary production on the shelf would decrease proportionately, and with it the ability of continental shelves to act as a sink for carbon. If, on the other hand, there was a significant increase in nutrients (from increased fertilizers and urban use), higher levels of productivity, and even eutrophication, in estuaries or coastal regions could be maintained. Over a longer term, sea-level rises could have pronounced effects on carbon. We are currently not in a position to estimate the exact magnitudes of these effects, but it seems likely that these changes will have a much larger impact on regional issues than on atmospheric $CO_2$.

## Vegetation Cover–Dust–Ocean Fertilization

A third possible feedback arising out of the interaction of the atmosphere-land-ocean system concerns the increasing evidence that micronutrients such as iron play an important role in regulating the strength of the ocean's biological pump, particularly in the high-nutrient low-chlorophyll (HNLC) regions. As a substantial fraction of the iron input into the surface ocean comes from the atmosphere, changes in this input could lead to changes in ocean productivity, export production, and eventually atmospheric $CO_2$. This mechanism has actually been proposed as a possible reason for the low atmospheric $CO_2$ concentration during the last glacial period (e.g., Martin 1990; see Joos and Prentice, Chapter 7, this volume). This feedback may also operate in the future. If degradation or desertification on land leads to increased dust inputs into oceanic HNLC regions, this dust input would stimulate oceanic productivity and thus increase oceanic $CO_2$ uptake. We estimate, however, that this effect will be relatively minor over the next 20 or 100 years in really reducing atmospheric $CO_2$, but this effect might lead to large changes on a regional scale.

It also has been proposed that atmospheric iron input plays a major role in controlling marine $N_2$-fixation and that changes in the iron input would therefore lead to changes in the total inventory of fixed nitrogen in the ocean, which could fuel higher levels of marine productivity throughout the ocean (Falkowski 1997; Broecker and Henderson 1998). It is unlikely, however, that this mechanism will lead to large changes in the ocean atmosphere distribution of inorganic carbon over the next 20 to 100 years.

## Summary and Conclusions

Our analysis of the vulnerability of the carbon pool on land and in the ocean reveals that a substantial amount of carbon is at risk of becoming mobilized and being released into the atmosphere (Figure 3.2, Figure 3.3, and Table 3.1). Recognizing that our estimates are quite uncertain, we find that on the order of several tens of Pg could be lost from the land and ocean carbon pool over the next 20 years and several hundred Pg over the course of this century. On land, we estimate that the maximum potential losses are substantially larger than the maximum potential sinks, suggesting that the overall sign of the feedbacks arising from carbon-climate-human interactions is positive—that is, accelerating climate change. For the ocean, we must differentiate between the ocean uptake that exists in the absence of climate change and the ocean sink that is due to climate feedbacks. The ocean uptake in the absence of any feedback is very large, on the order of several hundred Pg C. This uptake could be substantially offset by the feedbacks, since their net sign is likely positive as well.

Our results imply that the fraction of the anthropogenic $CO_2$ emissions that will remain in the atmosphere and force a climate change will increase in the future. As a consequence, the permissible anthropogenic emissions in a stabilization pathway are smaller in the presence of interactions between the carbon-climate-human systems. The

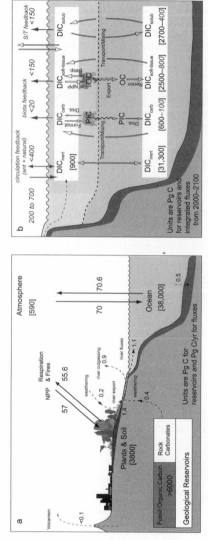
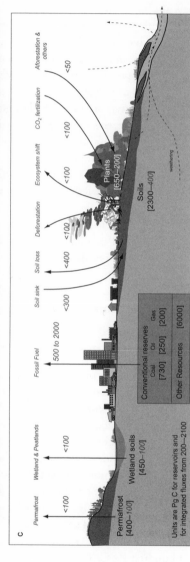

**Figure 3.3.** Global carbon cycle, its sink potential, and its vulnerability in the 21st century. (a) Preindustrial carbon cycle, (b) major ocean carbon pools and their vulnerability, and (c) major land pools and their vulnerability. The italic numbers shown in (b) and (c) represent our maximum estimates of the vulnerability. See also Table 3.1 and text. The oceanic changes in the pool sizes are larger than the induced air-sea $CO_2$ fluxes because of the buffering effect of seawater.

magnitudes of the feedbacks depend to a large degree on the state of the climate system. The feedbacks tend to get stronger with larger climate forcing, implying that the magnitude of the offsetting effect for the permissible emissions increases with the concentration of the atmospheric $CO_2$ stabilization target. A final observation that arises from the inspection of Table 3.1 is that the vulnerability of the global carbon cycle stems from the vulnerability of many different pools and processes. This finding highlights the need for a highly integrative and interdisciplinary approach for studying the global carbon cycle and the need to view Earth as a coupled system rather than as an entity that can be studied in parts.

Although we view our analysis as a first step toward identifying and quantifying these feedbacks, our estimates are preliminary at best. We hope, however, to encourage subsequent research that will improve our initial assessment.

## Note

1. The SRES A2 and B2 emission scenarios were developed by Nakicenovic et al. (2000) and represent two representative samples out of a large family of scenarios. The A2 story line describes a very heterogeneous world. The underlying theme is self-reliance and preservation of local identities. Fertility patterns across regions converge very slowly, which results in continuously increasing population. Economic development is primarily regionally oriented, and per capita economic growth and technological change are more fragmented and slower than in other storylines. The B2 story line describes a world in which the emphasis is on local solutions to economic, social, and environmental sustainability. It is a world with continuously increasing global population, at a rate lower than A2, and intermediate levels of economic development and of technological change. Although the scenario is also oriented toward environmental protection and social equity, it focuses on local and regional levels. Because of higher population and slow technological advances, the A2 scenario results in one of the highest levels of $CO_2$ emissions, whereas B2 represents an intermediate-level emission scenario.

## Literature Cited

Anisimov, O. A., F. E. Nelson, and A. V. Pavlov. 1999. Predictive scenarios of permafrost development under conditions of global climate change in the XXI century. *Earth Cryology* 3 (4): 15–25.

Archer, D., H. Kheshgi, and E. Maier-Reimer. 1997. Multiple timescales for neutralization of fossil fuel $CO_2$. *Geophysical Research Letters* 24 (4): 405–408.

Archer, D., G. Eshel, A. Winguth, and W. Broecker. 2000. Atmospheric $CO_2$ sensitivity to the biological pump in the ocean. *Global Biogeochemical Cycles* 14:1219–1230.

Armstrong, R. A., C. Lee, J. I. Hedges, S. Honjo and S. G. Wakeham. 2002. A new, mechanistic model for organic carbon fluxes in the ocean based on the quantitative association of POC with ballast minerals. *Deep-Sea Research* 49:219–236.

Bacastow, R., and C. D. Keeling. 1973. Atmospheric carbon dioxide and radiocarbon in the natural carbon cycle. II. Changes from A.D. 1700 to 2070 as deduced from a geochemical reservoir. Pp. 86–135 in *Carbon and the biosphere,* edited by G. M. Woodwell and E. V. Pecan. Springfield, VA: U.S. Department of Commerce.

Betts, R. A., P. M. Cox, S. E. Lee, and F. I. Woodward. 1997. Contrasting physiological and structural vegetation feedbacks in climate change simulations. *Nature* 387: 796–799.

Bonan, G. B., D. B. Pollard, and S. L. Thompson. 1992. Effects of boreal forest vegetation on global climate. *Nature* 359:716–718.

Bopp, L., P. Monfray, O. Aumont, J.-L. Dufresne, H. Le Treut, G. Madec, L. Terray, and J. C. Orr. 2001. Potential impact of climate change on marine export production. *Global Biogeochemical Cycles* 15 (1): 81–99.

Boyd, P. W., and S. C. Doney. 2003. The impact of climate change and feedback processes on the ocean carbon cycle. Pp. 157–193 in *Ocean biogeochemistry: The role of the ocean carbon cycle in global change,* edited by M. J. R. Fasham. Berlin: Springer Verlag.

Broecker, W. S., and G. M. Henderson. 1998. The sequence of events surrounding Termination II and their implications for the cause of glacial-interglacial $CO_2$ changes. *Paleoceanography* 13 (4): 352–364.

Caspersen, J. P., S. W. Pacala, J. C. Jenkins, G. C. Hurtt, P. R. Moorcroft, and R. A. Bird-sey. 2000. Contributions of land-use history to carbon accumulation in U.S. forests. *Science* 290:1148–1151.

Christensen, T., I. Prentice, J. Kaplan, A. Haxeltine, and S. Sitch. 1996. Methane flux from northern wetlands and tundra: An ecosystem source modelling approach. *Tellus* 48B:652–661.

Collatz, G. J., J. A. Berry, and C. Grivet. 1992. Coupled photosynthesis-stomatal conductance model for leaves of $C_4$ plants. *Australian Journal of Plant Physiology* 19:519–538.

Cox, P. M., R. A. Betts, C. D. Jones, S. A. Spall and I. J. Totterdell. 2000. Acceleration of global warming due to carbon-cycle feedbacks in a coupled climate model. *Nature* 408:184–187.

Cubasch, U., G. A. Meehl, G. J. Boer, R. J. Stouffer, M. Dix, A. Noda, C. A. Senior, S. Raper, and K. S. Yap. 2001. Projections of future climate change. Pp. 525–582 in *Climate change 2001: The scientific basis (Contribution of Working Group I to the Third Assessment Report of the Intergovernmental Panel on Climate Change),* edited by J. T. Houghton, Y. Ding, D. J. Griggs, M. Noguer, P. J. van der Linden, X. Dai, K. Maskell, and C. A. Johnson. Cambridge: Cambridge University Press.

DeFries, R. S., L. Bounoua, and G. J. Collatz. 2002. Human modification of the landscape and surface climate in the next fifty years. *Global Change Biology* 8 (5): 438–458.

Dickens, G. R., M. M. Castillo, and J. C. G. Walker. 1997. A blast of gas in the latest Paleocene: Simulating the first-order effects of massive dissociation of oceanic methane hydrate. *Geology* 25:259–262

Dore, J. E, R. Lukas, D. W. Sadler, and D. M. Karl. 2003. Climate-driven changes to the atmospheric $CO_2$ sink in the subtropical North Pacific Ocean. *Nature* 424:754–757.

Dufresne, J.-L., P. Friedlingstein, M. Berthelot, L. Bopp, P. Ciais, L. Fairhead, H. LeTreut, and P. Monfray. 2002. Effects of climate change due to $CO_2$ increase on land and ocean carbon uptake. *Geophysical Research Letters* 29 (10), doi:10.1029/2001GL013777.

Falkowski, P. G.1997. Evolution of the nitrogen cycle and its influence on the biological sequestration of $CO_2$ in the ocean. *Nature* 387:272–275.

Falkowski, P. G., and C. Wilson. 1992. Phytoplankton productivity in the North Pacific Ocean since 1900 and implications for absorption of anthropogenic $CO_2$. *Nature* 358:741–743.

Farquhar, G. D., S. von Caemmerer, and J. A. Berry. 1980. A biochemical model of photosynthetic $CO_2$ assimilation in leaves of $C_3$ species. *Planta* 149:78–90.

Field, C. B., F. S. Chapin III, P. A. Matson, and H. A. Mooney. 1992. Responses of terrestrial ecosystems to the changing atmosphere: A resource-based approach. *Annual Review of Ecology and Systematics* 23:201–235.

Friedlingstein, P., J.-L. Dufresne, P. M. Cox, and P. Rayner. 2003. How important is the positive feedback between climate change and the carbon cycle? *Tellus* (in press).

Gattuso, J.-P., M. Frankignoulle, I. Bourge, S. Romaine, and R. W. Buddemeier. 1998. Effect of calcium carbonate saturation of seawater on coral calcification. *Global and Planetary Change* 18:37–46.

Gloor, M., N. Gruber, J. L. Sarmiento, C. S. Sabine, R. A. Feely, and C. Rödenbeck. 2003. A first estimate of present and pre-industrial air-sea CO2 flux patterns based on ocean interior carbon measurements and models. *Geophysical Research Letters* 30 (1), doi:10.1029/2002GL015594.

Gregg, W. W., and M. E. Conkright. 2002. Decadal changes in global ocean chlorophyll. *Geophysical Research Letters* 29 (15), doi:10.1029/2002GL014689.

Gruber, N., and J. L. Sarmiento. 2002. Biogeochemical/physical interactions in elemental cycles. Pp. 337–399 in *The sea*, Vol. 12, *Biological-physical interactions in the sea,* edited by A. R. Robinson, J. J. McCarthy, and B. J. Rothschild. New York: John Wiley and Sons.

Houghton, R. 1999. The annual net flux of carbon to the atmosphere from changes in land use 1850–1990. *Tellus* 51B:298–313.

Houghton, J. T., G. J. Jenkins, and J. J. Ephraums, eds. 1990. *Climate change: The IPCC scientific assessment (Contribution of Working Group I to the first assessment report of the Intergovernmental Panel on Climate Change).* Cambridge: Cambridge University Press.

Houghton, R. A., J. L. Hackler, and K. T. Lawrence. 1999. The U.S. carbon budget: Contributions from land-use change. *Science* 285:574–578.

House, J. I., C. Prentice, and C. Le Quéré. 2002. Maximum impacts of future reforestation or deforestation on atmospheric $CO_2$. *Global Change Biology* 8 (11): 1047–1052.

Jackson, J. B. C., M. X. Kirby, W. H. Berger, K. A. Bjorndal, L. W. Botsford, B. J. Bourque, R. H. Bradbury, R. Cooke, J. Erlandson, J. A. Estes, T. P. Hughes, S. Kidwell, C. B. Lange, H. S. Lenihan, J. M. Pandolfi, C. H. Peterson, R. S. Steneck, M. J. Tegner, and R. R. Warner. 2001 . Historical overfishing and the recent collapse of coastal ecosystems. *Science* 293:629–638

Jones, C. D, P. M. Cox, R. L. H. Essery, D. L. Roberts, and M. J. Woodage. 2003. Strong carbon cycle feedbacks in a climate model with interactive $CO_2$ and sulphate aerosols. *Geophysical Research Letters* 30 (9): 1479, doi:10.1029/2003GL016867.

Joos, F., G.-K. Plattner, T. F. Stocker, O. Marchal, and A. Schmittner. 1999. Global warming and marine carbon cycle feedbacks on future atmospheric $CO_2$. *Science* 284:464–467.

Kammen, D. M., and D. M. Hassenzahl. 2001. *Why should we risk it? Exploring environ-*

*mental, health, and technological problem solving.* Princeton, NJ: Princeton University Press.

Klaas, C., and D. E. Archer. 2002. Association of sinking organic matter with various types of mineral ballast in the deep sea: Implications for the rain ratio. *Global Biogeochemical Cycles* 16 (4), doi:10.1029/2001GB001765.

Kleypas, J. A., R. W. Buddemeier, D. Archer, J.-P. Gattuso, C. Langdon, and B. N. Opdyke. 1999. Geochemical consequences of increased atmospheric carbon dioxide on coral reefs. *Science* 284:118–120.

Lal, R. 1999. Soil management and restoration for C sequestration to mitigate the greenhouse effect. *Progress in Environmental Science* 1:307–326

————. 2003. Global potential of soil C sequestration to mitigate the greenhouse effect. *Critical Reviews in Plant Sciences* 22:151–184.

Lal, R., J. M. Kimble, R. F. Follett, and C. V. Cole. 1998. *The potential of U.S. cropland to sequester C and mitigate the greenhouse effect.* Chelsea, MI: Ann Arbor Press.

Laws, E. A., P. Falkowski, E. A. Carpenter, and H. Ducklow. 2000. Temperature effects on export production in the open ocean. *Global Biogeochemical Cycles* 14 (4): 1231–1246.

Martin, J. H. 1990. Glacial-interglacial $CO_2$ change: The iron hypothesis. *Paleoceanography* 5 (1): 1–13.

Matear, R. J., and A. C. Hirst. 1999. Climate change feedback on the future oceanic $CO_2$ uptake. *Tellus* 51B:722–733.

McGuire, A. D., S. Sitch, J. S. Clein, R. Dargaville, G. Esser, J. Foley, M. Heimann, F. Joos, J. Kaplan, D. W. Kicklighter, R. A. Meier, J. M. Melillo, B. Moore III, I. C. Prentice, N. Ramankutty, T. Reichenau, A. Schloss, H. Tian, L. J. Williams, and U. Wittenberg. 2001. Carbon balance of the terrestrial biosphere in the twentieth century: Analyses of $CO_2$, climate and land-use effects with four process-based ecosystem models. *Global Biogeochemical Cycles* 15:183–206.

Nakicenovic, N., J. Alcamo, G. Davis, B. de Vries, J. Fenhann, S. Gaffin, K. Gregory, A. Grübler, T. Y. Jung, T. Kram, E. L. La Rovere, L. Michaelis, S. Mori, T. Morita, W. Pepper, H. Pitcher, L. Price, K. Raihi, A. Roehrl, H.-H. Rogner, A. Sankovski, M. Schlesinger, P. Shukla, S. Smith, R. Swart, S. van Rooijen, N. Victor, Z. Dadi. 2000. *IPCC special report on emissions scenarios.* Cambridge: Cambridge University Press.

Pacala, S. W., G. C. Hurtt, R. A. Houghton, R. A. Birdsey, L. Heath, E. T. Sundquist, R. F. Stallard, D. Baker, P. Peylin, P. Ciais, P. Moorcroft, J. Caspersen, E. Shevliakova, B. Moore, G. Kohlmaier, E. Holland, M. Gloor, M. E. Harmon, S.-M. Fan, J. L. Sarmiento, C. Goodale, D. Schimel, and C. B. Field. 2001. Convergence of land- and atmosphere-based U.S. carbon sink estimates. *Science* 292:2316–2320.

Parton, W. J., D. S. Schimel, C. V. Cole, and D. S. Ojima. 1987. Analysis of factors controlling soil organic matter levels in Great Plains grasslands. *Soil Science Society of America Journal* 51:1173–1179.

Plattner, G.-K., F. Joos, T. F. Stocker, and O. Marchal. 2001. Feedback mechanisms and sensitivities of ocean carbon uptake under global warming. *Tellus* 53B:564–592.

Raich, J. W., E. B. Rastetter, J. M. Melillo, D. W. Kicklighter, P. A. Steudler, B. J. Peterson, A. L. Grace, B. Moore III, and C. J. Vörösmarty. 1991. Potential net primary production in South America. *Ecological Applications* 1:399–429.

Raper, S. C. B., and U. Cubasch. 1996. Emulation of the results from a coupled general circulation model using a simple climate model. *Geophysical Research Letters* 23:1107–1110.

Riebesell, U., I. Zondervan, B. Rost, P. D. Tortell, R. Zebe, and F. M. M. Morel. 2000. Increased $CO_2$ decreases marine planktonic calcification. *Nature* 407:364—367.

Salati, E., and P. B. Vose. 1984. Amazon basin: A system in equilibrium. *Science* 225:129.

Sanderman, J., R. G. Amundson, and D. D. Baldocchi. 2003. Application of eddy covariance measurements to the temperature dependence of soil organic matter mean residence time. *Global Biogeochemical Cycles* 17:1061, doi:10.1029/2001GB001833.

Sarmiento, J. L., and C. Le Quéré. 1996. Oceanic $CO_2$ uptake in a model of century-scale global warming. *Science* 274:1346–1350.

Sarmiento, J. L., J. C. Orr, and U. Siegenthaler. 1992. A perturbation simulation of $CO_2$ uptake in an ocean general circulation model. *Journal of Geophysical Research* 97 (C3): 3621–3645.

Sarmiento, J. L., C. Le Quéré, and S. W. Pacala. 1995. Limiting future atmospheric carbon dioxide. *Global Biogeochemical Cycles* 9:121–137.

Sarmiento, J. L., T. M. C. Hughes, R. J. Stouffer, and S. Manabe. 1998. Simulated response of the ocean carbon cycle to anthropogenic climate warming. *Nature* 393:245–249.

Schimel, D. S., D. Alves, I. Enting, M. Heimann, F. Joos, D. Raynaud, T. Wigley, M. Prather, R. Derwent, D. Ehhalt, P. Fraser, E. Sanhueza, X. Zhou, P. Jonas, R. Charlson, H. Rohd, K. P. Sadasivan, Y. Shine, Y. Fouquart, V. Ramaswamy, S. Solomon, J. Srinivasan, D. Albritton, I. Isaken, M. Lal, and D. Wuebbles. 1996. Radiative forcing of climate change. Pp. 65–131 in *Climate change 1995: The science of climate change.* Cambridge: Cambridge University Press.

Schimel, D. S., J. I. House, K. A. Hibbard, P. Bousquet, P. Ciais, P. Peylin, B. H. Braswell, M. J. Apps, D. Baker, A. Bondeau, J. Canadell, G. Churkina, W. Cramer, A. S. Denning, C. B. Field, P. Friedlingstein, C. Goodale, M. Heimann, R. A. Houghton, J. M. Melillo, B. Moore III, D. Murdiyarso, I. Noble, S. W. Pacala, I. C. Prentice, M. R. Raupach, P. J. Rayner, R. J. Scholes, W. L. Steffen, and C. Wirth. 2001. Recent patterns and mechanisms of carbon exchange by terrestrial ecosystems. *Nature* 414:169–172.

Schlesinger, W. H., and J. Lichter. 2001. Limited carbon storage in soil and litter of experimental forest plots under increased atmospheric $CO_2$. *Nature* 411:466–469.

Schulze, E. D., D. Mollicone, F. Achard, G. Matteucci, S. Frederici, H. D. Eva, and R. Valentini. 2003. Making deforestation pay under the Kyoto protocol? *Science* 299:1669.

Shaw, M. R., E. S. Zavaleta, N. Chiariello, H. A. Mooney, and C. B. Field. 2002. Grassland responses to global environmental changes suppressed by elevated $CO_2$. *Science* 298:1987–1990.

Shukla, J., C. Nobre, and P. Sellers. 1990. Amazon deforestation and climate change. *Science* 247:1322–1325.

Sigman, D., and E. Boyle. 2000. Glacial/interglacial variations in carbon dioxide: Searching for a cause. *Nature* 407:859–869.

Smith, P., D. Powlson, J. U. Smith, P. Falloon, and K. Coleman. 2000. Meeting Europe's climate change commitments: Quantitative estimates of the potential for carbon mitigation by agriculture. *Global Change Biology* 6:525–539.

Smith, T. M., and H. H. Shugart. 1993. The transient response of terrestrial carbon storage to a perturbed climate. *Nature* 361:523–526.

Tortell, P. D., G. R. DiTullio, D. M. Sigman, and F. M. M. Morel. 2002. $CO_2$ effects on species composition and nutrient utilization in an Equatorial Pacific phytoplankton community. *Marine Ecology Progress Series* 236:37–43.

Watson, A., and J. Orr. 2003. Carbon dioxide fluxes in the global ocean. Pp. 123–143 in *Ocean biogeochemistry: The role of the ocean carbon cycle in global change,* edited by M. J. R. Fasham. Berlin: Springer Verlag.

Wigley, T. M. L., R. Richels, and J. A. Edmonds. 1996. Economic and environmental choices in the stabilization of atmospheric $CO_2$ concentrations. *Nature* 379:240–243.

Yi, C., P. Gong, M. Xu, and Y. Qi. 2001. The effects of buffer and temperature feedback on the oceanic uptake of $CO_2$. *Geophysical Research Letters* 28 (5): 751–754.

Zondervan, I., R. E. Zeebe, B. Rost, and U. Riebesell. 2001. Decreasing marine biogenic calcification: A negative feedback on rising atmospheric $pCO_2$. *Global Biogeochemical Cycles* 15 (2): 507–516.

# 4

# Scenarios, Targets, Gaps, and Costs

Jae Edmonds, Fortunat Joos, Nebojsa Nakicenovic,
Richard G. Richels, and Jorge L. Sarmiento

## The Technology "Gap"

With substantial increases in the global demand for energy services expected over the next century, energy technology holds the key to effective limitation of greenhouse gas emissions. Carbon-based fuels and their associated technologies supplied 88 percent of the world's energy in 1995. Over the coming decades, technologies that provide energy services but emit little or no $CO_2$ into the atmosphere will compete in the global marketplace with ever-improving carbon-emitting fossil-fuel technologies. How successfully they compete will determine future emissions.

Figure 4.1 illustrates the nature of the challenge. The middle curve depicts the carbon dioxide emissions associated with the Intergovernmental Panel on Climate Change (IPCC) central scenario, denoted IS92a. This IPCC reference case scenario is based on analysis of trends in global population, economic growth, land use, and energy systems. The underlying energy and land use changes produce carbon emissions that grow steadily throughout the century from 7 petagrams of carbon (PgC) per year in 1990 to almost 20 PgC per year in 2100.

The extent of technological improvements contained in the IS92a scenario is not always appreciated. They are substantial. To illustrate the extent of technological change incorporated in the IS92a scenario, we have computed carbon emissions that would be associated with a world having the same population and economic growth as IS92a but with energy technology held constant at its 1990 level. This calculation is not intended to be a realizable scenario but is rather intended to highlight the degree to which technologies that are expected to participate in energy systems consistent with the stabilization of greenhouse gas concentrations are already assumed to contribute. The difference between the upper and middle curves illustrates the technological improvement needed merely to achieve the IS92a emissions path with its corresponding impact on

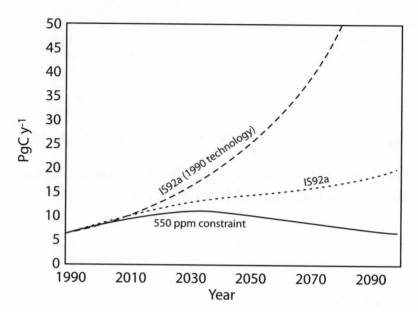

**Figure 4.1.** Carbon emissions, 1990–2100.

concentrations. The effects of energy intensity improvements and the deployment of non-carbon-emitting energy supply technologies are large. Power generation is 75 percent carbon free by the year 2100, and modern commercial biomass provides more energy than the combined global production of oil and gas in 1990.

Even under the advanced technology assumptions of IS92a, emissions will continue to grow. They will increase at a significantly slower rate than they would have without the technology developments envisioned by IS92a. Nevertheless, under the IS92a scenario, the concentration of carbon dioxide will rise to more than 700 parts per million (ppm) by the end of the 21st century—nearly triple the preindustrial level—and will continue rising.

The lower curve depicts an emissions path consistent with a 550-ppm concentration ceiling. This curve is depicted for illustrative purposes. While the Framework Convention on Climate Change commits the governments of the world to stabilize the concentration of greenhouse gases, it is silent as to which concentration. The concentration at which $CO_2$ is stabilized may turn out to be 350 ppm or 1,000 ppm.

Regardless of the concentration at which $CO_2$ is stabilized, the fact that $CO_2$ concentrations are determined by cumulative, not annual, global emissions implies that eventually global $CO_2$ emissions must peak and then begin a long-term, indefinite decline, as in this example.

We define the difference between carbon emissions that are anticipated to occur in a world that places no value on carbon and emissions required to stabilize at a specific concentration level as the "gap." Closing the gap means effecting a change in the technologies that are anticipated to come into use under the IS92a scenario. Substantial development of energy technologies is necessary to achieve the IS92a goals, and even greater development and deployment is needed to achieve stabilization. In the sections that follow, we explore how the size of the gap is affected by assumptions regarding (1) business as usual emissions, (2) the $CO_2$ concentration target, and (3) key components of the carbon cycle. We also show how the gap may be filled under alternative assumptions regarding technology cost and availability.

## Scenarios

We begin by examining reference scenarios that represent the range (but not necessarily the distribution) of scenarios found in the open literature and explore the role technology plays in shaping fossil-fuel carbon emissions. The main conclusion is that scenarios are generally optimistic about future development and deployment of technologies that provide energy services without carbon emissions. Yet, as with the IS92a scenario, assumed technology improvements do not guarantee that the concentration of atmospheric $CO_2$ will be stabilized in any given scenario. In most instances there remains a gap between emissions in the scenario and an emissions trajectory that would stabilize $CO_2$ concentrations. The size of this gap depends on the $CO_2$ stabilization concentration and the degree to which advanced energy technologies—both supply and energy efficiency—are assumed to have already displaced carbon-emitting energy technologies.

A wide array of scenarios have been developed that explore future energy, industrial, and land use carbon emissions. One of the most recent examinations is the IPCC Special Report on Emission Scenarios (SRES) (Nakicenovic et al. 2000). This document examines scenarios published in the open literature and develops an array of new global scenarios looking forward through the 21st century. (The many scenarios contained in the SRES are organized into four major groups labeled A1, A2, B1, and B2. The characteristics of each group are discussed in Colorplate 4, and the associated primary energy requirements are shown in Colorplate 6.) The literature is rich, containing scenarios that range from those with rapidly rising emissions to those in which emissions follow a pattern consistent with the stabilization of greenhouse gas concentrations. Some scenarios exhibit more rapid growth in emissions than IS92a, and some are slower. This range of anthropogenic $CO_2$ emissions for both fossil fuels and land use change is shown in Colorplate 6, from Nakicenovic et al. (2000).

The scenarios with lower energy requirements (Colorplate 6) represent sustainable futures with a transition to very efficient energy use and high degrees of conservation. Generally, these are also the scenarios in which energy sources with low carbon intensity play an important role. The scenarios with higher energy requirements generally rep-

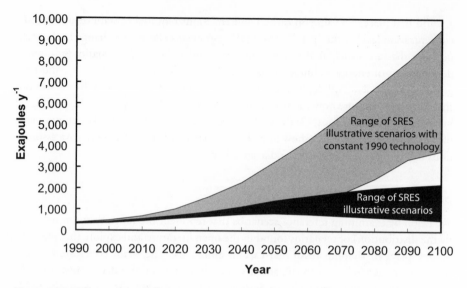

**Figure 4.2.** Effects of energy intensity improvements on energy demands in SRES illustrative scenarios.

resent futures with more rapid rates of economic growth or low rates of economic growth combined with high population.

## Energy Intensity

Like the IS92 scenarios before them, all of the SRES scenarios envision substantial improvements in technology. To illustrate the magnitude of the technological change assumed to occur over the course of the 21st century in the SRES scenarios, we compute the energy requirement for each scenario, given its population and gross world product, that would be associated with an energy intensity that remained unchanged from that of 1990 (see Figure 4.2).

World energy intensity is the ratio of energy to gross world product. Energy intensity improves in all SRES scenarios. That is, the amount of energy required to produce each dollar or yen or rupee of gross domestic product (GDP), after adjusting for inflation, is lower each decade than it was the decade before. Energy intensity declines as a result of many changes in the scenarios. These changes include improvements in the efficiency with which energy is used for a given process, shifts between processes, and the substitution of goods and materials with lower energy content for those with higher energy content.

Improvements in energy intensity are not new. They have been going on in some countries for a century. In the SRES scenarios, energy intensity improves at annual rates

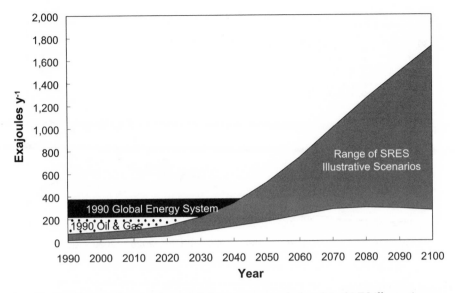

**Figure 4.3.** Range of non-carbon-emitting energy supply (gray) in SRES illustrative scenarios. Also shown are the magnitudes of the 1990 global energy system (black) and the 1990 global oil and gas system (dotted).

that range from less than 1 percent per year to more than 2 percent per year. By the year 2100 the cumulative effect is large. By the year 2100 primary energy demand in the SRES scenarios ranges from 55 percent to more than 90 percent lower than had no energy-intensity improvement occurred. Such is the effect of compounding.

Figure 4.2 shows the range of primary energy demands for the SRES illustrative scenarios and the corresponding range that would have been required assuming the same gross world product as in each of the SRES scenarios, but leaving energy intensity (energy per unit of gross world product) unchanged at 1990 levels. Without the improvements in energy intensity, it is hard to imagine how the global energy system could expand to produce the energy requirements of a constant-energy intensity-world. (This underscores the hypothetical nature of the constant-energy-intensity calculations and their purely illustrative character.)

## Non-Carbon Energy Supply

Scenarios that consider the future evolution of global energy systems not only have significant improvements in energy intensity as a characteristic, but also envision significant increases in the deployment of non-fossil energy supply.

Figure 4.3 shows the range of deployment in the SRES illustrative scenarios for non-carbon-emitting energy (i.e., solar, wind, nuclear, and biomass). The global energy system in 1990 produced 375 exajoules (EJ) of energy. In many of the SRES illustrative sce-

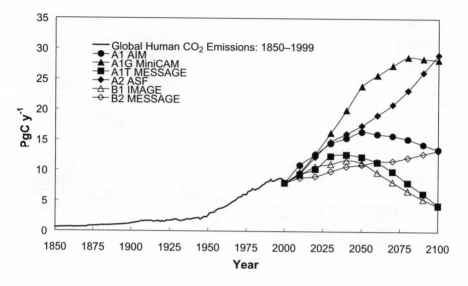

**Figure 4.4.** Global carbon dioxide emissions in billion tons of carbon (PgC) per year since 1850 to present, plus emissions trajectories for the six SRES illustrative scenarios.

narios, the deployment of non-carbon-emitting energy supply systems exceeds the size of the global energy system in 1990. In some cases the scale of the commercial biomass energy sector alone rivals or dwarfs the scale of the present global energy system.

## The Gap and $CO_2$ Stabilization

We have used the term "the gap" to refer to the difference between $CO_2$ emissions associated with reference scenarios, with their attendant technological progress, and emissions along a path that stabilizes the concentration of atmospheric $CO_2$. Historic emissions from 1850 through 1999 and the variety of SRES reference emissions paths are shown in Figure 4.4.

The Wigley, Richels, and Edmonds (1996, WRE) emissions paths (Figure 4.5) trace emissions trajectories that are consistent with stable atmospheric concentrations of $CO_2$ at five alternative levels. These paths exhibit the peak and decline pattern associated with the limit on cumulative emissions for atmospheric $CO_2$ concentrations. The higher the stabilization concentration, the later and higher is the peak in emissions. For most of the scenarios in the open literature, the dramatic advances in anticipated technology developments are insufficient to stabilize the concentration of atmospheric $CO_2$.

The magnitudes of the "gaps" between WRE trajectories for stabilization of atmospheric $CO_2$ concentrations at levels ranging from 350 ppm to 750 ppm are shown in Figure 4.6 and in Tables 4.1 and 4.2. The range of emissions reductions, relative to the

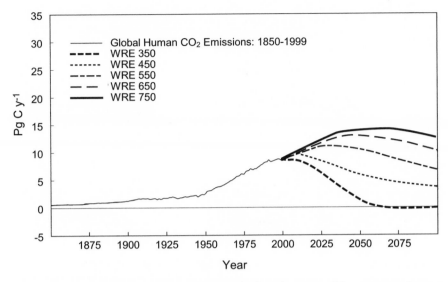

**Figure 4.5.** WRE $CO_2$ emissions trajectories for five alternative $CO_2$ concentrations.

reference case, that are required to attain a WRE concentration stabilization path is very wide (Figure 4.6). For some of the illustrative scenarios, events unfold in such a way as to stabilize the concentration of $CO_2$ at approximately 550 ppm without the need for explicit policies to limit greenhouse gas emissions. By construction, these scenarios assume that sustainable development is a priority and that policy measures ensure that non-carbon-emitting technologies are developed and deployed in preference to fossil energy technology. For most of the SRES reference scenarios, such good outcomes are not anticipated. Most of the scenarios require additional emissions reductions to stabilize the atmospheric concentration of $CO_2$.

## Closing the Gap

Numerous mitigation measures and policies need to be invoked in the IS92a and SRES scenarios to stabilize $CO_2$ concentration levels. There is no unique solution to closing the gap. In fact, regardless of the reference scenario, it is never the case that a single energy technology closes the gap. In all instances the gap is closed by the deployment of a suite of energy technologies. These technologies include familiar core technologies such as energy efficiency; other energy intensity improvements; production of solar, wind, nuclear, modern commercial biomass, and other renewable energy; and changes in land use practices such as afforestation, other forest management practices, and soil carbon management. Beyond that, technologies that are only minor components in the present

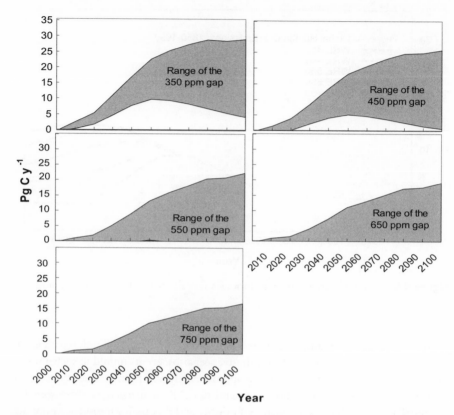

**Figure 4.6.** The "gap" for five alternative $CO_2$ concentrations. This panel shows for five alternative $CO_2$ concentrations the range of differences between reference emissions in the six SRES illustrative scenarios and the WRE emissions trajectory associated with stabilization of the concentration of $CO_2$ at the indicated level. This difference between anticipated emissions and emissions along a trajectory that stabilizes $CO_2$ concentrations is referred to as the "gap."

global energy system could become major components of the global energy system in the middle and latter half of the 21st century. These technologies could include carbon capture and disposal, hydrogen and advanced transportation systems, and biotechnology.

Many issues remain to be resolved before the technologies assumed to successfully deploy in the reference cases deploy as assumed. The development and deployment of advanced energy technologies raise other research questions.

As discussed elsewhere in this book, carbon capture and disposal could be a major technology in the 21st century. Its deployment would enable the continued employment of abundant fossil-fuel resources to provide energy services. But cost is an important

**Table 4.1.** The gap for the six SRES illustrative scenarios for atmospheric $CO_2$ concentrations ranging from 450 ppm to 750 ppm (PgC/year)

| Scenario | WRE 350 | | | WRE 450 | | | WRE 550 | | | WRE 650 | | | WRE 750 | | |
|---|---|---|---|---|---|---|---|---|---|---|---|---|---|---|---|
| | 2020 | 2050 | 2100 | 2020 | 2050 | 2100 | 2020 | 2050 | 2100 | 2020 | 2050 | 2100 | 2020 | 2050 | 2100 |
| A1 AIM | 5 | 15 | 13 | 4 | 10 | 10 | 2 | 6 | 7 | 1 | 4 | 3 | 1 | 3 | 1 |
| A1G MiniCAM | 6 | 23 | 28 | 4 | 18 | 25 | 2 | 13 | 21 | 2 | 11 | 18 | 1 | 10 | 16 |
| A1TMESSAGE | 3 | 11 | 4 | 1 | 6 | 1 | NA | NA | NA | NA | NA | NA | NA | NA | NA |
| A2 ASF | 5 | 16 | 29 | 3 | 12 | 25 | 1 | 7 | 22 | 1 | 5 | 19 | 1 | 4 | 17 |
| B1 IMAGE | 3 | 10 | 4 | 2 | 5 | 1 | NA | NA | NA | NA | NA | NA | NA | NA | NA |
| B2 MESSAGE | 2 | 10 | 13 | 0 | 5 | 10 | -2 | 0 | 7 | -2 | -2 | 3 | -2 | -3 | 1 |

*Note:* NA indicates not available.

**Table 4.2.** The gap for the six SRES illustrative scenarios for atmospheric $CO_2$ concentrations ranging from 450 ppm to 750 ppm (EJ/year)

| Scenario | WRE 350 | | | WRE 450 | | | WRE 550 | | | WRE 650 | | | WRE 750 | | |
|---|---|---|---|---|---|---|---|---|---|---|---|---|---|---|---|
| | 2020 | 2050 | 2100 | 2020 | 2050 | 2100 | 2020 | 2050 | 2100 | 2020 | 2050 | 2100 | 2020 | 2050 | 2100 |
| A1 AIM | 270 | 822 | 809 | 180 | 567 | 589 | 92 | 312 | 401 | 71 | 190 | 207 | 61 | 137 | 61 |
| A1G MiniCAM | 284 | 1,137 | 1,335 | 191 | 901 | 1,162 | 100 | 665 | 1,014 | 78 | 552 | 860 | 68 | 503 | 745 |
| A1TMESSAGE | 158 | 626 | 299 | 65 | 360 | 45 | NA | NA | NA | NA | NA | NA | NA | NA | NA |
| A2 ASF | 253 | 785 | 1,243 | 163 | 557 | 1,086 | 74 | 329 | 953 | 53 | 220 | 814 | 42 | 173 | 710 |
| B1 IMAGE | 166 | 491 | 200 | 79 | 262 | 27 | NA | NA | NA | NA | NA | NA | NA | NA | NA |
| B2 MESSAGE | 97 | 534 | 664 | 4 | 278 | 481 | -88 | 21 | 325 | -110 | -101 | 164 | -121 | -154 | 42 |

*Note:* NA indicates not available.

question. If the cost issue is successfully addressed, or if the value of carbon is high enough, carbon capture technologies could be deployed at very large scale. Cumulative capture amounting to hundreds of billions of tons of carbon over the course of the 21st century could occur if other technology issues are addressed. Disposal is also a critical question. Many potential reservoir classes exist, including depleted oil and gas wells, deep saline reservoirs, unminable coal seams, basalt formations, and oceans. Monitoring, verification, health, safety, and local environmental issues remain to be resolved before this technology can be deployed at a large scale.

Similarly, hydrogen systems could provide a major contribution to closing the gap. Hydrogen is an energy carrier that has the attractive property of exhausting water vapor when oxidized. It can be used directly in applications ranging from space heating to electric power generation to transport. But hydrogen is not a primary energy form. It is derived either from a hydrocarbon such as a fossil fuel or biomass, or from the splitting of water, $H_2O$, into its constituent parts, hydrogen and oxygen, usually using electricity. The use of hydrocarbons to provide a source of hydrogen raises the question of the disposition of the carbon, and hence may require carbon capture and disposal technology to contribute to closing the gap. Hydrocarbons derived from biomass could contribute to closing the gap by providing hydrocarbon feedstocks for hydrogen production without carbon capture and disposal technology, as the carbon contained in the biomass is obtained from the atmosphere. In combination with carbon capture and disposal, biomass could provide energy with an effectively negative carbon emission. The production of hydrogen using electricity raises both the question of cost and the question of how the electricity was produced.

Once hydrogen is produced, a system must be developed and deployed to cost-effectively utilize the fuel in the provision of energy services while simultaneously providing consumer amenities, and addressing other environment, health, and safety concerns. Fuel cells have attracted considerable attention in that they can convert hydrogen to electricity and heat, while producing only water vapor as exhaust. Fuel cells can be deployed in either stationary or mobile applications. Yet many questions remain to be addressed before either hydrogen systems or fuel cells are widely deployed. Economic issues loom large. The present fossil-fuel–based transportation system with the internal combustion engine has proved a highly cost-effective system for delivering transportation services. Furthermore, even if cost-competitive fuel cells are developed, they could have little or no effect on carbon emissions if they do not employ hydrogen as the fuel.

Both hydrogen and fuel cell technologies will require further research to address economic, technological, environmental, health, safety, and institutional questions.

Biotechnology could similarly play a significant part in closing the gap. Many illustrative scenarios assume that costs and performance will improve to the point where commercial biomass is a major component of the global energy system by the middle of the century. But, as with other technologies that are at an early stage of development, economic, technological, environmental, health, safety, institutional, and ethical questions abound. How is crop productivity to be improved? What other implications are implied for land use and land use emissions? Could genetically modified crops be used? How will this approach affect biodiversity?

Other biotechnology options also represent great potential contributions to closing the gap, while raising equally deep questions about technology, environmental impact, health, and safety considerations and ethics. The creation of new life forms to produce hydrogen from hydrocarbons or water, or to capture carbon from the air and store it in soils, for example, has extraordinary ethical implications in addition to technological challenges.

## Technology Development and Deployment

New technologies and policies play an important role in all scenarios, including baselines and stabilization scenarios. To meet the growing need for energy services, new technologies must deliver energy services in an ever more efficient, less polluting, and less costly manner. Technological change plays an important role in this process, along with other important developments such as new institutional arrangements, adequate investments in energy, capacity building and education, or free trade to mention just a few enabling developments. The need for technological change is even greater in the stabilization cases because additional technological measures are required to close the emissions gap. Thus, numerous new and advanced energy technologies will have to be developed and deployed during the next 100 years.

Large research and development (R&D) efforts are required to deliver technologies to fill the gap in addition to delivering the reference scenarios. R&D comes in many forms, ranging from basic scientific research to technology deployment. The knowledge base upon which future technologies will emerge will be created by curiosity-driven research, applied research, research in related fields, and by learning processes that begin once technologies begin to emerge in their first applications.

New technologies historically emerge after extensive experimentation and the accumulation of knowledge through experience. Technologies typically begin their existence in niche markets and are characterized by high costs and frequently inferior performance compared with the old ones in core applications. Dedicated development often, but not always, brings improvements. In economics this process is called technological learning or learning by doing. In engineering and business, one often refers to so-called cost buy-downs along a learning curve. It is only after the costs have been reduced and performance improved that widespread diffusion can take place and old technologies can be replaced by new ones. A rich literature that describes the enormous improvements and cost reductions that can be achieved with accumulated experience and deployment of new technologies, eventually resulting in superior performance and lower costs than older competitors. Gas turbines are an example of one technology that has been developed this way. It should be mentioned that these stages in the innovation chain are not intended to be linear or sequential; it is an interactive process. R&D is always required, and niche markets for experimentation are needed to advance even the mature technologies.

## The Cost of Closing the Gap

For those cases in which a gap exists—that is, for economies that are not on a course that will lead to the stabilization of $CO_2$ concentrations on their own—resources that would not otherwise be employed to reduce emissions must be diverted from some other human enterprise to the task of closing the gap. That is, there is an economic cost.

Many factors will determine the magnitude of that cost, including the scale of economic activity, the technical, political, and institutional ability of society to limit emissions wherever it is cheapest to do so, the set of technologies available to fill the gap, the stabilization concentration level, and the distribution of emissions in time.

The latter factor reflects the fact that cumulative, not annual, emission of carbon to the atmosphere determines atmospheric $CO_2$ concentrations. That is, the same concentration target can be achieved through a variety of emission pathways. This process is illustrated in Figure 4.7, which shows alternative concentration profiles leading to stabilization at 350, 450, 550, 650, and 750 ppm.

The choice of emission pathway can be thought of as a "carbon budget" allocation problem. In a first approximation, a concentration target defines an allowable amount of carbon to be emitted into the atmosphere between now and the date at which the target is to be achieved. The issue is how best to allocate the carbon budget over time.

Some insight into the characteristics of the least-cost mitigation pathway can be obtained from an exercise conducted under the Stanford Energy Modeling Forum. A group of modelers was asked to examine two alternative emission pathways for stabilizing concentrations at 450, 550, 650, and 750 ppm (see Figure 4.8). The solid and dashed lines are referred to as WG 1 and WRE, respectively, denoting their source (Houghton et al. 1995; Wigley et al. 1996).

Notice that for each model, global mitigation costs are less expensive under WRE. There are several reasons why the models tend to favor a more gradual departure from their reference path. First, energy-using and energy-producing capital stock (e.g., power plants, buildings, and transport) are typically long lived. The current system was put into place based upon a particular set of expectations about the future. Large emission reductions in the near term will require accelerated replacement. This replacement is apt to be costly. There will be more opportunity for reducing emissions cheaply once the existing capital stock turns over.

Second, the models suggest that there are currently insufficient low-cost substitutes, on both the supply and demand sides of the energy sector, for deep near-term cuts in carbon emissions. With the anticipated improvements in the efficiency of energy supply, transformation, and end-use technologies, such reductions should be less expensive in the future.

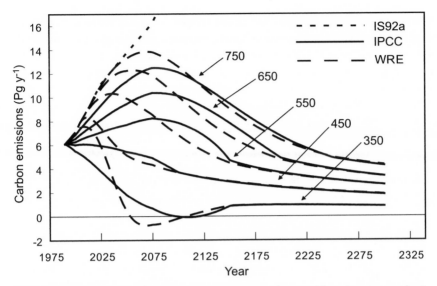

**Figure 4.7.** Comparison of emissions trajectories consistent with various atmospheric $CO_2$ concentrations. Developed by the IPCC (S350–S750) and by Wigley, Richels, and Edmonds (S350a–S750a) (Wigley et al. 1996).

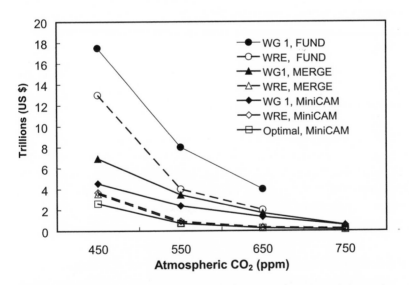

**Figure 4.8.** Relationship between present discounted costs for stabilizing the concentrations of $CO_2$ in the atmosphere at alternative levels.

Third, with the economy yielding a positive return on capital, future reductions can be made with a smaller commitment of today's resources. For example, assume a net real rate of return on capital of 5 percent per year. Further, suppose that it costs $50 to remove a ton of carbon, regardless of the year in which the reduction occurs. If we were to remove the ton today, it would cost $50. Alternatively, we could invest $19 today to have the resources to remove a ton of carbon in 2020.

The result that the lower-cost mitigation pathway tends to follow the baseline in the early years has been misconstrued by some as an argument for inaction. Wigley et al. (1996, 242), referring to their own work, argue that this is far from the case.

> We must stress that, even from the narrow perspective of a cost effectiveness analysis, our results should not be interpreted as suggesting a "do nothing" or "wait and see" policy. First, all stabilization targets still require future capital stock to be less carbon-intensive than under a BAU scenario. As most energy production and use technologies are long-lived, this has implications for current investment decisions. Second, new supply options typically take many years to enter the marketplace. To ensure sufficient quantities of low-cost, low-carbon substitutes in the future requires a sustained commitment to research, development and demonstration today. Third, any "no regrets" measures for reducing emissions should be adopted immediately. Last, it is clear from [Figure 4.7] that one cannot go on deferring emission reductions indefinitely, and that the need for substantial reductions in emissions is sooner the lower the concentration target.

Returning to Figure 4.8, note the "bend" in the cost curve as we move from a 550 to a 450 ppm concentration target. The reason is that even under WRE, a 450 ppm target requires an immediate departure from the baseline resulting in premature retirement of existing plant and equipment.

Finally, it should be noted that different emission pathways for achieving a given concentration target imply not only different mitigation costs, but also different benefits in terms of environmental impacts averted. These differences occur because of the differences in concentration in the years preceding the accomplishment of the target. It is therefore important to examine the environmental consequences of choosing one emission path over another.

## Uncertainties in Emission Allowance from Uncertainties in the Carbon Cycle–Climate System

The integrated assessment models described in previous sections estimate the carbon emissions that result from a given scenario of economic development and population growth. These emissions are then converted into atmospheric carbon dioxide partial pressures using reduced-form carbon cycle models. In the flip side to these emission scenarios, the stabilization scenarios calculated by the integrated assessment models also make use of the reduced-form carbon cycle models, but now to calculate permissible emissions, which are then translated into an energy supply from fossil fuels. The aim of this section is to examine how uncertainties in the global carbon cycle might affect both the growth rate in atmospheric carbon dioxide that results from a given emission scenario and the permissible emissions that result from a given concentration stabilization scenario.

We begin our discussion with the set of stabilization scenarios shown in Figure 4.9 and a detailed breakdown of two stabilization scenarios given in Table 4.3. Because of the dependence of asymptotic atmospheric carbon dioxide on cumulative rather than annual emissions, annual carbon emissions have to peak and eventually decline well below present levels (Figure 4.9a) in order to stabilize cumulative emissions (Figure 4.9b) and thus atmospheric $CO_2$. This is true irrespective of the stabilization target. Note, however, that a considerable amount of emissions can continue for many decades and even centuries beyond the peak, as the deep ocean gradually comes into play as a carbon sink.

A range of processes determine carbon uptake by the ocean and the terrestrial biosphere (Prentice et al. 2001; see chapters by Friedlingstein, Le Quéré and Metzl, and Greenblatt and Sarmiento in this volume) and thus allow carbon emissions for a given stabilization pathway. Factors that influence oceanic and terrestrial storage are the increase in atmospheric $CO_2$, the increase in radiative forcing by $CO_2$ and non-$CO_2$ greenhouse gases (GHGs) and other agents, the sensitivity of climate to radiative forcing, and the sensitivity of carbon cycle processes to climate change (Joos et al. 2001; Leemans et al. 2002). Terrestrial carbon storage depends on past and future land use (Hurtt et al. 2002; Leemans et al. 2002; Gitz and Ciais 2003), the sensitivity of soil respiration to soil temperature and moisture changes, and changes in structure and distribution of ecosystems (Cox et al. 2000; Cramer et al. 2001; Joos et al. 2001).

The role of increasing atmospheric $CO_2$ and of additional nutrient input into terrestrial ecosystems (Townsend et al. 1996) in enhancing primary productivity remains controversial. Recent reviews (House et al. 2003; Houghton et al. 2003) discuss the conflicting evidence for environmental growth enhancement (see also Hättenschwiler et al. 1997; Luo et al. 1999; De Lucia et al. 1999; Caspersen et al. 2000; Joos et al. 2002; Cowling and Field 2003).

For a range of plausible assumptions, it is found that the terrestrial biosphere could act as a strong carbon source or sink during this century (Cox et al. 2000; Joos et al. 2001; Hurtt et al. 2002). The ocean's physico-chemical-driven response is to sequester carbon released into the atmosphere. The rate-limiting step is surface-to-deep mixing of excess carbon associated with typical timescales ranging from decades for the upper ocean to many centuries for the abyss (Oeschger et al. 1975). Solubility of $CO_2$ decreases in warming waters, tending to reduce the uptake. Other factors that can

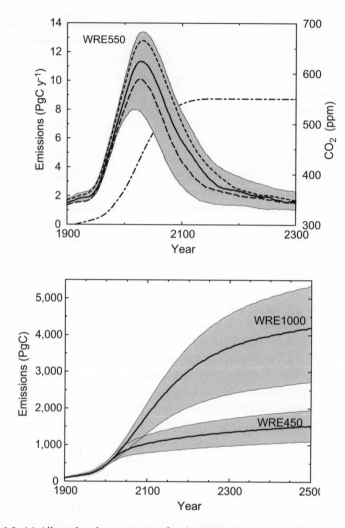

**Figure 4.9.** (a) Allowed carbon emission for the WRE 550 pathway where atmospheric $CO_2$ is stabilized at 550 ppm (dot-dash line, right axis) as simulated with the Bern CC model (Joos et al. 2001). The model's climate sensitivity, expressed as equilibrium temperature increase for a doubling of atmospheric $CO_2$, has been varied between 0°C (no climate feedbacks, dash), 2.5°C (standard case, solid), and 4.5°C (long-dash). The gray band provides an estimate of the overall uncertainty of the allowed emissions. The lower bounding has been calculated by phasing out $CO_2$ fertilization, the major terrestrial sink process in the model, after year 2000 and by setting slow ocean mixing rates. The upper bounding has been obtained by implementing no dependence of soil respiration rates on soil warming, thereby suppressing the major terrestrial source process in the model and by setting high ocean mixing rates. (b) Cumulative carbon emissions allowance for the two WRE stabilization pathways leading to stabilization at 450 ppm and 1,000 ppm.

**Table 4.3.** Allowed carbon emissions for WRE 450 and WRE 750 stabilization pathways

| Pathway | 2000–2020 (20 years) | 2020–2050 (30 years) | 2050–2100 (50 years) |
|---|---|---|---|
| *WRE 450 stabilization pathway* | | | |
| Standard | 189 | 217 | 393 |
| Bounding cases | 145 to +227 | 136 to 268 | 306 to 462 |
| Deviations due to: | | | |
| Climate feedback | | | |
| None | +21 | +36 | +51 |
| High | −23 | −36 | −34 |
| Ocean mixing | | | |
| Low | −17 | −23 | −26 |
| High | +12 | +18 | +19 |
| Land | | | |
| No $CO_2$ fertilization | −28 | −58 | −61 |
| Temperature-independent soil respiration rates | +26 | +33 | +50 |
| Bounding deviations | −44 to +38 | −81 to +51 | −87 to +69 |
| *WRE 750 stabilization pathway* | | | |
| Standard | 202 | 399 | 654 |
| Bounding cases | 155 to 235 | 271 to 468 | 396 to 801 |
| Deviations due to: | | | |
| Climate feedback | | | |
| None | +23 | +50 | +111 |
| High | −24 | −40 | −99 |
| Ocean mixing | | | |
| Low | −17 | −33 | −57 |
| High | +13 | +24 | +43 |
| Land | | | |
| No $CO_2$ fertilization | −30 | −96 | −200 |
| Temperature-independent soil respiration rates | +20 | +45 | +104 |
| Bounding deviations | −47 to +33 | −128 to +69 | −259 to +147 |

*Note:* Total allowed emission in PgC for the standard and bounding cases and deviations from the standard in PgC for the other cases, as obtained with the Bern CC model. The lower bounding case is obtained by combining low ocean mixing with the phasing out of $CO_2$ fertilization on land after year 2000, whereas the upper bounding case results from combining high ocean mixing with soil respiration rate coefficients that remain unchanged under global warming.

influence the oceanic uptake are a slowing of the circulation in response to changed heat and freshwater fluxes into the surface ocean, and changes in marine ecosystems and their productivity in response to changes in the physical and chemical environment (Maier-Reimer et al. 1996; Sarmiento et al. 1998; Joos et al. 1999; Matear and Hirst 1999; Plattner et al. 2001). Uncertainties in projections are thus associated both with uncertainties in basic carbon cycle processes, such as the rate of surface-to-deep mixing of excess carbon or fertilization mechanisms on land, and with uncertainties in carbon cycle–climate feedbacks, including a possible collapse of the formation of North Atlantic Deep Water or dieback of extant forests.

In the following sections, we explore these uncertainties in a quantitative framework, using primarily sensitivity studies that have been carried out within the framework of IPCC with the Bern carbon cycle–climate (CC) model (Joos et al. 2001; Prentice et al. 2001), but also including recent results from two three-dimensional Earth system models (Cox et al. 2000; Friedlingstein et al. 2000). The Bern CC model links a chemistry and radiative forcing module, the HILDA box-diffusion-type ocean model, the Lund-Potsdam-Jena (LPJ) dynamic global vegetation model (Sitch et al. 2003), and a substitute of the ECHAM3/LSG AOGCM. The three-dimensional models are computationally expensive, and simulations are not available for the stabilization scenarios. Figure 4.9 and Table 4.3 illustrate how uncertainties might affect estimates of emissions for various stabilization scenarios. The cumulative upper-to-lower-limit range of these uncertainties for a standard climate response of 2.5°C warming for a $CO_2$ doubling is on the order of 40 percent of the standard scenario for the next 20 years, then increases to 50 percent and greater for the 2020 to 2050 and 2050 to 2100 periods, and beyond.

The largest contribution to these uncertainties in all the scenarios arises from the parameterization of land processes. The LPJ model as used for the standard scenarios given in the top row of the tables includes both a representation of the effect of $CO_2$ fertilization on plant growth and a conventional temperature dependence of soil respiration rates (Sitch et al. 2003). These mechanisms have opposite effects. $CO_2$ fertilization leads to carbon uptake for rising atmospheric $CO_2$ concentrations, whereas global warming leads to higher soil respiration rates and a loss of soil carbon. The extent of stimulation of carbon storage in natural ecosystems by $CO_2$ has been a matter of controversy, and there is no proof that the biospheric sink on the global scale (see Chapter 1) is indeed primarily due to $CO_2$ fertilization, as implied by the LPJ results. Other processes, such as nitrogen fertilization (Schindler and Bayley 1993; Townsend et al. 1996), climate variations (Dai and Fung 1993; McGuire et al. 2001), and forest regrowth and woody encroachment (Pacala et al. 2001; Goodale et al. 2002; see also Nabuurs, Chapter 16, this volume), might be responsible for part or most of the terrestrial sink. If this were the case, then it would be unlikely that primary productivity would increase as a function of future $CO_2$ concentrations. A simple alternative hypothesis is that primary productivity remains close to its present level. Then the allowed carbon emissions would have to drop by a very large amount (row "no $CO_2$ fertilization" in tables), as the terrestrial biosphere turns into a source in the coming decades as soil carbon is lost. There is also conflicting evidence on the temperature dependence of soil and litter respiration (Trumbore et al. 1996; Giardina and Ryan 2000) over multiannual timescales. If soil and litter respiration rates are indeed independent of global warming, then carbon loss from litter and soil would be suppressed and allowed carbon emissions could be higher (row "temperature independent soil respiration rates"). Uncertainty in ocean circulation, represented in the Bern CC model by varying the ocean mixing, is also substantial, though not as great as uncertainties in land processes. Uncertainty in climate projections also translate into uncertainties in estimated carbon emission allowance. The climate feedback rows vary the equilibrium temperature for a $CO_2$ doubling from its standard value of 2.5°C to 0°C in the no climate feedback case and 4.5°C in the high climate feedback case. Projected temperature and precipitation changes affect photosynthesis and soil respiration rates on land and the solubility of $CO_2$ in the ocean. Climate change also affects vegetation dynamics, leading to forest expansion in high northern latitudes as well as to forest dieback in other regions. The results obtained with the three-dimensional Hadley and IPSL models summarized in Table 4.4 give another illustration of the climate feedback effect. Here the carbon emissions are fixed per the IPCC IS92a (Hadley) and A2 (IPSL) scenarios, and the models predict the resulting $CO_2$ in the atmosphere. The Hadley model land biosphere, which suffers from a massive loss of soil carbon in the climate feedback case, shifts from being a large net sink to a large net source when climate feedback is added. As a consequence, atmospheric $CO_2$ climbs to almost 1,000 ppm from the non-climate feedback result of 702 ppm. This represents a difference of almost 600 PgC in the carbon sinks between the two cases. By contrast, the IPSL model, which is more similar to the Bern model in its behavior, shows a difference of only 160 PgC in the carbon sinks between the two scenarios.

As discussed in Chapter 2, there exist vulnerabilities and possible "surprises" that are not readily included in state-of-the art models or that are possible but not necessarily likely. For example, an uncertainty not considered in the projections presented by IPCC and shown here in Figure 4.9 and Table 4.3 is the possibility of a collapse of the North Atlantic Deep Water formation and its impact on oceanic and terrestrial carbon uptake (see also Joos and Prentice, Chapter 7, this volume). Since the detection of rapid abrupt climate change in Greenland ice cores, European lake sediments, and sediments in the deep Atlantic (Oeschger et al. 1984; Broecker et al. 1985; Clark et al. 2002), concerns have been expressed that the formation of North Atlantic Deep Water may cease in response to global warming (Broecker 1987; Manabe and Stouffer 1993; Stocker et al. 2001). This would imply reduced ocean heat transport to the North Atlantic region with large consequences for the climate in Europe and the Northern Hemisphere. Model results suggest that the meridional overturning circulation may be vulnerable to future changes in the hydrological cycle and in sea surface temperature (Cubasch et al. 2001) and that North Atlantic Deep Water formation may even eventually cease in

**Table 4.4.** Impact of climate feedback on carbon sinks and global warming in the Hadley (Cox et al. 2000) and IPSL (Friedlingstein et al. 2000) simulations

| Period | Emissions (PgC) | Ocean uptake (PgC) | Land uptake (PgC) | $CO_2$ in final year (ppm) | Temperature change (°C) |
|---|---|---|---|---|---|
| *HADLEY model* | | | | | |
| With climate feedback | | | | | |
| 2001–2020 | 205 | 58 | 20 | 457 | 0.45 |
| 2021–2050 | 397 | 111 | –14 | 596 | 1.60 |
| 2051–2099 | 865 | 221 | –182 | 982 | 2.00 |
| No climate feedback | | | | | |
| 2001–2020 | 205 | 42 | 70 | 409 | |
| 2021–2050 | 397 | 81 | 134 | 494 | |
| 2051–2099 | 865 | 161 | 258 | 702 | |
| *IPSL model* | | | | | |
| With climate feedback | | | | | |
| 2001–2020 | 209 | 67 | 55 | 409 | 0.5 |
| 2021–2050 | 464 | 154 | 102 | 506 | 0.72 |
| 2051–2100 | 1,128 | 349 | 209 | 778 | 1.12 |
| No climate feedback | | | | | |
| 2001–2020 | 209 | 66 | 66 | 399 | |
| 2021–2050 | 464 | 145 | 139 | 484 | |
| 2051–2100 | 1,128 | 336 | 334 | 701 | |

*Note:* The IS92a emission scenario is applied in the Hadley simulations, and the SRES A2 scenario in the IPSL run.

response to anthropogenic forcing (Stocker and Schmittner 1997), similar to what happened frequently during the last glacial period. However, since such ocean circulation changes, and in particular large-scale reorganizations, are highly nonlinear processes involving thresholds, there are inherent limitations to the predictability of such phenomena (Knutti and Stocker 2002). Sensitivity experiments carried out with dynamical ocean-biogeochemical models (Sarmiento and Le Quéré 1996; Joos et al. 1999; Plattner et al. 2001) show that a collapse of North Atlantic Deep Water formation has the potential to substantially reduce oceanic carbon uptake as surface-to-deep mixing is slowed considerably on century timescales (Figure 4.10).

Uncertainties in the IPCC projections of emission allowance appear to be asymmetrically distributed around the best estimates. The chance that emission allowance is overestimated is higher than that it is underestimated. One reason is that these projections do

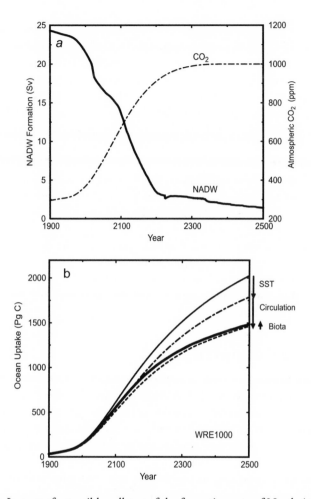

**Figure 4.10.** Impact of a possible collapse of the formation rate of North Atlantic Deep Water (NADW) on the oceanic carbon uptake, and hence the carbon emission allowance to meet a stabilization target (Joos et al. 1999). (a) Prescribed atmospheric $CO_2$ stabilization pathway (WRE1000) (dashed, right-hand scale) and the modeled decrease in NADW formation under global warming (solid). (b) Oceanic carbon uptake is reduced for the simulation with global warming and a collapse in NADW formation (solid) compared with a simulation without global warming (thin solid). The contribution of various mechanisms to the overall reduction has been quantified by additional simulations and is indicated by arrows. Only the reduction in carbon uptake due to the effect of sea surface warming on the $CO_2$ solubility (difference between thin solid and dash-dot, "SST") has been included in the IPCC projections. The slowed surface-to-deep ocean exchange leads to a further reduction in oceanic carbon uptake (difference between dash-dot and dash, "circulation"), whereas a more efficient nutrient utilization slightly enhances uptake of excess carbon (difference between dash and solid, "biota").

not fully account for the vulnerabilities of carbon reservoirs discussed previously and in Chapter 2, most of which are positive feedbacks. Another reason is that current terrestrial models (Cramer et al. 2001; McGuire et al. 2001) include relatively strong sink mechanisms driven primarily by $CO_2$ fertilization. A strong sink mechanism is consistent with the traditionally high land-use emission estimates (Fearnside 2000; McGuire et al. 2001; Houghton 2003) and the contemporary carbon budget (Chapter 1). If land-use emissions (Archard et al. 2002; DeFries et al. 2002), and thus the present terrestrial sink processes, are overestimated or if the sink decreases in the near future (Hurtt et al. 2002; Cowling and Field 2003), this would have by far the largest potential impact on emission allowance of all the processes considered (Table 4.3). Yet another reason is that additional warming by non-$CO_2$ GHGs, not included in the IPCC stabilization pathway calculations (Prentice et al. 2001), would likely lead to reduced ocean and terrestrial carbon uptake (Joos et al. 2001).

In summary, there exist a broad range of uncertainties in the projected emission allowance for a distinct $CO_2$ stabilization pathway. The impacts of these uncertainties on estimates of the cost required to close the gap between no climate policy scenarios and stabilization pathways and on optimal hedging strategies need yet to be explored and is a task for future research.

## Some Closing Comments

In this chapter we explored a range of scenarios for future $CO_2$ emissions drawing upon the extensive literature on this subject. Models of the socioeconomic system are coupled with those of the carbon cycle to determine future emissions under alternative assumptions about population and income growth, the cost and availability of existing and future energy-producing and -using technologies, and the key determinants of the carbon cycle.

Most scenarios suggest that in the absence of a constraint on emissions, atmospheric $CO_2$ concentrations will continue to rise well beyond current levels, highlighting a gap between business-as-usual $CO_2$ emissions and those required to stabilize concentrations at levels currently under consideration. To eliminate this gap will require the development and deployment of a new generation of energy technologies.

Of course, technology development is but one of the options for dealing with global climate change. As pointed out in Metz et al. 2001, climate policy requires a portfolio of responses. The challenge facing today's policy makers is to arrive at a prudent hedging strategy in the face of climate-related uncertainties. Among the options are

- immediate reductions of greenhouse gas emissions,
- investments in actions to assist human and natural systems in adapting to any climate change that should occur,
- continued research to reduce uncertainties about how much change will occur and what effects it will have, and

• R&D on energy supply and end-use technologies to reduce the costs of limiting greenhouse gas emissions.

The issue is not one of "either-or" but one of finding the right blend of options. Policy makers must decide how to divide greenhouse insurance dollars among these competing needs.

# Literature Cited

Archard, F., H. D. Eva, H. Stibig, P. Mayaux, J. Gallego, R. Richards, and J. Malingreau. 2002. Determination of deforestation rates of the world's humid tropical forest. *Science* 297:999–1002.

Broecker, W. S. 1987. Unpleasant surprises in the greenhouse. *Nature* 328:123–126.

Broecker, W. S., D. M. Peteet, and D. Rind. 1985. Does the ocean-atmosphere system have more than one stable mode of operation? *Nature* 315:21–25.

Caspersen, J. P., S. W. Pacala, J. C. Jenkins, G. C. Hurtt, P. R. Moorcroft, and R. A. Birdsey. 2000. Contributions of land-use history to carbon accumulation in U.S. forests. *Science* 290:1148–1151.

Clark, P. U., N. G. Pisias, T. F. Stocker, and A. J. Weaver. 2002. The role of the thermohaline circulation in abrupt climate change. *Nature* 415:863–869.

Cowling, S. A., and C. B. Field. 2003. Environmental control of leaf area production: Implication for vegetation and land-surface modeling. *Global Biogeochemical Cycles* 17, doi:10.1029/2002GB001915.

Cox, P. M., R. A. Betts, C. D. Jones, S. A. Spall, and I. J. Totterdell. 2000. Will carbon-cycle feedbacks accelerate global warming in the 21st century? *Nature* 408:184–187.

Cramer, W., A. Bondeau, F. I. Woodward, I. C. Prentice, R. A. Betts, V. Brovkin, P. M. Cox, V. Fisher, J. A. Foley, A. D. Friend, C. Kucharik, M. R. Lomas, N. Ramankutty, S. Sitch, B. Smith, A. White, and C. Young-Molling. 2001. Global response of terrestrial ecosystem structure and function to $CO_2$ and climate change: Results from six dynamic global vegetation models. *Global Change Biology* 7:357–373.

Cubasch, U., G. A. Meehl, G. J. Boer, R. J. Stouffer, M. Dix, A. Noda, C. A. Senior, S. Raper, and K. S. Yap. 2001. Projections of future climate change. Pp. 525–582 in *Climate change 2001: The scientific basis. (Contribution of Working Group I to the Third Assessment Report of the Intergovernmental Panel on Climate Change)*, edited by J. T. Houghton, Y. Ding, D. J. Griggs, M. Noguer, P. J. van der Linden, X. Dai, K. Maskell, and C. A. Johnson. Cambridge: Cambridge University Press.

Dai, A., and I. Y. Fung. 1993. Can climate variability contribute to the "missing" $CO_2$ sink? *Global Biogeochemical Cycles* 7:599–609.

DeFries, R. S., R. A. Houghton, M. C. Hansen, C. B. Field, D. Skole, and J. Townshend. 2002. Carbon emissions from tropical deforestation and regrowth based on satellite observations for the 1980s and 90s. *Proceedings of the National Academy of Sciences* 99:14256–14261.

De Lucia, E. H., J. G. Hamilton, S. L. Naidu, R. B. Thomas, J. A. Andrews, A. Finzi, M. Lavine, R. Matamala, J. E. Mohan, G. R. Hedrey, and W. H. Schlesinger. 1999. Net primary production of a forest ecosystem with experimental $CO_2$ enrichment. *Nature* 284:1177–1179.

Fearnside, P. M. 2000. Global warming and tropical land-use change: Greenhouse gas emissions from biomass burning, decomposition and soils in forest conversion, shifting cultivation and secondary vegetation. *Climatic Change* 46:115–158.

Friedlingstein, P., L. Bopp, P. Ciais, J.-L. Dufresne, L. Fairhead, H. LeTreut, P. Monfray, and J. Orr. 2000. Positive feedback of the carbon cycle on future climate change. *Note du Pôle de Modélisation* (Institut Pierre Simon Laplace) 19:1–15.

Giardina, C. P., and M. G. Ryan. 2000. Evidence that decomposition rates of organic carbon in mineral soil do not vary with temperature. *Nature* 404:858–861.

Gitz, V., and P. Ciais. 2003. Amplifying effects of land-use change on future atmospheric $CO_2$ levels. *Global Biogeochemical Cycles*, doi 10.1029/2002GB001963.

Goodale, C. L., M. J. Apps, R. A. Birdsey, C. B. Field, L. S. Heath, R. A. Houghton, J. C. Jenkins, G. H. Kohlmaier, W. Kurz, S. Liu, G.-J. Nabuurs, S. Nilsson, and A. Z. Shvidenko. 2002. Forest carbon sinks in the Northern Hemisphere. *Ecological Applications* 12:891–899.

Hättenschwiler, S., F. Miglietta, A. Raschi, and C. Körner. 1997. Morphological adjustments of mature *Quercus ilex* trees to elevated $CO_2$. *Acta Oecologica* 18:361–365.

Houghton, R. A. 2003. Revised estimates of the annual net flux of carbon to the atmosphere from changes in land use and land management, 1850–2000. *Tellus* 55B: 378–390.

Houghton, J. T., Y. Ding, D. J. Griggs, M. Noguer, P. J. van der Linden, X. Dai, K. Maskell, and C. A. Johnson, eds. 2001. *Climate change 2001: The scientific basis (Contribution of Working Group I to the Third Assessment Report of the Intergovernmental Panel on Climate Change)*. Cambridge: Cambridge University Press.

Houghton, J. T., L. G. M. Filho, J. Bruce, H. Lee, B. A. Callander, E. Haites, N. Harris, and K. Maskell, eds. 1995. *Climate change 1994: Radiative forcing of climate change and an evaluation of the IPCC IS92 emissions scenarios*. Cambridge: Cambridge University Press.

Houghton, R. A., F. Joos, and G. P. Asner. 2003. The effects of land use and management on the global carbon cycle. In *Land change science: Observing, monitoring, and understanding trajectories of change on the Earth's surface*, edited by G. Gutman. Kluwer, submitted, 2003.

House, J. I., N. Ramankutty, I. C. Prentice, M. Heimann, and R. A. Houghton. 2003. Reducing uncertainties in the global terrestrial $CO_2$ sink. *Tellus* 55B:345–363.

Hurtt, G. C., S. W. Pacala, P. R. Moorcroft, J. Caspersen, E. Shevliakova, R. A. Houghton, and B. Moore. 2002. Projecting the future of the U.S. carbon sink. *Proceedings of the National Academy of Sciences of the United States* 99 (3): 1389–1394.

Joos, F., G.-K. Plattner, T. F. Stocker, O. Marchal, and A. Schmittner. 1999. Global warming and marine carbon cycle feedbacks on future atmospheric $CO_2$. *Science* 284:464–467.

Joos, F., I. C. Prentice, S. Sitch, R. Meyer, G. Hooss, G.-K. Plattner, S. Gerber, and K. Hasselmann. 2001. Global warming feedbacks on terrestrial carbon uptake under the Intergovernmental Panel on Climate Change (IPCC) emission scenarios. *Global Biogeochemical Cycles* 15:891–907.

Joos, F., I. C. Prentice, and J. I. House. 2002. Growth enhancement due to global atmospheric change as predicted by terrestrial ecosystem models: Consistent with U.S. forest inventory data. *Global Change Biology* 8:299–303.

Knutti, R., and T. F. Stocker. 2002. Limited predictability of the future thermohaline circulation close to an instability threshold. *Journal of Climate* 15:179–186.

Leemans, R., B. Eickhout, B. Strengers, L. Bouwman, and M. Schaeffer. 2002. The consequences for the terrestrial carbon cycle of uncertainties in land use, climate and vegetation responses in the IPCC SRES scenarios. *Science in China* 45:126.

Luo, Y. Q., J. Reynolds, and Y. P. Wang. 1999. A search for predictive understanding of plant responses to elevated $CO_2$. *Global Change Biology* 5:143–156.

Maier-Reimer, E., U. Mikolajewicz, and A. Winguth. 1996. Future ocean uptake of $CO_2$: Interaction between ocean circulation and biology. *Climate Dynamics* 12:711–721.

Manabe, S., and R. J. Stouffer. 1993. Century-scale effects of increased atmospheric $CO_2$ on the ocean-atmosphere system. *Nature* 364:215–218.

Matear, R. J., and A. C. Hirst. 1999. Climate change feedback on the future oceanic $CO_2$ uptake. *Tellus* 51B (3): 722–733.

McGuire, A. D., S. Sitch, J. S. Clein, R. Dargaville, G. Esser, J. Foley, M. Heimann, F. Joos, J. Kaplan, D.W. Kicklighter, R. A. Meier, J. M. Melillo, B. Moore III, I. C. Prentice, N. Ramankutty, T. Reichenau, A. Schloss, H. Tian, L. J. Williams, and U. Wittenberg. 2001. Carbon balance of the terrestrial biosphere in the twentieth century: Analyses of $CO_2$, climate and land-use effects with four process-based ecosystem models. *Global Biogeochemical Cycles* 15:183–206.

Metz, B., O. Davidson, R. Swart, and J. Pan, eds. 2001. *Climate change 2001: Mitigation (The Contribution of Working Group III to the Third Assessment Report of the Intergovernmental Panel on Climate Change)*. Cambridge: Cambridge University Press.

Nakicenovic, N., J. Alcamo, G. Davis, B. de Vries, J. Fenhann, S. Gaffin, K. Gregory, A. Grubler, T. Y. Jung, T. Kram, E. L. La Rovere, L. Michaelis, S. Mori, T. Morita, W. Pepper, H. Pitcher, L. Price, K. Raihi, A. Roehrl, H.-H. Rogner, A. Sankovski, M. Schlesinger, P. Shukla, S. Smith, R. Swart, S. van Rooijen, N. Victor, and Z. Dadi. 2000. *IPCC special report on emissions scenarios*. Cambridge: Cambridge University Press.

Oeschger, H., U. Siegenthaler, U. Schotterer, and A. Gugelmann. 1975. A box diffusion model to study the carbon dioxide exchange in nature. *Tellus* 27:168–192.

Oeschger, H., J. Beer, U. Siegenthaler, B. Stauffer, W. Dansgaard, and C. C. Langway. 1984. Late glacial climate history from ice cores. Pp. 299–306 in *Climate processes and climate sensitivity*, edited by J. E. Hansen and T. Takahashi. Geophysical Monograph 29. Washington, D.C.: American Geophysical Union.

Pacala, S. W., G. C. Hurtt, D. Baker, P. Peylin, R. A. Houghton, R. A. Birdsey, L. Heath, E. T. Sundquist, R. F. Stallard, P. Ciais, P. Moorcroft, J. P. Caspersen, E. Shevliakova, B. Moore, G. Kohlmaier, E. Holland, M. Gloor, M. E. Harmon, S.-M. Fan, J. L. Sarmiento, C. L. Goodale, D. Schimel, and C. B. Field. 2001. Consistent land- and atmosphere-based U.S. carbon sink estimates. *Science* 292:2316–2320.

Plattner, G.-K., F. Joos, T. F. Stocker, and O. Marchal. 2001. Feedback mechanisms and sensitivities of ocean carbon uptake under global warming. *Tellus* 53B:564–592.

Prentice, I. C., G. D. Farquhar, M. J. R. Fasham, M. L. Goulden, M. Heimann, V. J. Jaramillo, H. S. Kheshgi, C. Le Quéré, R. J. Scholes, and D. W. R. Wallace. 2001. The carbon cycle and atmospheric carbon dioxide. Pp. 183–237 in *Climate change 2001: The scientific basis (Contribution of Working Group I to the Third Assessment Report of the Intergovernmental Panel on Climate Change)*, edited by J. T. Houghton, Y. Ding, D. J.

Griggs, M. Noguer, P. J. van der Linden, X. Dai, K. Maskell, and C. A. Johnson. Cambridge: Cambridge University Press.

Sarmiento, J. L., and C. Le Quéré. 1996. Oceanic carbon dioxide uptake in a model of century-scale global warming. *Science* 274:1346–1350.

Sarmiento, J. L., T. M. C. Hughes, R. J. Stouffer, and S. Manabe. 1998. Simulated response of the ocean carbon cycle to anthropogenic climate warming. *Nature* 393:245–249.

Schindler, W. D., and S. E. Bayley. 1993. The biosphere as an increasing sink for atmospheric carbon: Estimates from increased nitrogen deposition. *Global Biogeochemical Cycles* 7:717–733.

Sitch, S., B. Smith, I. C. Prentice, A. Arneth, A. Bondeau, W. Cramer, J. O. Kaplan, S. Levis, W. Lucht, M. T. Sykes, K. Thonicke, and S. Venevsky. 2003. Evaluation of ecosystem dynamics, plant geography and terrestrial carbon cycling in the LPJ dynamic global vegetation model. *Global Biogeochemical Cycles* 9:161–185.

Stocker, T. F., and A. Schmittner. 1997. Influence of $CO_2$ emission rates on the stability of the thermohaline circulation. *Nature* 388:862–865.

Stocker, T. F., G. K. C. Clarke, H. LeTreut, R. S. Lindzen, V. P. Meleshko, R. K. Mugara, T. N. Palmer, R. T. Pierrehumbert, P. J. Sellers, K. E. Trenberth, and J. Willebrand. 2001. Physical climate processes and feedbacks. Pp. 417–470 in *Climate change 2001: The scientific basis (Contribution of Working Group I to the Third Assessment Report of the Intergovernmental Panel on Climate Change)*, edited by J. T. Houghton, Y. Ding, D. J. Griggs, M. Noguer, P. J. van der Linden, X. Dai, K. Maskell, and C. A. Johnson. Cambridge: Cambridge University Press.

Townsend, A. R., B. H. Braswell, E. A. Holland, and J. E. Penner. 1996. Spatial, temporal patterns in terrestrial carbon storage due to the deposition of fossil fuel nitrogen. *Ecological Applications* 6:806–814.

Trumbore, S. E., O. A. Chadwick, and R. Amundson. 1996. Rapid exchange between soil carbon and atmospheric carbon dioxide driven by temperature change. *Science* 272:393–396.

Wigley, T. M. L., R. Richels, and J. A. Edmonds. 1996. Economic and environmental choices in the stabilization of atmospheric $CO_2$ concentrations. *Nature* 379: 240–243.

# 5

# A Portfolio of Carbon Management Options

Ken Caldeira, M. Granger Morgan, Dennis Baldocchi,
Peter G. Brewer, Chen-Tung Arthur Chen,
Gert-Jan Nabuurs, Nebojsa Nakicenovic, and
G. Philip Robertson

Continuation of current trends in fossil-fuel and land use is likely to lead to significant climate change, with important adverse consequences for both natural and human systems. This has led to the investigation of various options to reduce greenhouse gas emissions or otherwise diminish the impact of human activities on the climate system. Here, we review options that can contribute to managing this problem and discuss factors that could accelerate their development, deployment, and improvement.

There is no single option available now or apparent on the horizon that will allow stabilization of radiative forcing from greenhouse gases and other atmospheric constituents. A portfolio approach will be essential.

The portfolio contains two broad options:

- Reducing sources of carbon (or carbon equivalents) to the atmosphere (e.g., reduce dependence on fossil fuels, reduce energy demand, reduce releases of other radiatively active gases, limit deforestation)
- Increasing sinks of carbon (or carbon equivalents) from the atmosphere (e.g., augment carbon uptake by the land biosphere or the oceans over what would have occurred in the absence of active management)

A variety of options could make a significant contribution in the short term. These include: changing agricultural management practice to increase carbon storage and reduce non-$CO_2$ gas emission; improving appliances, lighting, motors, buildings, industrial processes, and vehicles; mitigating non-$CO_2$ greenhouse gas emissions from industry; reforestation; and geoengineering Earth's climate with stratospheric sulfate aerosols.

Longer-term options that could make a significant contribution include separating carbon from fossil fuels and storing it in geologic reservoirs or the ocean; developing large-scale solar and wind resources with long-distance electricity transmission and/or long-distance $H_2$ distribution and storage; ceasing net deforestation; developing energy-efficient urban and transportation systems; developing highly efficient coal technologies (e.g., integrated gasifier combined cycle, or IGCC, discussed later in this chapter); generating electricity from biomass, possibly with carbon capture and sequestration; producing transportation fuels from biomass; reducing population growth; and developing next-generation nuclear fission.

As long as we continue to use fossil fuels, there are relatively few places to put the associated carbon.

- If $CO_2$ is put directly into the atmosphere, about one-third stays in the atmosphere, causing climate change. Another one-third currently goes to the biosphere, but this sink will eventually saturate, leaving $CO_2$ to accumulate in the atmosphere, where it can cause climate to change. The remaining one-third quickly enters the ocean (and most of the increased atmospheric burden will end up in the ocean on longer timescales). This movement causes significant acidification of the biologically active surface waters before mixing and diluting in the deep ocean on timescales of centuries.
- If $CO_2$ is put directly into the deep ocean (through deep injection), most of it will stay there without first producing a substantial acidification of biologically more active surface waters, but risks to deep ocean biota are not well understood.
- If $CO_2$ is put into deep (>1 km) geological formations (through geologic sequestration) it may be effectively sequestered, but there is uncertainty about the available geological storage capacity, about how much of the injected carbon dioxide will stay in place and for how long, and what ecological and other risks may be associated if and when reservoirs leak.
- If $CO_2$ could be mineralized to a solid form of carbonate (or dissolved forms in the ocean), it could be effectively sequestered on geological timescales, but currently we do not know how to mineralize carbon dioxide or accelerate natural mineral weathering reactions in a cost-effective way.

In the short run (<20 years), management of emissions of non-$CO_2$ greenhouse gases and black carbon may hold as much or more potential to limit radiative forcing than management of carbon dioxide. Continued management of these non-$CO_2$ greenhouse gases and particulates will remain essential in the long run.

Management strategies must be regionally adaptive since sources, sinks, energy alternatives, and other factors vary widely around the world. In industrializing and industrialized countries, the largest sources of $CO_2$ are from fossil fuel. In less-industrialized countries the largest sources involve land use.

Technologies and approaches for achieving stabilization will not arise automatically though market forces. Markets can effectively convert knowledge into working solutions, but scientists do not currently have the knowledge to efficiently and effectively stabilize radiative forcing at acceptable levels. Dramatically larger investments in basic technology research, in understanding consequences of new energy systems, and in understanding ecosystem processes will be required to produce the needed knowledge. Such investments would enable creation of essential skills and experience with innovative pilot programs for technologies and options that could be developed, deployed, and improved to facilitate climate stabilization while maintaining robust economic growth.

# A Portfolio of Options

## *The Importance of a Portfolio Approach*

Stabilizing radiative forcing from greenhouse gases and other atmospheric constituents will require massive changes in the design and operation of the energy system, in the management of forests and agriculture, and in several other important human activities. No single technology or approach will be sufficient to accomplish these changes (Hoffert et al. 1998, 2002). Successful control of greenhouse gases will thus require the development of a portfolio of options, potentially including greater efficiency in the production and use of energy; expanded use of renewable energy technologies; technologies for removing carbon from hydrocarbon fuels and sequestering it away from the atmosphere; a mixture of changes in forestry, agricultural, and land use practices; a reduction in the emissions of the non-$CO_2$ greenhouse gases; and other approaches, some of which are currently very controversial, such as nuclear power and certain types of "geoengineering."

A failure to adopt a portfolio approach runs the risk of dramatically increasing the cost of controls and needlessly polarizing public discourse. Thus, for example, if the proponents of carbon capture and sequestration were incorrectly to suggest that these technologies could resolve all of the problems of limiting emissions, their claims would likely alienate members of the environmental community, who tend to be strong proponents of conservation and renewables, and might impede development of an understanding of this important option. At the same time, arguments that the entire problem could be solved by expanded use of conservation and renewables fail to recognize important technical, economic, and behavioral realities and could unnecessarily confuse the public debate.

Just as a mixed portfolio of solutions will be needed, so too a portfolio approach is needed in research and development. Not every nation need make substantial investments in every technology—indeed few, if any, can afford to do so. Across the world, however, it is essential that substantial investments be made in all promising technologies since there is considerable ambiguity about which ones will ultimately prove most useful, socially acceptable, and cost-effective. Indeed, because of the high diversity across the world's nations, peoples, and ecosystems, different mixes of options are likely to prove desirable in different locations.

It is important that experts remain cognizant of the considerable uncertainties that confront this field. It is incumbent upon the technical community to provide leadership to maintain a wide search for options. It would be a serious mistake if work on promising options were prematurely foreclosed by incomplete expert or public understanding or by short-term political or business agendas. Developing practical hands-on experience with many different technologies is essential. Equally important is the recognition that markets are good at commercializing existing intellectual capital, but they are generally not very good at making sustained investments in the basic research needed to develop this capital. Investments in basic technology and environmental research will be crucial for providing the intellectual capital that the world will need over the next century as it grows progressively more serious about addressing this problem.

## Overview of the Portfolio

Carbon emissions (C) can be represented as the product of gross domestic product (GDP) and carbon emissions per unit GDP (C/GDP), that is, $C = GDP \times (C/GDP)$. The growth rate of GDP today is roughly 2.5 percent per year. Stabilizing $CO_2$ emissions in a world whose GDP increases 2–3 percent per year requires comparable or greater percentage reductions in C/GDP. Stabilizing $CO_2$ concentrations ultimately requires making deep long-term cuts in $CO_2$ emissions (Houghton et al. 2001).

C/GDP can be expressed as the product of the amount of $CO_2$ (or $CO_2$-equivalents) emitted per unit of energy consumed (C/E) and the amount of energy consumed per unit GDP (E/GDP), that is, $(C/GDP) = (C/E) \times (E/GDP)$. Reduction in C/E can be accomplished by using renewable fuels (solar, wind, biomass, etc.), using fossil fuels with carbon sequestration, reducing C-equivalent emissions of non-$CO_2$ greenhouse gases, or using nuclear power or potential future sources such as fusion power. Reduction in E/GDP can be accomplished by developing, for example, more efficient appliances, vehicles, buildings, and industrial processes (i.e., device efficiency), and by developing, for example, urban centers that lend themselves to more efficient transportation systems (i.e., systems efficiency).

Table 5.1 shows energy system and biophysical options categorized by how quickly a significant fraction of each option's total potential could be realized (columns) and by the potential magnitude of $CO_2$ mitigation or its radiative equivalent (rows). This table identifies many options that we can begin deploying now and that can make a significant contribution over the coming decades. Research and development undertaken now can produce the additional options needed to stabilize climate on a longer timescale. In Table 5.2 we indicate the readiness of various options for deployment, as well the magnitude of carbon emissions that could be mitigated by each option. Furthermore, we indicate our subjective appraisal of the relative size of the research and development budget that should be allocated to each option. We believe that highest allocations should go to the most promising options that are limited now by unresolved, but tractable, scientific or technological issues (e.g., energy distribution systems that can facilitate large-scale wind and solar power, improved energy production efficiency, and fossil-fuel carbon capture

**Table 5.1.** Categorization of mitigation options by timescale to achieve a significant proportion of possible reductions (columns) and by potential magnitude of $CO_2$ equivalent impact on radiative forcing (rows)

|  | *Rapidly deployable*[a] | *Not rapidly deployable*[b] |
|---|---|---|
| *Minor contributor* ≤ 3% | • Biomass co-fire in coal-fired power plants<br>• Cogeneration (smallscale distributed)<br>• Expanded use of natural gas combined cycle<br>• Hydropower<br>• Wind without storage (=10% of electric grid)<br>• Niche options: wave and tidal, geothermal, smallscale solar | • Building-integrated photovoltaics<br>• Forest management/ fire suppression<br>• Ocean fertilization |
| *Major contributor* > 3% | • Carbon storage in agricultural soils (no-till cultivation, cover crops)<br>• Improved appliance, lighting, and motor efficiency<br>• Improved buildings<br>• Improved industrial processes<br>• Improved vehicle efficiency<br>• Non-$CO_2$ gas abatement from industrial sources including coal mines, landfills, pipelines<br>• Non-$CO_2$ gas abatement from agriculture including soils, animal industry<br>• Reforestation/land restoration<br>• Stratospheric sulfate aerosol geoengineering | • Biomass to hydrogen or electricity possibly with carbon capture and sequestration<br>• Biomass to transportation fuel<br>• Cessation of net deforestation<br>• Energy-efficient urban and transportation system design<br>• Fossil-fuel carbon separation with geologic or ocean storage<br>• Highly efficient coal technologies (e.g., IGCC)<br>• Large-scale solar (with $H_2$, long-distance transmission, storage)<br>• Next-generation nuclear fission<br>• Reduced population growth<br>• Wind (with $H_2$, long-distance transmission, storage)<br>• Speculative technologies (direct atmospheric scrubbing, space solar, fusion, exotic geoengineering, bioengineering) |

*Note:* Minor contributors are capable of contributing <0.2 PgC $y^{-1}$; major contributors> 0.2 PgC $y^{-1}$. The left column represents technologies that can achieve a significant fraction of their potential within a few decades. The right column represents technologies that could be available in the coming decades if research and development begin now.

[a]A significant fraction of option's potential could be achieved within a few decades.

[b]Unlikely to achieve a significant fraction of option's potential within a few decades.

**Table 5.2.** Magnitude of R&D needed driven by $CO_2$ mitigation needs

| Options | Magnitude of potential contribution | Longevity of energy source | Economic efficiency | Technical readiness | Relative size of preferred R&D allocation | Comments |
|---|---|---|---|---|---|---|
| *Rapidly deployable, minor potential* *≤ 3% (≤ 0.2 PgC y⁻¹) and ≤ 20 years* | | | | | | |
| Biomass co-fire in coal-fired power plants | • | ••• | ••• | ••• | • | Conversion is straightforward |
| Cogeneration (small-scale distributed) | • | ••• | ••• | ••• | • | Principal obstacles are regulatory |
| Expanded use of natural gas combined cycle | • | •• | ••• | ••• | • | Economic considerations are already driving adoption of this option; attractiveness depends on price and availability of natural gas |
| Hydropower | • | ••• | ≥•• | ••• | • | Siting, relicensing, ecosystem disruption |
| Wind | • | ••• | ≥•• | ••• | •• | Limits imposed by dispatch in power systems, resource availability |
| Niche options: wave and tidal, geothermal, small-scale solar | • | ••• | ≤•• | ••• | • | |

*Rapidly deployable, major potential*
*> 3% (> 0.2 Pg C $y^{-1}$) and = 20 years*

| Option | | | | | Comments |
|---|---|---|---|---|---|
| Carbon storage in agricultural soils (no-till cultivation, cover crops) | •• | •• | • | • | Verification, incentives, research into persistence of stored carbon; can be driven by nonclimate considerations |
| Improved appliance, lighting, and motor efficiency | •• | •• | •• | •• | Many possibilities, principal issue is incentives and public communication; can be driven by nonclimate considerations |
| Improved buildings | •• | •• | •• | •• | Zoning, codes, and construction practice are important, higher capital requirements; can be driven by nonclimate considerations |
| Improved industrial processes | •• | ≤••• | ••• | •• | Many possibilities, principal issue is incentives; can be driven by nonclimate considerations |
| Improved vehicle efficiency | •• | ≤••• | •• | •• | Regulatory environment more important than technology; can be driven by nonclimate considerations |
| Non-$CO_2$ gas abatement from industrial sources including coal mines, landfills, pipelines | •• | •• | • | • | Principal issues are regulation, cost, and incentives |
| Non-$CO_2$ gas abatement from agriculture including soils, animal industry | •• | •• | • | •• | Verification; Research on nitrogen cycling |
| Reforestation/land restoration | •• | •• | • | • | Verification; land competition; win-win possibilities |
| Stratospheric sulfate aerosol geoengineering | ••• | ••• | ••• | • | Issues of public acceptance, international law and unintended consequences |

*(continued)*

**Table 5.2.** *(continued)*

| Options | Magnitude of potential contribution | Longevity of energy source | Economic efficiency | Technical readiness | Relative size of preferred R&D allocation | Comments |
|---|---|---|---|---|---|---|
| *Not rapidly deployable, minor potential = 3% (= 0.2 PgCy⁻¹) and > 20 years* | | | | | | |
| Building-integrated photovoltaics | • | ••• | •• | •• | •• | Primary research issue is system cost reduction |
| Forest management/fire suppression | • | • | ••• | ••• | • | Win-win possibilities, but potential ecological costs, enhanced long-term fire vulnerability, questions about long-term effectiveness |
| Ocean fertilization | • | •• | | • | • | Issues include public acceptance, verification, efficacy, and unintended consequences |
| *Not rapidly deployable, major potential > 3% (> 0.2 PgCy⁻¹) and > 20 years* | | | | | | |
| Biomass to hydrogen or electricity possibly with carbon capture and sequestration | •• | ••• | •• | •• | •• | Large land requirements with potential landscape and ecological consequences; potential for negative net emissions |
| Biomass to transportation fuel | •• | ••• | ••• | •• | •• | Large land requirements with potential landscape ecological consequences |
| Cessation or possible reversal of net deforestation | •• | •• | ••• | ••• | • | Social/economic/political factors are limiting |
| Energy-efficient urban and transportation system design | •• | •• | ≤••• | ••• | • | Social/economic/political factors are limiting |

*Note: units in italic are $PgCy^{-1}$.*

| Option | Magnitude of potential | Readiness | R&D needs | Comments |
|---|---|---|---|---|
| Fossil-fuel carbon separation and transport with geologic or ocean storage | •• | •• | ••• | Unintended consequences, leakage, public acceptance; development of cost-effective $H_2$ use technologies |
| Highly efficient coal technologies (e.g., IGCC) | •• | •• | ••• | Need process improvements and cost reduction |
| Large-scale solar (with $H_2$, long-distance transmission, storage) | ••• | • | ••• | Principal research costs in low-cost cell design, energy storage, transport |
| Next-generation nuclear fission | •• | •• | •• | Principally limited by public acceptance |
| Reduced population growth | ••• | ••• | • | Social/economic/political factors are limiting |
| Wind (with $H_2$, long-distance transmission, storage) | •• | • | ••• | Principal research costs in storage and transmission not in turbine design |
| Speculative technologies (direct atmospheric scrubbing, space solar, fusion, exotic geoengineering, bioengineering) | ••• | ? | • | |

*Note:* Many important options have relatively low R&D needs. Description of symbols: Magnitude of potential: • = could be a minor contributor ($\leq$ 0.2 PgC y$^{-1}$), •• = could be a major contributor (> 0.2 PgC y$^{-1}$) but inadequate to be entire solution, ••• = could be entire solution (i.e., contribute > 20 PgC y$^{-1}$). Longevity: • = most of potential could be exhausted this century, •• = most of potential could be exhausted over this millennium, ••• = could be sustained for many millennia. Economic efficiency as measured by cost per ton C avoided as measured against current mix of energy production: • = $\geq$ US$150, •• = between US$150 and US$50, ••• = < US$50 and in a few cases negative. Readiness: • = not ready, •• = ready after additional R&D, ••• = ready. R&D needs: • = modest, •• = significant, ••• = major.

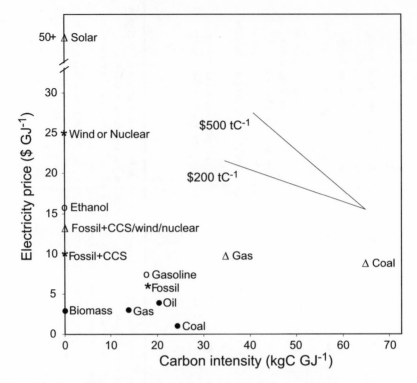

**Figure 5.1.** Cost of electricity presented as function of $CO_2$ emissions per unit energy produced. Lower carbon emission technologies are generally more expensive. Costs per ton of carbon avoided can be estimated from the slope of the line connecting the initial electricity generation technology to the lower carbon emission technology (redrawn from Keith and Morgan 2002). Symbols: • = primary energy, o = Liquid transportation fuel, Δ = electricity, * = hydrogen. CCS refers to carbon capture and storage.

and storage). Figure 5.1 illustrates estimated carbon emission avoidance costs associated with moving to lower-carbon-emissions energy systems.

In the following sections, we review opportunities to reduce energy demand, to improve energy production efficiency, to develop renewable energy sources, to capture and sequester carbon, to reduce the emissions of non-$CO_2$ greenhouse gases, and to mitigate climate change using fission power, geoengineering, and other options.

## Demand Reduction (Conservation, Efficiency)

Energy demand can be reduced in several ways. New technology can provide the same services with lower energy inputs than were previously required (e.g., higher-mileage automobiles, light-emitting diodes). Social preferences and industrial needs can change so that less energy is required to perform the mix of desired functions (e.g., reduced meat and greater vegetable content in diets, reduced use of energy-intensive transport and travel). While the population of the world as a whole will continue to grow in the coming decades, populations in some parts of the world have stabilized or even begun to shrink. Reduced rates of population growth result in fewer persons who must be served with a given level of energy, although the essential need for economic development in much of the world will continue to result in growing energy demand, even if the energy efficiency of these societies can be significantly increased.

There is considerable potential for progress in demand reduction. While new technology is often a necessary ingredient, regulation, pricing, tax policy, social norms, and other factors can be equally or more important.

Efficiency improvements can reduce energy demand by ~1 percent per year through this century (Lightfoot and Green 2002). Thus, efficiency improvements could reduce energy demand by some few tens of percent on the timescale of several decades and could more than halve energy demand by century's end. A more extensive discussion of these options can be found in Metz et al. (2001).

## Renewables and Noncarbon Energy Sources

Renewable energy sources produced by direct solar capture (photovoltaics), wind, hydro, and biomass are all forms of solar energy. Hence, all sources face a finite upper limit of available energy (Metz et al. 2001; Lightfoot and Green 2002), based on net flux density (e.g., average radiation at the ground of ~200 watts per square meter [W m$^{-2}$]).

### BIOMASS ENERGY

Biomass production is limited by the photosynthetic efficiency of conversion of solar energy, which on a canopy scale is capped at about 2–3 percent during the growing season. One must consider the phenology of the plant system, which may leave the landscape bare or sparse a considerable portion of the year, reducing annual mean photosynthetic conversion efficiency (Baldocchi and Valentini, Chapter 15, this volume). Additional energy inputs may be required for cultivation and for fertilizers in order to prevent soil degradation. Land availability is limited by competition with other needs. Furthermore, a significant portion of the terrestrial biosphere is nonarable, where scarce imported water would be needed to produce biomass.

### HYDROPOWER

Hydropower produces $CO_2$-free energy, but this option is suitable only in selected watersheds. There are environmental costs associated with the disruption of ecosystems, fisheries, and landscapes, and silting leads to a finite reservoir life (e.g., Aswan Dam in Egypt). Moreover, when dams are situated in remote locations, there are energy transmission losses. In many industrialized countries, potential hydropower resources are already largely exploited.

## WIND

All locations are not suitable for producing wind power. Wind power is a function of wind speed cubed (kinetic energy increases with wind speed squared and that energy is transported to the wind turbine at the wind speed). Best sites are near seashores (e.g., Denmark), high-wind areas in continental interiors (e.g., North Dakota), or hilltops (e.g., Altamont Pass, California). The jet stream, with energy fluxes of >10 kilowatts (kW) m$^{-2}$, represents a high-density resource, but one that is difficult to harvest. In isolated regions transmission losses reduce efficiencies. For many sites public acceptance is an important barrier (e.g., offshore in Cape Cod and the North Sea). Intermittency limits wind power, in the absence of energy storage and long-distance transmission technologies, to ~10 percent of base load power. Wind power can potentially provide a greater fraction of total power if coupled with improved energy storage and transmission systems (e.g., hydrogen, superconducting long-distance electricity transmission). At very large scale, wind energy could begin to extract a significant portion of kinetic energy from the boundary layer and thus potentially have adverse environmental consequences.

## SOLAR

Photovoltaics are increasing in efficiency, but to contribute significantly to climate stabilization, this technology must be implemented cost-effectively on a large scale. The most efficient photovoltaics from a physics perspective may not be the most efficient economically, as high-efficiency photovoltaics are expensive (Hoffert et al. 2002). Moreover, not all regions are sunny. The use of exotic materials may also be an eventual barrier to large-scale solar voltaic implementation. Because of the intermittency of solar power (and the day-night cycle), solar can only supply a major fraction of base load power if it is combined with a means of energy storage and effective transmission.

## Greenhouse Gas Capture and Sequestration

### LOW-CARBON USE OF FOSSIL FUEL

Hydrocarbon fuels contain both hydrogen and carbon. Today most hydrogen is made for industrial purposes by extracting it from natural gas. If a similar separation is performed on coal, and the $CO_2$ is not released to the atmosphere, the world's abundant supplies of coal could be used as a climate-neutral source of energy for many decades. This practice could buy time to develop new technologies that are completely independent of fossil fuels.

Carbon can be removed from coal or other fossil fuels either before, during, or after combustion (Herzog and Drake 1996; Freund and Ormerod 1997; Metz et al. 2001). The technology for removing carbon after combustion is more developed but is likely to be less economically attractive in the long run. Although the components of carbon capture and sequestration systems exist today at commercial scale, much additional research, development, and deployment will be required both to bring the costs down and to gain essential experience.

Four different technologies have been proposed to separate $CO_2$ after combustion: solvent absorption, adsorption on a solid, membrane separation, and cryogenic separation. All four of these technologies are under development. The use of amine as a solvent for absorption of $CO_2$ from flue gases and its separation through steam reforming is the most mature of these technologies. Amine scrubbing is currently used for producing $CO_2$ for soft drinks, in scrubbing $CO_2$ from air on nuclear submarines, and in many other industrial applications including $CO_2$ removal from natural gas in the Norwegian Sleipner field.

Scientists are exploring various approaches to increase the concentration of $CO_2$ in combustion gases, and thus improve the ease with which the $CO_2$ can be separated. One approach is to combust the fuel in relatively pure oxygen; another idea involves combusting methane in pure oxygen and using the resulting $CO_2$ to drive a turbine, resulting in virtually pure $CO_2$ streams suitable for storage. The combustion temperatures would be far higher than techniques in use today and would pose a challenge to materials science.

Carbon can also be removed either before or during combustion. In this case, there are two basic methods for carbon capture and removal. The first involves carbon separation from the fossil energy source before combustion. For example, one technology already in a relatively mature state of development is steam reforming of methane (natural gas) followed by a shift reaction that results in a mixture of $CO_2$ and $H_2$. After separation, the $CO_2$ could be stored, while the $H_2$ could be used as a very clean energy carrier for electricity production in gas turbines, in fuel cells, for heat production, or in chemical uses. The main challenge is cost reduction.

The second option for $CO_2$ separation and capture before combustion involves coal gasification by partial oxidation to make "syngas" (mainly CO and $H_2$) for combined-cycle turbines. Such schemes for electricity production from coal are called integrated gasifier combined cycle (IGCC) plants and have relatively low air pollutant emissions. Today power from IGCC cogeneration plants can often be competitive with power from coal steam-electric plants with stringent air pollution controls.

IGCC technology is one example of a set of technologies that can increase the efficiency with which fossil fuel can be converted to usable energy. IGCC and more conventional technologies, such as combined cycle natural gas, achieve improved efficiency by operating at a much higher temperature, thus increasing the thermodynamic efficiency of the process. There are other ways to increase energy conversion efficiency. For example, distributed small-scale electric power generation technology is much more effi-

**Table 5.3.** Estimates of storage capacity of geologic reservoirs

| Carbon storage reservoir | Capacity (PgC) |
|---|---|
| Basalt formations* | >100 |
| Deep saline reservoirs | 87–2,727 |
| Depleted gas reservoirs | 136–300 |
| Depleted oil reservoirs | 41–91 |
| Unminable coal seams | > 20 |

*Estimate for Columbia River basalt only.

*Sources:* Herzog et al. (1997), Freund and Ormerod (1997), McGrail et al. (2002).

cient than conventional central generation technology, because in addition to generating electricity, these technologies also supply heat for space conditioning in the buildings in which they are located. Similarly, advanced engine technologies and vehicle design can achieve dramatic improvements in the efficacy with which automobiles convert fuel into kilometers traveled. Together, options for increased conversion efficiency could make major contributions to reducing greenhouse gas emissions in both the short and long term and have the additional advantage of limiting air pollution and other undesirable consequences from fossil-fuel use.

Although they are often not included in discussions of advanced coal technology and of carbon capture and separation, it is important to remember that coal mining, processing, and transport have major land-use and adverse environmental consequences that will continue to grow as the use of coal expands. A full accounting of these and other technologies' costs should be factored into evaluations of different options.

Injection of Carbon Dioxide into Deep Geological Formations

Once carbon dioxide has been separated from fossil fuel, it must be sequestered away from the atmosphere. On land it can be injected into a variety of deep (>1 km) geological formations. Candidates include injection into spent gas fields, injection into depleted oil fields to stimulate additional (secondary) recovery, injection into deep briny aquifers, and injection into coal beds that are too deep for economic production. Table 5.3 provides a very preliminary estimate of reservoirs capacities.

In some parts of the world, large-volume deep injection of fluids already occurs. For example, the United States already injects a mass of fluids into geologic reservoirs larger than the mass of all $CO_2$ produced by U.S. power plants (Wilson and Keith 2002). These fluids are principally wastewaters from municipalities and oil and gas operations (and, to a lesser extent, $CO_2$ for secondary oil recovery).

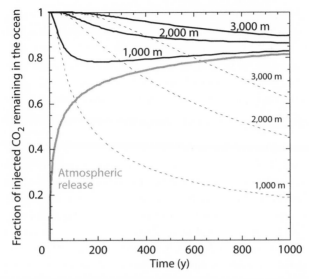

**Figure 5.2.** Representation of model results for direct injection of carbon into the ocean at three depths under two different boundary conditions (Caldeira et al. 2001). The solid lines represent the fraction of carbon that would be found in the ocean at any time after an initial injection at 1,000 m, 2,000 m, or 3,000 m depth. Most of the $CO_2$ added to the ocean through direct injection will remain in the ocean essentially forever. The dashed lines show all the carbon leaking out of the ocean on the timescale of several hundred years. This subtracts from the amounts shown in the solid lines the $CO_2$ that has leaked to the atmosphere but was then reabsorbed by the ocean. The solid lines represent the amount of carbon found in the ocean, but the dashed lines show the amount of $CO_2$ storage credited to ocean sequestration

We currently have a very limited understanding of how to accurately characterize and license geological reservoirs for $CO_2$, monitor possible leakage of $CO_2$ from reservoirs, and assess and deal with potential ecological and other risks (Wilson and Keith 2002).

DIRECT OCEANIC INJECTION OF CARBON DIOXIDE

The ocean is the principal sink for absorbing $CO_2$ from the atmosphere. This capacity arises from the alkalinity of seawater and the reaction of $CO_2$ with dissolved carbonate to form bicarbonate in the ocean waters. The ocean surface waters now absorb ~25 million tons of $CO_2$ per day. The ocean will eventually absorb up to 85 percent of all atmospheric fossil-fuel $CO_2$.

Most of the $CO_2$ added to the ocean through direct injection will remain in the ocean over geological time periods. An important issue is how to credit ocean sequestration for carbon in the ocean, taking into account the fact that most of the $CO_2$ released to the atmosphere would have entered the ocean on this timescale anyway (Figure 5.2).

Total fossil fuel $CO_2$ emissions to the atmosphere today are ~6 petagrams of carbon (PgC) $y^{-1}$, of which ~2 PgC $y^{-1}$ enter the ocean. If we were to permit atmospheric $CO_2$ levels to rise to ~ 600 parts per million (ppm), then surface ocean waters would experience an acidification of ~0.3 pH units. If all ocean waters were to experience ~0.3 pH reduction, then the quantity of $CO_2$ absorbed would be ~1,800 PgC.[1] Thus stabilization at an atmospheric $pCO_2$ of ~600 ppm would commit us to ~1,800 PgC of ocean carbon storage.

The ultimate capacity of the ocean is large, and direct injection of $CO_2$ into deep (~3,000 m) ocean waters (bypassing the atmospheric step with its attendant global warming) would in theory be possible in very large quantities (Herzog et al. 2001). Technological capability for this already exists, making it a high-capacity and quickly available option. Possibilities exist for making ocean sequestration more permanent and reducing pH consequences using the dissolution of carbonate minerals; furthermore, it may be possible to use carbonate neutralization as a flue gas separation process with long-term ocean storage (Kheshgi 1995; Caldeira and Rau 2000).

Significant barriers remain, however: There are a limited number of power plants close to deep ocean waters; public acceptance may not be forthcoming; the London Convention could prohibit this option; and the impact of increased $CO_2$ on deep-sea ecosystems is poorly understood. Research to investigate the potential and risks of ocean storage options is an important priority.

### ACCELERATED CHEMICAL WEATHERING

A basic approach to neutralizing carbon acidity is to dissolve carbonate or silicate minerals. Accelerated weathering of carbonate minerals could be used to store carbon in the ocean for tens of thousands of years (Kheshgi 1995; Caldeira and Rau 2000). Like deep ocean injection, this approach raises legal, social, and environmental issues associated with the use of the oceans as a disposal site. In contrast, accelerated weathering of silicate minerals could potentially lead to carbon storage as a solid carbonate mineral (Lackner 2002). The process is exothermic, but the kinetics are very slow. Thus, the potential for accelerated weathering to make a difference in the next 50 years must be regarded as speculative.

## Potential for Abating Non-$CO_2$ Greenhouse Gas Sources

Non-$CO_2$ greenhouse gases and black carbon are responsible for about 45 percent of total anthropogenic radiative forcing (Houghton et al. 2001), a magnitude that provides substantial opportunities for abatement. These gases include $CH_4$, $N_2O$, tropospheric $O_3$, and halocarbons (Prinn, Chapter 9, and Robertson, Chapter 29, both this volume).

The total anthropogenic flux of $CH_4$ is 344 Tg $CH_4$ $y^{-1}$, equivalent to 2.1 $PgC_{equiv}$ $y^{-1}$ (based on a 100-year GWP time horizon). About half of this flux (1.1 $PgC_{equiv}$) is of agricultural origin; the remainder is from energy extraction and production, industrial combustion, and landfills (Houghton et al. 2001; Robertson, Chapter 29, Table 29.3). There are multiple options for reducing the total $CH_4$ flux; for example, industrial sources including landfills have declined in the United States by 7 percent since 1990 largely owing to the economic value of methane recovery. If these efficiencies were applied more broadly, we might expect to achieve a 10 percent global flux reduction in the next 20 years and a 25 percent reduction in the years following.

The large agricultural $CH_4$ flux is mainly from enteric fermentation, rice cultivation, biomass burning, and livestock waste treatment. Current technology is available to abate 10–30 percent of emissions from confined animals via nutritional supplements, as is now common in U.S. feedlots and dairies. With adoption of best management practices, methane in rice can be reduced substantially; recent results suggest a potential for 50–80 percent abatement based on irrigation management and management for high yields (Robertson, Chapter 29). Likewise, technology is available now to abate most $CH_4$ from animal waste by storing waste in lagoons and generating power from the captured methane; such use has a net negative carbon cost. Modest abatement of these fluxes in the next 20 years and in some cases more aggressive abatement afterward, together with industrial savings, could generate a total abatement of 75 $PgC_{equiv}$ over 100 years.

$N_2O$ fluxes can also be mitigated. The total anthropogenic flux of $N_2O$ is 8.1 $TgN_2O\text{-}N\ y^{-1}$, equivalent to 1.0 $PgC_{equiv}\ y^{-1}$. More than 80 percent of this flux (0.9 $PgC_{equiv}$) is from agriculture; most of the rest is from the industrial production of adipic and nitric acids and combustion. Agricultural soils emit annually about half of the entire anthropogenic flux; waste handling in feedlots and dairies generates about 25 percent of the flux, and biomass burning and industry generates the remainder (Table 5.4).

Technology is now available to abate most of the industrial sources of $N_2O$, and $N_2O$ from waste handling could be largely comitigated with $CH_4$ waste management abatement (as described above). Agricultural soils are more problematic; soils emit $N_2O$ largely as a function of available soil N, and although it is easy to reduce soil N, it is difficult to do so without affecting yields. Better nitrogen placement and timing could reduce current fertilizer needs and therefore $N_2O$ flux by perhaps 20 percent using today's technology, which is the basis for our 20-year abatement estimate (Table 5.4); further mitigation using site-specific farming technologies, varietal improvements for plant nitrogen-use efficiency, and new forms of fertilizer and nitrification inhibitors could lead to 80 percent mitigation in 20–50 years. If so, total 100-year $N_2O$ abatement could reach 75 $PgC_{equiv}$.

Other non-$CO_2$ greenhouse gases, notably the halocarbons, the ozone precursors CO and $NO_x$, and black carbon, are also abatable. Although GWPs for the ozone precursors are at present uncertain and for black carbon unknown, best estimates suggest

**Table 5.4.** Potential for non-$CO_2$ greenhouse gas abatement and biosphere carbon storage

| Gas/source | Current flux strength ($TgC_{eq} y^{-1}$) | Feasible abatement rate ($TgC_{eq} y^{-1}$) 0–20 y | 20–100 y | >100 y | Total abatement ($PgC_{eq}$/100y) |
|---|---|---|---|---|---|
| $CH_4$ | | | | | |
| Industry including landfills | 1,016 | 100 | 250 | 250 | 22 |
| Agriculture | | | | | |
|   Enteric fermentation | 590 | 60 | 300 | 300 | 25 |
|   Rice cultivation | 251 | 160 | 160 | 160 | 16 |
|   Biomass burning | 213 | 20 | 50 | 50 | 4 |
|   Animal waste treatment | 88 | 75 | 75 | 75 | 8 |
|   Total agricultural $CH_4$ | 1,142 | | | | 53 |
| Total anthropogenic $CH_4$ | 2,158 | | | | 75 |
| $N_2O$ | | | | | |
| Industry, transport | 165 | 80 | 130 | 130 | 12 |
| Agriculture | | | | | |
|   Soils | 533 | 100 | 425 | 425 | 36 |
|   Animal waste treatment | 266 | 200 | 250 | 250 | 24 |
|   Biomass burning | 63 | 6 | 30 | 30 | 3 |
|   Total agricultural $N_2O$ | 862 | 306 | 705 | 705 | 63 |
| Total anthropogenic $N_2O$ | 1,027 | 386 | 835 | 835 | 75 |
| *Other* | | | | | |
| Halocarbons | 98 | 10 | 75 | 90 | 6 |
| Ozone precursors (CO, $NO_x$) | 1650 | 150 | 750 | 1500 | 63 |
| Black carbon | nd | | | | nd |
| Total other gases | 1748 | 160 | 825 | 590 | 69 |
| *Biosphere carbon storage* | | | | | |
| Agricultural soils | 0 | 300 | 500 | 0 | 46 |
| Reforestation/agroforestry | | | | | |
|   Improved management | 0 | 170 | 150 | 0 | 15 |
|   Agroforestry + afforestation | 0 | 200 | 100 | 0 | 12 |
|   Cessation of deforestation | 0 | 100 | 200 | 0 | 18 |
|   Total forestry | | 470 | 450 | 0 | 45 |
| Total biosphere C storage | | 770 | 950 | 0 | 91 |

*Note:* Current flux strengths are based on Houghton et al. (2001) and other sources (see Prinn, Chapter 9; Smith, Chapter 28; and Robertson, Chapter 29; all this volume). nd = currently not determinable because of GWP uncertainty.

a total source strength >1.6 $PgC_{equiv}$ $y^{-1}$, of similar importance to $CH_4$ and $N_2O$ (Prinn, Chapter 9, this volume). Mitigation potentials appear to be of a similar magnitude.

## Biological Sinks and Source Reduction

### FOREST MANAGEMENT AND OTHER ACTIVITIES IN THE BIOSPHERE

Forests cover $42 \times 10^{12}$ $m^2$ globally (Sabine et al., Chapter 2, this volume); some 21 percent ($\sim 7 \times 10^{12}$ $m^2$) can be considered managed in some direct manner. Forest carbon storage can be achieved by three principal means: (1) improve the management of currently forested areas, (2) expand the area currently forested via afforestation and agroforestry, and (3) reduce the rate of deforestation. All of these measures aim to adapt management to alter the balance between carbon fluxes into the system (photosynthesis) and fluxes out (plant and soil respiration and harvest), resulting in increased carbon stocks or avoided emissions.

Potential forest sequestration approaches 1 $PgC$ $y^{-1}$ (Watson et al. 2000); more realistic estimates of achievable sequestration are on the order of 0.17 $PgC$ $y^{-1}$ from improved management of existing forests and 0.2 $PgC$ $y^{-1}$ from establishment of new forests on formerly wooded and degraded lands (Watson et al. 2000). Financial costs are modest to high in Annex I countries (US\$3 to \$120 per ton C) and often small elsewhere (US\$0.2 to \$29 per ton C). Management measures to improve carbon storage in forestry include prolonging rotations, changing tree species, continuous-cover forestry, fire control, combined water storage with peat swamp afforestations, fertilization, thinning regimes, and mixed species rotations, among others. At some point management improvements will saturate forest carbon sinks; after 100 years of improved management, it is unlikely that any further net storage will occur.

Cessation of deforestation is another major means for improving biosphere carbon storage in forests. Complete cessation of the current 1.8 $PgC$ $y^{-1}$ is unrealistic for a variety of reasons but nevertheless offers the single largest potential for forest carbon sequestration. Tropical forests and peatlands are at particular risk. With respect to peatlands, it is important to ensure that water tables are maintained so that C continues to be sequestered, especially in the face of global warming at high latitudes.

Although reversing deforestation is a laudable goal, it will be difficult to implement, govern, or reinforce without tangible socioeconomic incentives. Verification also poses difficulties. Over the next 20 years we estimate that with proper incentives 0.1 $PgC$ $yr^{-1}$ could be saved from deforestation, and twice this amount in the period afterward. Once the biosphere is essentially deforested, of course, this potential savings becomes nil.

Despite large attention for carbon storage measures in the political discussions before and after the Conference of the Parties in Kyoto, only a limited number of example projects are underway. Together they affect only some tens of millions of hectares. There are a number of reasons for this: (1) the biosphere inherently shows a large natural dynamic, thus creating uncertainty and risk; (2) political discussions have been lengthy and outcomes uncertain; (3) biospheric sinks will eventually saturate in the future; (4) many other aims for land management exist (food, fiber, biodiversity, water storage), and all these aims have to be met in an integrated way; and (5) a verification infrastructure must be created and maintained.

## AGRICULTURAL CARBON CAPTURE IN SOILS

Some 40–90 PgC has been lost in agricultural soils since the onset of cultivation. Recapturing some portion of this carbon forms the basis for the soil carbon sequestration sink. The most optimistic estimates of soil carbon sequestration place potential rates of net carbon capture at 0.9 PgC $y^{-1}$, which would restore most of the lost carbon within 50–100 years; more measured estimates place the potential at 0.3–0.5 PgC $y^{-1}$ (Smith, Chapter 28, this volume).

Soil carbon can be built through a variety of agronomic techniques, usually based on increasing plant carbon inputs, slowing soil carbon decomposition rates, or (more commonly) both. Carbon inputs can be enhanced by growing more high-biomass crops, by leaving more crop biomass to decompose in situ, by increasing belowground net primary production (NPP), and by growing cover crops during portions of the year that the soil would otherwise remain fallow. Decomposition rates can be slowed by reducing tillage and by growing crops with low residue quality—that is, containing organic carbon that is less susceptible to microbial attack.

If financial incentives were sufficient, agronomic management for carbon storage could lead to the sequestration of 0.3 PgC $y^{-1}$ within the first 20 years of adoption and on average 0.5 PgC $y^{-1}$ for the following 80 years, providing a 100-year total of 46 PgC mitigation. After this period the soil sink will be essentially saturated and sequestration rates will be nil (Table 5.4).

## IRON FERTILIZATION OF THE OCEAN

Sinking organic matter transports carbon from the near-surface ocean to the deep ocean. It has been suggested that the addition of trace nutrients such as iron could enhance this sinking flux and thus act as a way to store additional carbon in the deep ocean, where it is relatively isolated from the atmosphere. Field experiments of iron addition to the ocean have yielded convincing evidence (e.g., Coale et al. 1998) that productivity can increase several fold within weeks, and there is some evidence that export of organic matter from the surface ocean increases after both experimental and natural iron additions (Bishop et al. 2002). Theory and experimental evidence suggest, however, that most of the exported carbon will return to the surface layers within decades.

It has been suggested that iron fertilization could represent an inexpensive carbon storage option; however, widely discussed cost estimates are questionable as they typically have been based on implausible assumptions including that exported carbon has high C/Fe ratios, that all of the added organic carbon exported from the euphotic zone is balanced by a corresponding $CO_2$ influx from the atmosphere, and that $CO_2$ taken up by the ocean through iron fertilization remains there for a long time. Model simulations involving extreme assumptions (e.g., complete phosphate utilization south of 31°S) indicate maximum sustained carbon uptake rates of < 1 PgC $y^{-1}$. Realizable sequestration potential is likely to be much less.

## Advanced Noncarbon Technologies

Advanced noncarbon technologies, such as nuclear fission or fusion, space solar power, and geoengineering, could potentially play an important role in climate stabilization. Several of these technologies are controversial, in early stages of development, or both. Until an option can be shown not to be viable, however, we should work to understand the option's potential benefits and drawbacks.

### NUCLEAR FISSION

Nuclear fission is an existing technology that could help stabilize climate. In some countries (e.g., France) nuclear power generates a substantial fraction of electricity, thus displacing $CO_2$ emissions that might otherwise occur. Fission involves generating electricity by splitting heavy atomic nuclei, most commonly $U^{235}$, into lighter atomic nuclei. Present nuclear reactor technology provides $CO_2$-free electricity while posing unresolved problems of waste disposal and nuclear weapons proliferation. The supply of fissile material, which depends on price, can be extended greatly through the use of breeder reactors; however, such reactors could greatly exacerbate nuclear weapons proliferation. Fission can potentially play a large role in providing carbon-free energy, if the issues of safety, waste disposal, weapons proliferation, resource availability, and public acceptance can be adequately addressed.

### NUCLEAR FUSION

Fusion involves generating electricity through the joining or fusing of light atomic nuclei to form heavier atomic nuclei. Fusion power holds the promise of a nearly inexhaustible source of climate-neutral energy. It is unlikely, however, that fusion power will be commercially available in the time frame needed to stabilize climate (Hoffert et al. 2002), although it may play a role in maintaining climate stability in future centuries.

### SPACE SOLAR POWER

Solar power satellites could be constructed to generate power in Earth orbit or on the moon and beam that power to the Earth (Hoffert et al. 2002). Advantages of the space environment include higher and more consistent solar fluxes (avoidance of clouds, day-night cycle, etc.). Currently, however, launch costs make this approach uneconomical. In general, space power options require very large scales before economies of scale can be realized. Furthermore, there are environmental and public health concerns.

Public resistance to beaming energy through the atmosphere to Earth's surface is likely. Space power will probably not be economically feasible during this century.

GEOENGINEERING

It has been suggested that climate change induced by anthropogenic $CO_2$ could be cost-effectively counteracted with geoengineering schemes designed to diminish the solar radiation incident on, or absorbed by, Earth's surface. Several schemes have been proposed; these schemes typically involve placing reflectors or other light scatterers in the stratosphere or in orbit between the Earth and Sun, diminishing the amount of solar radiation incident on the Earth (Keith 2000; Govindasamy and Caldeira 2000). Less exotic, biosphere-based geoengineering approaches are possible, although they may be impractical or expensive or have other undesirable consequences. For example, the albedo of large-scale forests could be increased through selective logging. Changing $C_4$ grasses to $C_3$ grasses could partition more available energy into latent heat rather than sensible heat, thus cooling the planetary surface.

There are serious ethical, environmental, legal, technical, and political concerns associated with intentional climate modification. For example, political tensions could be heightened if countries were to undertake geoengineering efforts without first obtaining international consensus. It has been suggested that geoengineering should be researched as an emergency backup strategy in case we needed to head off a truly threatening climate change catastrophe (e.g., runaway methane hydrate degassing [see Gruber et al., Chapter 3, this volume]). Any geoengineering scheme is likely to have negative consequences, which would need to be carefully studied before any serious consideration of deployment.

# Cross-Cutting Issues

## The Role of Uncertainty

Although we have tried in the preceding discussion to provide some rough quantification of the potential magnitude of the contribution that may be achieved through various management options, as well as the timescale on which it may be possible to achieve these contributions, it is important to recognize that there is great uncertainty about both the cost and the efficacy of many options. We do not view these uncertainties as a basis for delay or inaction (Caldeira et al. 2003). The evidence of a growing problem is sufficiently compelling that action is clearly needed today. Economic, business, and ecological theory suggests that when faced with large uncertainty, the best option is to invest in a broad portfolio that creates a diversity of future options.

Our ability to project population, per capita GDP, political and social revolutions, and so on, is quite limited (Nakicenovic and Swart 2000). For example, the widely used IS92a scenario of the Intergovernmental Panel on Climate Change (IPCC) was technologically overoptimistic but failed to anticipate reduced $CO_2$ emissions associated with the end of the Soviet Union; thus the IS92a scenario overestimated year 2000 $CO_2$ releases. Experts meeting a century ago could not have anticipated developments like world wars, jet travel, nuclear power (and its rejection in many places), the computer, and the Internet. We assume we are in an equally disadvantageous position regarding the prediction of technical, social, and political innovation likely to occur this century.

## The Essential Role of Research

In contrast to other important industrial sectors, such as microelectronics and pharmaceuticals, which invest 10 percent or more of gross revenues in research, the energy sector has long had among the lowest research and development (R&D) investments of any major industrial sector.[2] Although concern about the need to deal with the problem of climate change has begun to spark modest increases in private and public research investment, the magnitude of those investments remains dramatically low given the magnitude of the challenge the world faces.

One of the issues that has been consistently lacking in the national and international discourse on greenhouse gas (GHG) policy has been a focus on finding ways to divert a modest portion of the large monetary flows that will be involved in any serious management program into investments into research, development, and demonstration.

## Enabling Technologies

As shown in some of the comments in the right column of Table 5.2, a number of technologies, although not direct options for managing radiative forcing, will be essential for implementing some of the options we have identified. For example, if cost-effective carbon-free strategies for producing hydrogen fuel can be developed, then cost-effective strategies for compact hydrogen storage and conversion (e.g., fuel cells) become important. Similarly, energy storage technologies and/or highly efficient long-distance transmission of electric power (e.g., by superconducting cables) will be critical to making large-scale use of wind or photovoltaics a practical reality.

## Barriers to Implementation

Engineers frequently adopt the view that "if we build it (so that it is cheap and effective) they will come." The reality is that large-scale technology adoption and diffusion are often much more complicated and uncertain. Even when there are no major barriers to adoption, it may take several decades or more for a new technology to become widely used because old capital stock remains economically attractive and personnel are slow to understand and appreciate the benefits of new technology. Beyond this, in many cases large vested interests have a stake in sustaining old technologies. These interests often work actively to impede the introduction of new competing technologies. Frequently regulatory or similar barriers inhibit introduction. For example, in many countries electric utilities are granted exclusive service territories, making it illegal for a private entrepreneurial company to introduce microgrid systems built on small-scale combined heat and power distributed generation (which is more energy efficient than central station power). At the same time, traditional utility companies may see little or no incentive to invest in such technology.

Social acceptability can also play an important role in the rate at which a new technology is adopted. As already noted, social concerns are often not founded on a full understanding of a technology, its strengths and weaknesses, and those of available alternatives. In such cases the professional community has an obligation to provide leadership and keep important options open for development and evaluation. But technology proponents also have a long history of arrogantly ignoring legitimate public concerns, consequently fostering a climate of mistrust and hostility that makes rational public decision making difficult or impossible.

## Ancillary Benefits of $CO_2$ Stabilization Technologies

Probably the single greatest motivation for adopting a new technology is direct cost: If it is cheaper than existing technologies it is adopted quickly. Externalities rarely figure into private sector motivations unless encouraged with government incentives or regulatory structures. Nevertheless, a number of mitigation options have ancillary benefits that can substantially multiply their value to GHG mitigation per se.

Biosphere sequestration of carbon and mitigation of non-$CO_2$ fluxes are two areas that can have substantial societal value beyond GHG mitigation. Organic carbon sequestered in soil contributes to soil and hence ecosystem health, with benefits for soil and water conservation, nutrient storage, porosity, invertebrate biodiversity, plant health, and ground and surface water quality. Carbon sequestered in reestablished and regrowing forests has similar benefits for forested watersheds, in addition to abetting plant and animal biodiversity. Thus, organic carbon storage has important practical implications for drinking water quality, coastal fisheries, farmland quality, and flood protection.

Reducing the emission of non-$CO_2$ greenhouse gases also provides ancillary benefits. $N_2O$ suppression through the better management of nitrogen in cropping systems will help to keep exogenous nitrogen from environmental fates other than crop yields (e.g., air pollution). At present the amount of nitrogen fixed by anthropogenic means is close to that fixed biologically; because less than half of the fertilizer applied to cropping systems is taken up by the crop and the remainder is available to cause significant environmental harm. Likewise $CH_4$ capture from waste handling can provide energy savings for individual farms and perhaps rural communities, and composted waste applied to soils can substitute for synthetic fertilizer, with its economic and $CO_2$ manufacturing cost.

Additionally, industrial capture of carbon and higher carbon use efficiency in the industrial and transport sectors will lead to the emission of fewer industrial non-$CO_2$ greenhouse gases. Black carbon, while not a gas, is nonetheless responsible for about 7 percent of the radiative forcing attributable to anthropogenic sources, and cleaner power generation will reduce radiative forcings from this source. Likewise, lower emissions of the ozone precursors—namely $NO_x$, CO, and the non-methane volatile organic carbons (NMVOCs)—will potentially reduce concentrations of tropospheric ozone, responsible for about 12 percent of total anthropogenic radiative forcing (see Prinn, Chapter 9, and Robertson, Chapter 29). $NO_x$ reductions will attenuate both rainfall acidity and much of the unintentional nitrogen deposition now occurring over much of the Earth's surface (Holland et al. 1999).

## Conclusions

Stabilizing climate will require massive changes in the design and operation of the energy system, in the management of forests and agriculture, and in several other important human activities. Yet we do not currently have sufficient knowledge to efficiently and effectively make the changes necessary to stabilize climate at acceptable levels.

A portfolio of options is required because no single technology or approach will be sufficient, although over time specific options may assume dominant roles. The development of expanded use of conservation and renewables is important, as is fossil-fuel carbon capture and sequestration. The suggestion that just one of these options could solve the entire problem, however, fails to recognize technical and economic realities. Furthermore, desirable portfolio options will vary by location. A failure to adopt a portfolio approach runs the risk of delaying implementation and dramatically increasing the cost of controls.

Technologies and approaches for achieving stabilization will not arise automatically through market forces. Markets can effectively convert knowledge into working solutions, but the needed knowledge can only be developed with dramatically larger investments in basic technology research, in efforts to understand consequences of new energy systems, and in efforts to understand ecosystem processes.

The technical community must provide leadership in a wide-ranging search for options. Incomplete expert or public understanding or short-term political or business agendas cannot be allowed to short-circuit the search for promising solutions. This search must include pilot experiments to gain hands-on experience with many different technologies and approaches. This approach would enable creation of essential skills and experience needed to develop systems of energy production and use to develop land management options, and to promote climate stabilization and vigorous economic growth.

## Notes

1. $1 \text{ PgC} = 10^{15} \text{ gC} = 1 \text{ GtC} = 1{,}000 \text{ TgC} = 3.7 \text{ Gt CO}_2 = 8.3 \times 10^{13} \text{ mol C}$.
2. For example, the U.S. electricity industry invests on the order of 0.3 percent of gross sales in basic technology research.

## Literature Cited

Bishop, J. K. B., R. E. Davis, and J. T. Sherman. 2002. Robotic observations of dust storm enhancement of carbon biomass in the North Pacific. *Science* 298 (5594): 817–821.

Caldeira, K., and G. H. Rau. 2000. Accelerating carbonate dissolution to sequester carbon dioxide in the ocean: Geochemical implications. *Geophysical Research Letters* 27:225–228.

Caldeira, K., H. Herzog, and M. Wickett. 2001. Predicting and evaluating the effectiveness of ocean carbon sequestration by direct injection. In *First national conference on carbon sequestration*. Washington, DC: National Energy Technology Laboratory. http://www.netl.doe.gov/publications/proceedings/01/carbon_seq/p48.pdf.

Caldeira, K., A. K. Jain, and M. I. Hoffert. 2003.Climate sensitivity uncertainty and the need for non-$CO_2$-emitting energy sources. *Science* 299 (5615): 2052–2054.

Coale, K. H., K. S. Johnson, S. E. Fitzwater, S. P. G. Blain, T. P. Stanton, and T. L. Coley. 1998. IronEx-I, an in situ iron-enrichment experiment: Experimental design, implementation and results. *Deep-Sea Research Part II–Topical Studies in Oceanography* 45 (6): 919–945.

Freund, P., and W. G. Ormerod. 1997. Progress toward storage of carbon dioxide. *Energy Conversion and Management* 38 (Supplement): S199–S204.

Govindasamy, B., and K. Caldeira. 2000. Geoengineering Earth's radiation balance to mitigate $CO_2$-induced climate change. *Geophysical Research Letters* 27:2141–2144.

Herzog, H. J., and E. M. Drake. 1996. Carbon dioxide recovery and disposal from large energy systems. *Annual Review of Energy and the Environment* 21:145–166.

Herzog, H., E. Drake, and E. Adams. 1997. $CO_2$ capture, reuse, and storage technologies for mitigating global climate change: A white paper. DOE Order No. DE-AF22-96PC01257. MIT Energy Laboratory, MIT, Cambridge, MA. http://sequestration.mit.edu/pdf/WhitePaper.pdf.

Herzog, H., K. Caldeira, and E. Adams. 2001. Carbon sequestration via direct injection. Pp. 408–414 in *Encyclopedia of ocean sciences*, Vol. 1, edited by J. H. Steele, S. A. Thorpe, and K. K. Turekian. London: Academic Press.

Hoffert, M. I., K. Caldeira, A. K. Jain, E. F. Haites, L. D. D. Harvey, S. D. Potter, M. E. Schlesinger, S. H. Schneiders, R. G. Watts, T. M. Wigley, and D. J. Wuebbles. 1998. Energy implications of future stabilization of atmospheric CO2 content. *Nature* 395:881–884.

Hoffert, M. I., K. Caldeira, G. Benford, D. R. Criswell, C. Green, H. Herzog, J. W. Katzenberger, H. S. Kheshgi, K. S. Lackner, J. S. Lewis, W. Manheimer, J. C. Mankins, G. Marland, M. E. Mauel, L. J. Perkins, M. E. Schlesinger, T. Volk, and T. M. L. Wigley. 2002. Advanced technology paths to global climate stability: Energy for a greenhouse planet. *Science* 295:981–987.

Holland, E. A., F. J. Dentener, B. H. Braswell, and J. M. Sulzman. 1999. Contemporary and pre-industrial global reactive nitrogen budgets. *Biogeochemistry* 46:7–43.

Houghton, J. T., Y. Ding, D. J. Griggs, M. Noguer, P. J. van der Linden, X. Dai, K. Maskell, and C. A. Johnson, eds. 2001. *Climate change 2001: The scientific basis (Contribution of Working Group I to the Third Assessment Report of the Intergovernmental Panel on Climate Change)*. Cambridge: Cambridge University Press.

Keith, D. W. 2000. Geoengineering the climate: History and prospect. *Annual Review of Energy and Environment* 25:245–284.

Keith, D. W., and M. G. Morgan. 2002. Industrial carbon management: A review of the technology and its implications for climate policy. Aspen Global Change Institute. http://www.agci.org/cfml/programs/eoc/ASPEN/science/eoc00/eoc00.cfm.

Kheshgi, H. S. 1995. Sequestering atmospheric carbon dioxide by increasing ocean alkalinity. *Energy* 20:915–922.

Lackner, K. S. 2002. Carbonate chemistry for sequestering fossil carbon. *Annual Review of Energy and Environment* 27:193–232.

Lightfoot, H. D., and C. Green. 2002. *An assessment of IPCC Working Group III findings in climate change 2001: Mitigation of the potential contribution of renewable energies to atmospheric carbon dioxide stabilization*. Report No. 2002-5. Montreal: McGill Centre for Climate and Global Change Research.

McGrail, B. P., S. P. Reidel, and H. T. Schaef. 2002. Use and features of basalt formations for geologic sequestration. In *Sixth international conference on greenhouse gas control technologies,* edited by N. Matsumiya. Kyoto, Japan: Research Institute of Innovative Technology for the Earth.

Metz, B., O. Davidson, R. Swart, and J. Pan, eds. 2001. *Climate change 2001: Mitigation (Contribution of Working Group III to the Third Assessment Report of the Intergovernmental Panel on Climate Change)*. Cambridge: Cambridge University Press.

Nakicenovic, N., and R. Swart, eds. 2000. *IPCC special report on emissions scenarios*. Cambridge: Cambridge University Press.

Watson, R. T., I. R. Noble, B. Bolin, N. H. Ravindranath, D. J. Verardo, and D. J. Dokken. 2000. *Land use, land-use change, and forestry.* A special report of the Intergovernmental Panel on Climate Change. Cambridge: Cambridge University Press.

Wilson, E. J., and D. W. Keith. 2002. Geologic carbon storage: Understanding the rules of the underground. In *Sixth international conference on greenhouse gas control technologies,* edited by N. Matsumiya. Kyoto, Japan: Research Institute of Innovative Technology for the Earth.

# 6

# Interactions between $CO_2$ Stabilization Pathways and Requirements for a Sustainable Earth System

Michael R. Raupach, Josep G. Canadell, Dorothee C. E. Bakker, Philippe Ciais, Maria José Sanz, JingYun Fang, Jerry M. Melillo, Patricia Romero Lankao, Jayant A. Sathaye, E.-Detlef Schulze, Pete Smith, and JeffTschirley

Efforts to stabilize the atmospheric $CO_2$ concentration take place within a fully coupled Earth system in which there are major interactions between the carbon cycle, the physical climate, and human activities. Hence it is necessary to consider atmospheric $CO_2$ mitigation in the context of the Earth system as a whole and its long-term sustainability.

"Sustainability" and "sustainable development" have been defined in a number of ways. The Brundtland Report (World Commission on Environment and Development 1987) defined sustainable development as "development that meets the needs of the present, without compromising the ability of future generations to meet their own needs." In practice, definitions of sustainable development hinge on several key issues: what has to be sustained, what has to be developed, for how long, and with what trade-offs. It is necessary to consider economic, social, and environmental dimensions and to accommodate the different perspectives of each dimension as well as the interrelations between them. For the purposes of this chapter, we will follow the consensus of the United Nations Commission on Sustainable Development (1996), which identified the major challenges for global sustainable development as (1) combating poverty; (2) protecting the quality and supply of freshwater resources; (3) combating desertification and drought; (4) combating deforestation; (5) promoting sustainable agriculture and rural development; and (6) conserving biological diversity. To this list may be added the goal of the 1992 UN Framework Convention on Climate Change (UNFCCC) of preventing dangerous anthropogenic interference in the climate system, implying the need to stabilize the atmospheric concentrations of greenhouse gases and $CO_2$ in particular.

This chapter examines the challenge of achieving $CO_2$ stabilization in the context of the requirements for a sustainable Earth system, broadly defined as above. We have three objectives, each the topic of a major section. First, we sketch a systems framework for analyzing $CO_2$ stabilization pathways within the full range of carbon-climate-human interactions. Second, we identify the wider economic, environmental, and sociocultural implications of a large number of carbon management options in order to provide information about these implications for the systems analysis. Third, we consider a particular important case (land-based mitigation) in more detail. An overall synthesis concludes the chapter.

## A Systems Analysis

### The Carbon Cycle and the Stabilization Challenge

We begin with the familiar aggregate atmospheric $CO_2$ budget, written in the form

$$\frac{dC_A}{dt} = \underbrace{F_{Land} + F_{Ocean}}_{\substack{\text{Indirect human} \\ \text{influence}}} + \underbrace{F_{Foss} + F_{LULUC} + F_{Seq} + F_{Disp}}_{\substack{\text{Direct human} \\ \text{influence}}} \tag{1}$$

where $C_A$ is the mass of $CO_2$ in the atmosphere. The fluxes on the right-hand side, which change $C_A$, include the land-air and ocean-air fluxes ($F_{Land}$, $F_{Ocean}$); industrial (mainly fossil fuel) emissions ($F_{Foss}$); fluxes from land use and land use change ($F_{LULUC}$) excluding managed sequestration; managed terrestrial and ocean biological carbon sequestration ($F_{Seq}$); and engineered disposal of $CO_2$ in land or ocean reservoirs ($F_{Disp}$). Fluxes from the atmosphere (uptake to land and ocean pools) are negative. All these fluxes are influenced by human activities, in the sense that they are different from what they would have been without human intervention in the carbon cycle. We can, however, distinguish different levels of human influence. Some fluxes are linked with identifiable human actions as proximate causes: for instance, fossil-fuel burning, land clearing, or managed $CO_2$ sequestration. Others are influenced by human actions only through changes in the Earth system, such as rising atmospheric $CO_2$, rising average temperatures, changes in precipitation patterns, and changes in nutrient cycles. The distinction is not absolute, but it is important because the directly human-influenced fluxes are candidates for carbon management. Therefore we identify some fluxes in equation (1) ($F_{Foss}$, $F_{LULUC}$, $F_{Seq}$, $F_{Disp}$) as "directly influenced" by human activities and others ($F_{Land}$, $F_{Ocean}$) as "indirectly influenced," with the fluxes in former group being actually or potentially directly manageable.

Most of the fluxes in equation (1) depend on the atmospheric CO$_2$ concentration. These dependencies (often called "feedbacks") are both direct and indirect. The direct feedbacks occur through physical and biological mechanisms like CO$_2$ fertilization of carbon uptake into terrestrial and marine pools, while the indirect feedbacks occur through climate properties like temperature, light, precipitation, and ocean circulation, which influence the fluxes. Climate, in turn, is responding to a range of "climate forcings," including CO$_2$ and non-CO$_2$ greenhouse forcing, solar variability, and volcanogenic aerosols. The result is that the fluxes in equation (1) depend on the atmospheric CO$_2$ level ($C_A$) both directly through physics and biology and indirectly through climate. Through this dual set of dependencies, the behavior of the fluxes in response to rising $C_A$ can potentially change drastically as $C_A$ itself rises; for example, negative (stabilizing) feedbacks can be replaced by positive (destabilizing) feedbacks through mechanisms like a shift in the land-air flux toward CO$_2$ emission as respiration increases with warming, or a substantial change in the oceanic thermohaline circulation. Such possible changes in system behavior, or "vulnerabilities" in the carbon-climate-human system, are associated with nonlinearities or thresholds in the relationships between fluxes ($F$), atmospheric CO$_2$ ($C_A$), and climate (see Gruber et al., Chapter 3, this volume, for more discussion of vulnerabilities). Their significance here is that they have major implications for efforts to stabilize atmospheric CO$_2$.

At present, the largest of the directly human-influenced fluxes in equation (1) is the fossil-fuel emission ($F_{Foss}$). This can be expressed[1] as a product of five driving factors:

$$F_{Foss} = P\underbrace{\left(\frac{G}{P}\right)}_{g}\underbrace{\left(\frac{E_{Pri}}{G}\right)}_{e}\underbrace{\left(\frac{E_{Foss}}{E_{Pri}}\right)}_{f}\underbrace{\left(\frac{F_{Foss}}{E_{Foss}}\right)}_{i} = Pgefi \qquad (2)$$

where $P$ is population, $G$ is gross world economic product, $E_{Pri}$ is global primary energy,[2] and $E_{Foss}$ is the primary energy generated from fossil fuels. Key ratios among these variables are the per capita gross economic product ($g = G/P$), primary energy per unit economic product ($e = E_{Pri}/G$), fraction of primary energy from fossil fuel ($f = E_{Foss}/E_{Pri}$), and carbon intensity of energy generation from fossil fuel ($I = F_{Foss}/E_{Foss}$). We note that $ge = E_{Pri}/P$ is the per capita primary energy. The sum of the directly human-influenced fluxes in equation (1) now becomes

$$\begin{aligned} F_{DHI} &= F_{Foss} + F_{LULUC} + F_{Seq} + F_{Disp} \\ &= Pgefi + F_{LULUC} + F_{Seq} + F_{Disp} \end{aligned} \qquad (3)$$

Table 6.1 gives current global average values for quantities in equations (3) and (1). Of course, these averages mask huge regional differences (Romero Lankao, Chapter 19, this volume).

**Table 6.1.** Current (1990–1999) global average values for terms in the global carbon budget and the quantities $P$, $ge$, $f$, $i$

| Quantity | Description | Value | Unit | Reference |
|---|---|---|---|---|
| $DC_A/dt$ | Growth rate of atmospheric $CO_2$ | +3.3 | PgC y$^{-1}$ | Prentice et al. (2001) |
| $F$ | Indirect human-induced C flux to atmosphere | −4.7 | PgC y$^{-1}$ | Prentice et al. (2001) |
| $= F_{Ocean}$ | = ocean-air flux | −1.7 | | |
| $+F_{Land}$ | + land-air flux | −3.0 | | |
| $F_{DHI}$ | Direct-human-induced C flux to atmosphere | +8.0 | PgC y$^{-1}$ | Prentice et al. (2001) |
| $= F_{Foss}$ | = industrial (mainly fossil fuel) emission | +6.4 | | |
| $+F_{LULUC}$ | + disturbance flux from land use change | +1.7 | | |
| $+F_{Seq}$ | + managed terrestrial C sequestration | −0.1 | | |
| $+F_{Disp}$ | + engineered disposal of $CO_2$ | −0.0 | | |
| $E_{Pri}$ | Primary energy | 12 | TW | Hoffert et al. |
| | | $3.8 \times 10^{14}$ | MJ y$^{-1}$ | (2002) |
| $P$ | Population | $6.0 \times 10^9$ | Humans | |
| $ge = E_{Pri}/P$ | Per capita primary energy | 2.0 | kW human$^{-1}$ | Calculated using |
| | | $6.3 \times 10^4$ | MJ y$^{-1}$ human$^{-1}$ | $ge = E_{Pri}/P$ |
| $f$ | Fraction of energy from fossil fuel | 0.85 | — | Hoffert et al. (2002) |
| $i$ | Carbon intensity of energy generation from fossil fuel | 19.9 | gC MJ$^{-1}$ | Calculated using $F_{Foss} = Pgefi$ |

Of the eight quantities ($P$, $g$, $e$, $f$, $i$, $F_{LULUC}$, $F_{Seq}$, $F_{Disp}$) determining the directly human-influenced fluxes through equation (3), six are associated with strategies for carbon mitigation:

- $e$: energy conservation and efficiency (while maintaining economic well-being, $g$);
- $f$: use of non-fossil-fuel energy sources;
- $i$: more carbon-efficient energy generation from fossil fuels;
- $F_{LULUC}$: reduction of carbon emissions from land disturbance;
- $F_{Seq}$: managed sequestration of carbon in terrestrial or oceanic biological sinks; and
- $F_{Disp}$: engineered disposal of $CO_2$ in geological or oceanic repositories.

The other two quantities ($P$ and $g$) are not regarded here as carbon mitigation strategies. That is, we do not consider options for reducing fossil-fuel emissions by lowering population ($P$) or economic well-being ($g$). Nakicenovic (Chapter 11, this volume) discusses trends in both variables.

We turn now to the stabilization challenge. A stabilization trajectory for direct-human-induced emissions, $F_{Stab}(t)$, is a (non-unique) trajectory for the directly human-influenced carbon flux, $F_{DHI}(t) = F_{Foss} + F_{LULUC} + F_{Seq} + F_{Disp}$, such that the atmospheric $CO_2$ level ($C_A$) is eventually stabilized ($dC_A/dt = 0$) at a given target level (Wigley et al.

1996; Houghton et al. 2001, Figures 3.13 and 9.16). Examples are shown in Figure 6.1. Stabilization trajectories are determined by the feedbacks between the fluxes in equation (1), particularly $F_{Land}$ and $F_{Ocean}$, and the atmospheric CO$_2$ level ($C_A$). Stabilization trajectories have several important features: first, $F_{Stab}(t)$ increases initially and then peaks and declines to near zero in the far future (and for low stabilization CO$_2$ targets $F_{Stab}$ can actually become negative). Second, the lower the stabilization CO$_2$ target, the sooner and lower is the allowed emissions peak. Third, short-term trajectories of $F_{Stab}(t)$ (over the next decade) do not depend much on the selected stabilization CO$_2$ target. Fourth, it is crucial to note the implications of the vulnerabilities associated with possible positive feedbacks in the relationships between CO$_2$, fluxes, and climate. These vulnerabilities add considerable uncertainty to stabilization trajectories, especially when the stabilization CO$_2$ target is high. The higher the target, the greater is the magnitude of likely climate change and the greater the probability of carbon-climate feedbacks becoming important. In extreme cases, positive feedbacks could prevent stabilization from being achieved at all.

A quite different picture emerges from consideration of the direct-human-induced C flux $F_{DHI}$ (equation (3)), especially its fossil-fuel component. At present, fossil fuels represent about 85 percent of the 12 TW of primary energy generated by human societies (Table 6.1). Future human energy demand will increase substantially (though far from uniformly across regions and scenarios), in response to both increasing population ($P$) and increasing globally averaged per capita energy use ($E_{Pri}/P = ge$) associated with globally rising standards of living. If a significant fraction of this demand is met from fossil fuels, the trajectory for $F_{DHI}$ is likely to be far from that required for stabilization. Emissions scenarios (Nakicenovic et al. 2000; Edmonds et al., Chapter 4, this volume; Nakicenovic, Chapter 11, this volume) give an internally consistent set of assumptions for the time trajectories of the quantities $P$, $g$, $e$, $f$, $i$, $F_{LULUC}$, $F_{Seq}$, and $F_{Disp}$, which influence $F_{DHI}$. Under most (though not all) scenarios, $F_{DHI}(t)$ substantially exceeds the human-induced C flux required for stabilization, $F_{Stab}(t)$. The difference $F_{DHI}(t) - F_{Stab}(t)$ is called the carbon gap.

To give an idea of the relationship between stabilization requirements and possible trajectories of the terms in equation (3), Figure 6.1 shows how these terms evolve over the next century from their current values (Table 6.1) under simple growth assumptions for three cases (see Figure 6.1 caption). Case 1 approximates the well-known IS92a scenario (Houghton et al. 1996) and includes no mitigation through $F_{LULUC}$, $F_{Seq}$, and $F_{Disp}$ except for a decreasing land disturbance emission. It does include some "default" or "business-as-usual" mitigation through negative values for the growth rates of $f$ and $i$, as is commonly done in baseline scenarios (Edmonds et al., Chapter 4, this volume). Case 2 includes major mitigation through $F_{LULUC}$, $F_{Seq}$, and $F_{Disp}$ but is otherwise the same as Case 1. Case 3 includes major mitigation in all terms ($e$, $f$, $i$, $F_{LULUC}$, $F_{Seq}$, $F_{Disp}$) and produces a trajectory for $F_{DHI}$ that approximates a stabilization trajectory. This figure shows that major mitigation through sequestration and disposal (Case 2) cannot

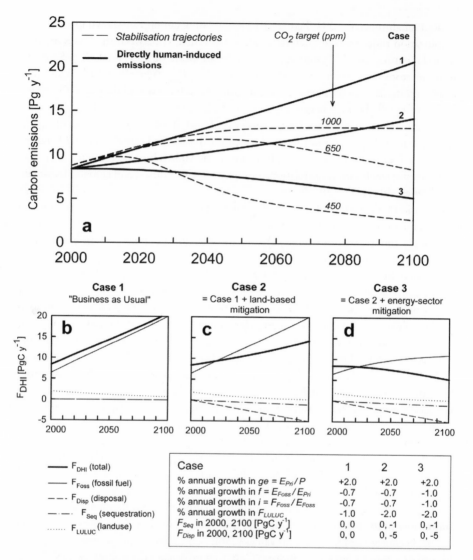

**Figure 6.1.** (a) Scenarios for the directly human-induced C flux ($F_{DHI}$) in three cases, compared with approximate stabilization trajectories ($F_{Stab}$) for stabilization at 450, 650, and 1000 ppm (Wigley et al. 1996). (b, c, d) Carbon fluxes contributing to $F_{DHI}$ for Cases 1, 2, 3, respectively. Cases assume that per capita primary energy $ge$ grows geometrically, $f$, $i$ and $F_{LULUC}$ decline geometrically, and $F_{Seq}$ and $F_{Disp}$ grow linearly, at constant rates shown in the inset table. Case 1 is a typical "business as usual" case approximating the IS92a scenario (see text). Case 2 includes major land-based mitigation. Case 3 additionally includes major energy-sector mitigation.

achieve stabilization unless there is also major mitigation in the energy sector (Case 3). Our intention here is not to construct self-consistent scenarios like those described in Edmonds et al. (Chapter 4, this volume), but only to indicate the orders of magnitude of the terms in equation (3), their trends under simple constraining assumptions, and how those trends relate to stabilization requirements.

## Trajectories of the Carbon-Climate-Human System as Emergent Properties

It might seem that closing the carbon gap is simply a matter of managing the future values of the quantities $P$, $g$, $e$, $f$, $i$, $F_{LULUC}$, $F_{Seq}$, and $F_{Disp}$, which determine the total direct-human-induced flux $F_{DHI}(t)$, so that its trajectory conforms with a stabilization trajectory $F_{Stab}(t)$. Of course, the real world is not like this "command and control" or "rational" decision-making model (Keely and Scoones 1999). All these quantities are internal variables in a coupled carbon-climate-human system and are constrained by numerous interactions with other variables. The future trajectory of the system, and of each of its components, is an emergent property—that is, a property of component interactions rather than the result of external control.

To explore these constraints and interactions, we consider the multiple effects of a portfolio of mitigation options or technologies. These options may include any of those listed after equation (3), together with abatement of non-CO$_2$ greenhouse gas emissions from agriculture and other sectors. The effect of any one mitigation technology can be described by its technical potential ($T$) and its uptake proportion ($u$), where $T$ is the maximum mitigation or carbon equivalent avoided emission (in tCeq per year) that can be achieved by the technology, subject only to biophysical constraints such as resource availability; and $u$ is a number between 0 and 1 determining how much of the technical potential is actually utilized, subject to additional economic, environmental, and sociocultural factors. The achieved mitigation from a portfolio of options will then be $u_1 T_1 + u_2 T_2 + \ldots$, where the subscripts refer to different technologies.

A range of constraints and driving factors influence the uptake of a mitigation technology, and the effort devoted to maximizing its technical potential. Briefly (pending more detailed discussion in the next subsection), these include:

- *climate factors:* the need to avoid dangerous anthropogenic interference in the climate system by minimizing CO$_2$ and other greenhouse gas emissions (the direct purpose of mitigation);
- *economic factors*, including the competitiveness of energy options, the material and energy intensities of economic growth, access to markets, and industrialization pathways;
- non-greenhouse *environmental factors*, including the provision and maintenance of ecosystem services such as clean air, clean water, and biodiversity; and

• *social, cultural, and institutional factors,* including consumption patterns, lifestyles, class structures, incentives, policy climates, and demographics at scales from local to global.

To account for these influences, we consider "utility functions" or "benefit functions" $U_{Clim}$, $U_{Econ}$, $U_{Env}$, and $U_{Soc}$ that quantify the effects of the portfolio of mitigation options in climate, economic, environmental, and sociocultural spheres. These functions may be either positive (net benefits) or negative (net costs). They reflect all aspects of societal well-being and are not confined to measurement in economic terms. They depend on the uptake proportions of the various mitigation options: thus, $U_{Econ} = U_{Econ}(u_k)$ (where the subscript $k$ distinguishes the various technologies) and similarly for $U_{Env}$ and $U_{Soc}$. A major part of the climate utility $U_{Clim}$ is clearly the total achieved mitigation $u_1 T_1 + u_2 T_2 + \ldots$, though there may be climate costs as well, for instance through adverse effects of land use change on regional climates (Betts et al. 2000) or increased $N_2O$ emissions from higher fertilizer use.

It is possible to consider the overall utility or net benefit from the portfolio of mitigation options, $U_{Total}$, accounting for benefits and costs in all spheres. This is a combination of the utilities $U_{Clim}$, $U_{Econ}$, $U_{Env}$, and $U_{Soc}$. There is a substantial literature from the discipline of welfare economics on the formulation and properties of such overall utility measures (see, for example, Brock et al. 2002). Here it suffices to assume that $U_{Total}$ is the weighted sum

$$U_{Total} = w_{Clim} U_{Clim}(u_k) + w_{Econ} U_{Econ}(u_k) + w_{Soc} U_{Soc}(u_k) + w_{Env} U_{Env}(u_k) \quad (4)$$

where the weight factors $w$ describe the relative importance that a society gives to each component of the overall utility of the portfolio of options. The uptake proportions $u_k$ can now be formally specified as the quantities $u_k(t)$, which maximize the overall utility, integrated over some time period, subject to several basic constraints: that energy supply meets demand and that requirements are met for other basic resources such as water, food, and land. It is also possible to regard the technical potentials $T_k$ as variable to maximize $U_{Total}$, to the extent that the technical potentials are influenced by societal choices such as investment in research and development. Thus, both $u_k$ and $T_k$ emerge as "control variables" in a constrained optimization.

This is merely an indicative analysis rather than a quantitative recipe. Even so, several features emerge. First, major variables determining the outcome are the weights $w$, which express the economic, environmental, sociocultural, and policy priorities emerging from societal institutions and structures. These weights are key "levers" influencing future trajectories of the energy system and more generally the carbon-climate-human system.

Second, diverse societies have different institutions, structures, and priorities. The analysis and the outcomes are therefore both regionally specific. A major point of intersection between regions is that they all share the atmosphere as a global commons and hence all pay a climate-change cost as a result of global fossil-fuel emissions. Even so, climate-change impacts and hence the nature of this cost are different among regions.

Third, the existence of benefits as well as costs is of great significance. Here are two examples: (1) a switch to clean fuels or renewables is likely to have benefits for regional air quality, and (2) more efficient transport systems in cities usually confer net social and urban-environment benefits. Such benefits in the non-greenhouse aspects of the overall utility are crucial in adding to the likelihood that real greenhouse mitigation will occur through changes in energy use.

Fourth, there are synergisms between different constraints (or different components of the overall utility in equation (4)). For example, if the technology for a massive switch to biofuels were available at reasonable cost, consequences would include declines in the availability of land for agriculture, biodiversity, and carbon stocks in forest ecosystems. This point is expanded later.

Finally, the uptake of mitigation options is constrained by technological and institutional inertia or contingent history: for example, it is not usually practical to change technologies at a rate faster than the turnover time of the infrastructure (see Caldeira et al., Chapter 5, this volume).

## Diversity of Influences on Carbon Mitigation Strategies

We now consider in more detail the diverse factors affecting the uptake of a particular carbon mitigation strategy. Figure 6.2 is a graphical representation of the factors influencing the difference between the technical potential of a particular mitigation strategy and the actual, achieved mitigation. The horizontal axis is the mitigation potential, measured by the total amount of carbon-equivalent greenhouse forcing avoided (in tCeq per year, for example), and the vertical axis is the price of carbon (in dollars per tCeq), a measure of the weight ascribed to carbon mitigation relative to other goals. The maximum mitigation that can be achieved by a strategy is its technical mitigation potential, based solely on biophysical estimates of the amount of carbon that may be sequestered or greenhouse gas emissions avoided, without regard to other environmental or human constraints. This amount is not dependent on the price of carbon and hence is a straight line. There is also a minimum mitigation achieved by baseline or business-as-usual technological development (as reflected, for example, in the family of scenarios appearing in the Intergovernmental Panel on Climate Change [IPCC] *Special Report on Emissions Scenarios* [SRES], Nakicenovic et al. 2000), which already include assumed mitigation such as through shifts toward non-fossil fuels). This baseline mitigation is likewise independent of the price of carbon. Between these limits, the actual, achieved mitigation is a price-dependent curve. It is less (often very much less) than the technical potential, because of a range of economic, environmental, and social drivers and constraints. In more detail than before, these include the following:

- *Economics:* Economic factors include (1) access to markets for carbon-relevant products; (2) the nature of those markets; (3) the influence of pathways of industrialization and urbanization on existing and new carbon-relevant economic sectors; (4) the

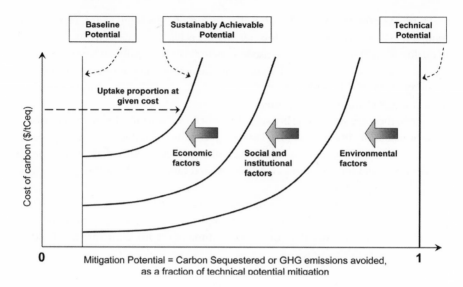

**Figure 6.2.** Effects of economic, environmental, and social-institutional factors on the mitigation potential of a carbon management strategy. The technical potential (independent of cost) is reduced a combination of economic factors (markets, trade, economic structures, urbanization, industrialization); environmental factors (need for land, water and other resources, waste disposal, property rights); and social and institutional factors (class structure, politics and formal policies, informal rules, lifestyles, attitudes, behaviour). The end result is a sustainably achievable mitigation potential for the carbon management strategy being considered. This depends on the cost of carbon, which is a measure of the weight ascribed to carbon mitigation relative to other goals. The uptake proportion for the strategy is the ratio of the sustainably achievable potential to the technical potential. The figure also shows a baseline potential, representing the extent to which the carbon management strategy is deployed in a "business-as-usual" scenario.

existence or absence of crisis-prone economic conditions; and (5) the indebtedness of many countries, especially in the developing world. Economic markets play an important role in governing access to resources and, used intelligently by governments, can provide important incentives to switch to lower-carbon energy portfolios.

• *Environmental requirements for other resources:* The need for resources to supply essentials, such as food, timber, and water, can reduce the estimated technical potential.

• *Environmental constraints:* Mitigation activities can incur environmental costs such as waste disposal and ecological impacts.

• *Social factors:* Differences in social factors between countries and between urban and rural locations exert strong influences on mitigation outcomes. On a personal level, class structure and lifestyles are often related on one hand to increasing consumption and use of carbon-relevant commodities as cultural symbols, such as cars, mobility, and travel to exotic places (Lebel, Chapter 20, this volume). On the other hand, lifestyles are linked to poverty and lack of access, for instance, to technical alternatives. On a public level, societal values and attitudes determine the level of support for carbon management strategies, through education and societal self-image (such as frontier, pro-modernization, pro-conservation).

- *Institutional factors:* Institutions determine the structure of incentives influencing any management option, both in terms of taxes, credits, subsidies, sectoral strategies, property rights regimes, and other formal components and in terms of the policy climate or informal policies within which management strategies are designed and implemented. Examples of policy climate include the level of performance, the presence or absence of corruption, and the extent and nature of powerful vested interests. To illustrate the last point, significant constraints can arise within both public and private sectors that affect the speed of technology deployment and the choices of alternative systems. Owners of existing energy technologies can use their considerable financial and technological influence to block the development or deployment of alternative systems. Similarly, government regulators may use their powers to control the flow of investment in mitigation technology or its application in their country in order to protect perceived national interests.

- *Institutional and timing aspects of technology transfer:* Some features of technology transfer systems, like the patenting system, do not allow all countries and sectors to gain access to the best available technology rapidly or at all. The timing of technology transfer is an issue, as many technological paradigms need 50–70 years to be completely established (Nakicenovic, Chapter 11, and Romero Lankao, Chapter 19, both this volume).

- *Demography:* The density, growth, migration patterns, and distribution of the population can form another constraint, especially in countries with high levels of social segregation. For instance, in regions with a high concentration of land in few hands, population pressure on land can pose an obstacle to a reforestation strategy.

Some of these constraints are price dependent, implying that a higher carbon price would increase the viability of the carbon management strategy. Hence, in general, the relationship between the sustainable mitigation potential and the price of carbon is a curve in Figure 6.2.

After considering these factors, the remaining amount of mitigated carbon emission or sequestration is the sustainably achievable potential at a given price of carbon. In the schematic analysis of the previous subsection, the price of carbon is a measure of the relative weights ($w$) between the climate cost and other components of the overall cost in equation (4), and the uptake proportion ($u$) is the ratio of the sustainably achievable potential to the technical potential. Figure 6.2 again emphasizes the importance of the weighting between carbon mitigation and other societal goals in determining the uptake of mitigation technologies and hence mitigation outcomes.

# Implications of Carbon Management Options

A large portfolio of carbon management options is needed to give flexibility in designing a practical pathway for achieving stabilization of atmospheric $CO_2$ (Caldeira et al., Chapter 5, this volume). As already emphasized, all carbon management options bring both positive and negative environmental, economic, and sociocultural impacts, in addition to greenhouse mitigation. In designing carbon management strategies it is necessary to take advantage of beneficial synergies between mitigation and other impacts.

Our discussion of the wider implications of mitigation is intended to supplement the more technically oriented contributions of Caldeira et al. (Chapter 5) and the engineering-oriented review of Hoffert et al. (2002), by providing some analysis of collateral effects. We consider four classes of impact corresponding to the four terms in equation (4): climate change and greenhouse, economic, environmental, and sociocultural. The technical options are organized into five categories identifiable with the six mitigation-oriented terms in equation (3): conservation and efficiency (combining the factors $e$ and $i$ into one category), non-fossil-fuel energy sources (factor $f$), land-based options including disturbance reduction and biological sequestration (terms $F_{LULUC}$ and $F_{Seq}$), biological sequestration in oceans (term $F_{Seq}$), and engineered $CO_2$ disposal (term $F_{Disp}$). We restrict discussion to options that are currently technically feasible, at least at moderate scales, omitting (for example) nuclear fusion and spaceborne solar power. Also, although the focus is on carbon mitigation toward $CO_2$ stabilization, we include discussion of other greenhouse gases where these are closely related to carbon mitigation options, for example in land use.

The results of our analysis are summarized in Table 6.2, where we have somewhat subjectively estimated the collateral costs and benefits for each strategy in each impact class as minor, moderate, or major.

## Conservation and Efficiency

A number of changes in technology, policy, and human behavior will reduce energy demands, with benefits or at most small costs for economic productivity. These influence the factor $e$ (primary energy per unit economic product) in equation (3). Examples include:

- *more efficient appliances* (light-emitting diodes, low-power computers, . . .);
- *more efficient indoor environments* (passive lighting, heating, cooling, insulation . . .);
- *urban microclimate design* to minimize energy demand in extreme weather conditions (for example, using trees to lessen both heating and cooling demands);
- *better urban planning* (improving public transport, shortening travel distances, . . .);
- *shifts toward diets* that require less energy inputs (shifting diets toward vegetarianism).

Technological changes that will reduce the carbon intensity of fossil-fuel energy generation (factor $i$ in equation (3)) include

**Table 6.2.** Assessment of positive and negative climate, economic, environmental, and sociocultural impacts associated with mitigation strategies

| Mitigation strategy | Climate change and greenhouse | Economic | Environ- mental | Socio- cultural |
|---|---|---|---|---|
| *Conservation and efficiency* | | | | |
| More efficient appliances | | | | |
| More efficient indoor environments | | | | |
| More efficient automotive transport | (+++) | (++) and (–) | (+++) and (–) | (++) and (–) |
| Better urban travel planning | | | | |
| Urban microclimate design | | | | |
| Better use of fossil fuels | | | | |
| Cogeneration | | | | |
| Changes to diets | | | | |
| *Non-fossil-fuel energy sources* | | | | |
| Hydropower | (++) and (–) | (+) and (–) | (++) and (––) | Developed: (–) Developing: (–––) |
| Solar power | (+++) | (–) | (+++) and (–) | ? |
| Wind power | (++) | (+) and (–) | (+++) and (–) | (+) and (–) |
| Bioenergy | (+++) and (––) | (+) and (–) | (+++) and (––) | (+) and (––) |
| Geothermal power | (+) | ? | (++) and (–) | ? |
| Nuclear energy | (+++) | (++) and (––) | (++) and (–––) | (–––) |
| *Land-based options* | | | | |
| Afforestation, reforestation, and land restoration | (++) and (–) | Incentives needed | (++) and (–) | (++) and (––) |
| Reduction of net deforestation | (+++) | Incentives needed | (++) | (++) and (––) |
| Forest management and fire suppression | (+) and (–) | (+) and (–) | (++) and (–) | ? |
| Changing agricultural management | (++) | (++) and (–) | (++) and (––) | ? |
| Non-CO$_2$ mitigation from land biosphere | (++) | (++) and (–) | (++) and (–) | ? |
| Bioengineering solutions | ? | (++) and (––) | (+) and (–––) | (––) |
| *Biological sequestration in the oceans* | | | | |
| Ocean fertilization | (++) and (––) | ? | (––) | (–––) |
| *CO$_2$ disposal on land and oceans* | | | | |
| C separation with ocean storage | (+++) and (––) | ? | (––) | (–––) |
| C separation with geological storage | (+++) and (––) | ? | (––) | (–) |

*Note:* Symbols (+), (++), and (+++) indicate minor, moderate, and major positive impacts (benefits); likewise, (–), (––), and (–––) indicate minor, moderate, and major negative impacts (costs). Where distinct benefits and costs occur, these are indicated separately. Impacts in the climate change and greenhouse area refer to technical potential, indicated as minor (< 0.3 PgC y$^{-1}$), moderate (0.3 to 1 PgC y$^{-1}$), and major (> 1 PgC y$^{-1}$). A question mark indicates that insufficient information is available to make a judgment.

- *more efficient automotive transport* (hybrid vehicles that electrically recover lost mechanical energy, use of smaller cars rather than sports utility vehicles in cities, . . .);
- *better use of fossil fuels* (natural gas, highly efficient coal combustion); and
- *cogeneration* (recovery and use of low-grade heat from electric power stations, usually with smaller and more distributed power stations).

The technical potential for gains in energy efficiency from these options are large, in many cases from tens to hundreds of percentage points on a sectoral basis, and in most cases the technology is readily deployable (Hoffert et al. 2002; Edmonds et al., Chapter 4, this volume). As a group, these options have significant environmental and socio-cultural benefits, including lower urban air pollution, more efficient urban transport networks, more congenial and livable cities, and improvements to population health. Some involve changes to lifestyles (such as reducing car sizes) that may be seen as socio-cultural drawbacks. They also involve some transfers of economic power (for instance, away from oil suppliers), although many oil companies, particularly in Europe, are turning this change to advantage by repositioning themselves as energy service providers, including renewables and conservation.

Because of the magnitude and generally rapid uptake time of these mitigation options, and also recognizing the significant collateral benefits and minimal costs, many future scenarios include significant continuing mitigation from this area. Although scenarios vary widely, it is common to assume improvements in energy efficiency in the order of 1 percent per year while maintaining economic growth of 2–3 percent per year (Edmonds et al., Chapter 4, this volume; Figure 6.1). A key question is whether these uptake rates can be improved.

## Non-Fossil-Fuel Energy Sources

These options reduce $f$ (the fraction of primary energy from fossil fuel) in equation (3).

### Hydropower

At present about 19 percent of the world's electricity is produced from hydropower (McCarthy et al. 2001). Although this source is reaching saturation in developed countries, continued deployment is likely in many developing countries. Hydropower is a renewable energy source with important benefits, such as flood control and regulation of river flows for agricultural, industrial, and urban use. From the power-engineering viewpoint, hydropower is fast to start in peak electricity consumption periods and can store surplus energy from other sources during low consumption periods, thus increasing overall system efficiency. There can, however, also be a range of negative sociocultural, environmental, economic, and ancillary greenhouse-gas consequences. During dam construction and first filling, few or many (up to millions) of people are displaced from their homes, often leading to impoverishment (World Commission on Dams 2000). Dams alter downstream nutrient cycling patterns, ecosystems, and food-web structures, reduce biodiversity in rivers, affect the viability of fishing industries, and accumulate sediment. WBGU (2003) suggests that at least 20 percent of global rivers should remain undammed. From the greenhouse-gas perspective, the global production of greenhouse gases from hydropower systems is relatively low, but lakes in vegetated areas can incur significant methane production in anoxic sediments (McCarthy et al. 2001).

## Solar Power

The solar option has low sociocultural and environmental impacts compared with many other energy systems. Available energy densities are moderate, about 15 W per square meter ($m^{-2}$) for electric power (Hoffert et al. 2002), so that current global electricity consumption could be met from a square less than 400 kilometers on a side (WBGU 2003). The resource is most concentrated in subtropical semiarid to arid areas, implying development opportunities in these areas, which are largely in the developing world. Extensive decentralization is also possible and is very efficient in some sectors (such as solar hot water), leading to substantial social benefits and microeconomic opportunities. On the cost side, large-scale solar energy deployment may involve significant land use changes with consequences for ecosystems; water supply may be an issue in arid regions; energy transport over large distances and across national borders is both a technical and an institutional hurdle; and environmental issues can arise in the production and disposal of photovoltaic cells (silica crystal with traces of heavy metals).

## Wind Power

Like solar energy, wind power is considered a clean energy. Its technology is rapidly deployable to many suitable regions around the world, making it the world's fastest growing energy source at present. Its energy density is comparable to solar, but areas for deployment are more restricted. Wind power has very limited negative effects on the environment. The largest concern is probably the visual impact of large-scale wind farms with turbines up to 200 meters tall on mountain ridges or in coastal regions with aesthetic and cultural values. Wind turbines, especially older models, can be noisy. A controversial impact is that on bird communities through killings by turbine blades.

## Bioenergy

An important option is the substitution of biofuels for fossil fuels, both for transport and for electricity generation from biofuels alone or with fossil fuels in co-fired plants. Biofuels have minimal net carbon emissions because atmospheric $CO_2$ is continuously recycled. The technical potential is large: up to 500 megahectares (Mha) of land (around 3 percent of the global land area) could be made available for biofuel crop production by 2100 (Watson et al. 2000b). This amount would produce in the range of 150–300 EJ $y^{-1}$ and displace 2–5 petagrams (Pg) C $y^{-1}$ of carbon emissions. Biofuel production offers many ancillary benefits: with good design, soil erosion can be reduced and water quality improved through increased vegetation cover. From a national viewpoint, dependence on imported oil can be reduced, rural investment increased, work opportunities created, and agricultural products diversified. There are few negative impacts when biomass sources are by-products (organic waste and residues) from agriculture and forestry, though transport infrastructure is an issue because of low energy densities ($0.3–0.6$ W m$^{-2}$). The large-scale production of dedicated biofuels can, however, have negative impacts, including competition for other land uses, primarily food production; soil sustainability (extraction of residues should not exceed 30 percent of total production to maintain soil fertility); poor resilience of monocultural plantations; and the implications for biodiversity and amenity (Watson et al. 2000b). The full greenhouse gas balance of biofuel production also needs to account for significant emissions from mechanization, transport, use of fertilizers, and conversion to usable energy. Taking into account some of these constraints, especially competition for other land uses, Cannell (2003) estimated the range $0.2–1$ PgC y$^{-1}$ to be a more conservative achievable mitigation capacity for C substitution using biofuels. This amount is about 15 percent of the potential mitigation described earlier. Other estimates are even lower (WBGU 2003).

## Geothermal Power

Geothermal power is a niche option with small mitigation potential but significant environmental benefits where available because of the clean energy yield. There are some environmental concerns over the depletion of geothermal resources.

## Nuclear Power

Nuclear energy is a non-fossil-fuel source with significant mitigation potential. Environmental concerns include waste and plant (old reactor) disposal and the potential for accidents. There is also a major sociocultural or political issue with weapons proliferation, especially where breeder reactors are used to create additional fissile material. This issue is too large to be explored here. It is notable that the UNFCCC Conference of Parties (COP 7 and 8) has agreed not to include nuclear power in the Kyoto Protocol flexible mechanisms.

## Terrestrial Biological Sequestration and Disturbance Reduction

In equation (3), these options reduce emissions from $F_{LULUC}$ and promote $CO_2$ uptake in the land component of $F_{Seq}$.

### Reforestation, Afforestation, and Land Restoration

The potential global carbon sequestration due to increased forest extension and other land uses is in the order of 1 PgC y$^{-1}$ by 2010 (Watson et al. 2000b). In the past few decades major plantation has occurred in many countries, often where previous deforestation took place. During the 1990s alone, the global area of plantations increased by

3.1 Mha y$^{-1}$ (Braatz 2001). Regionally, reforestation and afforestation in China during the past two decades is contributing 80 percent of the total current Chinese carbon sink (Fang et al. 2001), and the return of abandoned farmland to native vegetation is estimated to be responsible for 98 percent of the current sink in the eastern United States (Caspersen et al. 2000).

If done with proper regard for the ecology and history of human land use in a region, expansion of forest area usually brings a number of collateral benefits: preservation of biodiversity, decreased soil erosion and siltation, decreased salinization, and watershed protection with associated water supply and flooding regulation. These benefits will most likely come with little adverse effect on agricultural production, employment, economic well-being, and cultural and aesthetic values in countries with agricultural surpluses such as Europe and the United States. In developing countries, however, which are usually dependent on exports of raw materials and confronted with high levels of social segregation, diverting land for reforestation may have negative consequences if it is not adapted to the local environment, land use patterns, and institutional settings. In these regions, land for afforestation is usually created by logging primary forest (Schulze et al. 2003). This practice affects vulnerable communities such as small-holding farmers (Silva 1997). There are also some environmental concerns: reduction of runoff during dry periods (due to increased soil infiltration and transpiration) may have impacts downstream, for instance in farmlands and wetlands. In boreal regions, changes in albedo due to darker forest canopies may lead to positive climate forcing and regional warming (Betts 2000; Betts et al. 2000). Taking into account a set of institutional and socioeconomic constraints only, Cannell (2003) estimated that only 10–25 percent of the potential C sequestration could realistically be achieved by 2100.

### MANAGING WOOD PRODUCTS

Managing wood products is an important part of a terrestrial sequestration strategy. It involves recycling paper, making use of long-lasting wood products that can substitute for high fossil-carbon-content materials, and other steps to increase the residence time of carbon sequestered as utilized wood.

### REDUCTION OF NET DEFORESTATION AND EMISSIONS FROM LAND USE

Deforestation over the past 200 years has contributed 30 percent of the present anthropogenic increase of atmospheric CO$_2$. Current estimates of emissions from deforestation are 1.6 PgC y$^{-1}$, or about 20–25 percent of total anthropogenic emissions (Houghton et al. 2001). Therefore, reduction of deforestation has large carbon mitigation potential: a 3 to 10 percent reduction of deforestation in non–Annex I (developing) countries by 2010 would result in 0.053–0.177 PgC y$^{-1}$ carbon mitigation, equivalent to 1 to 3.5 percent of Annex I base-year emissions (Watson et al. 2000b). The environmental benefits of slowing down deforestation are numerous, including most of the benefits mentioned for reforestation, with maximum value for biodiversity conservation in pristine forests, particularly in the tropics. Although ending forest deforestation is a laudable goal, it has, however, proved difficult or impossible to implement in many regions unless there are tangible socioeconomic incentives and the other drivers of deforestation (such as markets and policy climate) are specifically addressed. This is especially so in developing countries, where most deforestation occurs (Lambin et al. 2001). Another unintended outcome of rewarding the preservation of carbon stocks in forests could be a more relaxed emphasis on reduction of energy emissions, the most important pathway for long-term $CO_2$ stabilization (Figure 6.1).

## FOREST MANAGEMENT AND FIRE SUPPRESSION

Changes in forest management could increase carbon stocks with an estimated global mitigation of 0.175 PgC $y^{-1}$ (Watson et al. 2000b). One management activity being suggested is fire suppression. Fire is a natural factor for large forest areas of the globe, and therefore an important component of the global carbon cycle. Fire is a major short-term source of atmospheric carbon, but it adds to a small longer-term sink (<0.1 PgC $y^{-1}$) through forest regrowth and the transfer of carbon from fast to inert pools including charcoal (Watson et al. 2000b; Czimczik et al. 2003). For instance, in the immense extent of tropical savanna and woodland (2.45 gigahectares [Gha], Schlesinger 1997), a 20 percent fire suppression would result in carbon storage of 1.4 tC $ha^{-1}$ $y^{-1}$ with associated mitigation of 0.7 PgC $y^{-1}$. Tilman et al. (2000) estimated that additional fire suppression in Siberian boreal forest and tropical savanna and woodland might conceivably decrease the rate of accumulation of atmospheric $CO_2$ by 1.3 PgC $y^{-1}$, or about 40 percent. The biggest problem with fire suppression, however, is that these estimated rates of carbon storage are not sustainable in the long term and that potential catastrophic fires in high biomass density stands could negate the entire mitigation project apart from some sequestration in charcoal.

## CHANGING AGRICULTURAL MANAGEMENT

Changes in agricultural management can restore large quantities of soil carbon lost since the onset of agriculture. The global potential for this strategy is estimated at 40–90 PgC. Estimates of potential soil C sequestration vary between 0.3–0.5 PgC $y^{-1}$ (Smith, Chapter 28, this volume) and 0.9 PgC $y^{-1}$ (Lal 2003), which in the best case would restore most of the lost carbon within 50 to 100 years. A large number of "stock-enhancing" practices bring environmental benefits: reduced erosion and pollution of underground and surface water, maintenance of biodiversity, and increased soil fertility (Smith, Chapter 28). However, the frequently proposed "win-win" hypothesis, according to which strategies aimed at increasing carbon accumulation will always bring other environmental benefits, needs to be verified on a case-by-case basis and with reference to sustainable development goals assigned to a region. For instance, increasing soil C stocks also leads to increasing soil organic nitrogen, which provides a source of mineralizable N, and therefore potential for increased $N_2O$ emissions, further enhanced by the use of nitrogen fertilizer (Robertson, Chapter 29, this volume). Also, activities such as the use of fertilizer and irrigation have an associated carbon cost that often exceeds the carbon benefit (Schlesinger 1999). Other management options such as zero or reduced tillage may reduce the carbon cost of production (owing to less on-farm diesel use, despite a carbon cost associated with the extra herbicide required), but the extra use of herbicide and pesticides may create further negative environmental impacts. Taking into account a number of environmental and socioeconomic con-straints, Freibauer et al. (2003) estimated that only 10 percent of the potential mitiga-tion in the agricultural sector in Europe by 2020 is realistically achievable.

## Non-$CO_2$ Mitigation from Land Biosphere

In the livestock sector, methane and $N_2O$ emissions can be reduced by better housing and manure management. Methane emissions from enteric fermentation can be reduced by engineering ruminant gut flora or by use of hormones, but in some coun-tries there are societal and legislative constraints on these technologies. $N_2O$ emissions from agriculture can be lessened by reducing N fertilization. This decline could be achieved without loss of productivity by better timing, spatial placement (precision farming), and selection of fertilizers (Smith et al. 1996). The total anthropogenic flux of $N_2O$ is 8.1 teragrams (Tg) $N_2O$-N $y^{-1}$, equivalent to 1.0 $PgC_{equiv}$ $y^{-1}$. More than 80 percent of this amount is from agriculture; most of the rest is from the industrial pro-duction of adipic and nitric acids and can be abated with available technology.

## Bioengineering Solutions

Genetically modified organisms (GMOs) were initially developed to increase food pro-duction, but also have possibilities for increasing biomass production through disease- and pest-resistant genes that promote higher productivity. This potential has possible implications for increasing both carbon sequestration and biofuel production. GMOs have still largely unknown ecological consequences, however. Major known hazards are increased weediness of GM plants, genetic drift of new genes to surrounding vegetation, development of pest resistance, and development of new viruses (Barrett 1997). Given the large uncertainty surrounding the ecological consequences of GMOs, this technol-ogy is banned in many countries.

A comment applicable to all of the options described involving terrestrial biological sequestration and disturbance reduction is that there are timescales associated with both biological sequestration and disturbance. In general, carbon losses to the atmosphere by land disturbance occur partly through delayed emissions long after the disturbance event (one to two decades). Gains from reforestation are quite slow, with similar time frames needed to rebuild carbon stocks. It follows that even if $F_{LULUC}$ were to stop, emissions from disturbed ecosystems would still continue for some time. On the other hand, if reforestation programs are to make a major contribution to closing the carbon gap, they must be implemented early enough to see their effect on stabilizing $CO_2$.

## Biological Sequestration in the Oceans

In equation (3), this option promotes $CO_2$ uptake in the ocean component of $F_{Seq}$.

### OCEAN FERTILIZATION

The efficiency and duration of carbon storage by ocean fertilization remain poorly defined and strongly depend on the oceanic region and fertilizer (iron, nitrogen, phosphorus) used (Bakker, Chapter 26, this volume). The maximum potential of iron fertilization has been estimated as 1 PgC yr$^{-1}$ by continuous fertilization of all oceanic waters south of 30°S (Valparaiso, Cape Town, Perth) for 100 years (Sarmiento and Orr 1991). This model study, however, probably strongly overestimates the potential for carbon storage by its assumption of complete nutrient depletion (Bakker, Chapter 26). Furthermore, roughly half of the stored carbon would be rapidly released to the atmosphere upon termination of the fertilization. Verification of actual C sequestration remains an issue.

Enhanced algal growth upon ocean fertilization of the surface oceans will decrease oxygen levels and may create anoxic (near-zero oxygen) conditions at intermediate depths (Fuhrman and Capone 1991; Sarmiento and Orr 1991), which will promote production of $N_2O$ and methane. The release of these potent greenhouse gases to the atmosphere will partly offset or even outweigh the reduction in radiative forcing by atmospheric $CO_2$ mitigation, especially for the equatorial Pacific Ocean (Jin and Gruber 2002; Jin et al. 2002; Bakker, Chapter 26). Increases in biological production may also result in the release of dimethyl sulphide (DMS) (Turner et al. 1996) and halocarbons (Chuck 2002) to the atmosphere, which will create powerful feedbacks on atmospheric chemistry and global climate.

Large-scale iron fertilization will profoundly change oceanic ecosystems, ocean biogeochemistry, and the composition of oceanic sediments (Watson et al. 2000a; Ducklow et al. 2003; Bakker, Chapter 26). A shift toward larger algal species was observed in iron fertilization experiments (Coale et al. 1996; Boyd et al. 2000). A comparison to mariculture and coastal seas suggests that harmful algal blooms may occur in intensively managed systems (Bakker, Chapter 26). Fertilization might affect fish stocks, and thence fisheries and other economic sectors.

The public is very concerned about large-scale fertilization manipulations of the surface ocean for carbon storage. International law with respect to fertilization is ambiguous.

## Engineered $CO_2$ Disposal on Land and Oceans

In equation (3), these options promote $CO_2$ uptake through $F_{Disp}$.

### CARBON SEPARATION WITH OCEAN STORAGE BY DEEP OCEAN INJECTION

The addition of pure liquid $CO_2$ in the deep ocean has potential for being a low-impact and highly effective mitigation option (Brewer, Chapter 27, this volume). There is still, however, little understanding of the potential effects of the instability of $CO_2$

deposits in the deep ocean and of the negative effects of the formation of $CO_2$-clathrates and a substantial lowering of pH on deep ocean biota. If numerous sequestration sites were concentrated in specific, rare deep ocean habitats, this practice could endanger biodiversity at these locations, but it seems likely the effects would be local. Further away a moderate pH decrease might be of similar magnitude as that expected from rising atmospheric $CO_2$ levels. In light of how little is known, deep ocean disposal may have unexpected effects on marine chemistry and ecosystems. Public opinion is skeptical about the technique. The London Convention (the Convention on the Prevention of Marine Pollution by Dumping of Wastes and Other Matter, 1972) is often cited as a treaty that could prohibit such a strategy.

CARBON SEPARATION WITH GEOLOGICAL STORAGE IN SEDIMENTS AND ROCKS

Given adequate technologies to capture $CO_2$, largely from industrial processes, $CO_2$ can be disposed of in exhausted oil and gas wells and in saline aquifers. This method is a relatively clean solution provided there are not $CO_2$ escapes, dissolution of host rock, sterilization of mineral resources, and unforeseen effects on groundwater (Metz et al. 2001, Section 3.8.4.4). At this stage, there are still large uncertainties regarding these possible environmental impacts, but it remains a promising option.

## Illustrative Case: Constraints on Land-Based Options

In this section we consider, as an illustrative case, the inclusion of land management as part of a portfolio of mitigation strategies to close the carbon gap defined earlier. Future scenarios such as the A1, A2, B1, and B2 scenario families (see Nakicenovic et al. 2000 and Figure 6.3) allocate different amounts of land for carbon mitigation strategies (such as cropland for biofuels), while also allowing for an increase in agricultural land to meet the food demands of a growing population. Globally, in 1990, 11 percent of the land surface (1,500 Mha) was under cropland. By 2100, cropland area is projected to increase to about 15 percent (A1), 17 percent (A2), 9–10 percent (B1), 16 percent (B2), and up to 24 percent under some A1 variants (Nakicenovic et al. 2000).

To illustrate the interactions between opportunities and constraints in carbon management, this section examines how each SRES scenario is likely to place pressure on land for food production as different amounts of land are allocated for carbon mitigation. It then asks two questions: (a) Are the areas of land required for mitigation realistic? (b) Under realistic environmental and sociocultural constraints on land use, what is the sustainably achievable potential for land-based options under each SRES scenario?

### Scenarios

The scenarios used are the six SRES marker scenarios, taken from four scenario families, A1, A2, B1, and B2. The main characteristics of each scenario family are summarized in Figure 6.3, which shows how the scenarios fall on two axes: one examining glob-

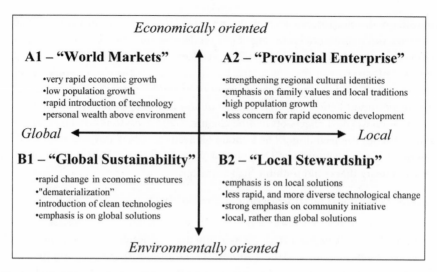

**Figure 6.3.** Characteristics of the main SRES marker scenario families (adapted from Smith and Powlson 2003)

alization versus regionalization and the other examining relative societal emphases on economic versus environmental considerations. Three variations of the A1 scenario (globalized, economically oriented) are considered here: A1FI (fossil fuels are used intensively), A1T (alternative technologies largely replace fossil fuels), and A1B (a balance between fossil fuels and alternative technologies). Further details are given in Edmonds et al. (Chapter 4, this volume); see also WBGU (2003). Key scenario parameters are given in Table 6.3a.

## Population Effects on Food Demand

We calculated 2100 food demand (assumed proportional to population growth) and the increase in productivity for each scenario (Table 6.3b). The area under crops in 1990 represents slightly over a third of the land that is theoretically estimated to be suitable for crop production. Although this estimated area may be optimistic (since some land is not well suited to permanent cropping, and other land will be removed from production by degradation), there is evidence that additional food can be generated sustainably to match population growth. This is supported by the fact that from 1961 to 1997, when human population doubled, agricultural land increased by only 11 percent. Nevertheless, the higher the food demand, the greater the pressure placed on land for agriculture, and less land may be available for other purposes, including carbon mitigation. This increased food demand could be met either by using more land for agriculture or by increasing cropping intensity and per-area productivity. Increased productivity can be obtained either (1) by increased fertilization and irrigation, subject to water availability, with associated carbon costs (Schlesinger 1999) and increased $N_2O$ emissions (Robertson, Chapter 29, this volume); or (2) by technological improvement through breeding or biotechnology.

The A1FI, A1B, A1T, and B1 scenarios require an increase in productivity of 0.3–0.4 percent $y^{-1}$ whereas the A2 and B2 scenarios require 0.9–1.1 percent $y^{-1}$. All these increases are in the range considered achievable within integrated assessment models; 1.5 percent $y^{-1}$ is assumed by Edmonds (personal communication, 2003). Although there is a limit to how much agricultural productivity can be increased, with yield increases slowing during the past two decades (Amthor 1998; Tilman et al. 2002), there is still capacity. The Food and Agriculture Organization of the United Nations (FAO) projects global aggregate crop production to grow at 1.4 percent $y^{-1}$ to 2030, down from 2.1 percent $y^{-1}$ over 1970–2000. Cereal yield growth, the mainstay of crop production growth, is projected to be 1.0 percent $y^{-1}$ in developing countries, compared with 2.5 percent $y^{-1}$ for 1961–1999 (Bruinsma 2003). We note also that future impacts of climate change on crop yields can be quite large. Leemans et al. (2002) estimate the effect of $CO_2$ fertilization on crop and biofuel yield and hence on land use demand, suggesting a 15 percent increase in the land use demand without $CO_2$ fertilization.

Increased intensity of cropping can lead to land degradation through salinization and erosion. Despite these environmental risks, irrigation is expected to play an important role in agricultural production growth in developing countries, where an estimated 40 Mha could come under irrigation, an expansion of 33 percent more than the land currently irrigated. Scenarios A2 and B2 require increased productivity close to potential growth rates, sustained until 2100.

In summary, scenarios with slow technology development and low per capita gross domestic product (GDP) will be less able to deliver food requirements on the existing land area and are likely to increase pressure on the land resource.

## Impact of Each Scenario on Possibilities for Meeting Food Demand

There are a number of logical qualitative trends for assessing whether food demand can be met by increased per-area productivity or by placing more land into production. Implications of possible trends in per capita GDP are:

- If per capita GDP is high, then (1) demand for more meat (protein) in the diet is high in developing countries, tending to increase pressure on land for food; and (2) there is capital available for increasing productivity, tending to decrease pressure on land for food.
- If per capita GDP is low, then the opposite trends ensue: (1) there is less protein in diet, decreasing pressure on land for food; and (2) less capital is available for increasing productivity, increasing pressure on land for food.

**Table 6.3a.** Scenario drivers taken from the SRES scenarios and total carbon gap

| Scenario | Total carbon gap ($PgC\,y^{-1}$) in 2100 at stabilization target of 450 ppm | Population in 2100 (billion) | World GDP in 2100 ($10^{12}$ 1990 US$) | Per capita GDP in 2100 ($10^3$ 1990 US$) | Income ratio in 2100 of Annex I to non–Annex I[a] | Technology |
|---|---|---|---|---|---|---|
| 1990 | — | 5.3 | 21 | 3.9 | 16.1 | — |
| A1FI | 25 | 7.0–7.1 | 522–550 | 73.5–78.6 | 1.5–1.6 | Rapid introduction of technology |
| A1B | 10 | 7.0–7.7 | 340–536 | 44.2–76.6 | 1.5–1.7 | Rapid introduction of technology |
| A1T | 1 | 7.0 | 519–550 | 74.1–78.6 | 1.6–1.7 | Rapid introduction of technology |
| A2 | 25 | 12.0–15.1 | 197–249 | 13.0–20.8 | 2.7–6.3 | Relatively (to A1) slower introduction of technology |
| B1 | 1 | 6.9–7.1 | 328–350 | 46.2–50.7 | 1.4–1.9 | Rapid introduction of technology |
| B2 | 10 | 10.3–10.4 | 199–255 | 19.1–24.8 | 2.0–3.6 | Less (relative to A1) rapid, and more diverse technological change |

*Sources:* (Nakicenovic et al. 2000; Table SPM-1a and story-line text) and total carbon gap from Chapter 4, this volume.

[a] Annex I = developed or industrialized nations; non–Annex I = developing countries.

**Table 6.3b.** Increase in food demand under each scenario

| Scenario | Increase in food demand relative to 1990 | Long-term average increase in per-year productivity required to meet food demand[a] |
|---|---|---|
| A1FI | 32–34% | 0.3% |
| A1B | 32–45% | 0.3–0.4% |
| A1T | 32–34% | 0.3% |
| A2 | 126–284% | 1.1% |
| B1 | 30–34% | 0.3% |
| B2 | 94–96% | 0.9% |

[a] calculated by % change by 2100 divided by 110 (years since 1990).

**Table 6.3c.** Impacts of scenarios on the pressure on land for food production

| Scenario | Per capita GDP impact on potential increase in productivity – pressure on land resource | Per capita GDP impact on change in diet – pressure on land resource | Technology – pressure on land resource | Carbon gap |
|---|---|---|---|---|
| A1FI | + | +++ | – – – | Small |
| A1B | + | +++ | – – | Moderate |
| A1T | + | +++ | – – – | Small |
| A2 | +++ | + | – | Large |
| B1 | ++ | ++ | – – – | Small |
| B2 | +++ | + | – | Large |

*Note:* + = more pressure on land for food, – = less pressure on land for food (relative to 1990).

Trends in technological development will be equally critical and can be summarized thus:

• If technology development and *effective* dissemination are high, there is more capacity to increase production, tending to decrease pressure on land for food, and vice versa.

A caveat is needed concerning thresholds. Since even the lowest change in per capita GDP by 2100 (A2) is more than three times greater than 1990 per capita GDP, this may already exceed the thresholds that lead to changes in food consumption patterns, thereby creating increased pressure on the land resource (Bruinsma 2003). Similarly, technology may cross a threshold allowing increases in productivity to be met with technological advances. Where these thresholds lie, or whether they exist at all, is unknown. Such threshold effects are not considered in this analysis.

To apply these trend indicators to estimate net pressure on land resources, we first use the A1FI scenario (global free market with intensive fossil-fuel use) as an example. Here food demand, driven by population growth, increases by 34 percent by 2100 (Table 6.3b). Per capita GDP is high, meaning that there is likely to be more demand for meat in the diet (pressure on land resource = +++), but also that capital will be available to increase per-area productivity (average of 0.3 percent per year required; pressure on land resource = +). Rapid introduction of new technologies means that technological advances, leading to higher per-area productivity, are favored to help meet the increased demand (pressure on land resource = –––). The first row in Table 6.3c shows the net consequences of these factors.

Applying the same reasoning, Table 6.3c shows how the parameters associated with each SRES marker scenario affect net pressure on land for food production. The highly regional scenarios A2 and B2 present the greatest pressure on land, since both have high food demand, low per capita GDP, and slow technological development and dissemination (hence low capacity to feed their high populations). Hence, scenarios A2 and B2 offer the least land for closing the carbon gap by using land-based options or for meeting other human needs.

Scenarios A1FI, A1B, A1T, and B1 present the lowest pressure on land for food production, because of relatively low population growth, high per capita GDP (leading to better ability to increase per-area productivity, even though diet would include more meat), and a high rate of technological development and dissemination. These scenarios also have the lowest, and therefore most realistically achievable, increases in productivity to meet increased food demand. Scenarios A1FI, A1B, A1T, and B1 are the most likely to offer land for helping to close the carbon gap and for providing land for other human needs.

The magnitude of the carbon gap for each SRES scenario has implications for the land available to close the gap. As shown in Table 6.3a, for a stabilization level of 450 parts per million (ppm), scenarios A1FI and A2 have the largest carbon gap by 2100 (25

PgC y$^{-1}$), A1T and B1 have the smallest gap (1 PgC y$^{-1}$), and A1B and B2 are intermediate (10 PgC y$^{-1}$). There would be little land available for carbon sequestration and biofuel cropping to help close these gaps for A2 and B2. Scenarios A1T and B1, however, have the smallest carbon gaps and are also the most likely to have land available for gap closure by land-based options. Scenarios A1FI and A1B are also more likely than A2 and B2 to have land available for gap closure, but the gap is 10–25 times larger than for A1T and B1. Hence, carbon sequestration and biofuel cropping may form part of the portfolio to close the carbon gap for A1FI or A1B, but the gap is much larger. Although the income ratio between developed and developing countries (Table 6.3a) decreases significantly in all scenarios between 1990 (16.1) and 2100 (1.5–6.3), scenarios A2 and B2 are the most likely to show regionally different impacts on land pressure, with low incomes in developing countries lessening the ability to meet food demand by improved management or technology.

In conclusion, this analysis suggests that given the constraints on land required for food production, land-based methods (biofuel cropping and carbon sequestration) for carbon gap closure show a sustainably achievable mitigation potential only under SRES scenarios A1T and B1. Under all other scenarios, land-based methods do not show a potential for contributing significantly to carbon gap closure.

## Concluding Summary

We have examined the challenge of achieving CO$_2$ stabilization in the context of a sustainable Earth system, by outlining a systems framework and then identifying the wider implications of carbon management options in economic, environmental, and sociocultural terms.

The systems analysis begins from the familiar global atmospheric carbon budget, and the stabilization requirement that direct-human-induced CO$_2$ emissions are low enough to allow land-air and ocean-air carbon fluxes to buffer atmospheric CO$_2$ at a constant future level. A range of strategies is available to keep net emissions within this constraint: (1) conservation of energy at end-use points, (2) use of non-fossil-fuel energy sources, (3) more carbon-efficient use of fossil fuels; (4) reduction of carbon emissions from land disturbance; (5) sequestration of carbon in terrestrial or oceanic biological sinks; and (6) engineered disposal of CO$_2$ in geological or oceanic repositories. Each strategy also has a range of impacts (benefits and costs), broadly in four classes: climatic, economic, environmental, and sociocultural.

The success of a suite of carbon mitigation strategies will be determined not only by the technical potential of each strategy (the amount of carbon emission that can be avoided, based on biophysical considerations alone), but also by the uptake rates of the various strategies. These rates are determined by the overall benefit-cost outcomes of the entire suite of strategies, judged not only against carbon-mitigation, but also against economic, environmental, and sociocultural criteria. Thus, the uptake rates (and the trajectories of all parts of the carbon-climate-human system) are emergent properties of the system, resulting from interactions among system components, rather than being imposable properties of isolated components. This situation has several important consequences: First, it is important to maintain a broad suite of mitigation options. Next, successful carbon mitigation depends on exploiting beneficial synergies between mitigation and other (economic, environmental, and sociocultural) goals. Third, societal choices play a major role in determining the relative weightings between these goals and thus the overall outcome.

An order-of-magnitude comparison of likely achievable mitigation potentials (Figure 6.1) shows that major mitigation through sequestration and disposal cannot achieve stabilization unless there is also major mitigation in the energy sector, which has a larger overall impact.

We have assessed a wide range of carbon management options for their economic, environmental, and sociocultural impacts, both positive and negative. The following sweeping (therefore imperfect) generalizations summarize our findings: strategies based on energy conservation and efficiency have broadly beneficial impacts and offer major achievable mitigation, as do strategies involving non-fossil-fuel energy (though with significant environmental and sociocultural negative implications in certain cases). Land-based options offer significant mitigation but with some significant negative impacts mainly by competing with other land uses such as food production. Ocean biological sequestration has major collateral concerns. $CO_2$ disposal in ocean and geological repositories has significant mitigation potential, but its side effects are still poorly known.

A more detailed analysis of a specific case, land-based mitigation through bioenergy and sequestration, has explored the implications for the pressure on land for food and other essentials under six SRES scenarios. These options offer a sustainably achievable mitigation potential only under SRES scenarios A1T and B1.

## Acknowledgments

We thank a number of colleagues, including Damian Barrett, Louis Lebel, Anand Patwardhan, Yoshiki Yamagata, and Oran Young, for comments on a draft of this chapter and discussions of the associated ideas. Their insights have been formative. We are indebted to Peter Briggs for assistance with figures and references. Two of us (Pep Canadell and Mike Raupach) express appreciation to the Australian Greenhouse Office and the Australian Greenhouse Science Program for funding support for the Canberra Office of the Global Carbon Project, which has made possible our participation in this project.

## Notes

1. Equation (2) is similar to the "Kaya identity" (Nakicenovic, Chapter 11, this volume): $F_{Foss} = P \times (G/P) \times (E/G) \times (F_{Foss}/E)$. To focus on mitigation, we break $(F_{Foss}/E)$ into the factors $f$ and $i$ defined above. Both equation (2) and the Kaya identity are examples of the "IPAT" model (Impact = Population × Affluence × Technology), where $g = G/P$ is affluence.

2. Primary energy $E_{Pri}$ is the total power generated by humankind, inclusive of waste heat in generation and transmission. It is conventionally called an "energy" although it is actually a power (energy per unit time) with units exajoules per year (EJ $y^{-1}$) or terawatt (TW) (1 EJ = $10^{18}$ J; 1 TW = $10^{12}$ W = 31.536 EJ $y^{-1}$).

## Literature Cited

Amthor, J. S. 1998. Perspective on the relative insignificance of increasing atmospheric $CO_2$ concentration to crop yield. *Field Crops Research* 58:109–127.

Barrett, K. 1997. Beyond boundaries: Converging disciplines, diverging interests: Meanings of responsibility in biotechnology research. University of British Columbia, Vancouver.

Betts, R. A. 2000. Offset of the potential carbon sink from boreal forestation by decreases in surface albedo. *Nature* 408:187–190.

Betts, R. A., P. M. Cox, and F. I. Woodward. 2000. Simulated responses of potential vegetation to doubled-$CO_2$ climate change and feedbacks on near-surface temperature. *Global Ecology and Biogeography* 9:171–180.

Board on Sustainable Development Policy Division and National Research Council. *Our common journey: A transition toward sustainability.* 1999. Washington, DC: National Academies Press.

Boyd, P. W., A. J. Watson, C. S. Law, E. R. Abraham, T. Trull, R. Murdoch, D. C. E. Bakker, A. R. Bowie, K. O. Buesseler, H. Chang, M. Charette, P. Croot, K. Downing, R. Frew, M. Gall, M. Hadfield, J. Hall, M. Harvey, G. Jameson, J. LaRoche, M. Liddicoat, R. Ling, M. T. Maldonado, R. M. McKay, S. Nodder, S. Pickmere, R. Pridmore, S. Rintoul, K. Safi, P. Sutton, R. Strzepek, K. Tanneberger, S. Turner, A. Waite, and J. Zeldis. 2000. A mesoscale phytoplankton bloom in the polar Southern Ocean stimulated by iron fertilization. *Nature* 407:695–702.

Braatz, S. M., ed. 2001. *State of the world's forests 2001.* Rome: Food and Agriculture Organization of the United Nations.

Brock, W. A., K.-G. Mäler, and C. Perrings. 2002. Resilience and sustainability: The economic analysis of nonlinear dynamic systems. Pp. 261–289 in *Panarchy: Understanding transformations in human and natural systems,* edited by L. H. Gundersen, and C. S. Holling. Washington, DC: Island Press.

Bruinsma, J., ed. 2003. *World agriculture: Towards 2015/2030—An FAO perspective.* Rome: Food and Agriculture Organization of the United Nations and Earthscan.

Cannell, M. G. R. 2003. Carbon sequestration and biomass energy offset: Theoretical, potential and achievable capacities globally, in Europe and the UK. *Biomass and Bioenergy* 24:97–116.

Caspersen, J. P., S. W. Pacala, J. C. Jenkins, G. C. Hurtt, P. R. Moorcroft, and R. A. Birdsey. 2000. Contributions of land-use history to carbon accumulation in US forests. *Science* 290:1148–1151.

Chuck, A. L. 2002. Biogenic halocarbons and light alkyl nitrates in the marine environment. University of East Anglia, Norwich.

Coale, K. H., K. S. Johnson, S. E. Fitzwater, R. M. Gordon, S. Tanner, F. P. Chavez, L. Ferioli, C. Sakamoto, P. Rogers, F. Millero, P. Steinberg, P. Nightingale, D. Cooper, W. P. Cochlan, M. R. Landry, J. Constantinou, G. Rollwagen, A. Trasvina, and R. Kudela. 1996. A massive phytoplankton bloom induced by an ecosystem-scale iron fertilization experiment in the equatorial Pacific Ocean. *Nature* 383:495–501.

Czimczik, C. I., C. M. Preston, M. W. I. Schmidt, and E. D. Schulze. 2003. How surface fire in Siberian Scots pine forests affects soil organic carbon in the forest floor: Stocks, molecular structure, and conversion to black carbon (charcoal). *Global Biogeochemical Cycles* 17, 10. 1029/2002GB001956.

Ducklow, H. W., J. L. Oliver, and W. O. Smith Jr. 2003. The role of iron as a limiting nutrient for marine plankton processes. In *Interactions of the major biogeochemical cycles: Global change and human impacts,* edited by J. M. Melillo, C. B. Field, and B. Moldan. SCOPE 61. Washington, DC: Island Press for the Scientific Committee on Problems of the Environment.

Fang, J. Y., A. P. Chen, C. H. Peng, S. Q. Zhao, and L. Ci. 2001. Changes in forest biomass carbon storage in China between 1949 and 1998. *Science* 292:2320–2322.

Freibauer, A., M. D. A. Rounsevell, P. Smith, and A. Verhagen. 2003. Carbon sequestration in European agricultural soils. *Soil Science Reviews* 1 (in press).

Fuhrman, J. A., and D. G. Capone. 1991. Possible biogeochemical consequences of ocean fertilization. *Limnology and Oceanography* 36:1951–1959.

Geist, H. J., and E. F. Lambin. 2001. *What drives tropical deforestation?* Land Use and Land-Cover Change (LUCC) Report Series 4. Louvain-la-Neuve, Belgium: LUCC International Project Office, University of Louvain.

Hoffert, M. I., K. Caldeira, G. Benford, D. R. Criswell, C. Green, H. Herzog, A. K. Jain, H. S. Kheshgi, K. S. Lackner, J. S. Lewis, H. D. Lightfoot, W. Manheimer, J. C. Mankins, M. E. Mauel, L. J. Perkins, M. E. Schlesinger, T. Volk, and T. M. L. Wigley. 2002. Advanced technology paths to global climate stability: Energy for a greenhouse planet. *Science* 298:981–987.

Houghton, J. T., L. G. Meira Filho, B. A. Callander, N. Harris, A. Kattenberg, and K. Maskell. 1996. *Climate change 1995: The science of climate change.* Cambridge: Cambridge University Press.

Houghton, J. T., Y. Ding, D. J. Griggs, M. Noguer, P. J. van der Linden, X. Dai, K. Maskell, and C. A. Johnson, eds. 2001. *Climate change 2001: The scientific basis (Contribution of Working Group I to the Third Assessment Report of the Intergovernmental Panel on Climate Change).* Cambridge: Cambridge University Press.

Jin, X., and N. Gruber. 2002. Offsetting the radiative benefit of ocean iron fertilization by enhancing ocean $N_2O$ emissions. *Eos, Transactions, American Geophysical Union* 83 (AGU Fall Meeting Supplement), Abstract U21A-0006.

Jin, X., C. Deutsch, N. Gruber, and K. Keller. 2002. Assessing the consequences of iron fertilization on oceanic $N_2O$ emissions and net radiative forcing. *Eos, Transactions, American Geophysical Union* 83 (Ocean Sciences Supplement), Abstract OS51F-10.

Keely, J., and I. Scoones. 1999. *Understanding environmental policy processes: A review.* IDS

Working Paper 89. Sussex, England: University of Sussex, Institute of Development Studies.

Lal, R. 2003. Global potential of soil carbon sequestration to mitigate the greenhouse effect. *Critical Reviews in Plant Sciences* 22:151–184.

Lambin, E. F., B. L. I. Turner, H. J. Geist, S. Agbola, A. Angelsen, J. W. Bruce, O. Coomes, R. Dirzo, G. Fischer, C. Folke, P. S. George, K. Homewood, J. Imbernon, R. Leemans, X. Li, E. F. Moran, M. Mortimore, P. S. Ramakrishnan, J. F. Richards, H. Skånes, W. Steffen, G. D. Stone, U. Svendin, T. Veldkamp, C. Vogel, and J. Xu. 2001. The causes of land-use and land-cover change: Moving beyond the myths. *Global Environmental Change: Human and Policy Dimensions* 11:261–269.

Leemans, R., B. Eickhout, B. Strengers, L. Bouwman, and M. Schaeffer. 2002. The consequences of uncertainties in land use, climate and vegetation responses on the terrestrial carbon. *Science in China (Life Sciences)* 45:126.

McCarthy, J. J., O. F. Canziani, N. A. Leary, D. J. Dokken, and K. S. White, eds. 2001. *Climate change 2001: Impacts, adaptation, and vulnerability (Contribution of Working Group II to the Third Assessment Report of the Intergovernmental Panel on Climate Change).* Cambridge: Cambridge University Press.

Metz, B., O. Davidson, R. Swart, and J. Pan, eds. 2001. *Climate change 2001: Mitigation (Contribution of Working Group II to the Third Assessment Report of the Intergovernmental Panel on Climate Change).* Cambridge: Cambridge University Press.

Nakicenovic, N., J. Alcamo, G. Davis, B. de Vries, J. Fenhann, S. Gaffin, K. Gregory, A. Grubler, T. Y. Jung, T. Kram, E. L. La Rovere, L. Michaelis, S. Mori, T. Morita, W. Pepper, H. Pitcher, L. Price, K. Raihi, A. Roehrl, H.-H. Rogner, A. Sankovski, M. Schlesinger, P. Shukla, S. Smith, R. Swart, S. van Rooijen, N. Victor, and Z. Dadi. 2000. *IPCC special report on emissions scenarios.* Cambridge: Cambridge University Press.

Prentice, C., G. D. Farquhar, M. J. R. Fasham, M. L. Goulden, M. Heimann, V. J.Jaramillo, H. S. Kheshgi, C. Le Quéré, R. J. Scholes, and D. W. R. Wallace. 2001. The carbon cycle and atmospheric carbon dioxide. In *Climate change 2001: The scientific basis (Contribution of Working Group I to the Third Assessment Report of the Intergovernmental Panel on Climate Change)*, edited by J. Houghton, Y. Ding, D. J. Griggs, M. Noguer, P. J. van der Linden, X. Dai, K. Maskell, and C. A. Johnson. Cambridge: Cambridge University Press.

Sarmiento, J. L., and J. C. Orr. 1991. 3-dimensional simulations of the impact of Southern-Ocean nutrient depletion on atmospheric CO2 and ocean chemistry. *Limnology and Oceanography* 36:1928–1950.

Schlesinger, W. H. 1997. *Biogeochemistry: An analysis of global change.* 2nd ed. San Diego: Academic Press.

———. 1999. Carbon and agriculture: Carbon sequestration in soils. *Science* 284:2095.

Schulze, E. D., D. Mollicone, F. Achard, G. Matteucci, S. Federici, H. D. Eva, and R. Valentini. 2003. Climate change: Making deforestation pay under the Kyoto Protocol? *Science* 299:1669.

Silva, E. 1997. The politics of sustainable development: Native forest policy in Chile, Venezuela, Costa Rica and Mexico. *Journal of Latin American Studies* 29:457–493.

Smith, P., and D. S. Powlson. 2003. Sustainability of soil management practices: A global perspective. In *Soil biological fertility: A key to sustainable land use in agriculture,* edited by L. K. Abbott and D. V. Murphy. Amsterdam: Kluwer (in press).

Smith, P., D. S. Powlson, and M. J. Glendining. 1996. Establishing a European soil organic matter network (SOMNET). Pp. 81–98 in *Evaluation of soil organic matter models using existing, long-term datasets,* edited by D. S. Powlson, P. Smith, and J. U. Smith. Berlin: Springer Verlag.

Tilman, D., P. Reich, H. Phillips, M. Menton, A. Patel, E. Vos, D. Peterson, and J. Knops. 2000. Fire suppression and ecosystem carbon storage. *Ecology* 81:2680–2685.

Tilman, D., K. G. Cassman, P. A. Matson, R. Naylor, and S. Polasky. 2002. Agricultural sustainability and intensive production practices. *Nature* 418:671–677.

Turner, S. M., P. D. Nightingale, L. J. Spokes, M. I. Liddicoat, and P. S. Liss. 1996. Increased dimethyl sulphide concentrations in sea water from in situ iron enrichment. *Nature* 383:513–517.

United Nations Commission on Sustainable Development. 1996. *Indicators of sustainable development: Framework and methodologies.* New York: United Nations.

Watson, A. J., D. C. E. Bakker, A. J. Ridgwell, P. W. Boyd, and C. S. Law. 2000a. Effect of iron supply on Southern Ocean $CO_2$ uptake and implications for glacial atmospheric $CO_2$. *Nature* 407:730–733.

Watson, R.T., I. R. Noble, B. Bolin, N. H. Ravindranath, D. J. Verardo, and D. J. Dokken. 2000b. *IPCC special report on land use, land-use change and forestry.* Cambridge: Cambridge University Press.

WBGU. 2003. *Welt im Wandel: Energiewende zur Nachhaltigkeit.* Heidelberg: Springer Verlag. Published in English as WBGU. 2004. *World in transition: Towards sustainable energy systems.* London: Earthscan.

Wigley, T. M. L., R. Richels, and J. A. Edmonds. 1996. Economic and environmental choices in the stabilization of atmospheric $CO_2$ concentrations. *Nature* 379:240–243.

World Commission on Dams. 2000. *Dams and development: A new framework for decision-making.* London: Earthscan.

World Commission on Environment and Development. 1987. *Our common future.* Oxford: Oxford University Press.

# PART II
# Overview of the Carbon Cycle

# 7

# A Paleo-Perspective on Changes in Atmospheric $CO_2$ and Climate

## Fortunat Joos and I. Colin Prentice

The goal of the United Nations Framework Convention on Climate Change is to stabilize atmospheric $CO_2$ and other greenhouse gases (GHGs) to prevent "dangerous" anthropogenic interference with climate. $CO_2$ is the most important of the anthropogenic GHGs in terms of radiative forcing, and its stabilization presents major challenges to society. To stabilize atmospheric $CO_2$ concentration at any given level, it is necessary to define trajectories of allowable $CO_2$ emissions. This calculation requires a thorough understanding of the carbon cycle—that is, the various terrestrial and oceanic processes that control the amount of $CO_2$ that stays in the atmosphere.

Timescales of carbon cycle processes range from months to centuries for carbon exchange between the atmosphere, land biosphere, and ocean to millennia for ocean-sediment interactions. Processes that operate on timescales longer than a few years cannot be investigated by experimental field studies, and long-term instrumental observations are missing for many variables. Paleodata (especially measurements on ice cores, marine and lake sediments, corals, tree rings, and historical documents) provide unique information on the magnitude of natural $CO_2$ variability, the processes regulating atmospheric $CO_2$ and their associated timescales (including multimillennial timescales), and the magnitude of carbon cycle–climate feedbacks.

The longest available ice-core records (from Vostok, Antarctica [Petit et al. 1999] and Dome Fuji, Antarctica [Kawamura et al. 2003]) demonstrate that atmospheric $CO_2$ concentration today is higher than at any time during (at least) the past 420,000 years.

The search for mechanisms driving the observed glacial-interglacial changes has led to the identification of a range of processes that contribute to the control of atmospheric $CO_2$ concentration and climate. The detection of abrupt, decadal-scale climatic changes influencing large regions has fueled the concern that anthropogenic GHG emissions may trigger future abrupt climate changes. Changes in $CO_2$ and climate during the Holocene, and during the last millennium of the Holocene, are less spectacu-

165

lar but provide important information on the linkages between the carbon cycle and climate.

## Glacial-Interglacial Variations

The Vostok and Dome Fuji ice-core records reveal that atmospheric $CO_2$ variations over the past four glacial cycles were confined to the range between ~ 180 parts per million (ppm) (around glacial maxima) and ~ 280 ppm (typical for interglacial periods) (Petit et al. 1999; Kawamura et al. 2003). Thus, preindustrial atmospheric $CO_2$ concentrations were consistently lower than today's value of 370 ppm and dramatically lower than the projected concentration range at year 2100 (450 to 1,100 ppm) reported by the Intergovernmental Panel on Climate Change (IPCC) (Joos et al. 2001; Prentice et al. 2001).

A success of the greenhouse theory, first established in the 19th century (Arrhenius 1896) and of today's climate models is that both the global warming over the industrial period (Houghton et al. 2001) and the widespread cold conditions of the last glacial maximum (LGM) (Ganopolski et al. 1998; Weaver et al. 1998; Kitoh et al. 2001; Kim et al. 2002; Hewitt et al. 2003; Shin et al. 2003) can be consistently explained by the radiative forcing due to changes in atmospheric GHG content and other factors. Detailed comparison of temperature proxies and $CO_2$ during the last glacial-interglacial transition, however, suggests that Antarctic temperature started to rise before atmospheric $CO_2$ (Figure 7.1). This finding is consistent with the view that natural $CO_2$ variations constitute a feedback in the glacial-interglacial cycle rather than a primary cause (Shackleton 2000). Changes in the Earth's orbit around the Sun are the pacemaker for glacial-interglacial cycles (Hays et al. 1976; Berger 1978), but these rather subtle orbital changes must be amplified by climate feedbacks in order to explain the large differences in global temperature and ice volume, and the relative abruptness of the transitions between glacial and interglacial periods (Berger et al. 1998; Clark et al. 1999).

Biogeochemical cycles play an important role for the amplification of orbital changes. Model simulations suggest that the direct radiative forcing by atmospheric $CO_2$ and $CH_4$ concentrations may have contributed up to half of the observed glacial-interglacial surface temperature difference at a global scale (Broccoli and Manabe 1987; Gallée et al. 1992; Shin et al. 2003). Other major factors involved in maintaining cold conditions during glacial periods include the water vapor feedback, the high albedo of the continental ice sheets (Broccoli and Manabe 1987; Hewitt and Mitchell 1997), the high albedo (especially when snow covered) of extensive nonforested regions at high latitudes (Gallée et al. 1992; Levis et al. 1999; Yoshimori et al. 2001; Wyputta and McAvaney 2002), and the reflection of shortwave radiation by the greatly enhanced atmospheric content of mineral dust (Claquin et al. 2003)—itself a consequence of reduced vegetation cover (Mahowald et al. 1999; Werner et al. 2002). It is plausible that such biogeochemical and biophysical feedbacks will amplify the direct anthropogenic greenhouse gas forcing, just as they have amplified orbital changes in the past.

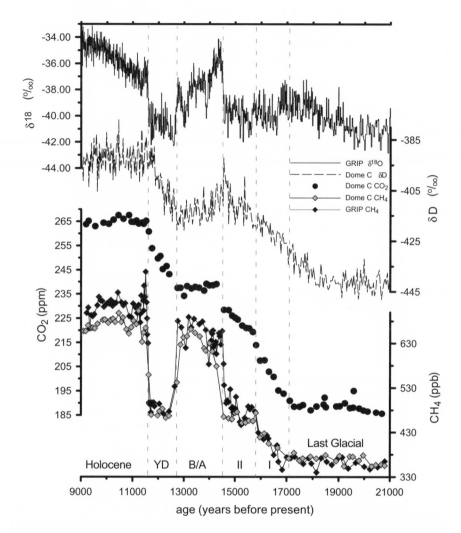

**Figure 7.1.** The evolution of proxies for local temperature in Greenland ($\delta^{18}O$, top line) and Antarctica ($\delta D$, dash) and the atmospheric concentration of the two greenhouse gases $CO_2$ (dots) and $CH_4$ (diamonds) over the last glacial-interglacial transition (Monnin et al. 2001). The records of Antarctic temperature and atmospheric $CO_2$ are highly correlated and can be divided into four phases as indicated by dashed vertical lines. The rapid (decadal-scale) variations in $CH_4$ at the beginning of the Bølling/Allerød (B/A) and at the beginning and end of the Younger Dryas (YD) are coeval with abrupt temperature changes in the North Atlantic region. Temperature changes in Antarctica are smaller and less abrupt than in Greenland, and temperature changes in Greenland and Antarctica are asynchronous. Measurements are from the EPICA ice core drilled at Dome Concordia, Antarctica, and from the GRIP ice core drilled at Summit, Greenland. The timescales of both cores have been synchronized to the GRIP scale.

The Earth's orbit around the Sun can be calculated with high precision for the future as well as the past (Berger 1978; Berger and Loutre 1991). Future climate changes can therefore be forecast on a multimillenial time scale with some confidence. Even under a natural $CO_2$ regime (i.e., with the global temperature-$CO_2$ correlation continuing as in the Vostok ice core), the next glacial period would not be expected to start within the next 50,000 years (Loutre and Berger 2000; Berger and Loutre 2002). The Holocene will thus last longer than the previous (Eemian) interglacial, which endured for only about 10,000 years. This difference is because orbital eccentricity was much higher in the Eemian than today. With low orbital eccentricity, the effects of precession are minimized, and extreme cold-northern-summer orbital configurations like that of the last glacial initiation at 115 kilo years before present (ka BP) do not occur. Sustained high atmospheric greenhouse concentrations, comparable to a mid-range $CO_2$ stabilization scenario, may lead to a complete melting of the Greenland ice cap (Church et al. 2001) and further delay the onset of the next glacial period (Loutre and Berger 2000).

The mechanistic explanation for the observed glacial-interglacial $CO_2$ variations remains a difficult attribution problem (Broecker and Henderson 1998; Archer et al. 2000). Benthic carbon-isotope evidence from marine sediments, and terrestrial carbon accounting based on pollen data from terrestrial sediments, suggest that terrestrial carbon storage was ~ 300–700 PgC less at the LGM than during the Holocene (Shackleton 1977; Curry et al. 1988; Duplessy et al. 1988; Van Campo et al. 1993; Bird et al. 1994; Crowley 1995; Peng et al. 1998; Beerling 1999), implying that the extra stored carbon during glacial periods must be in the ocean and not on land. Numerous oceanic processes have been identified that could have contributed to the low glacial concentrations of $CO_2$ (Archer et al. 2000; Sigman and Boyle 2000; Stephens and Keeling 2000). One explanation conceived in the 1980s (Knox and McElroy 1984; Sarmiento and Toggweiler 1984; Siegenthaler and Wenk 1984), and recently supported by nitrogen-isotope data (Francois et al. 1998; Crosta and Shemesh 2002), invokes a more efficient utilization of macronutrients in the Southern Ocean during glacial times, leading to higher rates of carbon export from the surface and thus to increased carbon storage at depth, reducing the equilibrium concentration of $CO_2$ at the ocean surface. This scenario could have been realized by an increased input of iron (which is a limiting micronutrient for phytoplankton, especially diatom, growth in the Southern Ocean today [Martin et al. 1994; Boyd et al. 2000]) provided by an enhanced supply and transport of mineral dust (Andersen et al. 1998; Mahowald et al. 1999), possibly in concert with a slower rate of surface-to-deep mixing in the Southern Ocean, in glacial time. This mechanism may account for up to half of the glacial-interglacial difference in atmospheric $CO_2$ concentration (Levèvre and Watson 1999; Watson et al. 2000; Bopp et al. 2003). It is unlikely that the iron hypothesis could account for more than half of the difference. The Antarctic ice-core records show that about half of the $CO_2$ lowering during the last glacial period took place before high dust fluxes appeared in the Antarctic (Petit et al. 1999; Röthlisberger et al. 2002). Similarly, substantial changes in CO$_2$ occurred during the LGM-Holocene transition after the high fluxes disappeared (Röthlisberger et al. 2002). The initial slow CO$_2$ decline during the first part of the last glacial requires a different explanation, possibly involving ocean-sediment interaction with a timescale of millennia (Archer et al. 1999). The subsequent lowering of atmospheric CO$_2$ by a further ~ 30 ppm occurred more rapidly, coeval with the increase of Antarctic dust.

Recent analysis of dust-storm data has shown that the natural component of the contemporary dust loading in the atmosphere does exceed 75 percent (Tegen et al. 2003), although a substantial fraction of the contemporary dust was formerly attributed to human activities such as overgrazing and construction (Tegen and Fung 1995). Future dust supply to the Southern Ocean and other high-nutrient low-chlorophyll regions could change through changes in vegetation cover and surface characteristics in response to changes in climate, atmospheric CO$_2$, and land use activities or through changes in atmospheric transport. The resulting changes in marine iron supply could provide a small feedback to increasing atmospheric CO$_2$.

## Abrupt Climatic Changes

Greenland ice-core records show abrupt, decadal-scale temperature changes locally of up to 16 K amplitude (Schwander et al. 1997; Severinghaus et al. 1998; Lang et al. 1999; Severinghaus and Brook 1999) during the last glacial period and the transition to the present interglacial. These are known as the Dansgaard Oeschger (D O) events (Dansgaard et al. 1982; Oeschger et al. 1984; Broecker et al. 1985; Clark et al. 2002) (Figure 7.2). Isotopic sedimentary and pollen records from lakes and marine sediments (Eicher et al. 1981; Ruddiman and McIntyre 1981; Bond et al. 1993; Yu and Eicher 1998; Ammann et al. 2000; Sánchez Goñi et al. 2000; Baker et al. 2001; Prokopenko et al. 2001; Sánchez Goñi et al. 2002; Tzedakis et al. 2002) and European and Asian loess records (Ding et al. 1999; Ye et al. 2000; Porter 2001; Rousseau et al. 2002) demonstrate that substantial effects of these abrupt climate changes extended over the North Atlantic region and beyond.

Concomitant temperature changes recorded in Antarctica are smaller, less abrupt, and asynchronous to the Northern Hemisphere changes (Indermühle et al. 2000). This asynchrony has been explained (Stocker and Johnsen 2003) by a reduction in the North Atlantic Deep Water formation rate and oceanic heat transport into the North Atlantic region (Stocker 2000), a phenomenon that, in combination with the large heat capacity of the Southern Ocean, produces cooling in the North Atlantic and warming in the Southern Hemisphere (Mikolajewicz 1996; Marchal et al. 1999a, b).

The lowest temperatures in Greenland are associated with the surging of large amounts of ice, recorded as "Heinrich events," which are characterized by the widespread appearance of ice-rafted debris in North Atlantic sediments (Bond et al. 1993; Bond and Lotti 1995). The ice release provides a plausible mechanism (Alley et al. 1999)

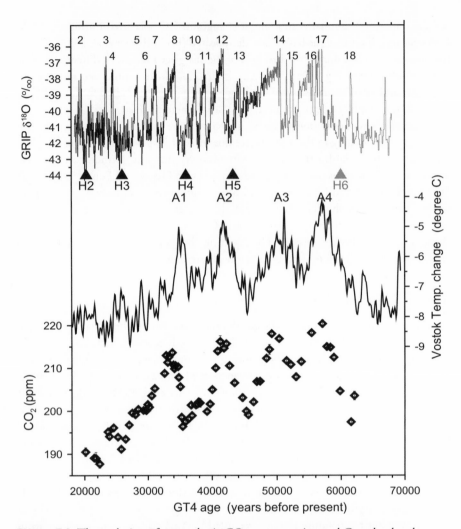

**Figure 7.2.** The evolution of atmospheric $CO_2$ concentration and Greenland and Antarctic temperature as indicated by $\delta^{18}O$, during the period from 70 ka BP to 20 ka BP (Indermühle et al. 2000). $\delta^{18}O$ values are from ice drilled at Summit during the Greenland Ice Core Project (GRIP) and Vostok, Antarctica, and $CO_2$ data are from Taylor Dome, Antarctica. All data are plotted on the GT4 chronology of Vostok. Dansgaard/Oeschger events 2 to 18 (warm interstadials in the Greenland $\delta^{18}O$ record), the Heinrich events H2 to H6 (filled triangles), recorded as ice-rafted debris in marine sediments in the North Atlantic, and the Antarctic warm periods A1 to A4 are indicated. The GRIP, Vostok, and Taylor Dome ice cores were synchronized based on methane measurements. The location of Heinrich events is based on the synchronization of the GRIP ice core to North Atlantic deep sea cores (Bond and Lotti 1995). The timescale is tentative for the first part of the record indicated by light gray in the $\delta^{18}O$ curve.

for a collapse of the thermohaline circulation (THC) as the meltwater input into the North Atlantic stabilizes the water column. Benthic carbon-isotope (Sarnthein et al. 1994) suggests that the deep Atlantic Ocean during the cold periods was filled with nutrient-rich water of Southern Hemisphere origin, and the atmospheric records of [14]C and [10]Be suggest that the ocean's ventilation was slowed (Hughen et al. 2000; Muscheler et al. 2000; Marchal et al. 2001).

A plausible explanation for the observed changes is the following sequence (Schmittner et al. 2002): Ice accumulates during the interstadial warm phases on the Laurentide ice sheet; this leads to an instability, triggering a massive iceberg discharge into the North Atlantic; the freshwater input causes a temporary breakdown of the THC, with cooling in the north and warming in the south; finally the THC resumes, perhaps in response to slow climatic changes in the Southern Ocean (Knorr and Lohmann 2003), when the freshwater perturbation has dissipated. This mechanism has been questioned by a high-resolution ice-core record that shows that the Antarctic cold reversal may have started earlier than the Bølling/Allerød (B/A) warming (Morgan et al. 2002), implying that the Antarctic cooling was not caused by the restarting of the THC in the Northern Hemisphere and thus suggesting a more important role for the tropics and Southern Hemisphere in controlling abrupt climate changes. A precise determination of hemispheric temperature relationships from the isotopic signatures in ice, however, remains challenging, as factors other than temperature, such as changes in the seasonality of precipitation, can affect the isotopic composition of the ice, and noise in the records may bias results.

Most current climate models, when forced with scenarios of increasing GHG concentrations, show a decrease in the THC and a reduction in the poleward heat transport by the North Atlantic (Cubasch et al. 2001). Idealized model simulations also suggest that the THC is subject to instabilities (as the palaeodata confirm) and that the North Atlantic Deep Water formation might collapse in a warming world (Bryan 1986; Stocker and Schmittner 1997) and remain in a collapsed state, owing to hysteresis, for many centuries. Unfortunately, the existence of critical thresholds and nonlinearities seriously limits our ability to predict the behavior of the THC (Knutti and Stocker 2002). Long-term ocean monitoring may be necessary if we are to reliably detect changes in the THC. Tracers related to the carbon cycle, such as dissolved $O_2$, are sensitive to circulation changes. Available observations suggest a decreasing inventory of dissolved $O_2$ in the ocean comparable to that predicted by modeling the effects of global warming on the ocean circulation (Plattner et al. 2001, 2002; Bopp et al. 2002). This effect is also present as an increase in atmospheric $O_2$ concentration, requiring a correction to the land-atmosphere partitioning of anthropogenic $CO_2$ (Bopp et al. 2002; Keeling and Garcia 2002; Plattner et al. 2002), estimated from observed atmospheric $CO_2$ and $O_2$ trends (Keeling and Shertz 1992; Prentice et al. 2001). An extended monitoring of dissolved $O_2$ has been proposed to detect changes in ocean circulation and to improve estimates of the carbon sinks (Joos et al. 2003b).

Despite the large fluctuations in climate, atmospheric $CO_2$ varied less than 20 ppm during the large Dansgaard/Oeschger events associated with Heinrich events and less than 10 ppm during the smaller Dansgaard/Oeschger events (Figure 7.2) (Stauffer et al. 1998; Indermühle et al. 2000). $CO_2$ variations were ~ 30 ppm around the Younger Dryas (YD) cold interval (Figure 7.1), during which the North Atlantic THC was greatly reduced (Sarnthein et al. 1994). Concomitant changes in methane and in $N_2O$ were about 100–200 ppb and ~30 ppb, respectively. Ocean model simulations reproduce the observed behavior of the coupled carbon-climate system during abrupt events of this type. Thus, when the North Atlantic THC is forced to collapse by imposing a meltwater pulse in an ocean model, the simulated consequences include a small (10 ppm) temporary increase in atmospheric $CO_2$, strong cooling in the North Atlantic region, a slight warming in the Southern Hemisphere, an increase in nutrient-rich water in the Atlantic, higher $^{14}C/^{12}C$ ratios in the atmosphere (Marchal et al. 1999a, b; Delaygue et al. 2003), and a 10 parts per billion (ppb) reduction in atmospheric $N_2O$ (Goldstein et al. 2003), that is, about a third of the observed $N_2O$ changes (Flückiger et al. 1999). Similarly, a relatively small positive feedback between atmospheric $CO_2$ and ocean circulation changes is found in global warming simulations in which the rate of North Atlantic Deep Water formation is reduced or even collapsed (Joos et al. 1999; Plattner et al. 2001). Thus, paleodata and model simulations agree that possible future changes in the North Atlantic Deep Water formation rate would have only modest effects on atmospheric $CO_2$. This finding does not, however, preclude the possibility that circulation changes in other ocean regions, in particular in the Southern Ocean, could have a larger impact on atmospheric $CO_2$ (Greenblatt and Sarmiento, Chapter 13, this volume).

This analysis does not consider the possible contribution of vegetation dieback to atmospheric $CO_2$. Vegetation dieback may have contributed to the observed changes in atmospheric $CO_2$ concentration during the YD and the Dansgaard/Oeschger events (Scholze et al. 2003) and could have larger effects if an abrupt collapse of the North Atlantic THC occurred during a warm-climate regime.

## The Holocene

Atmospheric $CO_2$ concentration varied slowly during the preindustrial Holocene (Figure 7.3). Between 11 and 8 ka, atmospheric $CO_2$ fell gradually from 265 to 260 ppm, then it increased again toward the preindustrial level of 280 ppm. The climate event at 8,200 yr BP, comparable to a small Dansgaard/Oeschger event and associated with a temperature drop of 1.5 to 4 K in the North Atlantic region and 4 to 8 K in central Greenland (Barber et al. 1999), left no trace in the available low-resolution $CO_2$ records (Figure 7.3). A preliminary set of twelve $\delta^{13}C$ measurements was used by Indermühle et al. (1999) to estimate possible causes of the observed long-term variations using a double deconvolution technique (Joos and Bruno 1998). This analysis suggested that the observed variations were caused mainly by uptake and subsequent release of ter-

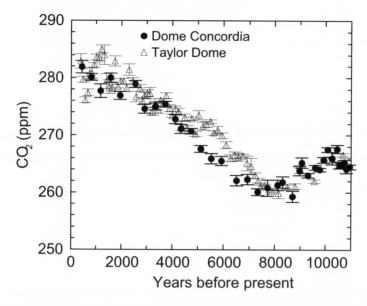

**Figure 7.3.** Holocene variations in atmospheric $CO_2$ concentration, from measurements on air entrapped in ice from Taylor Dome, Antarctica (Indermühle et al. 1999) and from Dome Concordia (Flückiger et al. 2002). The difference between the two cores, notably between 4 and 8 kyr, is most likely due to uncertainties in the age scales of the two cores (Stauffer et al. 2002).

restrial organic carbon (Indermühle et al. 1999). An alternative explanation (Broecker et al. 2001; Broecker and Clark 2003) is that the steady $CO_2$ rise after 8 ka is due to the $CaCO_3$ compensation mechanism, responding to the earlier extraction of carbon from the atmosphere by terrestrial uptake during the period after the LGM. This mechanism depends on ocean-sediment interactions and has an appropriately long time constants of 5 ka and longer (Archer et al. 1999). Unfortunately, uncertainties in the existing ice core $\delta^{13}C$ (Indermühle et al. 1999) and marine sediment data (Broecker and Clark 2003) do not allow us to discriminate reliably between the two hypotheses based on observations alone.

Other information sources provide circumstantial evidence. The benthic carbon isotope record (Curry et al. 1988; Duplessy et al. 1988) constrains the terrestrial uptake during the last glacial-interglacial transition and, hence, constrains the contribution by the $CaCO_3$ compensation mechanism to within 4 to 10 ppm (Joos et al. 2003a). Simulations with land biosphere models (François et al. 1999; Brovkin et al. 2002; Joos et al. 2003a; Kaplan et al. 2002) and pollen-based estimates (Adams and Faure 1998) yield ambiguous results ranging from a terrestrial release of 90 PgC to an uptake of 130 PgC during the late Holocene.

The Lund-Potsdam-Jena (LPJ) dynamic global vegetation model (Sitch et al. 2003) coupled to the atmosphere-ocean-sediment carbon cycle component of the Bern CC model has been used to compare the conflicting explanations for the Holocene $CO_2$ rise (Joos et al. 2003a). The model was forced with a 21 ka–long time series of "snapshot" simulations from the Hadley Center climate model (Kaplan et al. 2002) or the NCAR climate model and reconstructed atmospheric $CO_2$ (Monnin et al. 2001) during the transition until 11 ka BP. Afterward, atmospheric $CO_2$ was simulated. The entire Holocene ice-core $CO_2$ record is matched within a few ppm. A sensitivity analysis in combination with available observations suggests that $CaCO_3$ compensation, sea surface warming, and a terrestrial release, in response to land use and desertification, contributed about equally to the observed 20 ppm rise after 8 ka BP. The modeled $CO_2$ decrease during the early Holocene is, in agreement with earlier suggestions, mainly due to the establishment of boreal forest in formerly glaciated areas, partly compensated for by $CaCO_3$ compensation. Simulated LGM-Holocene terrestrial uptake is 800 PgC, slightly higher than the 300 to 700 PgC range derived from the benthic isotope record. These simulations took into account the dynamic land area changes caused by sea-level rise and ice retreat. Carbon sequestration by regrowth and soil establishment on formerly ice-covered areas is almost compensated for by carbon loss in response to sea-level rise and climate change on areas that were ice-free at the LGM. Increasing carbon storage on land in these simulations was principally a consequence of climate change (forced by orbital variations, $CO_2$, and ice extent), in combination with direct physiological effects of rising $CO_2$ concentration on plant productivity and water-use efficiency. These direct $CO_2$ effects were responsible for more than 80 percent of the simulated difference in global carbon storage between the Holocene and the LGM.

Existing uncertainties in the ice-core $\delta^{13}C$ and marine sediment records need to be reduced to further constrain the quantitative contributions of individual mechanisms to the Holocene $CO_2$ variations. Nevertheless, the observed Holocene $CO_2$ variations, combined with marine benthic $\delta^{13}C$ data (Shackleton 1977; Curry et al. 1988; Duplessy et al. 1988) are consistent with the timescale inferred for ocean-sediment interactions (Archer et al. 1999), and with a role for $CO_2$ fertilization in determining terrestrial carbon storage at concentrations of $CO_2$ within the natural range (180–280 ppm) (Esser and Lautenschlager 1994; Peng et al. 1998; Bennett and Willis 2000). One implication for the future evolution of atmospheric $CO_2$ is that the $CaCO_3$ compensation mechanism, which added $CO_2$ to the atmosphere during the preindustrial Holocene, will similarly remove atmospheric $CO_2$ emitted by human activities, but the timescale for this removal will be very long (multimillennia). The carbon sink mechanism due to $CO_2$ fertilization has additional experimental support for higher than present $CO_2$ concentrations (DeLucia et al. 1999; Luo et al. 1999), and is already accounted for in $CO_2$ stabilization scenario calculations (Joos et al. 2001; Prentice et al. 2001).

# The Last Millennium

Knowledge of natural climate variability and of the role of solar and volcanic forcing is important for the attribution of climate change during recent decades and centuries. Different reconstructions of forcing by past solar irradiance variations, however, based on cosmogenic isotopes ($^{14}C$, $^{10}Be$, $^{36}Cl$; Beer et al. 1994) and sunspot numbers (Hoyt and Schatten 1994) vary by up to a factor of five (Lean et al. 1995; Reid 1997; Bard et al. 2000; Crowley 2000). In addition, there is a growing literature arguing for an amplification of the climate impact of solar irradiance changes through a solar modulation of the cosmic ray flux, affecting, for example, low-level cloud cover on Earth, based on empirical correlations between climate change and solar activity over specific periods (e.g., Tinsley and Deen 1991; Shaviv and Veizer 2003). The proposed correlation, however, breaks down over the last years of the instrumental record, and cosmic ray flux and climate change proxies are unrelated for the period 20 to 60 ka BP, when very large changes in the cosmic ray flux (Laschamp event) occurred (Wagner et al. 2001).

The magnitude of Northern Hemisphere surface temperature variations is also debated (Jones et al. 1998; Mann et al. 1999; Briffa 2000; Crowley 2000; Huang et al. 2000; Beltrami 2002; Esper et al. 2002). For example, Esper et al. (2002) report that the decadal-average Northern Hemisphere surface temperature was about 1 K lower in the first half of the 17th century than at around year AD 1000 and the present climatological mean, whereas Mann et al. (1999) report that decadal-average temperatures varied within 0.4 K only.

Highly resolved ice-core $CO_2$ records of the past millennium provide a joint constraint on Northern Hemisphere temperature variability and the climate–carbon cycle feedback. Figure 7.4 illustrates the relationship between the variations of decadal to multidecadal Northern Hemisphere surface temperature, atmospheric $CO_2$ concentration, and the strength of the climate–carbon cycle feedback (as ppm $CO_2$ released per K temperature increase in the Northern Hemisphere). The strength of the feedback depends on several factors, including the change in solubility of $CO_2$ in seawater and the responses of productivity and heterotrophic respiration to temperature. The current best estimate of the actual $CO_2$ range during the past millennium (before the Industrial Revolution) is 6 ppm, based on emerging high-quality measurements (Siegenthaler and Monnin, pers. comm.). If we assume a climate–carbon cycle feedback of 12 ppm $K^{-1}$, as obtained for the coupled Bern CC-LPJ model (Gerber et al. 2003), the $CO_2$ range of 6 ppm then constrains the decadal-average Northern Hemisphere surface temperature variation to a range of 0.5 K during the past millennium (excluding the rise that has occurred during recent decades; Figure 7.4). Alternatively, if we accept the ranges of temperature variations reconstructed by Mann et al. (1999) and by Esper et al. (2002) as equally possible, then the $CO_2$ concentration range of 6 ppm constrains the climate–carbon cycle feedback to between 6 and 16 ppm $K^{-1}$ (for global mean surface temperature changes of less than ~1°C). This implies that the 0.6 ± 0.2°C increase in

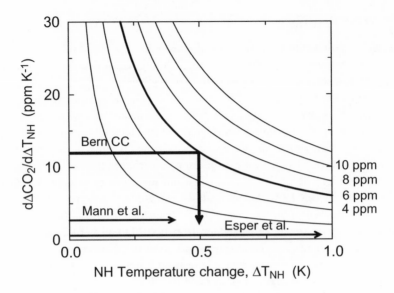

**Figure 7.4.** Relationship between Northern Hemisphere (NH) surface temperature change, climate–carbon cycle feedbacks, and variations in atmospheric $CO_2$. Isolines depict different ranges for $CO_2$ variation during the last millennium and are plotted against changes in decadal-average NH-temperature (horizontal axis) and the climate–carbon cycle feedback expressed as change in atmospheric $CO_2$ concentration per degree change in decadal-average NH surface temperature (vertical axis). The range of NH-temperature variations reconstructed by Mann et al. (1999) and Esper et al. (2002) are shown. The arrow labeled Bern CC depicts the NH temperature change obtained when combining the climate–carbon cycle feedback of the Bern CC-LPJ model (Gerber et al. 2003) with a $CO_2$ variability range of 6 ppm.

global mean surface temperature over the 20th century contributed little to the observed $CO_2$ increase over the same period. This data-based finding is somewhat in contrast with the model results of Dai and Fung (1993), which suggest climate change as a major driver of the terrestrial sink during the past century, but it is consistent with more recent model estimates that suggest a relatively small role of climate change in modulating the terrestrial sink over the past century (McGuire et al. 2001).

The temporal evolution of atmospheric $CO_2$ and Northern Hemisphere temperature in response to the reconstructed variability in solar irradiance and radiative forcing by volcanoes has been examined (Gerber et al. 2003). Modeled variations in atmospheric $CO_2$ and Northern Hemisphere mean surface temperature are compatible with reconstructions from different Antarctic ice cores and temperature proxy data for low solar forcing. Simulations where the magnitude of solar irradiance changes is increased toward the high estimates yield a mismatch between model results and the $CO_2$ measurements. This finding provides further evidence for modest changes in solar irradiance and global mean temperature over the past millennium and argues against a substantial amplification of the response of global or hemispheric annual mean temperature to solar changes by the suggested cosmic ray flux–cloud cover link or other solar-related mechanisms. The results imply that solar changes are not the dominant factor in the 20th-century warming.

## Acknowledgments

This chapter was written while one of us (FJ) was a visitor at the National Center for Atmospheric Research (NCAR), Boulder, CO. The hospitality and generous support by Dave Schimel and NCAR's Climate and Global Dynamics Division is appreciated. We wish to thank T. Blunier, S. Harrison, N. Mahowald, E. Monnin, and N. Rosenbloom for comments on this chapter and E. Monnin for help with figures.

## Literature Cited

Adams, J. M., and H. Faure. 1998. A new estimate of changing carbon storage on land since the last glacial maximum, based on global land ecosystem reconstruction. *Global and Planetary Change* 16/17:3–24.

Alley, R. B., P. U. Clark, L. D. Keigwin, and R. S. Webb. 1999. Making sense of millennial-scale climate change. Pp. 385–394 in *Mechanisms of millennial-scale global climate change,* edited by P. U. Clark, R. S. Webb, and L. D. Keigwin. Vol.112 of Geophysical Monograph. Washington, DC: American Geophysical Union.

Ammann, B., H. Birks, S. Brooks, U. Eicher, U. Grafenstein, W. Hofmann, G. Lemdahl, J. Schwander, K. Tobolski, and L. Wick. 2000. Quantification of biotic responses to rapid climatic changes around the Younger Dryas: A synthesis. *Palaeogeography, Palaeoclimatology, Palaeoecology* 159:313–347.

Andersen, K. K., A. Armengaud, and C. Genthon. 1998. Atmospheric dust under glacial and interglacial conditions. *Geophysical Research Letters* 25:2281–2284.

Archer, D., H. Kheshgi, and E. Maier-Reimer. 1999. Dynamics of fossil fuel CO$_2$ neutralization by marine CaCO$_3$. *Global Biogeochemical Cycles* 12:259–276.

Archer, D. A., A. Winguth, D. Lea, and N. Mahowald. 2000. What caused the glacial/interglacial atmospheric pCO$_2$ cycles? *Reviews of Geophysics* 38:159–189.

Arrhenius, S. 1896. On the influence of carbonic acid in the air upon the temperature of the ground. *The London, Edinburgh and Dublin Philosophical Magazine and Journal of Sciences* 41:237–276.

Baker, P. A., C. A. Rigsby, G. O. Seltzer, S. C. Fritz, T. K. Lowenstein, N. P. Bacher, and C. Veliz. 2001. Tropical climate changes at millennial and orbital timescales on the Bolivian Altiplano. *Nature* 409:698–701.

Barber, D. C., A. Dyke, C. Hillaire-Marcel, A. E. Jennings, J. T. Andrews, M. W. Kerwin, G. Bilodeau, R. McNeely, J. Southon, M. D. Morehead, and J.-M. Gagnon. 1999. Forcing of the cold event of 8,200 years ago by catastrophic drainage of Laurentide lakes. *Nature* 400:344–348.

Bard, E., G. Raisbeck, F. Yiou, and J. Jouzel. 2000. Solar irradiance during the last 1200 years based on cosmogenic nuclides. *Tellus* 52B:985–992.

Beer, J., F. Joos, C. Lukasczyk, W. Mende, J. Rodriguez, U. Siegenthaler, and R. Stellmacher. 1994. [10]Be as an indicator of solar variability and climate. Pp. 221–223 in *The solar engine and its influence on terrestrial atmosphere and climate,* edited by E. Nesme-Ribes. Heidelberg: Springer Verlag.

Beerling, D. J. 1999. New estimates of carbon transfer to terrestrial ecosystems between the last glacial maximum and the Holocene. *Terra Nova* 11:162–167.

Beltrami, H. 2002. Climate from borehole data: Energy fluxes and temperatures since 1500. *Geophysical Research Letters* 29, doi:10.1029/2002GL015702.

Bennett, K. D., and K. J. Willis. 2000. Effect of global atmospheric carbon dioxide on glacial-interglacial vegetation change. *Global Ecology and Biogeography* 9:355–361.

Berger, A. 1978. Long-term variations of daily insolation and Quaternary climatic changes. *Journal of the Atmospheric Sciences* 35:2362–2367.

Berger, A., and M. F. Loutre. 1991. Insolation values for the climate of the last 10 million years. *Quaternary Science Reviews* 10:297–317.

———. 2002. An exceptionally long interglacial ahead? *Science* 297:1287–1288.

Berger, A., M. F. Loutre, and H. Gallée. 1998. Sensitivity of the LLN climate model to the astronomical and $CO_2$ forcings over the last 200 ky. *Climate Dynamics* 14: 615–629.

Bird, M. I., J. Lloyd, and G. D. Farquhar. 1994. Terrestrial carbon storage at the LGM. *Nature* 371:566.

Bond, G. C., and R. Lotti. 1995. Iceberg discharges into the North Atlantic on millennial time scales during the last glaciation. *Science* 267:1005–1010.

Bond, G., W. S. Broecker, S. J. Johnsen, J. McManus, L. Labeyrie, J. Jouzel, and G. Bonani. 1993. Correlations between climate records from North Atlantic sediments and Greenland ice. *Nature* 365:143–147.

Bopp, L., C. Le Quéré, M. Heimann, A. C. Manning, and P. Monfray. 2002. Climate-induced oceanic oxygen fluxes: Implications for the contemporary carbon budget. *Global Biogeochemical Cycles* 16, doi:10.1029/2001GB001445.

Bopp, L., K. E. Kohlfeld, C. Le Quéré, and O. Aumont. 2003. Dust impact on marine biota and atmospheric $CO_2$ in glacial periods. *Paleoceanography* 18 (2):art. no. 1046.

Boyd, P. A., A. J. Watson, C. S. Law, E. R. Abraham, T. Trull, R. Murdoch, D. C. E. Bakker, A. R. Bowie, K. O. B., H. Chang, M. Charette, P. Croot, K. Downing, R. F., M. Gall, M. Hadfield, J. Hall, M. Harvey, G. Jameson, J. LaRoche, M. Liddicoat, R. Ling, M. T. Maldonado, R. M. McKay, S. Nodder, S. Pickmere, R. Pridmore, S. Rintoul, K. Safi, P. Sutton, R. Strzepek, K. Tanneberger, S. Turner, A. Waite, and J. Zeldis. 2000. A mesoscale phytoplankton bloom in the polar Southern Ocean stimulated by iron fertilization. *Nature* 407:695–702.

Briffa, K. R. 2000. Annual climate variability in the Holocene: Interpreting the message of ancient trees. *Quaternary Science Reviews* 19:87–105.

Broccoli, A. J., and S. Manabe. 1987. The influence of continental ice, atmospheric $CO_2$, and land albedo on the climate of the last glacial maximum. *Climate Dynamics* 1:87–99.

Broecker, W. S., and E. Clark. 2003. Holocene atmospheric $CO_2$ increase as viewed from the sea floor. *Global Biogeochemical Cycles* 17, doi:10.1029/2002GB001985.

Broecker, W. S., and G. M. Henderson. 1998. The sequence of events surrounding Termination II and their implications for the cause of glacial-interglacial CO$_2$ changes. *Paleoceanography* 13:352–364.

Broecker, W. S., D. M. Peteet, and D. Rind. 1985. Does the ocean-atmosphere system have more than one stable mode of operation? *Nature* 315:21–25.

Broecker, W. S., J. Lynch-Stieglitz, E. Clark, I. Hajdas, and G. Bonani. 2001. What caused the atmosphere's CO$_2$ content to rise during the last 8000 years? *Geochemistry Geophysics Geosystems* 2, art. no. 2001GC000177.

Brovkin, V., M. Hofmann, J. Bendtsen, and A. Ganopolski. 2002. Ocean biology could control atmospheric $\delta^{13}$C during glacial-interglacial cycle. *Geochemistry Geophysics Geosystems* 3, art. no. 1027.

Bryan, F. 1986. High-latitude salinity effects and interhemispheric thermohaline circulations. *Nature* 323:301–304.

Church, J. A., J. M. Gregory, P. Huybrechts, M. Kuhn, K. Lambeck, M. T. Nhuan, D. Qin, and P. L. Woodworth. 2001. Changes in sea level. Pp. 639–693 in *Climate change 2001: The scientific basis (Contribution of Working Group I to the Third Assessment Report of the Intergovernmental Panel on Climate Change)*, edited by J. T. Houghton, Y. Ding, D. Griggs, M. Noguer, P. van der Linden, X. Dai, K. Maskell, and C. A. Johnson. Cambridge: Cambridge University Press.

Claquin, T., C. Roelandt, K. Kohfeld, S. Harrison, I. Tegen, I. Prentice, Y. Balkanski, G. Bergametti, M. Hansson, N. Mahowald, H. Rodhe, and M. Schulz. 2003. Radiative forcing of climate by ice-age atmospheric dust. *Climate Dynamics* 20:193–202; doi:10.1007/s00382-002-0269-1.

Clark, P. U., R. A. Alley, and D. Pollard. 1999. Northern Hemisphere ice-sheet influences on global climate change. *Science* 286:1104–1111.

Clark, P. U., N. G. Pisias, T. F. Stocker, and A. J. Weaver. 2002. The role of the thermohaline circulation in abrupt climate change. *Nature* 415:863–869.

Crosta, X., and A. Shemesh. 2002. Reconciling down core anticorrelation of diatom carbon and nitrogen isotopic ratios from the Southern Ocean. *Paleoceanography* 17, art. no. 1010.

Crowley, T. J. 1995. Ice age terrestrial carbon changes revisited. *Global Biogeochemical Cycles* 9:377–389.

———. 2000. Causes of climate change of the last 1000 years. *Science* 289:270–277.

Cubasch, U., G. A. Meehl, G. J. Boer, R. J. Stouffer, M. Dix, A. Noda, C. A. Senior, S. Raper, and K. S. Yap. 2001. Projections of future climate change. Pp. 525–582 in *Climate change 2001: The scientific basis (Contribution of Working Group I to the Third Assessment Report of the Intergovernmental Panel on Climate Change)*, edited by J. T. Houghton, Y. Ding, D. Griggs, M. Noguer, P. van der Linden, X. Dai, K. Maskell, and C. A. Johnson. Cambridge: Cambridge University Press.

Curry, W. B., J. Duplessy, L. Labeyrie, and N. J. Shackleton. 1988. Changes in the distribution of $\delta^{13}$C of deepwater $\Sigma$CO$_2$ between the last glaciation and the Holocene. *Paleoceanography* 3:317–341.

Dai, A., and I. Y. Fung. 1993. Can climate variability contribute to the "missing" CO$_2$ sink? *Global Biogeochemical Cycles* 7:599–609.

Dansgaard, W., H. B. Clausen, N. Gundestrup, C. U. Hammer, S. F. Johnsen, P. M. Kristinsdottir, and N. Reeh. 1982. A new Greenland deep ice core. *Science* 218:1273–1277.

Delaygue, G., T. F. Stocker, F. Joos, and G.-K. Plattner. 2003. Simulation of atmospheric radiocarbon during abrupt oceanic circulation changes: Trying to reconcile models and reconstructions. *Quaternary Science Reviews* 22:1647–1658.

DeLucia, E. H., J. G. Hamilton, S. L. Naidu, R. B. Thomas, J. A. Andrews, A. Finzi, M. Lavine, R. Matamala, J. E. Mohan, G. R. Hedrey, and W. H. Schlesinger. 1999. Net primary production of a forest ecosystem with experimental $CO_2$ enrichment. *Nature* 284:1177–1179.

Ding, Z. L., J. Z. Ren, S. L. Yang, and T. S. Liu. 1999. Climate instability during the penultimate glaciation: Evidence from two high- resolution loess records, China. *Journal of Geophysical Research* 104:20123–20132.

Duplessy, J. C., N. J. Shackleton, R. G. Fairbanks, L. Labeyrie, D. Oppo, and N. Kallel. 1988. Deepwater source variations during the last climate cycle and their impact on the global deepwater circulation. *Paleoceanography* 3:343–360.

Eicher, U., U. Siegenthaler, and S. Wegmüller. 1981. Pollen and isotope analysis on late and post-glacial sediments of the Tourbiere de Chirens (Dauphine, France). *Quaternary Research* 15:160–170.

Esper, J., E. R. Cook, and F. H. Schweingruber. 2002. Low-frequency signals in long tree-ring chronologies for reconstructing past temperature variability. *Science* 295:2250–2253.

Esser, G., and M. Lautenschlager. 1994. Estimating the change of carbon in the terrestrial biosphere from 18000 BP to present using a carbon cycle model. *Environmental Pollution* 83:45–53.

Flückiger, J., A. Dällenbach, T. Blunier, B. Stauffer, T. F. Stocker, D. Raynaud, and J.-M. Barnola. 1999. Variations in atmospheric $N_2O$ concentrations during abrupt climatic changes. *Science* 285:227–230.

Flückiger, J., E. Monnin, B. Stauffer, J. Schwander, T. F. Stocker, J. Chappellaz, D. Raynaud, and J.-M. Barnola. 2002. High resolution Holocene $N_2O$ ice core record and its relationship with $CH_4$ and $CO_2$. *Global Biogeochemical Cycles* 16, doi:10.1029/2001GB001417.

Francois, R., M. A. Altabet, E. Yu, D. M. Sigman, M. P. Bacon, M. Frank, G. Bohrmann, G. Bareille, and L. D. Labeyrie. 1998. Contribution of Southern Ocean surface water stratification to low atmospheric $CO_2$ concentrations during the last glacial period. *Nature* 389:929–935.

François, L. M., Y. Goddéris, P. Warnant, G. Ramstein, N. de Noblet, and S. Lorenz. 1999. Carbon stocks and isotopic budgets of the terrestrial biosphere at mid-Holocene and last glacial maximum times. *Chemical Geology* 159:163–189.

Gallée, H., J. P. Van Ypersele, T. F. I. Marsiat, C. Tricot, and A. Berger. 1992. Simulation of the last glacial cycle by a coupled, sectorially averaged climate-ice sheet model. 2. Response to insolation and $CO_2$ variations. *Journal of Geophysical Research* 97:151713–151740.

Ganopolski, A., S. Rahmstorf, V. Petoukhov, and M. Claussen. 1998. Simulation of modern and glacial climates with a coupled global model of intermediate complexity. *Nature* 391:351–356.

Gerber, S., F. Joos, P. P. Brügger, T. F. Stocker, M. E. Mann, S. Sitch, and M. Scholze. 2003. Constraining temperature variations over the last millennium by comparing simulated and observed atmospheric $CO_2$. *Climate Dynamics* 20:281–299.

Goldstein B., F. Joos, and T. F. Stocker. 2003. A modeling study of oceanic nitrous oxide during the Younger Dryas cold period. *Geophysical Research Letters* 30, doi:10.1029/2002/GL016418.

Hays, J. D., J. Imbrie, and N. J. Shackleton. 1976. Variations in the Earth's orbit: Pacemaker of the ice ages. *Science* 194:1121–1132.

Hewitt, C. D., and J. F. B. Mitchell. 1997. Radiative forcing and response of a GCM to ice age boundary conditions: Cloud feedback and climate sensitivity. *Climate Dynamics* 13:821–834.

Hewitt, C. D., R. J. Stouffer, A. J. Broccoli, J. F. B. Mitchell, and P. J. Valdes. 2003. The effect of ocean dynamics in a coupled GCM simulation of the Last Glacial Maximum. *Climate Dynamics* 20:203–218; doi:10.1007/s00382-002-0272-6.

Houghton, J. T., Y. Ding, D. Griggs, M. Noguer, P. van der Linden, X. Dai, K. Maskell, and C. A. Johnson, eds. 2001. *Climate change 2001: The scientific basis (Contribution of Working Group I to the Third Assessment Report of the Intergovernmental Panel on Climate Change)*. Cambridge: Cambridge University Press.

Hoyt, D. V., and K. H. Schatten. 1994. The one hundredth year of Rudolf Wolf's death: Do we have the correct reconstruction of solar activity? *Geophysical Research Letters* 21:2067–2070.

Huang, S., H. N. Pollack, and P. Shen. 2000. Temperature trends over the past five centuries reconstructed from borehole temperatures. *Nature* 403:756–758.

Hughen, K. A., J. R. Southon, S. J. Lehman, and J. T. Overpeck. 2000. Synchronous radiocarbon and climate shifts during the last deglaciation. *Science* 290:1951–1954.

Indermühle, A., T. F. Stocker, F. Joos, H. Fischer, H. Smith, M. Wahlen, B. Deck, D. Mastroianni, J. Tschumi, T. Blunier, R. Meyer, and B. Stauffer. 1999. Holocene carbon-cycle dynamics based on $CO_2$ trapped in ice at Taylor Dome, Antarctica. *Nature* 398:121–126.

Indermühle, A., E. Monnin, B. Stauffer, T. F. Stocker, and M. Wahlen. 2000. Atmospheric $CO_2$ concentration from 60 to 20 kyr BP from the Taylor Dome ice core, Antarctica. *Geophysical Research Letters* 27:735–738.

Jones, P. D., K. R. Briffa, T. P. Barnett, and S. F. B. Tett. 1998. High-resolution palaeoclimatic records for the last millennium: Interpretation integration and comparison with General Circulation Model control-run temperatures. *The Holocene* 8:455–471.

Joos, F., and M. Bruno. 1998. Long-term variability of the terrestrial and oceanic carbon sinks and the budgets of the carbon isotopes $^{13}C$ and $^{14}C$. *Global Biogeochemical Cycles* 12:277–295.

Joos, F., G.-K. Plattner, T. F. Stocker, O. Marchal, and A. Schmittner. 1999. Global warming and marine carbon cycle feedbacks on future atmospheric $CO_2$. *Science* 284:464–467.

Joos, F., I. C. Prentice, S. Sitch, R. Meyer, G. Hooss, G.-K. Plattner, S. Gerber, and K. Hasselmann. 2001. Global warming feedbacks on terrestrial carbon uptake under the Intergovernmental Panel on Climate Change (IPCC) emission scenarios. *Global Biogeochemical Cycles* 15:891–907.

Joos, F., S. Gerber, I. C. Prentice, B. Otto-Bliesner, and P. Valdes. 2003a. Transient simulations of Holocene atmospheric carbon dioxide and terrestrial carbon since the Last Glacial Maximum. *Global Biogeochemical Cycles* (submitted).

Joos, F., G.-K. Plattner, T. F. Stocker, A. Körtzinger, and D. W. R. Wallace. 2003b. Trends in marine dissolved oxygen: Implications for ocean circulation changes and the carbon budget. *Eos, Transactions, American Geophysical Union*, 84:197–201.

Kaplan, J. O., I. C. Prentice, W. Knorr, and P. J. Valdes. 2002. Modeling the dynamics of terrestrial carbon storage since the Last Glacial Maximum. *Geophysical Research Letters* 29, doi:10.1029/2002GL015230.

Kawamura, K., T. Nakazawa, S. Aoki, S. Sugawara, Y. Fujii, and O. Watanbe. 2003. Atmospheric $CO_2$ variations over the last three glacial-interglacial climatic cycles deduced from the Dome Fuji deep ice core, Antarctica, using a wet extraction technique. *Tellus* 55B:126–137.

Keeling, R. F., and H. Garcia. 2002. The change in oceanic $O_2$ inventory associated with recent global warming. *Proceedings of the National Academy of Sciences of the USA* 99:7848–7853.

Keeling, R. F., and S. R. Shertz. 1992. Seasonal and interannual variations in atmospheric oxygen and implications for the global carbon cycle. *Nature* 358:723–727.

Kim, S. J., G. M. Flato, G. J. Boer, and N. A. McFarlane. 2002. A coupled climate model simulation of the Last Glacial Maximum. 1. Transient multi-decadal response. *Climate Dynamics* 19:515–537.

Kitoh, A., S. Murakami, and H. Koide. 2001. A simulation of Last Glacial Maximum with a coupled atmosphere-ocean GCM. *Geophysical Research Letters* 28:2221–2224.

Knorr, G., and G. Lohmann. 2003. Southern Ocean origin for the resumption of the Atlantic thermohaline circulation during deglaciation. *Nature* 424:532–536.

Knox, F., and M. McElroy. 1984. Change in atmospheric $CO_2$: Influence of the marine biota at high latitude. *Journal of Geophysical Research* 89:4629–4637.

Knutti, R., and T. F. Stocker. 2002. Limited predictability of the future thermohaline circulation close to an instability threshold. *Journal of Climate* 30:179–186.

Lang, C., M. Leuenberger, J. Schwander, and S. Johnsen. 1999. 16°C rapid temperature variation in Central Greenland 70,000 years ago. *Science* 286:934–937.

Lean, J., J. Beer, and R. Bradley. 1995. Reconstruction of solar irradiance since 1610: Implication for climate change. *Geophysical Research Letters* 22:3195–3198.

Levèvre, N., and A. J. Watson. 1999. Modeling the geochemical cycle of iron in the oceans and its impact on atmospheric $CO_2$ concentrations. *Global Biogeochemical Cycles* 13:727–736.

Levis, S., J. A. Foley, and D. Pollard. 1999. $CO_2$, climate, and vegetation feedbacks at the Last Glacial Maximum. *Journal of Geophysical Research* 104:31191–31198.

Loutre, M. F., and A. Berger. 2000. Future climate changes: Are we entering an exceptionally long interglacial? *Climatic Change* 46:611–90.

Luo, Y., J. Reynolds, and Y. Wang. 1999. A search for predictive understanding of plant responses to elevated $CO_2$. *Global Change Biology* 5:143–156.

Mahowald, N., K. E. Kohlfeld, M. Hansson, Y. Balkanski, S. P. Harrison, I. C. Prentice, M. Schulz, and H. Rhode. 1999. Dust sources and deposition during the Last Glacial Maximum and current climate: A comparison of model results with paleodata from ice cores and marine sediments. *Journal of Geophysical Research* 104:15895–15816.

Mann, M. E., R. S. Bradley, and M. K. Hughes. 1999. Northern Hemisphere temperatures during the past millennium: Inferences, uncertainties, and limitations. *Geophysical Research Letters* 26:759–762.

Marchal, O., T. F. Stocker, and F. Joos. 1999a. On large-scale physical and biogeochemical responses to abrupt changes in the Atlantic thermohaline circulation. Pp. 263–284 in *Mechanisms of millennial-scale global climate change,* edited by P. U. Clark, R. S.

Webb, and L. D. Keigwin. Vol. 112 of Geophysical Monograph. Washington, DC: American Geophysical Union.

Marchal, O., T. F. Stocker, F. Joos, A. Indermühle, T. Blunier, and J. Tschumi. 1999b. Modelling the concentration of atmospheric CO$_2$ during the Younger Dryas climate event. *Climate Dynamics* 15:341–354.

Marchal, O., T. F. Stocker, and R. Muscheler. 2001. Atmospheric radiocarbon during the Younger Dryas: Production, ventilation, or both? *Earth and Planetary Science Letters* 185:383–395.

Martin, J. H., K. H. Coale, K. S. Johnson, S. E. Fitzwater, R. M. Gordon, S. J. Tanner, C. N. Hunter, V. A. Elrod, J. L. Nowicki, T. L. Coley, R. T. Barber, S. Lindley, A. J. Watson, K. V. Scoy, C. S. Law, M. I. Liddicoat, R. Ling, T. Stanton, J. Stockel, C. Collins, A. Anderson, R. Bidigare, M. Ondrusek, M. Latasa, F. J. Millero, K. Lee, W. Yao, J. Y. Zhang, G. Friederich, C. Sakamoto, F. Chavez, K. Buck, Z. Kolber, R. Greene, P. Falkowski, S. W. Chisholm, F. Hoge, R. Swift, J. Yungel, S. Turner, P. Nightingale, A. Hatton, P. Liss, and N. W. Tindale. 1994. Testing the iron hypothesis in ecosystems of the equatorial Pacific Ocean. *Nature* 371:123–129.

McGuire, A. D., S. Sitch, J. S. Clein, R. Dargaville, G. Esser, J. Foley, M. Heimann, F. Joos, J. Kaplan, D. W. Kicklighter, R. A. Meier, J. M. Melillo, B. Moore. III, I. C. Prentice, N. Ramankutty, T. Reichenau, A. Schloss, H. Tian, L. J. Williams, and U. Wittenberg. 2001. Carbon balance of the terrestrial biosphere in the twentieth century: Analyses of CO$_2$, climate and land-use effects with four process-based ecosystem models. *Global Biogeochemical Cycles* 15:183–206.

Mikolajewicz, U. 1996. *A meltwater-induced collapse of the "conveyor belt" thermohaline circulation and its influence on the distribution of $\delta^{14}C$ and $\delta^{18}O$ in the oceans.* Technical Report 189. Hamburg: Max-Planck-Institut für Meteorologie.

Monnin, E., A. Indermühle, A. Däallenbach, J. Flückiger, B. Stauffer, T. F. Stocker, D. Raynaud, and J. Barnola. 2001. Atmospheric CO$_2$ concentrations over the last glacial termination. *Science* 291:112–114.

Morgan, V., M. Delmotte, T. van Ommen, J. Jouzel, J. Chapellaz, S. Woon, V. Masson-Delmotte, and D. Raynau. 2002. Relative timing of deglacial climate events in Antarctica and Greenland. *Science* 297:1862–1865.

Muscheler, R., J. Beer, G. Wagner, and R. C. Finkel. 2000. Changes in deep-water formation during the Younger Dryas event inferred from [10]Be and [14]C records. *Nature* 408:567–570.

Oeschger, H., J. Beer, U. Siegenthaler, and B. Stauffer. 1984. Late glacial climate history from ice cores. Pp. 299–306 in *Climate processes and climate sensitivity*, edited by J. E. Hansen and T. Takahashi. Vol. 29 of Geophysical Monograph. Washington, DC: American Geophysical Union.

Peng, C. H., J. Guiot, and E. Van Campo. 1998. Estimating changes in terrestrial vegetation and carbon storage: Using palaeoecological data and models. *Quaternary Science Reviews* 17:719–735.

Petit, J. R., J. Jouzel, D. Raynaud, N. I. Barkov, J.-M. Barnola, I. Basile, M. Bender, J. Chappellaz, M. Davis, G. Delaygue, M. Delmotte, V. M. Kotlyakov, M. Legrand, V. Y. Lipenkov, C. Lorius, L. Pépin, C. Ritz, E. Saltzman, and M. Stievenard. 1999. Climate and atmospheric history of the past 420,000 years from the Vostok ice core, Antarctica. *Nature* 399:429–436.

Plattner, G.-K., F. Joos, T. F. Stocker, and O. Marchal. 2001. Feedback mechanisms and sensitivities of ocean carbon uptake under global warming. *Tellus* 53B:564–592.

Plattner, G.-K., F. Joos, and T. F. Stocker. 2002. Revision of the global carbon budget due to changing air-sea oxygen fluxes. *Global Biogeochemical Cycles* 16:1096, doi:10.1029/2001GB001746.

Porter, S. C. 2001. Chinese loess record of monsoon climate during the last glacial-interglacial cycle. *Earth Science Reviews* 54:115–128.

Prentice, I. C., G. D. Farquhar, M. J. Fasham, M. I. Goulden, M. Heimann, V. J. Jaramillo, H. S. Kheshgi, C. LeQuéré, R. J. Scholes, and D. W. R. Wallace. 2001. The carbon cycle and atmospheric $CO_2$. Pp. 183–237 in *Climate change 2001: The scientific basis (Contribution of Working Group I to the Third Assessment Report of the Intergovernmental Panel on Climate Change)*, edited by J. T. Houghton, Y. Ding, D. Griggs, M. Noguer, P. van der Linden, X. Dai, K. Maskell, and C. A. Johnson. Cambridge: Cambridge University Press.

Prokopenko, A. A., D. F. Williams, E. B. Karabanov, and G. K. Khursevich. 2001. Continental response to Heinrich events and Bond cycles in sedimentary record of Lake Baikal, Siberia. *Global and Planetary Change* 28:217–226.

Reid, G. C. 1997. Solar forcing and global climate change since the mid-17th century. *Climatic Change* 37:391–405.

Röthlisberger, R., R. Mulvaney, E. W. Wolff, M. Hutterli, M. Bigler, S. Sommer, and J. Jouzel. 2002. Dust and sea salt variability in central East Antarctica (Dome C) over the last 45 kyrs and its implications for southern high-latitude climate. *Geophysical Research Letters* 29, doi:10.1029/2002GL015186.

Rousseau, D. D., P. Antoine, C. Hatte, A. Lang, L. Zoller, M. Fontugne, D. Ben Othman, J. M. Luck, O. Moine, M. Labonne, I. Bentaleb, and D. Jolly. 2002. Abrupt millennial climatic changes from Nussloch (Germany) upper Weichselian eolian records during the last glaciation. *Quaternary Science Reviews* 21:1577–1582.

Ruddiman, W. F., and A. McIntyre. 1981. The North Atlantic Ocean during the last glaciation. *Palaeogeography, Palaeoclimatology, Palaeoecology* 35:145–214.

Sánchez Goñi, M. F., J. Turon, F. Eynaud, and S. Gendreau. 2000. European climatic response to millennial-scale changes in the atmosphere-ocean system during the last glacial period. *Quaternary Research* 54:394–403.

Sánchez Goñi, M. F., I. Cacho, J. Turon, J. Guiot, F. J. Sierro, J. Peypouquet, J. Grimalt, and N. J. Shackleton. 2002. Sinchroneity between marine and terrestrial responses to millennial scale climatic variability during the last glacial period in the Mediterranean region. *Climate Dynamics* 19:95–105.

Sarmiento, J. L., and R. Toggweiler. 1984. A new model for the role of the oceans in determining atmospheric $pCO_2$. *Nature* 308:621–624.

Sarnthein, M., K. Winn, S. J. A. Jung, J. C. Duplessy, L. Labeyrie, H. Erlenkeuser, and G. Ganssen. 1994. Changes in East Atlantic deepwater circulation over the last 30,000 years: Eight time slice reconstruction. *Paleoceanography* 99:209–268.

Schmittner, A., M. Yoshimori, and A. J. Weaver. 2002. Instability of glacial climate in a model of ocean-atmosphere-cryosphere system. *Science* 295:1489–1493.

Scholze, M., W. Knorr, and M. Heimann. 2003. Modelling terrestrial vegetation dynamics and carbon cycling for an abrupt climate change event. *The Holocene* 13:319–326.

Schwander, J., T. Sowers, J. Barnola, T. Blunier, B. Malaizé, and A. Fuchs. 1997. Age scale of the air in the summit ice: Implication for glacial-interglacial temperature change. *Journal of Geophysical Research* 102:19483–19494.

Severinghaus, J. P., and E. J. Brook. 1999. Abrupt climate change at the end of the last glacial period inferred from trapped air in polar ice. *Science* 286:930–934.

Severinghaus, J. P., T. Sowers, E. J. Brook, R. B. Alley, and M. L. Bender. 1998. Timing of abrupt climate change at the end of the Younger Dryas interval from thermally fractionated gases in polar ice. *Nature* 391:141–146.

Shackleton, N. J. 1977. Carbon-13 in Uvigerina: Tropical rainforest history and the equatorial Pacific carbonate dissolution cycles. Pp. 401–428 in *The fate of fossil fuel CO$_2$ in the ocean*, edited by N. R. Andersen and A. Malahoff. New York: Plenum.

———. 2000. The 100,000-year ice-age cycle identified and found to lag temperature, carbon dioxide, and orbital eccentricity. *Science* 289:1897–1902.

Shaviv, N. J., and J. Veizer. 2003. Celestial driver of Phanerozoic climate? *GSA Today* 13:4–10.

Shin, S., Z. Liu, B. Otto-Bliesner, E. C. Brady, J. E. Kutzbach, and S. P. Harrison. 2003. A simulation of the Last Glacial Maximum climate using the NCAR-CCSM. *Climate Dynamics* 20:127–151; doi:10.1007/s00382-002-0260-x.

Siegenthaler, U., and T. Wenk. 1984. Rapid atmospheric CO$_2$ variations and ocean circulation. *Nature* 308:624–626.

Sigman, D. M., and E. A. Boyle. 2000. Glacial/interglacial variations in atmospheric carbon dioxide. *Nature* 407:859–869.

Sitch, S., B. Smith, I. C. Prentice, A. Arneth, A. Bondeau, W. Cramer, J. O. Kaplan, S. Levis, W. Lucht, M. Sykes, K. Thonicke, and S. Venevsky. 2003. Evaluation of ecosystem dynamics, plant geography and terrestrial carbon cycling in the LPJ dynamic global vegetation model. *Global Change Biology* 9:161–185.

Stauffer, B., T. Blunier, A. Dällenbach, A. Indermühle, J. Schwander, T. F. Stocker, J. Tschumi, J. Chappellaz, D. Raynaud, C. U. Hammer, and H. B. Clausen. 1998. Atmospheric CO$_2$ and millennial-scale climate change during the last glacial period. *Nature* 392:59–62.

Stauffer, B., J. Flückiger, E. Monnin, J. Schwander, J.-M. Barnola, and J. Chappellaz. 2002. Atmospheric CO$_2$, CH$_4$ and N$_2$O records over the past 60,000 years based on the comparison of different polar ice cores. *Annals of Glaciology* 35:202–208.

Stephens, B. B., and R. F. Keeling. 2000. The influence of Antarctic sea ice on glacial-interglacial CO$_2$ variations. *Nature* 404:171–174.

Stocker, T. F. 2000. Past and future reorganisations in the climate system. *Quaternary Science Reviews* 19:301–319.

Stocker, T. F., and S. J. Johnsen. 2003. A minimum thermodynamic model for the bipolar seesaw. *Paleoceanography* (submitted).

Stocker, T. F., and A. Schmittner. 1997. Influence of CO$_2$ emission rates on the stability of the thermohaline circulation. *Nature* 388:862–865.

Tegen, I., and I. Fung. 1995. Contribution to the atmospheric mineral aerosol load from land surface modification. *Journal of Geophysical Research* 100:18707–18726.

Tegen, I., S. P. Harrison, K. E. Kohfeld, and S. Engelstaedter. 2003. Soil dust aerosol emission trends: How important is anthropogenic dust? *Science* (submitted).

Tinsley, B. A., and G. W. Deen. 1991. Apparent tropospheric response to MeV-GeV particle flux variations: A correlation via electrofreezing of supercooled water in high-level clouds? *Journal of Geophysical Research* 12:22283–22296.

Tzedakis, P. C., I. T. Lawson, M. R. Frogley, G. M. Hewitt, and R. C. Preece. 2002. Buffered tree population changes in a Quaternary refugium: Evolutionary implications. *Science* 247:2044–2047.

Van Campo, E., J. Guiot, and C. Peng. 1993. A data-based re-appraisal of the terrestrial carbon budget at the Last Glacial Maximum. *Global and Planetary Change* 8:189–201.

Wagner, G., D. M. Livingstone, J. Masarik, R. Muscheler, and J. Beer. 2001. Some results relevant to the discussion of a possible link between cosmic rays and the Earth's climate. *Journal of Geophysical Research* 106:3381–3387.

Watson, A. J., D. C. E. Bakker, A. J. Ridgwell, P. W. Boyd, and C. S. Law. 2000. Effect of iron supply on southern ocean $CO_2$ uptake and implications for glacial atmospheric $CO_2$. *Nature* 407:730–733.

Weaver, A. J., M. Eby, A. F. Fanning, and E. C. Wiebe. 1998. Simulated influence of carbon dioxide, orbital forcing and ice sheets on the climate of the Last Glacial Maximum. *Nature* 394:847–853.

Werner, M., I. Tegen, S. P. Harrison, K. E. Kohfeld, I. C. Prentice, Y. Balkanski, H. Rodhe, and C. Roelandt. 2002. Seasonal and interannual variability of the mineral dust cycle under present and glacial climate conditions. *Journal of Geophysical Research-Atmospheres* 107 (D24): art. no. 4744.

Wyputta, U., and B. J. McAvaney. 2002. Influence of vegetation changes during the last glacial maximum using the BMRC atmospheric general circulation model. *Climate Dynamics* 17:923–932.

Ye, W., G. R. Dong, Y. J. Yuan, and Y. J. Ma. 2000. Climate instability in the Yili region, Xinjiang during the last glaciation. *Chinese Science Bulletin* 45:1604–1609.

Yoshimori, M., M. Reader, A. Weaver, and N. McFarlane. 2001. On the causes of glacial inception at 116 ka BP. *Climate Dynamics*, doi:10.1007/s00382-001-0186-8.

Yu, Z., and U. Eicher. 1998. Abrupt climate oscillations during the last deglaciation in Central North America. *Science* 282:2235–2238.

# 8

# Spatial and Temporal Distribution of Sources and Sinks of Carbon Dioxide

Martin Heimann, Christian Rödenbeck, and Manuel Gloor

Ever since the importance of the $CO_2$ gas for the radiative balance of the atmosphere was recognized, the mechanisms, both natural and anthropogenic, that control its concentration have been an active area of study. The possibility of a direct anthropogenic influence on atmospheric $CO_2$ concentration from the emissions of $CO_2$ from fossil-fuel burning and cement production was discussed by Arrhenius more than 100 years ago (Arrhenius 1896, 1903). Interestingly, Arrhenius predicted an atmospheric increase of only about 6 parts per million (ppm) over the 20th century, contrasting strongly with the 75 ppm that have been reconstructed from ice core measurements. Arrhenius's low prediction was partly due to assumed constant anthropogenic emissions of only 0.7 petagrams of carbon per year ($PgC\ y^{-1}$). In addition, he assumed perfect equilibration between the atmosphere and the entire ocean, hence neglecting the finite time needed to mix the anthropogenic excess $CO_2$ into the interior of the ocean.

Reliable atmospheric measurements of $CO_2$ in the Southern and Northern Hemispheres were started about 45 years ago (Keeling 1960). Early calculations with simple ocean box models calibrated against observations of natural and bomb-produced radiocarbon showed that a significant fraction of the anthropogenic $CO_2$ is taken up by carbon sinks, primarily in the oceans and possibly also on land (Keeling 1973; Oeschger et al. 1975). Indeed, in the late 1970s, it appeared that the global fossil-fuel carbon emissions were almost balanced by the atmospheric increase and the calculated ocean uptake (Siegenthaler and Oeschger 1978). Thereafter, the recognition of significant additional emissions from changes in land use, in particular deforestation in the tropics (Woodwell et al. 1978), started the quest for the "missing sink" to balance the global budget. Additional information from ice cores on the variability of $CO_2$ in the past and the evidence of changes in the atmospheric stable isotope ratios subsequently demonstrated

187

that changes in terrestrial sources and sinks also play a role and may constitute a sizable fraction of the "missing sink."

Since this first phase of carbon cycle research, many new techniques and observations have provided a wealth of heterogeneous information on the global carbon cycle. Many of the different identified sources and sinks of atmospheric $CO_2$ remain relatively elusive, however, and the processes that control them, especially the terrestrial exchange fluxes, are still poorly understood. In order to improve this knowledge, many recent efforts have been directed at estimating the regional patterns of carbon exchange and their variability, to better identify and quantify individual source processes and to assess their driving mechanisms. The establishment of the Kyoto Protocol has added a second motivation for the regional quantification of sources and sinks of $CO_2$. Independent and scientifically credible estimates of regional carbon balances must be established to corroborate carbon source and sink estimates as reported by individual countries.

In the sections that follow, we briefly summarize the existing methods to establish regional carbon budgets, comment on the global carbon balance, look at the direct anthropogenic imprint, and present a summary of regional estimates based on a recent inversion study (Rödenbeck et al. 2003).

# Methods

Generalizing, one may distinguish four main approaches to estimating regional distribution and variability of surface-air carbon fluxes, which, in practice, may also be combined in various ways.

## *Atmospheric Inverse Modeling*

During the 1980s atmospheric monitoring programs operated by several agencies started to expand and provided an improved spatial and temporal characterization of the global-scale atmospheric $CO_2$ concentration variations, which reflect the large-scale distribution of sources and sinks at the surface. The determination of the latter necessitates a model of the atmospheric transport that must be run in an "inverse" mode. This approach is commonly referred to as the "top-down" approach.

Most conspicuous was the recognition of a significant sink in the northern mid- to high latitudes inferred from observations of the meridional gradient of atmospheric $CO_2$ using two- and three-dimensional atmospheric transport models (Pearman and Hyson 1980; Keeling et al. 1989; Tans et al. 1989, 1990).

The substantial extension of the monitoring network during the 1990s, together with improved atmospheric models and mathematical methods, has led to the establishment of atmospheric inverse modeling as a means to infer location, magnitude, and temporal variability of surface sources and sinks of $CO_2$ (Enting et al. 1995; Kaminski et al. 1999; Rayner et al. 1999; Bousquet et al. 2000; Gurney et al. 2002; Rödenbeck et al. 2003).

## Process-Based Surface Modeling

Parallel to the advancement of the top-down approach has been the development of "bottom-up," spatially explicit, process-based models for the ocean initiated by Maier-Reimer and Hasselmann (1987) (see also Greenblatt and Sarmiento, Chapter 13, this volume, for a summary of the state of the art up to now) and the terrestrial biosphere initiated by Esser (1987) (see also McGuire et al. 2001 for a summary of the state of the art up to now). These models simulate the spatial and temporal evolution of sources and sinks as driven by the key natural and anthropogenic drivers of the carbon cycle: atmospheric $CO_2$ concentration, observed variations of climate (temperature, precipitation, surface winds, radiation, buoyancy fluxes, etc.), and historical patterns of land use. The most comprehensive, global simulation experiments to date have been undertaken through the Ocean Carbon Model Intercomparison Project (OCMIP; Orr et al. 2001), and the Carbon Cycle Model Linkage Project (CCMLP; McGuire et al. 2001).

## Extrapolations of In Situ Observations

In addition to the top-down and bottom-up modeling, several key datasets of regional and global extent have been painstakingly compiled. These also permit an assessment of location, magnitude, and variability of surface sources and sinks. Air-sea $CO_2$ partial pressure difference measurements allow the estimation of local air-sea fluxes (e.g., Takahashi et al. 2002). Carbon inventory changes, observed from repeated oceanic surveys or inferred by back-calculations of in situ inorganic carbon concentration measurements, allow basinwide estimates of ocean carbon uptake rates (e.g., Gruber 1998; Sabine et al. 1999, 2002; Gloor et al. 2003). On land, repeated forest inventories from national forestry statistics allow an estimation of changing carbon stocks in living biomass. Direct flux measurements above the vegetation by the eddy correlation technique provide in situ net carbon exchanges, and remote sensed information on land cover and land properties (e.g., Normalized Divided Vegetation Index) may be related to carbon fluxes. All these direct observations of carbon fluxes or carbon stock changes may, in principle, be used to estimate regional-scale carbon fluxes by means of scaling and extrapolation techniques (e.g., Pacala et al. 2001; Janssens et al. 2003; Papale and Valentini 2003).

## Carbon Cycle Tracers

In addition to the net carbon flux of interest, additional information can be extracted from tracers that are tightly coupled to the carbon cycle. Among these are carbon and oxygen isotope tracers and measurements of changes in the atmospheric oxygen content. Observations of these tracers in the atmosphere pose global and regional constraints on the contributions of differently "labeled" surface-air carbon fluxes. For example, radiocarbon may be used to track fossil fuel $CO_2$ (Levin and Kromer 1997), and the $^{13}C/^{12}C$ stable isotope ratio, at least on the global scale, provides constraints on terrestrial and oceanic $CO_2$ (Ciais et al. 1995; Francey et al. 1995; Keeling et al. 1995, Piper et al. 2001). Observations of the concurrent trends in the $O_2/N_2$ ratio and $CO_2$ permit a separation of oceanic and terrestrial carbon uptake (Keeling and Shertz 1992; Keeling et al. 1993, 1996; Battle et al. 2000).

## Consistency

The quantitative reconciliation of estimates by the different methods outlined for a continental region, such as Europe (Janssens et al. 2003) or the United States (Pacala et al. 2001), or for ocean basins, such as the North Atlantic or the equatorial Pacific Ocean, is difficult. One major obstacle arises from the fact that different methods, in general, measure different fluxes or carbon flows between the major reservoirs. For example, the top-down approach estimates the net surface-air $CO_2$ flux in a particular region, whereas an estimate for the same region based on extrapolating eddy flux measurements will not necessarily take into account disturbance factors (fire, harvest) or lateral carbon flows in or out of the target area (e.g., by wood products or carbon transport through rivers). Similarly, net ocean-atmosphere fluxes determined by the top-down approach will not directly correspond to an ocean carbon stock inventory change, since the net sea-air flux also includes a carbon flow induced by land-ocean transport of carbon through rivers (Sarmiento and Sundquist 1992). A second difficulty is induced by the varying timescales of the different approaches, necessitating averaging or interpolations in time for a consistent comparison.

## The Carbon Budget and Global Partitioning between Atmosphere, Land, and Ocean

The global fossil-fuel emissions of $CO_2$ and the net carbon balances of the atmosphere, land, and ocean over the past two decades are now rather well established through several independent observation and model-based methods (Prentice et al. 2002). The estimated accuracy of the net land and ocean uptake is on the order of 0.5–0.8 PgC y$^{-1}$. Less accurate is the quantification of individual components of the terrestrial carbon balance, in particular the emissions from changes in land use and land management and the corresponding uptake by natural terrestrial processes, such as fertilization by $CO_2$ or nutrients and changes in climate. The main issues in reconciling estimates of these different components have been assessed by House et al. (2003). An update on the global carbon budget is provided by Sabine et al. (Chapter 2, this volume).

From a scientific as well as an observational viewpoint, it is also important to determine the temporal variability underlying these global, decadal average quantities in the carbon budget. A recent compilation by Le Quéré et al. (2003) compares the various top-down and bottom-up methods for the ocean uptake. By difference from the atmospheric increase and estimates of the annual fossil-fuel emissions, the net land-atmosphere flux can also be inferred. Colorplate 7 shows the resulting interannual variability for the global ocean and terrestrial budgets (displayed as anomalies), using the different methods. In the panels on the right side, the time series have been smoothed to show only the decadal trends.

The main conclusion from this global analysis is that the terrestrial interannual and decadal variability is about five times larger than the variability of the oceanic fluxes. This result implies that the main interannual features in the atmospheric record, which are clearly correlated with large-scale climatic variations, such as events related to El Niño–Southern Oscillation and the climate anomaly after the Pinatubo volcanic eruption, are primarily of terrestrial origin.

## The Anthropogenic Perturbation

The direct anthropogenic imprint on the global carbon cycle through the emissions of $CO_2$ from fossil-fuel use and cement production is relatively well characterized based on energy production and energy use statistics (Andres et al. 2000). In addition to increasing the global concentration of $CO_2$, the anthropogenic perturbation has a direct effect on atmospheric $CO_2$ through the generation of a concentration gradient between the two hemispheres. One of the most compelling arguments for an anthropogenic cause of the current rise in atmospheric $CO_2$ is provided by the observations of this gradient from the two longest direct atmospheric records available: Mauna Loa station on Hawaii and the South Pole (Keeling and Whorf 2003). Both stations are reasonably representative for their respective hemispheres. Plotting this concentration difference against the difference in fossil-fuel emissions in the two hemispheres shows a remarkable linearity over the full range of the 41-year record (Figure 8.1). This behavior is exactly what would be expected if all non-fossil-fuel $CO_2$ sources and sinks were distributed uniformly over the entire earth (Keeling et al. 1989; Fan et al. 1999). The scatter of the points around the regression line indicates the level of variability of the interhemispheric asymmetry of the non-fossil-fuel sources. If most of the interannual variability is caused by terrestrial sources, the scatter reflects the Northern Hemisphere, extra-tropical, terrestrial sources. Tropical sources, because of their approximate meridional symmetry, would not strongly affect the interhemispheric concentration difference. A quantitative analysis, however, requires the use of an inverse modeling framework (to be discussed later in this chapter).

Extrapolating the concentration difference back to zero emissions leaves a small offset, with the South Pole concentration about 0.6 ppm higher than at Mauna Loa. The origin of this offset, originally noticed by Bacastow and Keeling (1981), is not clear. It may be the result of an interhemispherically asymmetric configuration of natural sources and sinks, or it may be caused by rectification effects generated through sea-

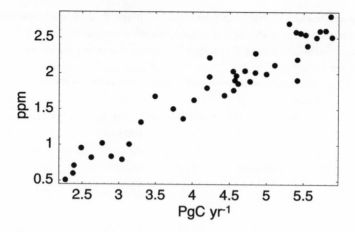

**Figure 8.1.** Annual mean atmospheric $CO_2$ concentration difference between the Mauna Loa and the South Pole stations (*y* axis, observations from Keeling et al. 1995, updated) shown against interhemispheric difference in annual fossil-fuel emissions (*x* axis, data from Andres et al. 2000, updated).

sonally varying transport patterns and seasonally varying sources (Hyson et al. 1980; Heimann et al. 1986; Keeling et al. 1989; Denning et al. 1995; Fan et al. 1999).

## Top-Down Regional Estimates

### General Considerations

Inferring the distribution of sources and sinks of $CO_2$ and their temporal variations in a consistent way from atmospheric concentration observations by means of a model of atmospheric transport constitutes an inverse problem of considerable complexity. Early attempts to address this problem were made as soon as the first reliable atmospheric $CO_2$ observations became available (Bolin and Keeling 1963). At that time, however, comprehensive meteorological observations of the tropospheric transport were not available, hence the atmospheric concentration measurements were also used to deduce the large-scale strength of meridional atmospheric mixing, in addition to the carbon sources.

With the increasing density of the observation network after 1980 and the emergence of three-dimensional atmospheric transport models based on realistic meteorological data from climate models or weather forecast models (Fung et al. 1983), it became possible to address the inversion problem in more detail. A comprehensive description of the mathematical methods involved has been given by Enting (2002). In general, the domain of interest is split into a series of spatiotemporal source patterns. Using the atmospheric transport model, a "base function" is computed for each of these source patterns, defined as the atmospheric response at the observation locations and times, subsequent to a unit input emitted from the source pattern. Finally, a linear combination of the base functions is determined that optimally matches the measurements at the observation points. The determined weights of the base functions then constitute the source strengths of the different source patterns.

The inversion problem raises a series of important difficulties:

1. Current atmospheric transport models are not perfect.
2. The observational network is very sparse—that is, there exist only ≈100 monitoring stations worldwide, a small number compared with the heterogeneity of the terrestrial or oceanic carbon sources. Furthermore, at some stations the sampling frequency is low, and there are often gaps in the observations.
3. Technically, the "inversion" of the atmospheric transport model is not trivial and requires much larger computing resources than running the model in the forward mode.
4. Individual measurements are often not representative of the appropriate temporal and spatial scale of the transport model.
5. Individual concentration observations are of limited accuracy and precision, and observations from different monitoring networks are often not easily comparable because of differences in measurement techniques and uses of different standards.

The most serious limitation of the top-down approach follows from the limited number of observations and the need to adequately represent sources and sinks with a relatively high spatial and temporal resolution. Indeed, a very large number of possible surface source-sink configurations are in principle consistent with the atmospheric observations. An example is provided by Kaminski and Heimann (2001), which demonstrates that an unrealistic sink of 2 PgC $y^{-1}$ over Europe may be perfectly compatible with the entire observations from the global monitoring station network. The atmospheric observations alone are thus not sufficient to uniquely determine the sources at the surface of the Earth.

Formally, the atmospheric inversion problem constitutes an ill-posed mathematical problem, which must be regularized by means of a priori information on spatial and temporal variability of sources and sinks including a priori estimates of their uncertainty. This prior information may be used to define a smaller number of spatiotemporally more complex base functions in order to solve for fewer unknowns compared with the number of observations (Enting et al. 1995; Fan et al. 1998; Rayner et al. 1999; Bousquet et al. 2000; Gurney et al. 2002). Alternatively, one may perform a high-resolution inversion by using the prior information in a Bayesian approach to regularize the problem (Kaminski et al. 1999; Rödenbeck et al. 2003).

To assess the impact of the different atmospheric transport models and the various possible methodologies on the determination of sources and sinks of $CO_2$, the international project TRANSCOM was established with support by the Global Analysis,

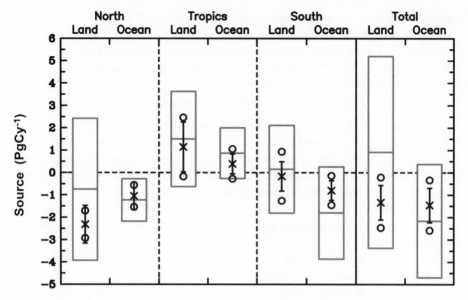

**Figure 8.2.** Annual mean latitudinal land-ocean breakdown of non-fossil-fuel carbon sources as determined in the TRANSCOM inversion intercomparison study of Gurney et al. 2002. Gray boxes: a priori fluxes and uncertainties. Crosses: inferred mean fluxes, average of 16 models or model variants. Vertical error bars indicate between-model variations (1 sigma); open circles show the mean estimated uncertainty across all models. Fluxes are representative for the five-year interval 1992–1996, determined from 76 atmospheric stations from the Cooperative Atmospheric Data Integration Project (2000) database.

Integration and Modeling task force of the International Geosphere-Biosphere Program. Several carefully defined model intercomparison studies have been conducted in TRANSCOM, including forward simulations of the terrestrial seasonal cycle and the fossil-fuel $CO_2$, (Law et al. 1996) and of the inert, anthropogenic tracer $SF_6$ (Denning et al. 1999). Later intercomparison experiments have explored the impact of different transport models based on annual mean global inversions (Gurney et al. 2002). As an illustration, Figure 8.2 shows the meridional land-ocean breakdown of the net non-fossil-fuel $CO_2$ sources aggregated in three global latitude bands separated at 30°N and 30°S, inferred by the 16 transport models or model variants included in the study of Gurney et al. (2002). These fluxes are representative for the five-year period 1992–1996. The light gray boxes show the specified a priori sources and their uncertainty (± 1 standard deviation). The crosses indicate the mean of the different model inversions performed using the observations over 1992–1996 from 76 stations included in the GLOBALVIEW-2000 data set (Cooperative Atmospheric Data Integration Project 2000). The vertical error bars reflect one standard deviation of the spread of the model averages ("between-model" error), while the circles indicate the average of the individual posterior source errors estimated by the different models (average "within-model" error). The Northern Hemisphere estimates exhibit substantial $CO_2$ uptake, especially on land, corroborating the existence of the northern extra-tropical $CO_2$ sink inferred by the earlier studies. The tropics, in particular the tropical land areas, are much less well constrained by the inversions, reflecting the poor atmospheric station coverage, but also a relatively larger spread of the simulated transport by the models in these regions as shown by the larger "between-model" error. Overall, tropical land and oceans appear as $CO_2$ sources in the inversion. The inferred estimates of the Southern Ocean fluxes are consistently less negative than the specified prior fluxes. This finding indicates a bias in air-sea flux field of Takahashi et al. (2002), which was used as prior ocean flux estimate in the inversion. The study of Gurney et al. (2002) represents the most comprehensive assessment of the impact of different atmospheric transport models on atmospheric inversions. On the other hand, the employed methodology using a time-independent mean inversion with a limited number of large-scale regions severely reduces the number of possible inversion solutions, as already discussed (see also the discussion of the aggregation error in global inversions by Kaminski et al. 2001). Therefore the error estimates shown in Figure 8.2 most likely are too optimistic.

## Results from a Time-Dependent, High-Resolution Global Inversion

In the following, we illustrate the current large-scale pattern of $CO_2$ sources and sinks and its variability, using results from a recent inversion study (Rödenbeck et al. 2003). In this study, special care was taken to address several of the shortcomings of previous studies. The atmospheric transport is based on the time-varying meteorology from the National Centers for Environment Prediction (NCEP) analyses over the past 20 years. This approach allows exact time synchronization between model simulation and observations by sampling of the model at the same time and location as the observations.

Atmospheric stations were selected for long-term continuous homogeneity of the records in order to eliminate spurious changes induced by data gaps. Because the density of observing stations has increased significantly over the past 20 years, four main inversions were run with different numbers of stations and correspondingly different inversion periods. Comparisons between these inversions over the overlapping time intervals permit an assessment of the robustness of large-scale source features with respect to the density of observation stations. The inversion used the high-resolution approach of Kaminski et al. (1999) by determining monthly sources on a global grid of approximately 8° latitude by 10° longitude, while the atmospheric transport is simulated using the TM3 transport model (Heimann and Körner 2003) with a resolution of approximately 4° latitude by 5° longitude and 19 layers in the vertical dimension.

Over the oceans, the prior information on the sources was specified from the large-scale patterns of the air-sea fluxes inferred from oceanic data by Gloor et al. (2003), modified regionally by the observed seasonal pattern (Takahashi et al. 2002). Over land the a priori source pattern was specified from the mean seasonal cycle of net $CO_2$ fluxes as simulated by the Lund-Potsdam-Jena (LPJ) terrestrial biosphere model (Sitch et al. 2000). In addition, the fossil-fuel emission distribution was specified from the EDGAR database (Olivier and Berdowski 2001). The atmospheric observations were obtained exclusively from the U.S. National Oceanographic and Atmospheric Administration-Climate Monitoring and Diagnostics Laboratory network (Conway and Tans 1999) to eliminate possible biases caused by unknown calibration offsets between networks.

Colorplate 8 displays the net surface-air flux averaged over July 1995–June 2000, inferred from an inversion with 35 stations (shown as black triangles). Shown is the total surface-atmosphere $CO_2$ flux including land, ocean, and fossil-fuel emissions (a) and the ocean and terrestrial fluxes only (b). Clearly, the total surface-atmosphere flux is dominated by the pattern of fossil-fuel emissions from the three major emission regions in the Northern Hemisphere: eastern North America, Europe, and eastern Asia. The mean pattern of the natural sources shows the expected $CO_2$ outgassing in the equatorial oceans, especially in the Pacific, the uptake in the North Atlantic Ocean, and a weak source in and around the Antarctic. In the tropics the terrestrial carbon balance exhibits sinks in the central tropical regions in Amazonia, Africa, and Indonesia.

The northern extra-tropics are dominated by a conspicuous source region in Eastern Europe and moderate sinks elsewhere. This pattern reflects the average over five years and may not be entirely representative of the longer-term source-sink distribution. In particular, the source in Eastern Europe may be a transient phenomenon. It is primarily generated by the stations in Hungary and at the Black Sea coast. It may, however, also reflect biases in the model transport or too-small industrial emissions estimates specified for this area. The inferred patterns near industrial areas may also be biased by neglecting the influence of carbon monoxide from fossil-fuel burning, which may be more significant on regional than on global scales (Enting and Mansbridge 1991). Interestingly, the European source pattern also appears in regional inversions for Europe based on the much denser regional observation network of the AEROCARB project (Rivier et al. 2003).

It must be realized that the impressive regional details shown in Colorplate 8a and b are not determined by the atmospheric data but are to a large extent provided by the a priori flux fields. The extent to which this flux field is modified by the inversion is indicated by Colorplate 8c, showing the error reduction defined as:

$$\text{error reduction} = (\text{posterior error} - \text{prior error})/(\text{posterior error})$$

and is expressed in percentage (Kaminski et al. 1999). The map in Colorplate 8c demonstrates the sobering fact that the employed network of stations reduces the error over most regions of the globe by only a modest amount. Only in the regions of the footprint of the monitoring stations does the error reduction reach values above 30 percent. Even though the posterior errors are partly correlated the error reduction increases when integrating over larger areas (Rödenbeck et al. 2003). Nevertheless, the small error reduction achievable with the present in situ observation network clearly demonstrates the need to enhance the measurements in undersampled areas and to develop methods to measure the $CO_2$ concentration from space (Houweling et al. 2003).

As already noted in the inversion study by Bousquet et al. (2000), the longer-term time-averaged source-sink patterns are less robustly determined than the temporal variability. The temporal pattern of the non-fossil-fuel sources and sinks averaged over five continental areas and three oceanic latitude bands is given in Figure 8.3. The different line styles indicate different inversion setups, with the shorter records including a larger number of atmospheric stations. The gray shading indicates time intervals of high values of the Multivariate El Niño Southern Oscillation Index (MEI) (Wolter and Timlin 1993). The prior fluxes included in these inversions do not contain any time dependence except the climatological seasonal cycle. Hence, the temporal trends resulting from the inversion are induced entirely by the temporal variability in the atmospheric concentration records. The variability of the terrestrial fluxes is clearly dominated by the fluxes from South America, which also show the strongest correlation with the ENSO climate fluctuation. The oceanic fluxes in the equatorial region, dominated by the Pacific Ocean, exhibit the expected reduced outgassing during warm El Niño periods.

## Concluding Remarks

The great challenge is relating top-down estimates to processes at the surface. On the one hand, this correlation would provide confidence in the inversion results, and on the other hand, it would potentially help to quantify individual source flux mechanisms. Such analyses, however, have not yet been performed in a comprehensive way. One approach is based on trying to identify surface variables that potentially drive carbon fluxes. Examples include features that can be seen with remote sensing, such as vegetation indexes, fire counts, or climate variables. As an example, we show in Figure 8.4 time series of the monthly anomalous net non-fossil-fuel $CO_2$ flux for two regions, as inferred from the atmospheric inversion described above (solid line), and compare it with fire counts detected from space, aggregated over the same spatial region (dashed line, arbitrary scale). The degree of correspondence between the time series indicates that fire is closely linked to anomalous $CO_2$ fluxes from the terrestrial biosphere. Because this is not a quantitative comparison, the correspondence may mean that similar driving conditions exist for enhanced $CO_2$ fluxes and for fires. Alternatively, the flux anomaly may be caused by the fires themselves, as inferred from other studies (Langenfelds et al. 2002). Indeed a preliminary comparison with modeled carbon emissions from fire (van der Verf et al. 2003) shows that these may account for up to 50 percent of the anomalies inferred by the inversion (Rödenbeck et al. 2003).

Regional comparisons of top-down fluxes with bottom-up model simulation results have shown encouraging results in particular areas (Bousquet et al. 2000; Janssens et al. 2003). As soon as carbon flux balances of smaller regions (such as Europe) are

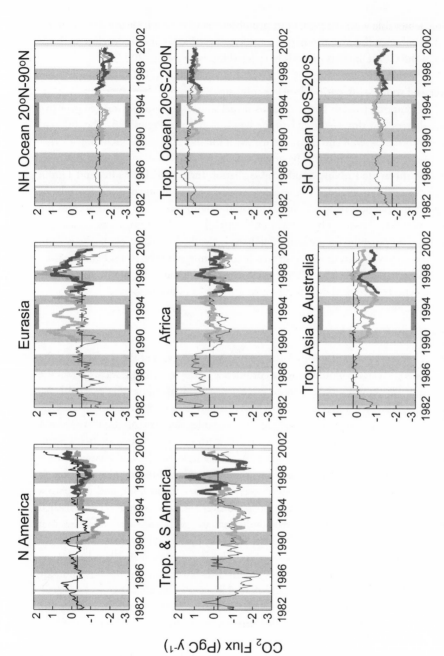

**Figure 8.3.** Time series of surface-atmosphere fluxes integrated over five continental areas (left) and over three oceanic latitudinal bands (right column) (Rödenbeck et al. 2003). The three curves reflect inversions using a different number of observing stations (black: 11 stations, light gray: 19 stations, dark gray: 35 stations). Units: PgC y[-1].

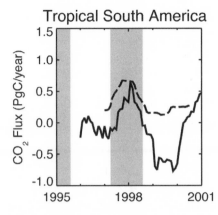
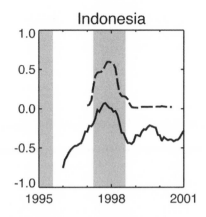

**Figure 8.4.** Time series of anomalous non-fossil-fuel net land-atmosphere flux from the inversion (solid line, PgC y$^{-1}$) compared with fire counts compiled by the European Space Agency (dashed line, arbitrary scale), both aggregated over two continental areas (Rödenbeck et al. 2003).

addressed, however, additional carbon fluxes not normally simulated by the ecosystem process models must be included. These include the source-sink patterns of lateral carbon transports by carbon-containing trade products (Tschirley and Servin, Chapter 21, this volume) and through erosion and rivers. In addition, air-sea carbon fluxes in marginal seas and from ocean shelves have to be taken into account (Chen, Chapter 18, this volume). It still remains to be shown that the planned efforts for intense carbon cycle observations in particular regions (e.g., North America [North American Carbon Observation Plan; Wofsy and Harris 2002] or Europe [CarboEurope; Schulze 2003]), together with detailed regional atmospheric modeling, will indeed provide consistent patterns of carbon sources and sinks inferred by the different approaches. Ultimately, the multitude of different data streams on the state of the carbon cycle from the atmosphere, oceans, and terrestrial surface might be merged into a regional or global data assimilation system, similar to the ones employed in current weather forecast models. The development of such a system, however, constitutes a huge challenge for the coming years.

## Literature Cited

Andres, R. J., G. Marland, T. Boden, and S. Bischof. 2000. Carbon dioxide emissions from fossil fuel consumption and cement manufacture, 1751–1991, and an estimate of their isotopic composition and latitudinal distribution. Pp. 53–62 in *The carbon cycle*, edited by T. M. L. Wigley and D. S. Schimel. Cambridge: Cambridge University Press.
Arrhenius, S. 1896. On the influence of carbonic acid in the air upon the temperature of

the ground. *The London, Edinburgh and Dublin Philosophical Magazine and Journal of Sciences* 41:237–276.

———. 1903. *Lehrbuch der kosmischen Physik.* Leipzig: Hirzel.

Bacastow, R. B., and C. D. Keeling. 1981. Atmospheric carbon dioxide concentration and the observed airborne fraction. Pp. 103–112 in *Carbon cycle modelling,* edited by B. Bolin. SCOPE 16. Chichester, UK: John Wiley and Sons.

Battle, M., M. L. Bender, P. P. Tans, J. W. C. White, J. T. Ellis, T. Conway, and R. J. Francey. 2000. Global carbon sinks and their variability inferred from atmospheric $O_2$ and $\delta^{13}C$. *Science* 287 (5462): 2467–2470.

Bolin, B., and C. D. Keeling. 1963. Large-scale atmospheric mixing as deduced from the seasonal and meridional variations of carbon dioxide. *Journal of Geophysical Research* 68:3899–3920.

Bousquet, P., P. Peylin, P. Ciais, C. Le Quéré, P. Friedlingstein, and P. P. Tans. 2000. Regional changes in carbon dioxide fluxes of land and oceans since 1980. *Science* 290 (5495): 1342–1346.

Ciais, P., P. P. Tans, M. Trolier, J. W. C. White, and R. J. Francey. 1995. A large northern-hemisphere terrestrial $CO_2$ sink indicated by the $^{13}C/^{12}C$ ratio of atmospheric $CO_2$. *Science* 269 (5227): 1098–1102.

Conway, T. J., and P. P. Tans. 1999. Development of the $CO_2$ latitude gradient in recent decades. *Global Biogeochemical Cycles* 13 (4): 821–826.

Cooperative Atmospheric Data Integration Project–Carbon Dioxide. 2000. GLOBALVIEW-$CO_2$. Boulder, CO: National Oceanic and Atmospheric Administration CMDL. CD-ROM (also available on Internet via anonymous FTP to ftp.cmdl.noaa.gov, Path: ccg/co2/GLOBALVIEW).

Denning, A. S., I. Y. Fung, and D. Randall. 1995. Latitudinal gradient of atmospheric $CO_2$ due to seasonal exchange with land biota. *Nature* 376 (6537): 240–243.

Denning, A. S., M. Holzer, K. R. Gurney, M. Heimann, R. M. Law, P. J. Rayner, I. Y. Fung, S. M. Fan, S. Taguchi, P. Friedlingstein, Y. Balkanski, J. Taylor, M. Maiss, and I. Levin. 1999. Three-dimensional transport and concentration of SF6: A model intercomparison study (TransCom 2). *Tellus* 51B (2): 266–297.

Enting, I. G. 2002. *Inverse modeling of atmospheric constituents.* Cambridge: Cambridge University Press.

Enting, I. G., and J. V. Mansbridge 1991. Latitudinal distribution of sources and sinks of $CO_2$: Results of an inversion study. *Tellus* 43B (2): 156–170.

Enting, I. G., C. M. Trudinger, and R. J. Francey. 1995. A synthesis inversion of the concentration and $^{13}C$ of atmospheric $CO_2$. *Tellus* 47B:35–52.

Esser, G. 1987. Senstitivity of global carbon pools and fluxes to human and potential climatic impacts. *Tellus* 39B:245–260.

Fan, S., M. Gloor, J. Mahlman, S. Pacala, J. Sarmiento, T. Takahashi, and P. Tans. 1998. A large terrestrial carbon sink in North America implied by atmospheric and oceanic carbon dioxide data and models. *Science* 282 (5388): 442–446.

Fan, S. M., T. L. Blaine, and J. L. Sarmiento. 1999. Terrestrial carbon sink in the Northern Hemisphere estimated from the atmospheric $CO_2$ difference between Mauna Loa and the South Pole since 1959. *Tellus* 51B (5): 863–870.

Francey, R. J., P. P. Tans, C. E. Allison, I. G. Enting, J. W. C. White, and M. Trolier. 1995. Changes in oceanic and terrestrial carbon uptake since 1982. *Nature* 373 (6512): 326–330.

Fung, I., K. Prentice, E. Matthews, J. Lerner, and G. Russell. 1983. 3-dimensional tracer model study of atmospheric $CO_2$: Response to seasonal exchanges with the terrestrial biosphere. *Journal of Geophysical Research–Oceans and Atmospheres* 88 (NC2): 1281–1294.

Gloor, M., N. Gruber, J. Sarmiento, C. L. Sabine, R. A. Feely, and C. Rödenbeck. 2003. A first estimate of present and preindustrial air-sea $CO_2$ flux patterns based on ocean interior carbon measurements and models. *Geophysical Research Letters* 30, doi:10.1029/2002GL015594.

Gruber, N. 1998. Anthropogenic $CO_2$ in the Atlantic Ocean, 1998. *Global Biogeochemical Cycles* 12:165–191.

Gurney, K. R., R. M. Law, A. S. Denning, P. J. Rayner, D. Baker, P. Bousquet, L. Bruhwiler, Y. H. Chen, P. Ciais, S. Fan, I. Y. Fung, M. Gloor, M. Heimann, K. Higuchi, J. John, T. Maki, S. Maksyutov, K. Masarie, P. Peylin, M. Prather, B. C. Pak, J. Randerson, J. Sarmiento, S. Taguchi, T. Takahashi, and C. W. Yuen. 2002. Towards robust regional estimates of $CO_2$ sources and sinks using atmospheric transport models. *Nature* 415 (6872): 626–630.

Heimann, M., and S. Körner. 2003. *The global atmospheric tracer model TM3: Model description and user's manual, release 3.8a.* Technical Report No. 5. Jena, Germany: Max-Planck-Institute for Biogeochemistry.

Heimann, M., C. D. Keeling, and I. Y. Fung. 1986. Simulating the atmospheric carbon dioxide distribution with a three-dimensional tracer model. Pp. 16–49 in *The changing carbon cycle: A global analysis,* edited by J. Trabalka and D. E. Reichle. New York: Springer Verlag.

House, J. I., I. C. Prentice, N. Ramankutty, R. A. Houghton, and M. Heimann. 2003. Reconciling apparent inconsistencies in estimates of terrestrial $CO_2$ sources and sinks. *Tellus* 55B:345–363.

Houweling, S., F.-M. Breon, I. Aben, C. Rödenbeck, M. Gloor, M., Heimann, and P. Ciais. 2003. Inverse modeling of $CO_2$ sources and sinks using satellite data: A synthetic inter-comparison of measurement techniques and their performance as a function of space and time. *Atmospheric Chemistry and Physics* Discussion 3, 5237–5274.

Hyson, P., P. J. Fraser, and G. I. Pearman. 1980. A two-dimensional transport simulation-model for trace atmospheric constituents. *Journal of Geophysical Research–Oceans and Atmospheres* 85 (NC8): 4443–4455.

Janssens, I. A., A. Freibauer, P. Ciais, P. Smith, G. J. Nabuurs, G. Folberth, B. Schlamadinger, R. W. A. Hutjes, R. Ceulemans, E. D. Schulze, R. Valentini, and A. J. Dolman. 2003. Europe's terrestrial biosphere absorbs 7 to 12% of European anthropogenic $CO_2$ emissions. *Science* 300 (5625): 1538–1542.

Kaminski, T., and M. Heimann. 2001. Inverse modeling of atmospheric carbon dioxide fluxes. *Science* 294 (5541): 259.

Kaminski, T., M. Heimann, and R. Giering. 1999. A coarse grid three-dimensional global inverse model of the atmospheric transport. 2. Inversion of the transport of $CO_2$ in the 1980s. *Journal of Geophysical Research–Atmospheres* 104 (D15): 18555–18581.

Kaminski, T., P. J. Rayner, M. Heimann, and I. G. Enting. 2001. On aggregation errors in atmospheric transport inversions. *Journal of Geophysical Research–Atmospheres* 106 (D5): 4703–4715.

Keeling, C. D. 1960. The concentration and isotopic abundances of carbon dioxide in the atmosphere. *Tellus* 12:200–203.

————. 1973. Carbon cycle. Pp. 251–329 in *Chemistry of the lower atmosphere,* edited by S. I. Rasool. New York: Plenum Press.

Keeling, R. F., and S. R. Shertz. 1992. Seasonal and interannual variations in atmospheric oxygen and implications for the global carbon cycle. *Nature* 358:723–727.

Keeling, C. D., and T. P. Whorf. 2003. Monthly and annual $CO_2$ concentration observed at Manua Loa and the South Pole. Oak Ridge, TN: Carbon Dioxide Information and Analysis Center (CDIAC), Oak Ridge National Laboratory. http://cdiac.esd.ornl.gov/home.html.

Keeling, C. D., S. C. Piper, and M. Heimann. 1989. A three dimensional model of atmospheric $CO_2$ transport based on observed winds. 4. Mean annual gradients and interannual variations. Pp. 305–363 in *Aspects of climate variability in the Pacific and the Western Americas,* edited by D. H. Peterson. Washington, DC: American Geophysical Union.

Keeling, R. F., R. P. Najjar, M. L. Bender, and P. P. Tans. 1993. What atmospheric oxygen measurements can tell us about the global carbon cycle. *Global Biogeochemical Cycles* 7:37–67.

Keeling, C. D., T. P. Whorf, M. Wahlen, and J. Vanderplicht. 1995. Interannual extremes in the rate of rise of atmospheric carbon-dioxide since 1980. *Nature* 375 (6533): 666–670.

Keeling, R. F., S. C. Piper, and M. Heimann. 1996. Global and hemispheric $CO_2$ sinks deduced from changes in atmospheric $O_2$ concentration. *Nature* 381:218–221.

Langenfelds, R., R. Francey, B. C. Pak, L. P. Steele, J. Lloyd, C. M. Trudinger, and C. A. Allison. 2002. Interannual growth rate variations of atmospheric $CO_2$ and its d13C, $H_2$, $CH_4$, and CO between 1992 and 1999 linked to biomass burning. *Global Biogeochemical Cycles* 16 (3): 1048.

Law, R. M., P. J. Rayner, A. S. Denning, D. Erickson, I. Y. Fung, M. Heimann, S. C. Piper, M. Ramonet, S. Taguchi, J. A. Taylor, C. M. Trudinger, and I. G. Watterson. 1996. Variations in modeled atmospheric transport of carbon dioxide and the consequences for $CO_2$ inversions. *Global Biogeochemical Cycles* 10 (4): 783–796.

Le Queré, C., O. Aumont, L. Bopp, P. Bousquet, P. Ciais, R. Francey, M. Heimann, C. D. Keeling, R. F. Keeling, H. Kheshgi, P. Peylin, S. C. Piper, I. C. Prentice, and P. J. Rayner. 2003. Two decades of ocean $CO_2$ sink and variability. *Tellus* 55B:649–656.

Levin, I., and B. Kromer. 1997. Twenty years of atmospheric $CO_2$-$^{14}C$ observations at Schauinsland station, Germany. *Radiocarbon* 39 (2): 205–218.

Maier-Reimer, E., and K. Hasselmann. 1987. Transport and storage of $CO_2$ in the ocean: An inorganic ocean-circulation carbon cycle model. *Climate Dynamics* 2:63–90.

McGuire, A. D., S. Sitch, J. S. Clein, R. Dargaville, G. Esser, J. Foley, M. Heimann, F. Joos, J. Kaplan, D. W. Kicklighter, R. A. Meier, J. M. Melillo, B. Moore, I. C. Prentice, N. Ramankutty, T. Reichenau, A. Schloss, H. Tian, L. J. Williams, and U. Wittenberg. 2001. Carbon balance of the terrestrial biosphere in the twentieth century: Analyses of $CO_2$, climate and land use effects with four process-based ecosystem models. *Global Biogeochemical Cycles* 15 (1): 183–206.

Oeschger, H., U. Siegenthaler, U. Schotterer, and A. Gugelmann. 1975. Box diffusion-model to study carbon-dioxide exchange in nature. *Tellus* 27 (2): 168–192.

Olivier, J. G. J., and J. J. M. Berdowski. 2001. Global emissions sources and sinks. Pp. 33–78 in *The climate system,* edited by J. Berdowski, R. Guicherit and B. J. Heij. Lisse, the Netherlands: A. A. Balkema/Swets and Zeitlinger.

Orr, J. C., E. Maier-Reimer, U. Mikolajewicz, P. Monfray, J. L. Sarmiento, J. R.

Toggweiler, N. K. Taylor, J. Palmer, N. Gruber, C. L. Sabine, C. Le Quéré, R. M. Key, and J. Boutin. 2001. Estimates of anthropogenic carbon uptake from four three-dimensional global ocean models. *Global Biogeochemical Cycles* 15:43–60.

Pacala, S. W., G. C. Hurtt, D. Baker, P. Peylin, R. A. Houghton, R. A. Birdsey, L. Heath, E. T. Sundquist, R. F. Stallard, P. Ciais, P. Moorcroft, J. P. Caspersen, E. Shevliakova, B. Moore, G. Kohlmaier, E. Holland, M. Gloor, M. E. Harmon, S. M. Fan, J. L. Sarmiento, C. L. Goodale, D. Schimel, and C. B. Field. 2001. Consistent land- and atmosphere-based US carbon sink estimates. *Science* 292 (5525): 2316–2320.

Papale, D., and R. Valentini. 2003. A new assessment of European forest carbon exchanges by eddy fluxes and artificial neural network spatialization. *Global Change Biology* 9:525–535.

Pearman, G. I., and P. Hyson. 1980. Activities of the global biosphere as reflected in atmospheric $CO_2$ records. *Journal of Geophysical Research–Oceans and Atmospheres* 85 (NC8): 4457–4467.

Piper, S. C., C. D. Keeling, and E. F. Stewart. 2001. Exchanges of atmospheric $CO_2$ and $^{13}CO_2$ with the terrestrial biosphere and oceans from 1978 to 2000. II. A three-dimensional tracer inversion model to deduce regional fluxes. SIO Reference Series no. 01–07. La Jolla, CA: Scripps Institution of Oceanography.

Prentice, I. C., G. D. Farquhar, M. J. R. Fasham, M. L. Goulden, M. Heimann, V. J. Jaramillo, H. S. Keshgi, C. Le Quéré, R. J. Scholes, and D. W. R. Wallace. 2002. The carbon cycle and atmospheric carbon dioxide. Pp. 183–237 in *Climate change 2001: The scientific basis (Contribution of Working Group I to the Third Assessment Report of the Intergovernmental Panel on Climate Change)*, edited by J. T. Houghton, Y. Ding, D. J. Griggs, M. Noguer, P. J. van der Linden, X. Dai, K. Maskell, and C. A. Johnson. Cambridge: Cambridge University Press.

Rayner, P. J., I. G. Enting, R. J. Francey, and R. Langenfelds. 1999. Reconstructing the recent carbon cycle from atmospheric $CO_2$, delta C-13 and $O_2/N_2$ observations. *Tellus* 51B (2): 213–232.

Rivier, L., P. Bousquiet, J. Brandt, P. Ciais, C. Geels, M. Gloor, M. Heimann, U. Karstens, P. Peylin, and C. Rödenbeck. 2003. Estimation of the regional sources and sinks of $CO_2$ with a focus on Europe using both regional and global atmospheric models. *Journal of Geophysical Research* (in preparation).

Rödenbeck, C., S. Houweling, M. Gloor, and M. Heimann. 2003. Atmospheric $CO_2$ source history over the last 20 years inferred from atmospheric data and models. *Atmospheric Chemistry and Physics* 3, 1919–1964. http://www.copernicus.org/EGU/acp/acp/3/1919/acp-3-1919.pdf.

Sabine, C. L., R. M. Key, K. M. Johnson, F. J. Millero, J. L. Sarmiento, D. W. R. Wallace, and C. D. Winn. 1999. Anthropogenic $CO_2$ inventory of the Indian ocean. *Global Biogeochemical Cycles* 13:179–198.

Sabine, C. L., R. A. Feely, R. M. Key, J. L. Bullister, F. J. Millero, K. Lee, T.-H. Peng, B. Tilbrook, T. Ono, and C. S. Wong. 2002. Distribution of anthropogenic $CO_2$ in the Pacific Ocean. *Global Biogeochemical Cycles* 16 (4): 1083, doi:10.1029/2001GB001639.

Sarmiento, J. L., and E. T. Sundquist. 1992. Revised budget for the oceanic uptake of anthropogenic carbon-dioxide. *Nature* 24 (356): 589–593.

Schulze, E.-D. 2003. Assessment of the European terrestrial carbon balance: Proposal for an integrated project. Jena, Germany: Max-Planck-Institute for Biogeochemistry.

Siegenthaler, U., and H. Oeschger. 1978. Predicting future atmospheric carbon-dioxide levels. *Science* 199 (4327): 388–395.

Sitch, S., I. C. Prentice, B. Smith, W. Cramer, J. Kaplan, W. Lucht, M. Sykes, K. Thonicke, and S. Venevsky. 2000. LPJ: A coupled model of vegetation dynamics and the terrestrial carbon cycle. In *The role of vegetation dynamics in the control of atmospheric CO content*, edited by S. Sitch. Lund, Sweden: Lund University.

Takahashi, T., S. C. Sutherland, C. Sweeney, A. Poisson, N. Metzl, B. Tilbrook, N. Bates, R. Wanninkhof, R. A. Feely, C. Sabine, J. Olafsson, and Y. Nojiri. 2002. Global sea-air $CO_2$ flux based on climatological surface ocean $pCO_2$, and seasonal biological and temperature effects. *Deep-Sea Research II* 49:1601–1623.

Tans, P. P., T. J. Conway, and T. Nakazawa. 1989. Latitudinal distribution of the sources and sinks of atmospheric carbon-dioxide derived from surface observations and an atmospheric transport model. *Journal of Geophysical Research–Atmospheres* 94 (D4): 5151–5172.

Tans, P. P., I. Y. Fung, and T. Takahashi. 1990. Observational constraints on the global atmospheric $CO_2$ budget. *Science* 247 (4949): 1431–1438.

van der Werf, G. R., J. T. Randerson, G. J. Collatz, and L. Giglio. 2003. Carbon emissions from fires in tropical and subtropical ecosystems. *Global Change Biology* 9:547–562.

Wofsy, S. C., and R. C. Harriss. 2002. *The North American Carbon Program (NACP)*. Report of the NACP Committee of the U.S. Interagency Carbon Cycle Science Program. Washington, DC: U.S. Global Change Research Program.

Wolter, K., and M. S. Timlin. 1993. Monitoring ENSO in COADS with a seasonally adjusted principal component index. Pp. 52–57 in *Proceedings of the 17th climate diagnostics workshop*. University of Oklahoma, Norman, OK.

Woodwell, G. M., R. H. Whittaker, W. A. Reiners, G. E. Likens, C. C. Delwiche, and D. B. Botkin. 1978. Biota and world carbon budget. *Science* 199 (4325): 141–146.

# 9

# Non-CO$_2$ Greenhouse Gases

## Ronald G. Prinn

The atmosphere contains a large number of anthropogenic greenhouse gases besides carbon dioxide, which, because of their rising concentrations, have collectively contributed an amount of added radiative forcing comparable to that of CO$_2$ since preindustrial times. Many of these non-CO$_2$ gases (e.g., CH$_4$, N$_2$O, CF$_2$Cl$_2$, SF$_6$) are emitted at the surface and contribute directly to this forcing. They are characterized by atmospheric lifetimes of decades to millennia (I define lifetime here as the amount of the gas in the global atmosphere divided by its global rate of removal). Other non-CO$_2$ gases (e.g., isoprene, terpenes, NO, CO, SO$_2$, (CH$_3$)$_2$S), most of which are also emitted at the surface, contribute indirectly to this forcing through production of either tropospheric ozone (which is a powerful greenhouse gas) or tropospheric aerosols (which can directly absorb sunlight or reflect it back to space, or indirectly change the reflection properties of clouds). This second group of climatically important non-CO$_2$ gases is characterized by much shorter lifetimes (hours to months).

To aid the handling of the long-lived non-CO$_2$ gases in the policy processes under the United Nations Framework Convention on Climate Change (UNFCCC), scientists have calculated so-called global warming potentials (GWPs). These dimensionless GWPs are intended to relate the time-integrated radiative forcing of climate by an emitted unit mass of a non-CO$_2$ trace gas to the forcing caused by emission of a unit mass of CO$_2$. The GWP concept has difficulties because the removal mechanisms for many gases (including CO$_2$ itself) have complex interactions involving chemical and/or biological processes, and because the time period (e.g., decade, century) over which one integrates the instantaneous radiative forcing of a gas to compute its GWP is somewhat arbitrary.

A listing of the major non-CO$_2$ greenhouse gases in the atmosphere along with their concentrations, temporal trends, emissions, lifetimes, and GWPs is given in Table 9.1, based on Prather et al. (2001). From this table it is evident that the radiative forcing by many non-CO$_2$ gases is far greater per unit emitted mass than CO$_2$, a characteristic that significantly offsets their much lower emissions relative to CO$_2$. To illustrate this further,

**Table 9.1.** Chemically reactive greenhouse gases and their precursors: Abundances, trends, budgets, lifetimes, and GWPs (based on Prather et al. 2001)

| Chemical species | Formula | Abundance[a] (ppt) 1998 | 1750 | Trend, 1990s (ppt/yr)[a] | Annual emission, late 1990s | Lifetime (years) | 100-year GWP[b] |
|---|---|---|---|---|---|---|---|
| Methane | $CH_4$ (ppb) | 1,745 | 700 | 7.0 | 600 Tg | 8.4/12 [c] | 23 |
| Nitrous oxide | $N_2O$ (ppb) | 314 | 270 | 0.8 | 16.4 TgN | 120/114 [c] | 296 |
| Perfluoromethane | $CF_4$ | 80 | 40 | 1.0 | ~15 Gg | >50,000 | 5,700 |
| Perfluoroethane | $C_2F_6$ | 3.0 | 0 | 0.08 | ~2 Gg | 10,000 | 11,900 |
| Sulphur hexafluoride | $SF_6$ | 4.2 | 0 | 0.24 | ~6 Gg | 3,200 | 22,200 |
| HFC-23 | $CHF_3$ | 14 | 0 | 0.55 | ~7 Gg | 260 | 12,000 |
| HFC-134a | $CF_3CH_2F$ | 7.5 | 0 | 2.0 | ~25 Gg | 13.8 | 1,300 |
| HFC-152a | $CH_3CHF_2$ | 0.5 | 0 | 0.1 | ~4 Gg | 1.40 | 120 |
| CFC-11 | $CFCl_3$ | 268 | 0 | -1.4 | | 45 | 4,600 |
| CFC-12 | $CF_2Cl_2$ | 533 | 0 | 4.4 | | 100 | 10,600 |
| CFC-13 | $CF_3Cl$ | 4 | 0 | 0.1 | | 640 | 14,000 |
| CFC-113 | $CF_2ClCFCl_2$ | 84 | 0 | 0.0 | | 85 | 6,000 |
| CFC-114 | $CF_2ClCF_2Cl$ | 15 | 0 | <0.5 | | 300 | 9,800 |
| CFC-115 | $CF_3CF_2Cl$ | 7 | 0 | 0.4 | | 1,700 | 7,200 |
| Carbon tetrachloride | $CCl_4$ | 102 | 0 | -1.0 | | 35 | 1,800 |
| Methyl chloroform | $CH_3CCl_3$ | 69 | 0 | -14 | | 4.8 | 140 |
| HCFC-22 | $CHF_2Cl$ | 132 | 0 | 5 | | 11.9 | 1,700 |
| HCFC-141b | $CH_3CFCl_2$ | 10 | 0 | 2 | | 9.3 | 700 |
| HCFC-142b | $CH_3CF_2Cl$ | 11 | 0 | 1 | | 19 | 2,400 |
| Halon-1211 | $CF_2ClBr$ | 3.8 | 0 | 0.2 | | 11 | 1,300 |
| Halon-1301 | $CF_3Br$ | 2.5 | 0 | 0.1 | | 65 | 6,900 |
| Halon-2402 | $CF_2BrCF_2Br$ | 0.45 | 0 | ~0 | | {<20 | |

*Other chemically active*
*gases directly or indirectly*
*affecting radiative forcing*

| | | | | | | | |
|---|---|---|---|---|---|---|---|
| Tropospheric ozone | $O_3$ (DU) | 34 | 25 | ? | | 0.01–0.05 | — |
| Tropospheric $NO_x$ | $NO + NO_2$ | 5–999 | ? | ? | ~52 TgN | <0.01–0.03 | — |
| Carbon monoxide | CO (ppb)[d] | 80 | ? | 6 | ~2,800 Tg | 0.08–0.25 | d |
| Stratospheric water | $H_2O$ (ppm) | 3–6 | 3–5 | ? | | 1–6 | — |

[a] All abundances are tropospheric molar mixing ratios in ppt ($10^{-12}$) and trends are in ppt $y^{-1}$ unless superseded by units on line (ppb = $10^{-9}$, ppm = $10^{-6}$). Where possible, the 1998 values are global, annual averages and the trends are calculated for 1996 to 1998.

[b] GWPs refer to the 100-year horizon values.

[c] Species with chemical feedbacks that change the duration of the atmospheric response; global mean atmospheric lifetime is given first followed by perturbation lifetime. Values are taken from the IPCC second assessment report (SAR) (Prather et al. 1995; Schimel et al. 1995) updated with data from Kurylo and Rodriguez (1999) and Prinn and Zander (1999) and new OH-scaling. Uncertainties in lifetimes have not changed substantially since the SAR.

[d] CO trend is very sensitive to the time period chosen. The value listed for 1996 to 1998, +6 ppb $y^{-1}$, is driven by a large increase during 1998. For the period 1991 to 1999, the CO trend was –0.6 ppb $y^{-1}$. CO is an indirect greenhouse gas.

**Table 9.2.** Current global emissions of non-$CO_2$ greenhouse gases expressed as equivalent amounts of carbon (Ceq) in $CO_2$ using GWPs with a 100-year time horizon

| Chemical species | Equivalent emissions (PgCeq $y^{-1}$) |
| --- | --- |
| Methane ($CH_4$) [a] | 3.8 |
| Nitrous oxide ($N_2O$) | 2.1 |
| Kyoto flourine gases [b] | 0.1 |
| Carbon monoxide (CO) [a,c] | 1.5 |
| Nitrogen oxides (NO, $NO_2$) [d] | 0.15 |

*Note:* Emissions and GWPs from Table 9.1 unless otherwise noted.

[a]These equivalent emissions do not include the $CO_2$ produced from oxidation of these gases (0.45 PgC $y^{-1}$ from $CH_4$ and 1.2 PgC $y^{-1}$ from CO).

[b]Specifically $CF_4$, $C_2F_6$, $SF_6$, $CHF_3$, $CF_3CH_2F$, and $CH_3CHF_2$ all regulated under the Kyoto Protocol. This list does not include the many ozone-depleting halocarbons already regulated under the Montreal Protocol.

[c]Assumes a GWP of 2, which is uncertain by at least a factor of 2 (Ramaswamy et al. 2001). CO generally produces $O_3$ and depletes OH.

[d]Assumes a GWP of 5 for surface emissions, which is also highly uncertain. For aircraft emissions (which are a small fraction of the total), the GWP is about 450 (Ramaswamy et al. 2001). Assumes all emissions of $NO_x$ are NO with $NO_2$ subsequently produced in the atmosphere. $NO_x$ usually produces $O_3$ and OH.

Table 9.2 shows the current emissions of major non-$CO_2$ gases converted into equivalent emissions of $CO_2$ using their GWPs. These equivalent emissions do not include the $CO_2$ produced from oxidation of $CH_4$ (0.45 PgC $y^{-1}$) and CO (1.2 PgC $y^{-1}$), which is usually included in the $CO_2$ carbon cycle. It is very clear that the total equivalent $CO_2$ emissions in Table 9.2 are comparable to the actual total global $CO_2$ emissions. Hence, these non-$CO_2$ gases are very important in all climate change discussions, and methods to lower their emissions need to be carefully evaluated (Robertson, Chapter 29, this volume).

In this chapter the life cycles of key gases (or groups of gases) involved in climate are reviewed followed by a critical discussion of their GWPs and a summary of major unsolved problems. Aerosols are not addressed but have been extensively reviewed by the IPCC (Penner et al. 2001).

## Methane

Methane is an important greenhouse substance because of its significant positive trend, its strong absorption of infrared radiation, and the location of its absorption bands at many wavelengths where $CO_2$ and $H_2O$ do not absorb (so-called window regions).

The major sources of methane are biological (basically methanogens operating in oxygen-poor environments like natural wetlands, rice paddies, digestive systems of termites and cattle, animal waste treatment, sewage, landfills, etc.). Another (indirect) biological source is biomass burning. Many of these biological sources (rice, cattle, biomass burning, etc.) are governed largely by human activity. Also under dominant human influence is another significant CH$_4$ source, namely escape of this gas during mining of coal and natural gas and leakages in natural gas distribution systems. The fossil sources can be differentiated from the biological sources because of the different carbon isotopic signatures of each, which show that roughly 20 percent of global total methane emissions come from fossil sources. The total methane source is estimated to be about 600 Tg y$^{-1}$ with 60 percent due to human activities (Kurylo et al. 1999; Prather et al. 2001).

The major sink for methane is its reaction with the OH free radical:

$$OH + CH_4 \rightarrow H_2O + CH_3$$

Using estimates of global OH levels based on analysis of global concentrations and emissions of the industrial chemical CH$_3$CCl$_3$ leads to an estimated methane loss rate due to OH of around 500 Tg y$^{-1}$ (Prather et al. 2001; Prinn et al. 2001). Note that because CH$_4$ is a significant global sink for OH, the levels of OH decrease when CH$_4$ increases, yielding a longer lifetime for methane. About 40 Tg y$^{-1}$ of CH$_4$ is destroyed in the stratosphere, and 20–40 Tg y$^{-1}$ currently accumulates in the atmosphere (producing the observed trend). The remaining 20–40 Tg y$^{-1}$ sink needed to balance the total source is probably methanotrophs in soils.

Although these numbers give the impression that the CH$_4$ budget is well understood, it is not. The estimates of individual CH$_4$ source strengths are very uncertain (up to a factor of two), and the reasons for the general decrease in the CH$_4$ trend in recent decades and the large year-to-year variations in this trend continue to be debated (see the section "Unsolved Problems").

## Nitrous Oxide

Nitrous oxide is important both as a source of NO, which destroys ozone in the stratosphere, and as a potent greenhouse gas like CH$_4$. It has risen quite steadily in the atmosphere at about 0.23 percent y$^{-1}$ (3.5 TgN y$^{-1}$) since 1978 (Prinn et al. 2000).

The major sources of N$_2$O involve nitrogen bacteria in soils (tropical, temperate) and the ocean (open, coastal). Additional sources involve nylon and nitric acid production, cattle and their feedlots, and atmospheric oxidation of NH$_3$. Like CH$_4$, the magnitudes of these individual sources are very uncertain. Best estimates of the total of these sources by various authors are around 15–17 TgN y$^{-1}$ (Prather et al. 2001).

The major sink for N$_2$O is photodissociation, and, to a much lesser extent, reaction with O($^1$D), in the stratosphere:

$$N_2O + \text{ultraviolet} \rightarrow N_2 + O$$

$$N_2O + O(^1D) \rightarrow NO + NO$$

Model studies imply a total stratospheric sink of around 12.6 TgN $y^{-1}$ (Prather et al. 2001), which, combined with the above rate of atmospheric accumulation, suggests an approximately balanced $N_2O$ budget. As for $CH_4$, however, the uncertainties in individual source and sink estimates are large, and hence the task of defining (for example) the precise reasons for the observed steady rise in $N_2O$ remains to be accomplished.

## Halocarbons and $SF_6$

Current concerns about the atmospheric trends in chlorofluorocarbons (CFCs), hydrochlorofluorocarbons (HCFCs), hydrofluorocarbons (HFCs), hydrochlorocarbons (HCCs), chlorocarbons, perfluorocarbons (PFCs), and sulfur hexafluoride ($SF_6$) are based on either their deleterious effects on the ozone layer (where many of them are sources of destructive Cl and ClO), and /or their radiative forcing of climate. Their GWPs are generally extremely large (Table 9.1), so that even very small emissions of these gases relative to $CO_2$ can have a nonnegligible climate impact.

The budgets of this wide-ranging set of gases, whose sources with few exceptions are almost exclusively industrial, have been recently reviewed (Kurylo and Rodriguez 1999; Prinn and Zander 1999; Prather et al. 2001; Montzka and Fraser 2002), and there is not space here to discuss all of the details. The total amount of chlorine contained in all the major chlorine-containing halocarbons peaked at 3.7 parts per billion (ppb) at the surface around 1992–1994 owing to the decreases in emissions of these gases mandated by the Montreal Protocol for the Protection of the Ozone Layer (Montzka et al. 1999; Prinn et al. 2000). Presuming these decreases continue in the future, attention to climate impacts and regulation of these gases in the Kyoto Protocol of the FCCC will increasingly be on halogen-containing gases not covered by the Montreal Protocol (specifically HFCs, PFCs, and $SF_6$; see Tables 9.1 and 9.2).

The major sources for all of these halogen-containing gases are anthropogenic. They are used primarily as refrigerant fluids, foam-blowing agents, or solvents. The primary sinks for CFCs, PFCs, and $SF_6$ are in the upper atmosphere (photodissociation, ionospheric reactions), while HFCs and HCFCs are destroyed by reaction with OH in the troposphere as well as being photodissociated in the stratosphere.

The most abundant HFCs and PFCs are, respectively, $CHF_3$ (a byproduct of production of the refrigerant $CHF_2Cl$) and $CF_4$ (a byproduct of aluminum production). Also rapidly rising are the HFC species $CF_3CH_2F$ (a common refrigerant), the PFC compound $C_2F_6$ (aluminum industry), and $SF_6$ (leaked from transformers, etc.). The latter gas, with one of the highest known GWPs (22,200) and a 6 percent $y^{-1}$ rate of increase, is indicative of the importance of this set of non-$CO_2$ gases.

# Nonmethane Hydrocarbons, CO, NO$_x$, and O$_3$

The ability of the troposphere to chemically transform and remove trace gases depends on complex chemistry driven by the relatively small flux of energetic solar ultraviolet radiation that penetrates through the stratospheric ozone layer (see, e.g., Ehhalt 1999). This chemistry is also driven by emissions of NO, CO, and hydrocarbons (RH) and leads to the production of ozone, which is a potent greenhouse gas. Globally, about 3.4–4.6 Pg y$^{-1}$ of ozone are produced in the troposphere with a lifetime of 28–37 days (Ehhalt 1999; Lelieveld and Thompson 1999; Prinn 2003). The most important cleansing chemical in the troposphere, however, is the hydroxyl free radical (OH), and a key measure of the capacity of the atmosphere to oxidize trace gases injected into it is the local concentration of hydroxyl radicals (Prinn 2003).

As noted earlier, the OH radical serves as the principal sink for the greenhouse gases CH$_4$, HFCs, and HCFCs. It removes about 3.65 Pg y$^{-1}$ of trace gases from the atmosphere (Ehhalt 1999; Prinn 2003). The principal catalyzed reactions creating OH and O$_3$ in the troposphere are:

$$O_3 + \text{ultraviolet} \rightarrow O_2 + O(^1D)$$
$$O(^1D) + H_2O \rightarrow 2OH$$
$$\text{Net effect: } O_3 + H_2O \rightarrow O_2 + 2OH$$

and

$$OH + CO \rightarrow H + CO_2$$
$$H + O_2 \rightarrow HO_2$$
$$HO_2 + NO \rightarrow OH + NO_2$$
$$NO_2 + \text{ultraviolet} \rightarrow NO + O$$
$$O + O_2 \rightarrow O_3$$
$$\text{Net effect: } CO + 2O_2 \rightarrow CO_2 + O_3$$

and

$$OH + RH \rightarrow R + H_2O$$
$$R + O_2 \rightarrow RO_2$$
$$RO_2 + NO \rightarrow RO + NO_2$$
$$NO_2 + \text{ultraviolet} \rightarrow NO + O$$
$$O + O_2 \rightarrow O_3$$
$$\text{Net effect: } OH + RH + 2O_2 \rightarrow RO + H_2O + O_3$$

In theoretical models, the global production of tropospheric $O_3$ by the pathway beginning with CO is typically about twice that beginning with RH. For most environments, these catalytic processes pump the majority of the OH and $O_3$ into the system. If the concentration of $NO_2$ gets too high, however, its reaction with OH to form $HNO_3$ ultimately limits the OH concentration.

How stable is the cleansing capability of the troposphere? If emissions of gases that react with OH, such as $CH_4$, CO, and $SO_2$, are increasing then, keeping everything else constant, OH levels should decrease. Conversely, increasing $NO_x$ emissions from combustion should increase tropospheric $O_3$ (and thus the primary source of OH), as well as increase the recycling rate of $HO_2$ to OH (the secondary source of OH). Also, if the oceans are increasing in temperature, one would expect increased water vapor in the lower troposphere. Because water vapor is part of the primary source of OH, climate warming also increases OH. This increase could be lowered or raised if changes in cloud cover accompanying the warming lead to more or less reflection of ultraviolet back toward space. Rising temperature also increases the rate of reaction of $CH_4$ with OH, thus lowering the lifetimes of both chemicals. Opposite conclusions apply if trace gas emissions increase or the climate cools. Finally, decreasing stratospheric ozone can also increase tropospheric OH, as discussed later in this chapter. Determining whether OH concentrations, and hence the cleansing capacity of the atmosphere, are changing is a major focus of recent research (Thompson 1992; Wang and Jacob 1998; Prinn 2003).

The ozone in the stratosphere is maintained by a distinctively different set of chemical reactions than in the troposphere. Stratospheric ozone is produced and destroyed at a massive global rate of about 120 Pg $y^{-1}$ (Warneck 1988). The fundamental stratospheric ozone source is photodissociation of $O_2$. The ozone sink is driven by chemicals like water vapor, chlorofluorocarbons, and nitrous oxide that are transported from the troposphere to the stratosphere, where they produce chlorine-, nitrogen-, and hydrogen-carrying free radicals, which catalytically destroy ozone.

There is an important link between stratospheric chemistry and climate. The precursor gases for the destructive free radicals in the stratosphere are greenhouse gases, as is stratospheric ozone. Therefore, while increases in the concentrations of the source gases will increase radiative forcing of warming, these increases will, at the same time, lead to decreases in stratospheric ozone, lowering the radiative forcing. As in the troposphere, there are important feedbacks between chemical processes and climate, through the various greenhouse gases. It is not just manmade chemicals that can change stratospheric ozone. Changes in natural emissions of $N_2O$ and $CH_4$, and changes in climate that alter $H_2O$ flows into the stratosphere, undoubtedly led to changes in the thickness of the ozone layer in the past.

There is also a strong link between the thickness of the stratospheric ozone layer and concentrations of OH in the troposphere. Because $O(^1D)$ can only be produced from $O_3$ at wavelengths less than 310 (nanometers) nm, and the current ozone layer effectively absorbs all incoming ultraviolet at less than 290 nm, a very narrow window of

290–310 nm radiation is driving the major oxidation processes in the troposphere. Decreasing stratospheric ozone, as in recent decades, can therefore increase tropospheric OH by increasing the flux of radiation less than 310 nm reaching the troposphere to produce $O(^1D)$ (Madronich and Granier 1992).

## Global Warming Potentials

As mentioned earlier, GWPs provide a tool for ranking the potential for a given mass of an emitted greenhouse gas to influence climate, relative to the same emitted mass of $CO_2$. Formally the GWP is defined as follows:

$$GWP = \int_0^T I(gas) \, M(gas) dt \Big/ \int_0^T I(CO_2) \, M(CO_2) dt$$

Here $I(gas)$ is the instantaneous radiative forcing by the gas at time $t$. $I(gas)$ depends on the basic molecular properties of the gas and on general atmospheric composition (distribution of gases, clouds, aerosols). $M(gas)$ is the mass of added gas still remaining at time $t$ [initially $M(gas) = M(CO_2)$]. It depends on the lifetime of the gas, which in turn depends on both the amounts of the gas itself and of other gases. Finally, $T$ is the time horizon for integration (gases with lifetimes longer or shorter than $CO_2$ have GWPs increasing or decreasing, respectively, with increasing $T$).

At least three major difficulties are inherent in the definition of GWPs, and these difficulties limit their usefulness. The first is that the lifetime of a gas (and hence $M(gas)$) may not simply be a function of its own concentration. For example, the lifetime of $CH_4$ depends on OH, which depends not only on $CH_4$, but also on the levels of $NO_x$, CO, and other hydrocarbons. The OH levels also depend on ultraviolet fluxes, which in turn depend on levels of gases that deplete stratospheric ozone like $N_2O$ and the CFCs.

The second is that the time horizon $T$ is arbitrary. It should logically be long enough to correspond to long-term climatic (as opposed to weather) variations, but short enough to have relevance to the development of human activity. There is also the matter of the timing of emission reductions of various gases relative to the particular target year chosen for assessing the climate changes avoided by these reductions (the year 2100 is often used as this target year). The Kyoto Protocol chooses $T = 100$ years. But the effects of changing this assumption can be profound. For example, the GWP of $CH_4$ is 23 for $T = 100$ years but increases to 62 for $T = 20$ years and decreases to 7 for $T = 500$ years (the lifetime of $CH_4$ is about 10 times less than $CO_2$). Conversely the GWP of $SF_6$ rises from 15,100 to 22,200 to 32,400 for $T = 20, 100,$ and 500 years, respectively (the lifetime of $SF_6$ is about 30 times greater than $CO_2$). As an example of the combined effects of these first two shortcomings, Reilly et al. (1999) have shown that the use of 100-year GWPs to convert non-$CO_2$ emissions into equivalent $CO_2$ emissions leads to much greater model-projected warming between 1990 and 2100 than a similar model run that uses actual non-$CO_2$ gas emissions and includes all of their chemical and radiative interactions. Specifically, $CH_4$ emissions reductions are significantly undervalued using $T = 100$ years.

Finally, the GWP definition depends upon $M(CO_2)$ and hence upon all the assumptions regarding oceanic and terrestrial $CO_2$ sinks. The many uncertainties in the carbon cycle discussed elsewhere in this volume therefore carry over into all of the GWP calculations.

## Unsolved Problems

Here is a partial list of outstanding issues regarding the budgets of non-$CO_2$ gases and their climatic effects relative to $CO_2$:

1. For methane, the magnitudes of its wide range of surface sources as functions of position, time, and process need better quantification. What drives the lowering of the trend in $CH_4$? Is it driven by steady increases in OH (Krol et al. 1998; Karlsdottir and Isaksen 2000) or by near-constant long-term OH and emissions, leading to an approach to a (zero-trend) steady state (Dlugokencky et al. 1998; Prinn et al. 2001).

2. For nitrous oxide, similar questions remain concerning its surface sources. What causes the steady increase in $N_2O$ since 1978 (Prinn et al. 2000)?

3. For halocarbons and $SF_6$, there is a continued need for improved measurements and industrial emission estimates to reconcile the global budgets of many species.

4. For nonmethane hydrocarbons, CO, $NO_x$, and tropospheric $O_3$, there is a need to better understand the roles of these species in controlling OH levels and hence the lifetimes of $CH_4$, HFCs, and HCFCs (Prinn 2003). For this purpose, improved emission and /or trend measurements of hydrocarbons, $NO_x$, and $O_3$ over the globe are essential. For stratospheric $O_3$, there is a need to improve knowledge of its role in controlling the lifetimes of trace gases through modulation of ultraviolet fluxes to the troposphere.

5. For both methane and nitrous oxide, the combined roles of climate change and ecosystem (soil, wetland) changes on the surface fluxes of these gases need better definition. We expect their emissions to increase with increased warming, soil wetting, N deposition, and soil carbon (Hall and Matson 1999; Prinn et al. 1999). Is this occurring?

6. Methods for lowering emissions of these non-$CO_2$ gases need to be carefully evaluated and included along with $CO_2$ in climate policy (Robertson, Chapter 29, this volume).

7. For ranking the climatic effects of non-$CO_2$ gases relative to $CO_2$, there is a need to improve the GWP concept (Manne and Richels, Chapter 25, this volume). It is probable that a single number will not be adequate and that the use of models that explicitly include the interactions missed by the GWPs will be necessary.

# Acknowledgments

This work was supported by the National Science Foundation (Grant ATM-0120468) and by the federal and industry sponsors of the MIT Joint Program on the Science and Policy of Global Change.

# Literature Cited

Dlugokencky, E. J., K. A. Masarie, P. M. Lang, and P. P. Tans. 1998. Continuing decline in the growth rate of the atmospheric methane burden. *Nature* 393:447–450.

Ehhalt, D. H. 1999. Gas phase chemistry of the troposphere. Pp. 21–109 in *Global aspects of atmospheric chemistry,* edited by R. Zellner. Topics in Physical Chemistry 6. Springer-Darmstadt-Steinkopff.

Hall, S. J., and P. A. Matson. 1999. Nitrogen oxide emissions after N additions in tropical forests. *Nature* 400:152–155.

Karlsdottir, S., and I. Isaksen. 2000. Changing methane lifetime: Possible cause for reduced growth. *Geophysical Research Letters* 27:93–97.

Krol, M., P. J. van Leeuwen, and J. Leileveld. 1998. Global OH trend inferred from methyl chloroform measurements. *Journal of Geophysical Research* 103:10697–10711.

Kurylo, M. J., and J. M. Rodriguez. 1999. Short-lived ozone-related compounds. Pp. 2.1–2.56 in *Scientific assessment of ozone depletion: 1998.* Report 44. Geneva: World Meteorological Organization, Global Ozone Research and Monitoring Project.

Lelieveld, J., and A. Thompson. 1999. Tropospheric ozone and related processes. Pp. 8.1–8.41 in *Scientific assessment of ozone depletion: 1998.* Report 44. Geneva: World Meteorological Organization, Global Ozone Research and Monitoring Project.

Madronich, S., and C. Granier. 1992. Impact of recent total ozone changes on tropospheric ozone photodissociation, hydroxyl radicals, and methane trends. *Geophysical Research Letters* 19:465–467.

Montzka, S. A., J. H. Butler, J. W. Elkins, T. M. Thompson, A. D. Clarke, and L. T. Lock. 1999. Present and future trends in the atmospheric burden of ozone-depleting halogens. *Nature* 398:690–694.

Montzka, S., and P. Fraser. 2003. Controlled substances and other source gases. Pp. 1.1–1.83 in *Scientific assessment of ozone depletion: 2002.* Report 47. Geneva: World Meteorological Organization, Global Ozone Research and Monitoring Project.

Penner, J., M. Andreae, H. Annegarn, L. Barrie, J. Feichter, D. Hegg, A. Jayaraman, R. Leaitch, D. Murphy, J. Nganga, and G. Pitari. 2001. Aerosols: Their direct and indirect effects. Pp. 289–348 in *Climate change 2001: The scientific basis (Contribution of Working Group I to the Third Assessment Report of the Intergovernmental Panel on Climate Change),* edited by J. T. Houghton, Y. Ding, D. J. Griggs, M. Noguer, P. J. van der Linden, X. Dai, K. Maskell, and C. A. Johnson. Cambridge: Cambridge University Press.

Prather, M., R. Derwent, D. Ehhalt, P. Fraser, E. Sanhueza, and X. Zhou. 1995. Other trace gases and atmospheric chemistry. Pp. 73–126 in *Climate change 1994: The science of climate change,* edited by J. T. Houghton, L.G. Meira Filho, J. Bruce, H. Lee, B.A. Callander, E. Haites, N. Harris, and K. Maskell. Cambridge: Cambridge University Press.

Prather, M., D. Ehhalt, F. Dentener, R. Derwent, E. Dlugokencky, E. Holland, I. Isaksen, J. Katima, V. Kirchhoff, P. Matson, P. Midgley, and M. Wang. 2001. Atmospheric chemistry and greenhouse gases. Pp. 239–288 in *Climate change 2001: The scientific basis (Contribution of Working Group I to the Third Assessment Report of the Intergovernmental Panel on Climate Change)*, edited by J. T. Houghton, Y. Ding, D. J. Griggs, M. Noguer, P. J. van der Linden, X. Dai, K. Maskell, and C. A. Johnson. Cambridge: Cambridge University Press.

Prinn, R. G., J. Huang, R. F. Weiss, D. M. Cunnold, P. J. Fraser, P. G. Simmonds, A. McCulloch, C. Harth, P. Salameh, S. O'Doherty, R. H. J. Wang, L. Porter, and B. R. Miller. 2001. Evidence for substantial variations of atmospheric hydroxyl radicals in the past two decades. *Science* 292:1882–1888.

Prinn, R. G., H. D. Jacoby, A. P. Sokolov, C. Wang, X. Xiao, Z. L. Yang, R. S. Eckaus, P. H. Stone, A. D. Ellerman, J. M. Melillo, J. Fitzmaurice, D. W. Kicklighter, G. L. Holian, and Y. Liu. 1999. Integrated global system model for climate policy assessment: Feedbacks and sensitivity studies. *Climatic Change* 41:469–546.

Prinn, R., R. Weiss, P. Fraser, P. Simmonds, D. Cunnold, F. Alyea, S. O'Doherty, P. Salameh, B. Miller, J. Huang, R. Wang, D. Hartley, C. Harth, L. Steele, G. Sturrock, P. Midgley, and A. McColloch. 2000. A history of chemically and radiatively important gases in air deduced from ALE/GAGE/AGAGE. *Journal of Geophysical Research* 105:17751–17792.

Prinn, R. G., and R. Zander. 1999. Long-lived ozone-related compounds. *Scientific assessment of ozone depletion: 1998*. Report 44. Geneva: World Meteorological Organization, Global Ozone Research and Monitoring Project.

Prinn, R. 2003. The cleansing capacity of the atmosphere. *Annual Review of Environment & Resources* 28:29–57.

Ramaswamy, V., O. Boucher, J. Haigh, D. Hauglustine, J. Haywood, G. Myhre, T. Nakajima, G. Y. Shi. and S. Solomon. 2001. Radiative forcing of climate change. Pp. 349–416 in *Climate change 2001: The scientific basis (Contribution of Working Group I to the Third Assessment Report of the Intergovernmental Panel on Climate Change)*, edited by J. T. Houghton, Y. Ding, D. J. Griggs, M. Noguer, P. J. van der Linden, X. Dai, K. Maskell, and C. A. Johnson. Cambridge: Cambridge University Press.

Reilly, J., R. G. Prinn, J. Harnisch, J. Fitzmaurice, H. D. Jacoby, D. Kicklighter, J. Melillo, P. H. Stone, A. P. Sokolov, and C. Wang. 1999. Multi-gas assessment of the Kyoto Protocol. *Nature* 401:549–555.

Schimel, D., D. Alves, I. Enting, M. Heimann, F. Joos, D. Raynaud, T. Wigley, M. Prather, R. Derwent, D. Ehhalt, P. Fraser, E. Sanhueza, X. Zhou, P. Jonas, R. Charlson, H. Rodhe, S. Sadasivan, K. P. Shine, Y. Fouquart, V. Ramaswamy, S. Solomon, J. Srivinasan, D. Albritton, R. Derwent, I. Isaksen, M. Lal, and D. Wuebbles. 1995. Radiative forcing of climate change. Pp. 65–131 in *Climate change 1994: The science of climate change (Contribution of Working Group I to the second sssessment report of the Intergovernmental Panel on Climate Change)*, edited by J. T. Houghton, L. G. Meira Filho, B. A. Callander, N. Harris, A. Kattenberg, and K. Maskell. Cambridge: Cambridge University Press.

Thompson, A. M. 1992. The oxidizing capacity of the Earth's atmosphere: Probable past and future changes. *Science* 256:1157–1165.

Wang, Y., and D. J. Jacob. 1998. Anthropogenic forcing on tropospheric ozone and OH since pre-industrial times. *Journal of Geophysical Research* 103:31123–31135.

Warneck, P. 1988. *Chemistry of the natural atmosphere*. New York: Academic Press.

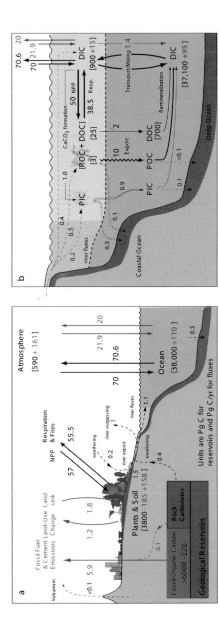

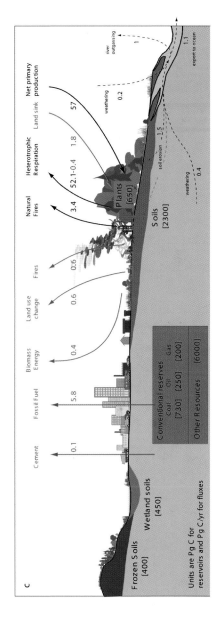

**Colorplate 1.** The current carbon cycle, showing (a) the globe, plus detailed views for (b) the oceans, and (c) the land. Stocks of carbon are in Pg (1 Pg = $10^{15}$ g = $10^9$ ton). Annual fluxes are in PgC y$^{-1}$. Background or preanthropogenic stocks and fluxes are in black. The human perturbations to the stocks and fluxes are in red. The values are averages over the 1980s and 1990s. Sources for the individual values are discussed throughout the chapter.

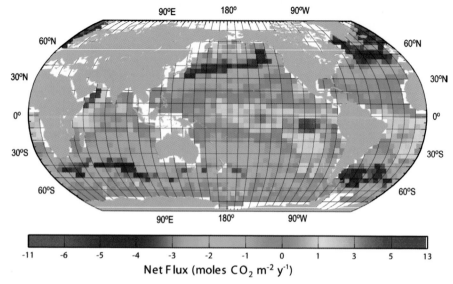

**Colorplate 2.** Mean annual net air-sea $CO_2$ flux for 1995 (moles C $m^{-2}$ $y^{-1}$), with negative values representing ocean sinks and positive values representing carbon sources to the atmosphere. Figure adapted from Takahashi et al. (2002).

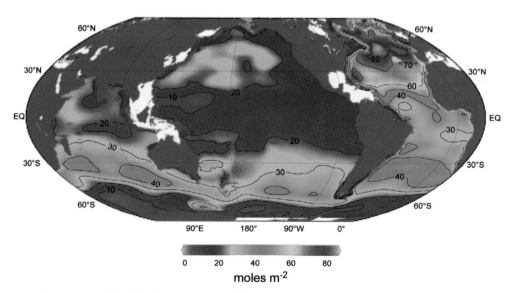

**Colorplate 3.** Total column inventory of anthropogenic $CO_2$ in the oceans (moles C $m^{-2}$) based on estimates using the $\Delta C^*$ technique. Indian, Pacific, and Atlantic data are from Sabine et al. (1999), Sabine et al. (2002), and Lee et al. (2003), respectively.

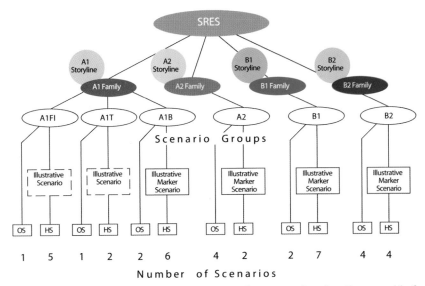

**Colorplate 4.** Characteristics of SRES scenarios. The A1 storyline describes a world of rapid economic growth. Global population peaks in mid-century. Introduction of new technologies is rapid, and regional differences in per capita income are substantially reduced. The three A1 groups emphasize fossil-intensive (A1Fl), non-fossil (A1T), and balanced energy sources (A1B). The A2 storyline describes a heterogeneous world emphasizing preservation of local identities. Global population increases continuously. Economic development and technological change are more fragmented and slower than in other storylines. The B1 storyline has the same population trajectory as A1, but with rapid transition to a service and information economy. It emphasizes reductions in material intensity, clean technologies, and global solutions to economic, social, and environmental sustainability, including improved equity. The B2 storyline describes a world that emphasizes local solutions to economic, social, and environmental sustainability. Global population increases continuously at a rate lower than A2. Economic development is intermediate. Technological change is less rapid and more diverse than in B1 and A1.

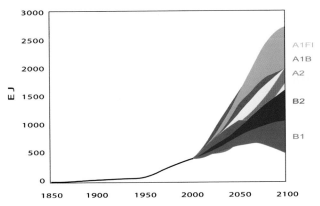

**Colorplate 5.** Global primary energy requirements since 1850 and in the IPCC SRES scenarios to 2100 in EJ per year.

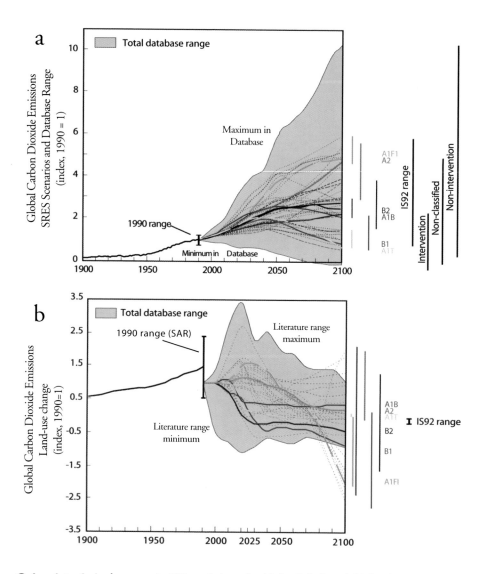

**Colorplate 6.** Anthropogenic $CO_2$ emissions for (a) fossil fuels and (b) land use change, from Nakicenovic et al. (2000). SAR is the IPCC second assessment report.

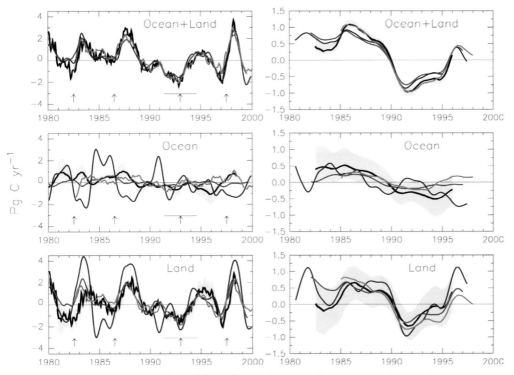

**Colorplate 7.** Interannual variability of anomalous global ocean-atmosphere and land-atmosphere $CO_2$ fluxes, estimated by various methods, as compiled by Le Quéré et al. (2003): (1) time-dependent atmospheric $CO_2$ inversion (black lines, Bousquet et al. 2000), (2) ocean model simulation driven by observed climate (red lines), (3) and (4) $^{13}C$-$CO_2$ double deconvolutions using different atmospheric $^{13}C$ records and atmospheric inversion models (blue lines: Keeling et al. 1995, updated; green lines: update of Rayner et al. 1999). Left panels: anomalous fluxes low pass filtered with a cut-off frequency of one year, right panels: anomalous fluxes low pass filtered with a cut-off frequency of five years. The gray shading indicates the range of eight different sensitivity inversions from the Bousquet et al. (2000) study, documenting the robustness of the atmospheric inversion–based source flux estimates. Arrows indicate El Niño events, the thin horizontal line indicates the time period following the Mt Pinatubo volcanic eruption. The anomalous fluxes were derived by subtracting the 1985–1995 mean.

A Posteriori Fluxes + Fossil Emissions, Average July 1995 - June 2000 [gC/m2/yr]

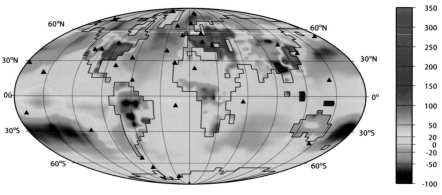

A Posteriori Fluxes, Average July 1995 - June 2000 [gC/m2/yr]

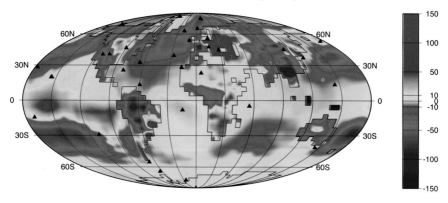

Uncertainty Reduction, Average July 1995 - June 2000 [%]

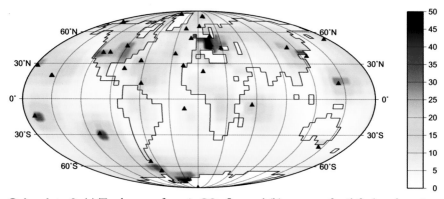

**Colorplate 8.** (a) Total net surface-air $CO_2$ flux and (b) net non-fossil-fuel surface-air flux inferred from the atmospheric inversion average July 1995–June 2000, based on 35 stations (black triangles) from the NOAA-CMDL network. Units: g C m$^{-2}$ y$^{-1}$ (Rödenbeck et al. 2003). (c) Average uncertainty reduction on the prior fluxes induced by the atmospheric observations (in percentage).

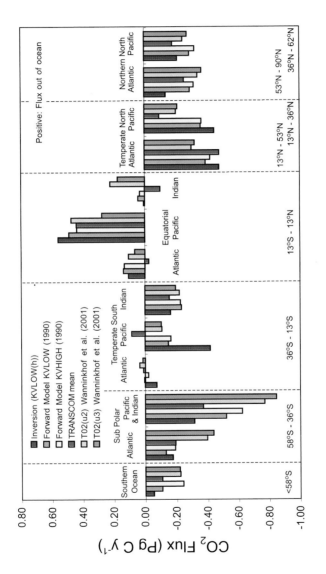

**Colorplate 9.** Regional estimates of air-sea $CO_2$ fluxes (Figure 2c from Gloor et al. 2003) based on ocean tracer inversion and forward models (Gloor et al. 2003), atmospheric inversions (labeled TRANSCOM; Gurney et al. 2002), and ocean measurements (labeled T02; Takahashi et al. 2002).

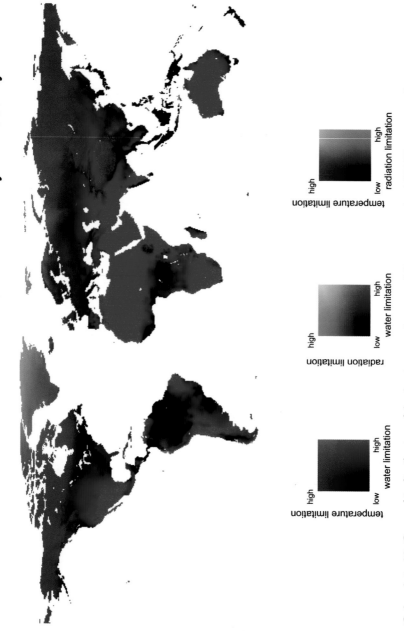

**Colorplate 10.** Geographic distribution of the dominant environmental factors governing NPP. This Figure refers to NPP on an annual basis and does not consider seasonal variations in environmental controls of NPP. After Churkina and Running (1998).

# 10

# Climate–Carbon Cycle Interactions

Pierre Friedlingstein

On centennial timescales the climate is largely controlled by the composition of the atmosphere. Over the past 150 years, the concentration of greenhouse gases (GHGs) rose to levels that the Earth has not experienced for at least 500,000 years. Atmospheric $CO_2$, the major anthropogenic GHG, increased from 280 parts per million (ppm) before 1850 to 370 ppm today (Etheridge et al. 1996). The increase in atmospheric $CO_2$ was caused by the imbalance between emissions from fossil-fuel combustion and deforestation on the one hand and uptake by the land and oceans on the other.

Uptake by land and oceans is driven by biogeochemical processes such as photosynthesis and physical processes such as oceanic circulation. These processes are influenced by climate and atmospheric $CO_2$. There is, therefore, a feedback loop between the climate system and the carbon cycle. Changes in atmospheric $CO_2$ will drive a climate change that will affect the ability of land and oceans to absorb atmospheric $CO_2$. If climate change, induced by enhanced atmospheric $CO_2$, leads to an increase in carbon uptake, the feedback is negative. Conversely, a positive feedback occurs if the climate change leads to a reduction of land and ocean $CO_2$-uptake (Friedlingstein et al. 2003). On annual to decadal timescales, this feedback is negligible. Climate anomalies such as El Niño–Southern Oscillation (ENSO) have a strong impact on land and ocean $CO_2$ fluxes and therefore on atmospheric $CO_2$ (Bousquet et al. 2000), but the induced $CO_2$ anomalies, on the order of a few ppm, are too small to cause appreciable climate feedback. On longer timescales, the story is different. Over the past four glacial cycles (400,000 years), ice-core data show a strong correlation between temperature and atmospheric $CO_2$ (Petit et al. 1999). Atmospheric $CO_2$ fell to 200 ppm during glacial periods and reached 280 ppm during the interglacials. Recent studies based on both conceptual models and careful analyses of leads and lags in paleo-records show that atmospheric $CO_2$ and Antarctic temperature are almost in phase (temperature is believed to precede $CO_2$ by about 100 years), and they both lead major changes in ice sheets by a couple of thousand years (Petit et al. 1999; Shackleton 2000). The change in atmospheric $CO_2$ contributes significantly to glacial-interglacial temperature differences. Model studies show that the 80 ppm difference in atmospheric $CO_2$ may explain up to half of the observed temperature change.

In the context of future climate change, there is a potentially important coupling between the change in the climate and the functioning of the carbon cycle. Although large uncertainty remains in future climate predictions, several patterns are robust. For a given greenhouse gas emission scenario, models predict global warming ranging from 1.3°C to 4.5°C by the end of the 21st century (Cubasch et al. 2001). Projected warming is generally greater on land than on the ocean. All models predict an increase in global mean precipitation (from 1.3 to 6.8 percent for the SRES A2 scenario), although the spatial pattern is model dependent. Several simulations show an increase in precipitation in the tropics (especially over the oceans) and decreased rainfall in the sub-tropics. Because of changes in precipitation and evaporation, soil moisture, which is a crucial quantity for vegetative activity, may also experience changes. In general, models predict increased soil moisture in high northern latitudes in winter but drier soils in summer. Most models show a decrease in the strength of the meridian circulation caused by a reduced equator-to-pole temperature gradient combined with a precipitation-induced decrease in salinity at high latitudes. These changes in surface temperature, soil moisture, salinity, and ocean circulation are important because they control the major processes driving the land and ocean exchange of $CO_2$ with the atmosphere.

On land, biosphere activity, which is mainly driven by the climate, exerts a strong control over the fluxes of water, energy, and $CO_2$. There are, therefore, numerous feedbacks between the land biogeochemistry and the atmosphere. This is not generally the case for the ocean, where biological activity has little control over sea-air fluxes of water and energy. A recent study, however, showed that major blooms of phytoplankton have the potential to greatly reduce light penetration into the ocean, leading to a local enhancement of sea surface temperature (Murtugudde et al. 2002). Land-atmosphere feedbacks operate at different scales, from local (e.g., diurnal coupling between the vegetation and surrounding atmosphere) to continental (e.g., Amazon deforestation) and global (e.g., carbon cycle). An extended review of local-scale feedbacks can be found in Raupach (1998).

In the following, I concentrate on global-scale feedbacks caused by changes in atmospheric $CO_2$, climate, or both (Figure 10.1).

## Climate–Carbon Cycle Feedbacks Driven by Atmospheric $CO_2$

Although this chapter concentrates on feedbacks through the global carbon cycle, it is worth mentioning two pioneering studies on interactions between change in climate, atmospheric $CO_2$, and changes in ecosystem functioning in a coupled mode. Increased atmospheric $CO_2$ affects plant physiology. Stomata conductance decreases with increasing $CO_2$ concentration (Field et al.1995), typically reducing water loss without affecting $CO_2$ uptake. Sellers et al. (1996) used an atmospheric Global Circulation Model

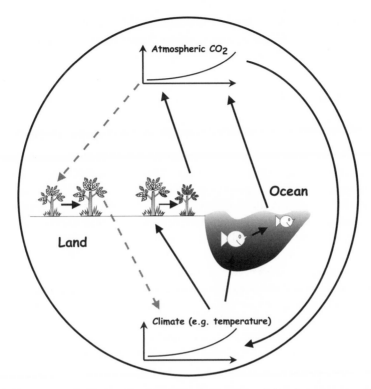

**Figure 10.1.** Global feedbacks between the carbon cycle and the climate system. The dashed arrows represent the impact of increased $CO_2$ on plant biogeochemistry and the feedback on the climate system caused by changes in water and energy fluxes. The solid arrows represent the impact of climate change on the land and ocean carbon cycles, the impact on the growth rate of atmospheric $CO_2$, and the feedback on the climate system.

(GCM) coupled to a land surface scheme to compare this physiological effect to the standard radiative (GHG) effect in a $2 \times CO_2$ experiment. They showed that a decrease in stomata conductance induces an increase in air temperature, amplifying the changes caused by the radiative effect. In the tropics, this process leads to an additional 0.4°C warming on top of the 1.7°C warming caused by the radiative effect.

Shortly after Sellers et al. (1996), Betts et al. (1997) confirmed the stomata feedback but noted a second mechanism that acts in the opposite direction. An increase in $CO_2$ can enhance plant productivity leading to increased leaf biomass and therefore higher plant transpiration. They found that the two effects (decreased stomata conductance and increased leaf biomass) tended to cancel each other, leading to a net effect on surface temperature lower than that reported by Sellers et al. Another potentially important structural feedback that has received little attention is the change in root distribution.

Climate change and increase in $CO_2$ have the potential to affect allocation to roots (Friedlingstein et al. 1999), which in turn, affects climate through changes in plant transpiration (Kleidon et al. 2000).

## Climate–Carbon Cycle Feedbacks Driven by Climate

Both land and ocean carbon cycles are primarily driven by the state of the climate. Several "offline" carbon cycle studies, driven by but not coupled to the climate, have investigated the impact of future climate changes on the carbon cycle. These generally agree that climate change may reduce the uptake of $CO_2$ by the ocean or the land biosphere.

Melillo et al. (1993) performed one of the earliest studies, in which a terrestrial model was forced by both future $CO_2$ and climate. Since that time, many terrestrial models have been run under simulated future climates at regional (Vemap Members 1995; White et al. 2000) or global scales (King et al. 1997; Cao and Woodward 1998; Cramer et al. 2001; Berthelot et al. 2002). Climate change generally causes a reduction in net biosphere uptake. This negative impact is generally greatest in the tropics, where higher temperatures and increased drought reduce plant productivity but increase heterotrophic respiration. In an extreme case, this process leads to a severe dieback of tropical forest, which further increases the carbon loss from the biosphere (Cramer et al. 2001). Temperate and high-latitude regions often increase carbon stocks in response to simulated climate change (Cao and Woodward 1998; White et al. 2000; Berthelot et al. 2002). The patterns of ecosystem response to climate change, shown in one recent study, project that tropical forest expands (Joos et al. 2001). Boreal trees, however, are partly replaced by temperate trees and grasses because of the warm winters and hot summers. This pattern may reflect a difference in climate forcing, with increased soil moisture in the tropics. Ocean studies generally show that the effect of climate change on ocean circulation and sea surface temperature tends to reduce the global ocean carbon uptake because stratification of the upper ocean increases as circulation slows (Maier-Reimer et al. 1996; Sarmiento and Le Quéré 1996; Sarmiento et al. 1998; Joos et al. 1999; Matear and Hirst 1999; Friedlingstein et al. 2001).

These studies were extremely important because they highlighted the potential for positive feedbacks from the coupling between the climate system and the carbon cycle. Reduced carbon uptake for both land and ocean should, with realistic coupling, translate into a higher atmospheric $CO_2$ concentration.

Two recent studies demonstrate the importance of this feedback using fully coupled climate–carbon cycle models. The Hadley (Cox et al. 2000) and IPSL (Dufresne et al. 2002) groups used Ocean-Atmosphere General Circulation Models (OAGCMs) to calculate future climate changes in response to an Intergovernmental Panel on Climate Change (IPCC) scenario of $CO_2$ emissions. Atmospheric $CO_2$ was calculated as the difference between the imposed emissions and the land and ocean carbon uptake, which was a function of the changing climate.

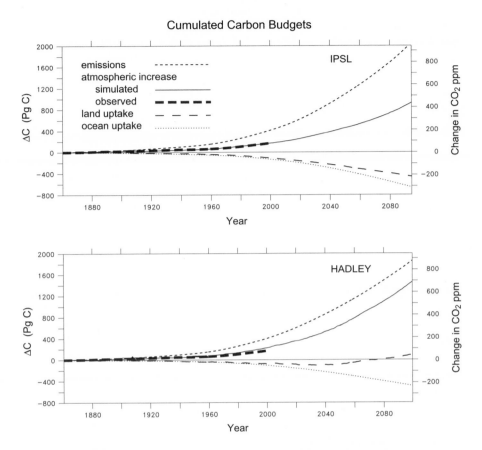

**Figure 10.2.** Cumulative carbon budgets for the IPSL (top) and Hadley (bottom) coupled simulations. All numbers are differences from the preindustrial steady state. For the Hadley model, atmospheric $CO_2$ is higher and the land sink is much smaller, becoming a source after 2080. For the IPSL model, ocean uptake is much larger.

Both models simulated a positive climate-carbon feedback. The Hadley model feedback was much stronger than from the IPSL. Without climate impacts on the carbon cycle, both models simulate an atmospheric $CO_2$ concentration of 700 ppm. With the climate impact, the Hadley model reaches a $CO_2$ level in 2100 of 980 ppm, while the IPSL model reaches 780 ppm (Figure 10.2). Although the sensitivity of the two models is very different, both attribute the positive feedback to the land biosphere. Both climate models simulate a substantial land warming associated with a general drying in the tropics. This causes reduced tropical ecosystem productivity (leading in the Hadley simulations to replacement of tropical forests by savannas). In addition, the increased temperature leads to accelerated loss of soil carbon from increased respiration. This effect is limited to the tropics in the IPSL results but is worldwide for the Hadley output.

In both simulations the magnitude of ocean carbon uptake is largely unaffected by the climate change. This contrast with the previous offline studies reflects another aspect of the coupling. In both simulations, the climate-induced increase in atmospheric $CO_2$, caused by reduced land carbon uptake, leads to enhanced geochemical uptake for both land and ocean (Dufresne et al. 2002). For the ocean, this enhancement is large enough to counterbalance the direct climate-induced reduction of the ocean sink. This additional (negative) feedback is not accounted in offline studies. Mathematically, (1) climate change is proportional to $CO_2$, (2) carbon uptake is proportional to $CO_2$, and (3) carbon uptake is proportional to climate change, one can write:

$$\Delta T_{cou} = 1/(1 - g)*\Delta T_{unc} \tag{1}$$

$$\Delta C_{cou} = 1/(1 - g)*\Delta C_{unc} \tag{2}$$

where $\Delta T_{cou}$ ($\Delta C_{cou}$) and $\Delta T_{unc}$ ($\Delta C_{unc}$) are respectively the coupled and uncoupled changes in temperature (atmospheric $CO_2$), and $g$ is the gain of the climate-carbon feedback (Friedlingstein et al. 2003). This gain is a function of the sensitivities of climate to $CO_2$, carbon cycle to $CO_2$, and carbon cycle to climate change. The sensitivity of the IPSL ocean carbon cycle to atmospheric $CO_2$ is much larger than that of the Hadley model. Further, the sensitivity of the Hadley land carbon cycle to climate change is much larger than that in the IPSL model. This latter difference explains most of the difference between the simulations.

In summary, large uncertainties remain in the carbon cycle, especially in the coupling of the climate and the carbon cycle. The few fully coupled studies to date reveal large gaps in our understanding of soil carbon dynamics, large-scale vegetation dynamics, and ocean circulation (especially in the southern oceans). Process studies based on manipulative experiments and long-term monitoring of carbon cycle components are critical for filling this gap. With our current knowledge, it seems unlikely that the future carbon cycle will act as a strong negative feedback on global warming.

## Literature Cited

Berthelot, M., P. Friedlingstein, P. Ciais, P. Monfray, J.-L. Dufresne, H. LeTreut, and L. Fairhead. 2002. Global response of the terrestrial biosphere to $CO_2$ and climate change using a coupled climate-carbon cycle model. *Global Biogeochemical Cycles* 10:1029/2001GB001827.

Betts, R. A., P. M. Cox, S. E. Lee, and F. I. Woodward. 1997. Contrasting physiological and structural vegetation feedbacks in climate change simulations. *Nature* 387:796–799.

Bousquet, P., P. Peylin, P. Ciais, C. Le Quéré, P. Friedlingstein, and P. P. Tans. 2000.

Regional changes in carbon dioxide fluxes of land and oceans since 1980. *Science* 290:1342–1346.

Cao, M., and F. I. Woodward. 1998. Dynamic responses of terrestrial ecosystem carbon cycling to global climate change. *Nature* 393:249–252.

Cox, P., R. Betts, C. Jones, S. Spall, and I. Totterdell. 2000. Acceleration of global warming due to carbon-cycle feedbacks in a coupled climate model. *Nature* 408:184–187.

Cramer, W., A. Bondeau, F. Woodward, I. Prentice, R. Betts, V. Brovkin, P. Cox, V. Fisher, J. Foley, A. Friend, C. Kucharik, M. Lomas, N. Ramankutty, S. Sitch, B. Smith, A. White, and C. Young-Molling. 2001. Global response of terrestrial ecosystem structure and function to $CO_2$ and climate change: Results from six dynamic global vegetation models. *Global Change Biology* 7:357–373.

Cubasch, U., G. A. Meehl, G. J. Boer, R. J. Stouffer, M. Dix, A. Noda, C. A. Senior, S. Raper, and K. S. Yap. 2001. Projections of future climate change. Pp. 525–582 in *Climate change 2001: The scientific basis (Contribution of Working Group I to the Third Assessment Report of the Intergovernmental Panel on Climate Change)*, edited by J. T. Houghton, Y. Ding, D. J. Griggs, M. Noguer, P. J. van der Linden, X. Dai, K. Maskell, and C. A. Johnson. Cambridge: Cambridge University Press.

Dufresne, J.-L., P. Friedlingstein, M. Berthelot, L. Bopp, P. Ciais, L. Fairhead, H. LeTreut, and P. Monfray. 2002. Effects of climate change due to $CO_2$ increase on land and ocean carbon uptake. *Geophysical Research Letters* 29, 10.1029/2001GL013777.

Etheridge, D., L. Steele, R. Langenfelds, R. Francey, J.-M. Barnola, and V. Morgan. 1996. Natural and anthropogenic changes in atmospheric $CO_2$ over the last 1000 years from air in antarctic ice and firn. *Journal of Geophysical Research* 101:4115–4128.

Field, C. B., R. B. Jackson, and H. A. Mooney. 1995. Stomatal responses to increased $CO_2$: Implications from the plant to the global scale. *Plant, Cell and Environment* 18:1214–1225.

Friedlingstein, P., G. Joel, C. Field, and I. Fung. 1999. Toward an allocation scheme for global terrestrial carbon models. *Global Change Biology* 5:755–770.

Friedlingstein, P., L. Bopp, P. Ciais, J.-L. Dufresne, L. Fairhead, H. Le Treut, P. Monfray, and J. Orr. 2001. Positive feedback between future climate change and the carbon cycle. *Geophysical Research Letters* 28:1543–1546.

Friedlingstein, P., J.-L. Dufresne, P. M. Cox, and P. Rayner. 2003. How positive is the feedback between climate change and the carbon cycle? *Tellus* 55:692–700.

Joos, F., G.-K. Plattner, T. F. Stocker, O. Marchal, and A. Schmittner. 1999. Global warming and marine carbon cycle feedbacks on future atmospheric $CO_2$. *Science* 284:464–467.

Joos, F., I. C. Prentice, S. Sitch, R. Meyer, G. Hooss, G.-K. Plattner, S. Gerber, and K. Hasselmann. 2001. Global warming feedbacks on terrestrial carbon uptake under the Intergovernmental Panel on Climate Change (IPCC) emission scenarios. *Global Biogeochemical Cycles* 15:891–908.

King, A., W. Postand, and S. Wullschleger. 1997. The potential response of terrestrial carbon storage to changes in climate and atmospheric $CO_2$. *Climatic Change* 35:199–227.

Kleidon, A., K. Fraedrich, and M. Heimann. 2000. A green planet versus a desert world: Estimating the maximum effect of vegetation on the land surface climate. *Climatic Change* 44:471–493.

Maier-Reimer, E., U. Mikolajewicz, and A. Winguth. 1996. Future ocean uptake of $CO_2$: Interaction between ocean circulation and biology. *Climate Dynamics* 12:711–721.

Matear, R., and A. Hirst. 1999. Climate change feedback on the future oceanic $CO_2$ uptake. *Tellus* 51B:722–733.

Melillo, J. M., A. D. McGuire, D. W. Kicklighter, B. Moore, C. J. Vorosmarty, and A. L. Schloss. 1993. Global climate change and terrestrial net primary production. *Nature* 363:234–240.

Murtugudde, R., J. Beauchamps, C. R. McClain, M. Lewis, and A. Busalacchi. 2002. Effects of penetrative radiation on the upper tropical ocean circulation. *Journal of Climate* 15:470–486.

Petit, J., J. Jouzel, D. Raynaud, N. I. Barkov, J.-M. Barnola, I. Basile, M. Bender, J. Chappellaz, M. Davis, G. Delaygue, M. Delmotte, V. M. Kotlyakov, M. Legrand, V. Y. Lipenkov, C. Lorius, L. Pepin, C. Ritz, E. Saltzman, and M. Stievenard. 1999. Climate and atmospheric history of the past 420,000 years from the Vostok ice core, Antarctica. *Nature* 399:429–436.

Raupach, M. 1998. Influences of local feedbacks on land-air exchanges of energy and carbon. *Global Change Biology* 4:477–494.

Sarmiento, J. L., and C. Le Quéré. 1996. Oceanic carbon dioxide uptake in a model of century-scale global warming. *Science* 274:1346–1350.

Sarmiento, J. L., T. M. C. Hughes, R. J. Stouffer, and S. Manabe. 1998. Simulated response of the ocean carbon cycle to anthropogenic climate warming. *Nature* 393:245–249.

Sellers, P. J., D. A. Randall, G. J. Collatz, J. Berry, C. Field, D. A. Dazlich, and C. Zhang. 1996. A revised land-surface parameterization (SiB2) for atmospheric GCMs. I. Model formulation. *Journal of Climate* 9:676–705.

Shackleton, N. 2000. The 100,000-year ice-age cycle identified and found to lag temperature, carbon dioxide, and orbital eccentricity. *Science* 289:1897–1902.

Vemap Members. 1995. Vegetation/ecosystem modeling and analysis project: Comparing biogeography and biogeochemistry models in a continental-scale study of terrestrial ecosystem responses to climate change and $CO_2$ doubling. *Global Biogeochemical Cycles* 9:407–437.

White, A., M. Cannel, and A. Friend. 2000. The high-latitude terrestrial carbon sink: A model analysis. *Global Change Biology* 6:227–245.

# 11

# Socioeconomic Driving Forces of Emissions Scenarios

Nebojsa Nakicenovic

Numerous factors, sometimes called driving forces, influence future greenhouse gas (GHG) emissions. Both future emissions and the evolution of their underlying driving forces (e.g., rate of technology changes, prices) are highly uncertain. This is the reason scenarios are used to describe how the future may develop, based on a coherent and internally consistent set of assumptions about key relationships, driving forces, and the emissions outcomes. The emissions scenarios in the literature encompass a wide range of different future developments that might influence GHG sources and sinks, such as alternative structures of energy systems and land use changes (Nakicenovic et al. 2000). Clearly, demographic and economic developments play a crucial role in determining emissions. Another driving force, at least as important, is technological change. Other factors are diverse, and it is not possible to devise a simple scheme that accounts for all the factors considered and included in the scenarios. They range from human resources such as education, institutional frameworks, and lifestyles to natural resource endowments, capital vintage structures, and international trade patterns.

For this chapter, I choose a simple scheme captured by the so-called Kaya identity. This hypothetical identity represents the main emissions driving forces as multiplicative factors (Yamaji et al. 1991):

$$CO_2 = (CO_2/E) \times (E/GWP) \times (GWP/P) \times P,$$

where $E$ represents energy consumption, $GWP$ the gross world product, and $P$ population. Changes in $CO_2$ emissions can be described by changes in these four factors or driving forces.

This type of identity was originally used to assess different impacts. The underlying method of analysis is called IPAT: Impact = Population × Affluence × Technology. The IPAT identity states that environmental impacts (e.g., emissions) are the product of the

225

level of population, affluence (income per capita), and the level of technology deployed (emissions per unit of income).

The advantage of the Kaya identity, and the IPAT approach in general, is that it is simple and facilitates at least some standardization in the comparison and analysis of historical developments and many diverse scenarios of future emissions. An important caveat is that the driving forces that determine emissions are not independent of each other, even though they are treated as such in the identity. In fact, in many scenarios they explicitly depend on each other. For example, based on historical experience, scenario builders often assume that high rates of economic growth lead to high capital turnover. This favors more advanced and more efficient technologies, resulting in lower energy intensities and thus lower emissions. Often a weak inverse relationship is assumed between population and economic growth. Thus, the scenario ranges for these main driving forces are not necessarily independent of each other. Also note that these kinds of relationships tend to partially compensate for or offset emissions consequences of a wide range of individual driving forces, for example, higher economic growth rates are associated with lower population growth and lower emissions intensities.

In the following I present the historical developments and scenario ranges for each of the factors in the Kaya identity representing the main (energy-related) emissions driving forces: population, GWP, energy consumption (total consumption and energy intensity), and carbon emissions (total and intensity). This approach has been used often, to review historical developments and scenarios in the literature (Ogawa 1991; Parikh et al. 1991; Nakicenovic et al. 1993; Parikh 1994) and in particular by the Intergovernmental Panel on Climate Change (IPCC) (Alcamo et al. 1995; Nakicenovic et al. 2000; Morita et al. 2001). This chapter is based on the findings reported in Nakicenovic (1996) and Nakicenovic et al. (1998, 2000). I begin with global $CO_2$ emissions, represented as a "dependent variable" in the Kaya identity, and then analyze the other factors, represented as "independent variables" (main emissions scenario driving forces). This sequence corresponds to the way the main scenario driving forces are represented in the Kaya identity, without implying a priori any causal relationships among the driving forces themselves or between the driving forces and the resulting $CO_2$ emissions.

The scenarios in the literature cover a wide range of future GHG emissions. There are more than 500 emissions scenarios in the literature (Nakicenovic et al. 2000). Most of them are documented in a database (Morita and Lee 1998) that is accessible at www-cger.nies.go.jp/cger-e/db/ipcc.html.

## Carbon Dioxide Emissions

The span of $CO_2$ emissions across all scenarios in the database is indeed large, ranging in 2100 from 10 times the current emissions all the way to negative emissions (due to carbon sequestration and enhanced carbon sinks, which are included in some scenarios). There are many possible interpretations of this large range and many good reasons

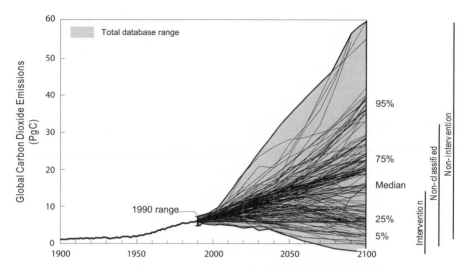

**Figure 11.1.** Global carbon emissions: Historical development and scenarios. Emissions are indexed to 1990, when actual global energy-related $CO_2$ emissions were about 6 PgC. The figure includes 232 scenarios. Three vertical bars on the right-hand side indicate the ranges for scenarios with emissions control such as mitigation measures (labeled "Intervention"), for scenarios that may or may not include controls ("Non-classified"), and for those without controls ("Non-intervention") (Morita and Lee 1998; Nakicenovic et al. 1998, 2000).

it should be so large. Common to all of them is the high uncertainty about how the main driving forces such as population growth, economic development, and energy production, conversion, and end use might unfold during the next century.

Figure 11.1 shows the global, energy-related, and industrial $CO_2$ emission paths from the database as "spaghetti" curves for the period to 2100 against the background of the historical emissions from 1900 to 1990. Historical emissions increased continuously, whereas the emissions scenarios cover a wider range of alternative developments. Future emissions are normalized against an index on the vertical axis rather than as absolute values, because of large differences in the values assumed for the base year, 1990. These sometimes arise from genuine differences among the scenarios (e.g., different data sources, definitions), different base years assumed in the analysis, or alternative calibrations.

Historically, gross $CO_2$ emissions have increased at an average rate of about 1.7 percent per year ($y^{-1}$) since 1900 (Nakicenovic 1996); if that trend continues, global emissions would double during the next three to four decades and increase more than sixfold by 2100. Many scenarios in the database describe such a development. Even by 2030, however, the range around this possible doubling of global emissions is very large.

The highest scenarios have emissions four times the 1990 level by 2030, whereas the lowest are barely above half the current emissions. This divergence continues so that the highest scenarios envisage a 10-fold increase of global emissions in 2100. These high emissions levels would no doubt lead to an alarming increase in atmospheric $CO_2$ concentrations and could cause significant global climate change. The median scenarios lead to about a 3-fold emissions increase over the same time period or to about 16 PgC. This is somewhat lower than the medians from the earlier scenarios comparisons (see Alcamo et al. 1995). Nevertheless, the median emissions path would more than double atmospheric concentrations of $CO_2$ to approximately 750 parts per million (ppm) by 2100. (The current value is about 370 ppm.) The distribution of emissions, however, is asymmetric with long "tails." The thin emissions tail that extends above the 95th percentile (i.e., between the 6- and 10-fold increase of emissions by 2100 compared with 1990) includes only a few scenarios. A number of scenarios in the low range below the 5th percentile are consistent with stabilizing concentrations at relatively benign levels of 450 to 550 ppm. All scenarios that are consistent with stabilization paths reach maximum emissions and decline to below current emissions levels by the end of the 21st century.

The spaghetti curves in Figure 11.1 do not include all of the global emissions paths in the database, but are a representative sample of scenarios, selected to be discernable as individual trajectories. It should be noted that not all emissions paths in Figure 11.1 increase monotonically; some oscillate from high rates of increase to periods of declining emissions; other paths cross each other. What is important is that, on average, there is a strong trend toward continuous increases in emissions, with most of the scenarios clustering between a small decrease in global emissions over the next century to more than a 4-fold increase. It is also important to note that the distribution of scenarios is asymmetric with long tails.

## Population Projections

Population is one of the fundamental driving forces of future emissions. Consequently, all emissions scenarios require some kind of population assumptions. Nevertheless, relatively few of the underlying population projections are reported in the literature, perhaps because they are exogenous inputs in most of the models used to formulate emissions scenarios.

Three main research groups project long-term global population: United Nations (UN 1998), World Bank (Bos and Vu 1994), and the International Institute for Applied Systems Analysis (IIASA) (Lutz et al. 1997). The so-called central, or median, population projections lead to less than a doubling of current global population to some 10 billion people by 2100 compared with 6 billion in 2000. In recent years the central population projections for the year 2100 have declined somewhat. The latest long-term UN (2002) medium-low and medium-high projections show a range of 5.2 to 16.2 bil-

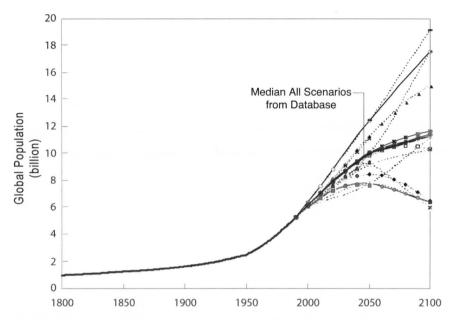

**Figure 11.2.** Global population projections: Historical development and scenarios. The figure includes 59 scenarios (Durand 1967; Demeny 1990; Morita and Lee 1998; Nakicenovic et al. 1998, 2000).

lion people by 2100, with the medium scenario at 9.5 billion. The IIASA central estimate for 2100 is also 8.4 billion, with 95 percent of projections above 7.3 and below 12 billion (Lutz et al. 2001).

Figure 11.2 contrasts the global population projections across the scenarios in the literature with the historical trend. The long-term historical population growth rate has averaged about 1 percent $y^{-1}$ during the past two centuries and at about 1.3 percent $y^{-1}$ since 1900. Currently, the world's population is increasing at about 2 percent $y^{-1}$. The global population projections envision a slowing population growth in the future. The most recent doubling of the world population took approximately 40 years. Even the highest population projections in Figure 11.2 require 70 years or more for the next doubling, while roughly half of the scenarios are well below these levels during the 21st century. The lowest average population growth across all projections is 0.1 percent per year, the highest is 1.2 percent per year, and median is about 0.7 percent per year.

Of the 428 scenarios documented in the database (Morita and Lee 1998), only 59 report their underlying population projections. The range is from more than 6 to about 19 billion in 2100, with the central or median estimates in the range of 10 billion. Thus, the population assumptions in the emissions scenarios appear to be broadly consistent with the recent population projections, with the caveat that only a few underlying projections have been reported in the literature. The population projections across the emission scenarios are not evenly distributed across the whole range. Instead, they are grouped into three clusters. The middle cluster is representative of the central projections, with the range of about 9 billion to fewer than 12 billion people by 2100. The other two clusters mark the highest and the lowest population projections available in the literature with about 6 billion on the low end and between 14 and 19 billion at the high end.

Despite the large range of future global populations across alternative projections, the variation by 2100, compared with the base year, is the smallest of all scenario-driving forces considered in this comparison. Compared with 1990 global population, the proportional increase varies from less than one to less than four.

## Gross World Product (GWP)

Economic development is a fundamental prerequisite for achieving an increase in living standards. It is thus not surprising that the assumptions about economic development are also among the most important determinants of emissions levels in the scenarios. At the same time, the prospects for economic growth are among the most uncertain determinants of future emissions. This uncertainty is reflected in the very wide range of economic development paths assumed in the scenarios. Figure 11.3 shows the future increase in GWP compared with the historical experience since 1950 (measured at market exchange rates).

The historical GWP growth rate has been about 4 percent $y^{-1}$ since 1950. In the scenarios the average growth rates to 2100 range from 1.1 percent $y^{-1}$ to 3.2 percent $y^{-1}$, with the median value of 2.3 percent $y^{-1}$. This translates into a GWP in 2100 that varies from 3.5 to more than 32 times the 1990 GWP. The 1990 and 2000 GWP were about US\$20 trillion and US\$35 trillion, respectively. These translate into a range of about US\$70 to more than US\$640 trillion (in 1990 dollars) by 2100. The full range of GWP development includes a few noticeable outliers, while the rest of the scenarios are grouped much more closely, compressing the range to a proportional increase of about 7 to 17 times compared to 1990.

## Gross World Product per Capita

The scenarios in the literature portray a weak, slightly negative relationship between population and economic growth. Scenarios that lead to very high GWP are generally associated with central to low population projections, while high population projections do not lead to the highest GWP scenarios. This important tendency tends to lead to higher development in scenarios with completion of the population transition during the 21st century. Scenarios with continuous population growth tend to lag in economic development, magnifying current poverty and deprivation in the world. At the same

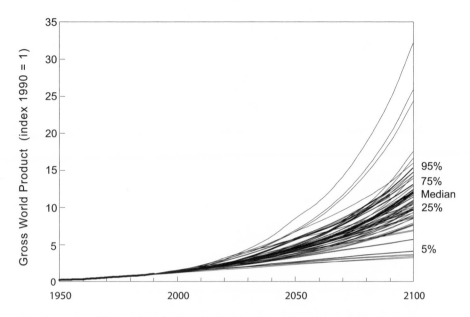

**Figure 11.3.** Global economic development measured by gross world product (GWP) measured at market exchange rates: Historical development and scenarios. The figure includes 167 scenarios. Because of the relatively large differences in the base-year GWP across scenarios, the GWP paths are plotted as an index and spliced to historical data in 1990 (United Nations 1993a, b; Morita and Lee 1998; Nakicenovic et al. 2000).

time, this tendency "compresses" the range of economic development, because scenarios with high population tend to have lower per capita economic development, and ones with low population tend to have higher per capita GWP.

Figure 11.4 illustrates some of the relationships between population and GWP in the scenarios. It compares only 39 scenarios, because the information about population assumptions is available for only a few scenarios. In most of them, a global population transition is achieved during the next century and stabilization occurs at a population between 10 and 12 billion people. Generally, this is associated with relatively high levels of economic development, in the range from US$200 to US$500 trillion (in 1990 dollars).

Scenarios on the lower end of this scale are collectively labeled the "mid-range cluster." One example (IS92f) leads to 18 billion people by 2100 with comparatively low economic growth. The two highest scenarios are labeled as the "extra-high-growth" cases. On the other side of the scale are two scenarios with low population projections (about 6 billion people by 2100). They are representing the "great transitions" scenarios, because they lead to relatively high economic growth with low global populations[1].

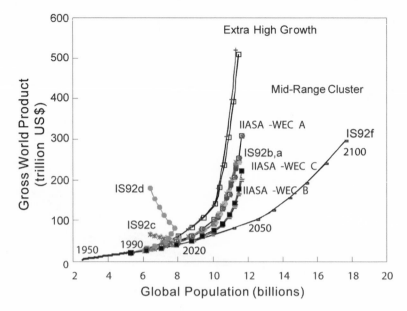

**Figure 11.4.** GWP and population growth: Historical development and scenarios. The figure includes 39 scenarios. Some of the scenarios are identified by name (Nakicenovic et al. 1998, 2000).

Recently, demographers have given a lot of attention to such low population scenarios, because they are consistent with continuous declines in fertility rates across the world.

## Primary Energy Requirements

Primary energy requirements are another of the fundamental determinants of GHG emissions. Clearly, high primary energy levels tend to result in high GHG emissions. More important for emissions, however, is the structure of energy systems. High carbon intensities of energy—namely high shares of fossil energy sources, especially coal—lead to scenarios with very high carbon emissions. Here, the primary energy paths of different scenarios and energy carbon intensity are treated in turn.

Figure 11.5 shows the primary energy consumption paths across the scenarios, plus the historical development since 1900. Because of relatively large differences in the base-year values, the primary energy consumption paths are plotted as an index and spliced to the historical data in 1990. In 1990 primary energy was about 360 exajoules (EJ). It was about 400 EJ (10 petagrams [Pg] oil$_{equiv.}$) in 2000.

On average, global primary energy consumption has increased at more than 2 percent y$^{-1}$ (fossil energy alone has risen at almost 3 percent y$^{-1}$) since 1900. In the sce-

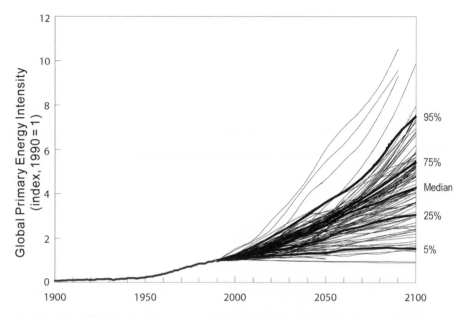

**Figure 11.5.** Global primary energy requirements: Historical development and scenarios. The figure includes 153 scenarios. Because of the relatively large differences in the base-year primary energy requirements across scenarios, the primary energy paths are plotted as an index and spliced to historical data in 1990 (Morita and Lee 1998; Nakicenovic et al. 1998).

narios, the average growth rates to 2100 range from 0.1 percent $y^{-1}$ to 2.4 percent $y^{-1}$, with a median value of 1.3 percent $y^{-1}$.

The full range includes a few noticeable outliers (Figure 11.5), especially some scenarios without emissions control (non-control) or mitigation measures, which fall toward the high end of energy consumption levels. The rest of the scenarios are grouped more closely, compressing the range to a factor increase of about one to eight times compared with 2000.

## Primary Energy Intensities

Most of the scenarios in the literature portray a clear relationship between primary energy and GWP. In all scenarios, economic growth outpaces the increase in primary energy requirements, leading to substantial reductions in energy intensities (the ratio of primary energy requirements to GWP). This tendency across scenarios is the cumulative result of technological and structural changes. Individual technologies improve or are replaced by new ones—inefficient technologies are retired in favor of more efficient

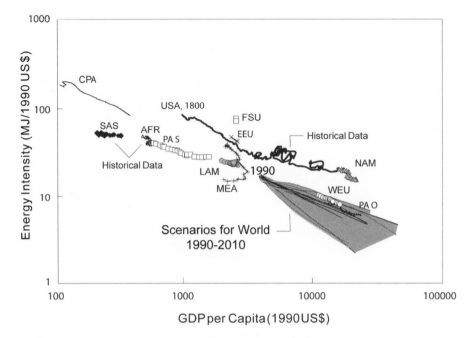

**Figure 11.6.** Primary energy intensity versus GDP per capita measured at market exchange rates: Regional historical developments for USA, North America (NAM), Latin America and Caribbean (LAM), Western Europe (WEU), Eastern Europe (EEU), former Soviet Union (FSU), Middle East and North Africa (MEA), Sub-Saharan Africa (AFR), centrally planned Asia and China (CPA), South Asia (SAS), other Pacific Asia (PAS), and Pacific OECD (PAO). Future global developments of energy intensities across scenarios are shown as a range (IEA 1993; World Bank 1993; Morita and Lee 1998; Nakicenovic et al. 1998).

ones. Also, the structure of the energy system and patterns of energy services change. All of these factors tend to reduce the aggregate amount of primary energy needed per unit of GWP. All other things being equal, the more rapid is the economic growth, the higher is the turnover of capital, and the greater is the resulting decline in energy intensity. These long-term relationships between energy and economic development are reflected in the majority of scenarios and are consistent with historical experience across a range of alternative development paths in different countries.

Figure 11.6 shows the historical relationship since 1970 between energy intensity and gross domestic product (GDP) per capita for the major world regions. This shorter-term historical record of development is contrasted with the experience of the United States since 1800. In all cases economic development is associated with a reduction in energy intensity; the levels of energy intensities in developing countries today are generally comparable to that of the now industrialized countries when they had the same per capita

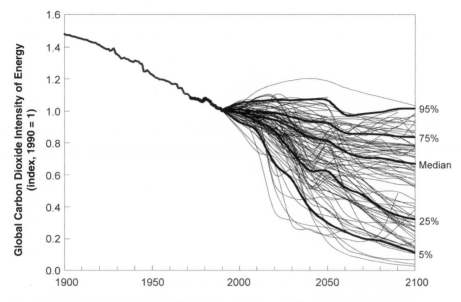

**Figure 11.7.** Global decarbonization of primary energy: Historical development and scenarios. The carbon intensities of the scenarios are shown as indexes spliced into the base year 1990 (Nakicenovic 1996; Morita and Lee 1998).

GDP. The historical experiences illustrate that different countries and regions can follow different development paths with some persistent differences in energy intensities, even at similar levels of per capita GDP.

Global energy intensities diverge across scenarios in the literature. They are contained within a "cone" that opens from the base year 1990, into the future (Figure 11.6). The cone delineates alternative developments across scenarios, with the same general relationship that higher levels of economic development are associated with lower energy intensities. This pattern clearly illustrates the persistent inverse relationship between economic development and energy intensity, across the wide range of scenarios with numerous other differences.

## Decarbonization of Primary Energy

Decarbonization denotes the declining average carbon intensity of primary energy over time. Although the decarbonization of the world's energy system is comparatively slow (0.3 percent $y^{-1}$) (Figure 11.7), the trend has persisted throughout the past two centuries (Nakicenovic 1996). The overall tendency toward lower carbon intensities is due to the continuous replacement of energy sources with high carbon content, such as coal, by those with low carbon content, such as natural gas; intensities are, however, currently increasing in some developing regions. The carbon intensity at the global level has decreased by a third since 1900. This decline is primarily associated with the replacement of coal, which was the dominant energy source at the beginning of the 20th century and accounted for almost 60 percent of all primary energy during the 1920s. Oil and natural gas substituted for coal during the 20th century. Zero-carbon sources such as hydropower and nuclear energy had a much less important role in reducing carbon intensities, because their shares in total global primary energy were relatively small.

The median of all the scenarios indicates the continuation of the historical trend, with a decarbonization rate of about 0.4 percent $y^{-1}$. At this rate, the global carbon intensity would decline by some 40 percent by the end of the 21st century.

All of the scenarios below the median would lead to higher rates of decarbonization. The highest rates (up to 3.3 percent $y^{-1}$) occur in scenarios that envision a complete transition in the energy system away from carbon-intensive fossil fuels. Generally, these are scenarios that describe more sustainable development paths with lower energy intensities as well as significant contributions of renewables. Some of the scenarios in this low range include large shares of nuclear energy and significant decarbonization of fossil energy sources through carbon sequestration and disposal. In such scenarios, carbon is removed from fossil energy sources, primarily natural gas and coal, either pre- or postcombustion, and stored over geological timescales. The most promising storage options are depleted oil and gas fields and deep aquifers, while deep ocean storage possibilities continue to be highly controversial (Brewer, Chapter 27, this volume).

The scenarios that are most intensive in use of fossil fuels lead to practically no reduction in carbon intensity. Some of the scenarios above the median actually anticipate significant increases in carbon intensity, particularly over the next few decades. They include cases where developing countries rich in domestic coal resources, including India and China, as well as other now more developed countries, such as the United States, remain on or return to carbon-intensive development paths. Coal technologies are assumed to improve in these scenarios, leading to much cleaner and less environmentally intrusive development paths, but with extremely high GHG emissions.

Figure 11.8 illustrates the relationships between energy intensities of GWP and carbon intensities of energy across the scenarios in the literature. Scenarios that unfold horizontally are pure decarbonization cases with little structural change in the economy; scenarios that unfold vertically indicate reduction of energy intensity of economic activities, with little change in the energy system. Most scenarios stay away from these extremes and develop a fan-shaped pattern—marked by both decarbonization and declining energy intensity (Figure 11.8).

The fan-shaped graph illustrates the wide range in the policies and structures of the global energy systems among scenarios. For example, even the scenarios that achieve high levels of decarbonization can take alternative development paths. In some, decarbonization is achieved largely through energy-efficiency improvements, while in others,

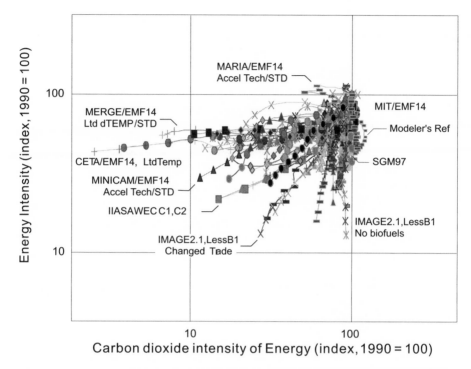

**Figure 11.8.** Global decarbonization and decline of energy intensity (1990–2100). Both intensities are shown on logarithmic scales. The starting point is the base year 1990, normalized to an index (1990 = 100) for both intensities. Some of the scenarios are identified by name (Morita and Lee 1998; Nakicenovic et al. 1998).

decarbonization is mainly the result of lower carbon intensity, due to vigorous substitution of fossil energy sources.

## Literature Cited

Alcamo, J., A. Bouwman, J. Edmonds, A. Grübler, T. Morita, and A. Sugandhy. 1995. An evaluation of the IPCC IS92 emission scenarios. Pp. 233–304 in *Climate change 1994: Radiative forcing of climate change and an evaluation of the IPCC IS92 emission scenarios*, edited by J. T. Houghton, L. G. Meira Filho, J. P. Bruce, H. Lee, B. A. Callander, and E. F. Haites. Cambridge: Cambridge University Press.

Bos, E., and M. T. Vu. 1994. *World population projections: Estimates and projections with related demographic statistics, 1994–1995 edition*. Washington, DC: World Bank.

Demeny, P. 1990. Population. In *The Earth as transformed by human action: Global and regional changes in the biosphere over the past 300 years*, edited by B. L. Turner II, W. C.

Clark, R. W. Kates, J. F. Richards, J. T. Mathews, and W. B. Meyer. Cambridge: Cambridge University Press.

Durand, J. D. 1967. The modern expansion of world population. *Proceedings of the American Philosophical Society* 111 (3): 136–159.

Gallopin, G., A. Hammond, P. Raskin, and R. Swart. 1997. Branch points: Global scenarios and human choice. A Resource Paper of the Global Scenario Group. PoleStar Series 7. Division of Environment and Human Settlements, United Nations Economic Commission for Latin America and the Caribbean, Santiago, Chile.

Houghton, J. T., L. G. Meira Filho, J. P. Bruce, H. Lee, B. A. Callander, and E. F. Haites, eds. 1995. *Climate change 1994: Radiative forcing of climate change and an evaluation of the IPCC IS92 emission scenarios.* Cambridge: Cambridge University Press.

IEA (International Energy Agency). 1993. *Energy statistics and balances for OECD and non-OECD countries, 1971–1991.* Paris: Organisation for Economic Co-operation and Development (OECD).

Lutz, W., W. Sanderson, and S. Scherbov. 1997. Doubling of world population unlikely. *Nature* 387 (6635): 803–805.

————. 2001. The end of world population growth. *Nature* 412:543–545.

Morita, T., and H.-C. Lee. 1998. *IPCC SRES Database, Version 0.1.* Emission Scenario Database prepared for IPCC Special Report on Emission Scenarios. http://www-cger.nies.go.jp/cger-e/db/ipcc.html.

Morita, T., J. Robinson, A. Adegbulugbe, J. Alcamo, D. Herbert, E. L. La Rovere, N. Nakicenovic, H. Pitcher, P. Raskin, K. Riahi, A. Sankovski, V. Sokolov, B. de Vries, and D. Zhou. 2001. Greenhouse gas emission mitigation scenarios and implications. In *Climate change 2001: Mitigation (Contribution of Working Group III to the Third Assessment Report of the Intergovernmental Panel on Climate Change)*, edited by B. Metz, O. Davidson, R. Swart, and J. Pan. Cambridge: Cambridge University Press.

Nakicenovic, N. 1996. Freeing energy from carbon. *Daedalus* 125 (3): 95–112. Also published as pp. 74–88 in *Technological trajectories and the human environment*, edited by J. A. Ausubel and H. D. Langford. Washington, DC: National Academies Press.

Nakicenovic, N., A. Grübler, A. Inaba, S. Messner, S. Nilsson, Y. Nishimura, H.-H. Rogner, A. Schaefer, L. Schrattenholzer, M. Strubegger, J. Swisher, D. Victor, and D. Wilson. 1993. Long-term strategies for mitigating global warming. *Energy* 18 (5): 401–609.

Nakicenovic, N., N. Victor, and T. Morita. 1998. Emissions scenarios database and review of scenarios. *Mitigation and Adaptation Strategies for Global Change* 3 (2–4): 95–120.

Nakicenovic, N., J. Alcamo, G. Davis, B. de Vries, J. Fenhann, S. Gaffin, K. Gregory, A. Gruebler, T. Y. Jung, T. Kram, E. L. La Rovere, L. Michaelis, S. Mori, T. Morita, W. Pepper, H. Pitcher, L. Price, K. Riahi, A. Roehrl, H.-H. Rogner, A. Sankovski, M. Schlesinger, P. Shukla, S. Smith, R. Swart, S. van Rooijen, N. Victor, and D. Zhou. 2000. *Emissions scenarios (Special Report of the Intergovernmental Panel on Climate Change).* Cambridge: Cambridge University Press.

Ogawa, Y. 1991. Economic activity and greenhouse effect. *Energy Journal* 12 (1): 23–34.

Parikh, J. K. 1994. North-south issues for climate change. *Economic and Political Weekly,* November 5–12, 2940–2943.

Parikh, J. K., K. S. Parikh, S. Gokarn, J. P. Painuly, B. Saha, and V. Shukla. 1991. *Consumption patterns: The driving force of environmental stress.* Report prepared for the

United Nations Conference on Environment and Development (UNCED), IGIDR-PP-014. Mumbai, India: Indira Gandhi Institute for Development Research.

United Nations. 1993a. *Macroeconomic data systems, MSPA Data Bank of World Development Statistics, MSPA Handbook of World Development Statistics.* MEDS/DTA/1 and 2 June. New York: UN.

———. 1993b. *UN MEDS Macroeconomic Data Systems, MSPA Data Bank of World Development Statistics.* MEDS/DTA/1 MSPA-BK.93. New York: Long-Term Socioeconomic Perspectives Branch, Department of Economic and Social Information and Policy Analysis, UN.

———. 1998. *World population projections to 2150.* New York: UN Department of Economic and Social Affairs, Population Division.

———. 2002. *World population prospects.* New York: UN Department of Economic and Social Affairs, Population Division. http://esa.un.org.

World Bank. 1993. *World development report 1993.* New York: Oxford University Press.

Yamaji, K., R. Matsuhashi, Y. Nagata, and Y. Kaya. 1991. An integrated system for $CO_2$/energy/GNP analysis: Case studies on economic measures for $CO_2$ reduction in Japan. In *Workshop on $CO_2$ reduction and removal: Measures for the next century, 19–21 March 1991.* Laxenburg, Austria: International Institute for Applied Systems Analysis (IIASA).

# PART III
## The Carbon Cycle of the Oceans

# 12

# Natural Processes Regulating the Ocean Uptake of $CO_2$

Corinne Le Quéré and Nicolas Metzl

## Natural $CO_2$ Fluxes

The oceans hold 50 times more carbon than the atmosphere. The large carbon content of the ocean results from carbon chemistry. When $CO_2$ dissolves in the ocean, it reacts with water and carbonate ($CO_3^{2-}$) to form bicarbonate ($HCO_3^-$) ions according to the following equilibrium reaction:

$$CO_2 + H_2O + CO_3^{2-} \leftrightarrow 2HCO_3^- \tag{1}$$

The sum of $CO_2$, $HCO_3^-$, and $CO_3^{2-}$ is called dissolved inorganic carbon (DIC). Ninety-one percent of DIC is in the form of $HCO_3^-$, 8 percent in $CO_3^{2-}$, and 1 percent in $CO_2$. Only $CO_2$ can exchange with the atmosphere.

At large scale, it is now well established that the ocean takes up $CO_2$ mostly in temperate and high latitudes, whereas the ocean outgases $CO_2$ generally in the tropics (thick full line, Figure 12.1). The air-sea $CO_2$ flux distribution is controlled by $CO_2$ solubility and by biological and physical processes.

$CO_2$ is more soluble in cold than in warm water. When water is heated, $CO_2$ is outgassed to the atmosphere. This process occurs mostly in equatorial regions, where upwelling brings cold water from the deep ocean into contact with the atmosphere (dashed line, Figure 12.1). Conversely, the solubility effect creates $CO_2$ uptake at mid-latitudes, where currents transport warm surface waters from the low latitudes. Most of the intermediate and deep waters of the ocean were last in contact with the atmosphere at high latitudes. Thus, $CO_2$ at depth has equilibrated with the atmosphere at cold temperatures, which creates a concentration of carbon 5 percent higher at depth than at the surface (Murnane et al. 1999). This process is called the solubility pump.

Marine plankton take up carbon and nutrients. When they die or excrete detritus, the

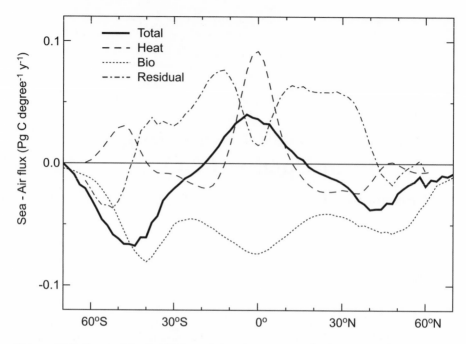

**Figure 12.1.** Sea-air $CO_2$ flux by latitude band (PgC degree$^{-1}$ y$^{-1}$). Negative values indicate a flux from the atmosphere to the ocean. The total $CO_2$ flux was estimated by Takahashi et al. (2002) based on ocean $pCO_2$ observations. The contribution of the biological export production was estimated using SeaWiFS chlorophyll observations averaged over 1997–2001, Behrenfeld and Falkowski's (1997) primary production, and Laws et al.'s (2000) ef-ratio. The contribution of the heat flux was estimated using NCEP heat fluxes averaged over 1985–1995 with the formulation of Murnane et al. (1999). The biology and heat contributions were smoothed with a 5° running mean corresponding to the grid used by Takahashi et al. (2002). Contributions from the biology and heat flux assume an infinite gas exchange. The residual flux is estimated from the total minus the biological and heat contributions. It represents ocean transport and anthropogenic uptake as well as the error associated with the other estimates.

dead tissues either remain in the water as dissolved organic carbon (DOC), which is transported by currents, or aggregate into particulate organic carbon (POC), which sinks and entrains carbon as it falls (dotted line, Figure 12.1). DOC and POC are remineralized back to DIC by living organisms and ultraviolet degradation. Sinking POC and DOC create a flux of carbon from the surface to the deep ocean, where it is isolated from the atmosphere for decades to centuries. The one-way flux of carbon from surface to depth is called export production. Deepwaters, rich in carbon, are transported back to the surface by physical processes (currents and mixing). The entire cycle of carbon consisting of export production balanced by physical transport is called the biological pump.

Some plankton species grow shells of calcium carbonate ($CaCO_3$), which alter the carbon chemistry of surface waters. When $CaCO_3$ shells sink to the deep ocean, the surface concentration of $CO_3^{2-}$ decreases, which drives the equilibrium reaction in equation 1 toward the left and releases $CO_2$, thus having the opposite effect on $CO_2$ from the biological pump. This is called the biological counterpump (or carbonate pump). Only about 10 percent of the export production, however, is associated with $CaCO_3$ production, and the overall effect of the two biological pumps is to increase the concentration of carbon at depth by 10 percent compared with the surface ocean (Murnane et al. 1999).

Plankton growth is limited by the availability of light, nutrients, and by grazing. Unlike the case for terrestrial plants, high $CO_2$ does not enhance plankton growth, because carbon is already abundant in the ocean. Nutrients are abundant in the intermediate and deep ocean, but they are depleted over most of the ocean surface, where growth takes place. The Southern Ocean and the North and Equatorial Pacific Oceans, however, have high nutrients but low biological productivity. They are called high-nutrient, low-chlorophyll (HNLC) regions. Their low productivity is generally explained by the low availability of iron (Bakker, Chapter 26, this volume). Iron is needed by phytoplankton for chlorophyll synthesis and nitrate reduction. The absence of iron thus limits productivity. In situ experiments in HNLC regions have shown that fertilizing the ocean with iron can enhance marine productivity. The actual impact on export production and the persistence of the potential $CO_2$ sink, however, are uncertain (Bakker, Chapter 26).

## Variability

The partial pressure of $CO_2$ in seawater ($pCO_2$) and $CO_2$ fluxes vary regionally and globally when sea surface temperature (SST), biological export production, or ocean circulation varies. Thanks to numerous oceanographic cruises conducted since the 1960s, the seasonal variations are now relatively well established at regional and global scales (see a synthesis in Takahashi et al. 2002). Figure 12.2 presents an example of in situ observations made in the South Indian Ocean in January and August 2000. The large seasonal variations of $pCO_2$ are mostly controlled by temperature in the subtropics, but by biological production and winter mixing at high latitudes (<50°S), leading to opposite seasonality in these two regions.

Interannual variability of $pCO_2$ and air-sea $CO_2$ fluxes is not so well known at the global scale. A one-degree warming causes a 4 percent increase in ocean $pCO_2$. Year-to-year variations in global SST measured by satellite are on the order of ±0.25°C, potentially driving $CO_2$ fluxes of ±0.8 PgC y$^{-1}$. Climate models forced by different scenarios predict that surface temperature will warm by 1.4° to 5.8°C in the next century, thus reducing the uptake capacity of the ocean.

Changes in biological export production translate almost directly into changes in

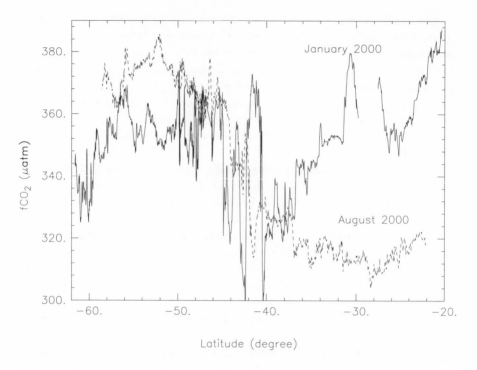

**Figure 12.2.** Meridional distribution of ocean fCO$_2$ measured in January 2000 (solid line) and August 2000 (dashed line) in the southwestern Indian Ocean (OISO cruises). fCO$_2$ is the pCO$_2$ corrected for the non-ideal behavior of CO$_2$ as a gas. The seasonal amplitude of fCO$_2$ in the Southern Ocean, south of the Polar Front (50°S), is not as high as in the subtropics and frontal zone but is large enough to move from a summer CO$_2$ sink to a winter CO$_2$ source (Metzl et al. 2001). These data highlight the meso-scale variability (highs and lows on 1- to 10-km scale) in the frontal zone (40–45°S) when primary production is enhanced during austral summer.

CO$_2$ fluxes. Year-to-year changes in global export production computed from satellite data (Figure 12.1) are on the order of ±0.4 PgC y$^{-1}$. It is difficult to say how biological export production will change in the future. Recent studies have highlighted extreme regime shifts in marine ecosystems, showing either decadal variations (Chavez et al. 2003) or trends (Beaugrand et al. 2002). We cannot yet estimate the impact of such ecosystem shifts on CO$_2$ fluxes. For this we would need to understand and quantify the impact of different plankton functional types on CO$_2$ and to estimate the response of these types under different climatic conditions. Research is very active in this area. Karl et al. (1999) showed that ocean stratification favors the presence of phytoplankton types that can fix atmospheric nitrogen but that lead to low export production. Riebesell et al. (2000) showed that the acidification of the ocean from elevated $CO_2$ concentration hinders the formation of $CaCO_3$ shells by phytoplankton, which has two effects: it alters carbon chemistry and increases surface $CO_2$ uptake, but reduces the export capacity of POC because $CaCO_3$ shells are associated with high export production (Klaas and Archer 2002). These effects have begun to be included in global carbon models (Bopp et al. 2001; Boyd and Doney 2002), but there is a long way to go before it is possible to rely on such model results.

The impact of changes in ocean circulation is also very difficult to estimate, because of the lack of large-scale observations. We rely mostly on ocean models to estimate this term. Much of the mixing in the ocean occurs on space scales of 1 meter (m) to 10 kilometers (km), one order of magnitude smaller than the resolution of current global carbon models. Upwelling is driven by wind divergence. Winds can now be measured from satellites, but the uncertainties in the estimates and in the upwelling produced by models are large. Coastal upwelling is also poorly resolved in global models. Ice melting induces ocean stratification and would also have impacts on the $CO_2$ flux. All these effects can be estimated with current ocean models, but the reliability of these estimates is debatable. Heat content observations show that models can reproduce trends in observed climate but not its natural variability (Levitus et al. 2001; Bopp et al. 2002).

The impact of changes in circulation on $CO_2$ fluxes depends on the timescale of interest. For interannual variability, a reduction in the rate of exposure of intermediate waters, either from stratification or reduced upwelling, would cause a $CO_2$ flux anomaly from the atmosphere to the ocean, because intermediate waters are richer in DIC than surface waters. Over longer timescales, we must consider the impact of changes in circulation on anthropogenic $CO_2$ as well. Anthropogenic $CO_2$ uptake would be reduced if the ocean stratifies, opposite to the effect for natural DIC. The net effect depends on the size and duration of the changes and can be estimated most reliably with a model (for model estimates, see Greenblatt and Sarmiento, Chapter 13, this volume).

## Anthropogenic $CO_2$ Fluxes in a Steady State Ocean

When atmospheric $CO_2$ increases, it is absorbed by the oceans at a rate that depends on the rate of increase of atmospheric $CO_2$, compared with the rate of ocean mixing. It takes only about one year for the surface ocean to equilibrate with the atmosphere, but it takes a few years to a few centuries to mix the surface ocean with intermediate and deepwaters. As a result, the storage of anthropogenic carbon in the ocean currently reaches only ~15 percent of its potential maximum (Figure 12.3). The oceans have the potential to absorb 70–85 percent of the anthropogenic $CO_2$, but this will take several centuries after the emissions occur (Archer et al. 1997).

The fraction of atmospheric $CO_2$ that can be absorbed by the ocean diminishes with high $CO_2$. As more anthropogenic $CO_2$ dissolves in the ocean, less $CO_3^{2-}$ is available to form $HCO_3^-$ (equation 1). Thus more $CO_2$ stays in its soluble gas form, which can

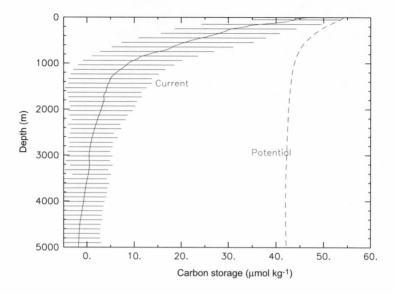

**Figure 12.3.** Global mean current inventory (solid line) and potential maximum (dashed line) of anthropogenic carbon for an atmospheric $CO_2$ concentration of 368 ppm. The difference between the two curves is caused by the long mixing time of the ocean. The current storage is calculated using an average of anthropogenic DIC estimates around the globe. Horizontal lines represent one standard deviation. The analysis of observations was provided by R. M. Key based on data referenced in Sabine et al. (1999, 2002) and Lee et al. (2003).

exchange with the atmosphere. This effect is large and easily predictable (Greenblatt and Sarmiento, Chapter 13).

The global uptake of carbon by the oceans can be estimated using several independent methods: atmospheric measurements of $CO_2$, augmented with $\delta^{13}C$, $O_2/N_2$, or the spatial distribution of atmospheric $CO_2$; direct measurements of ocean carbon, surface $\delta^{13}C$, or related tracers in the ocean; inversion of ocean tracers; and global ocean models. Recent estimates of uptake range between 1.5 and 2.8 PgC $y^{-1}$ for 1980–2000, representing one-quarter to one-half of the annual fossil-fuel emissions (Table 12.1). For the entire industrial period until 1994, the total uptake of carbon is estimated to be 111 ± 13 PgC, based on observations (Lee et al. 2003), or about one-quarter of the total anthropogenic emissions (fossil fuel plus land use; House et al. 2002). Model results are in the range of the observations.

Although the results cited are all coherent with one another and converge toward a value of 1.9 PgC $y^{-1}$ for 1980–2000 (Prentice et al. 2001), the uncertainty in the mean value is still ±0.7 PgC $y^{-1}$. The method based on atmospheric $CO_2$ and $O_2/N_2$, developed in the 1990s, offered promise for reducing this uncertainty. This method assumes,

**Table 12.1.** Estimates of mean ocean $CO_2$ uptake ($PgC\ y^{-1}$) from various methods and recent time periods (based on Prentice et al. 2001, Table 3.4, with additional studies)

| Time period and study | Method | Uptake |
|---|---|---|
| *1970–1990* | | |
| Quay et al. (1992) | Ocean $^{13}C$ inventory | $2.1 \pm 0.8$ |
| Heimann and Maier-Reimer (1996) | Ocean $^{13}C$ inventory | $2.1 \pm 0.9$ |
| *1980–1989* | | |
| Prentice et al. (2001) | Atmospheric $CO_2$ and $O_2/N_2$ | $1.9 \pm 0.6$ |
| Bopp et al. (2002); Le Quéré et al. (2003) | Atmospheric $CO_2$ and $O_2/N_2$ [a] | $1.8 \pm 0.8$ |
| Plattner et al. (2002) | Atmospheric $CO_2$ and $O_2/N_2$ [a] | $1.7 \pm 0.6$ |
| Battle et al. (1996) | $O_2/N_2$ in Antarctic firn ice[b] | $1.8 \pm 1.0$ |
| Orr et al. (2001) | Model intercomparison | $1.85 \pm 0.35$ |
| McNeil et al. (2003) | Ocean CFC inventory | $1.6 \pm 0.4$ |
| *1985–1995* | | |
| Gruber and Keeling (2001) | Ocean $^{13}C$ inventory | $1.5 \pm 0.9$ |
| *1990–1996* | | |
| Bopp et al. (2002) | Atmospheric $CO_2$ and $O_2/N_2$ [a] | $2.3 \pm 0.7$ |
| *1994* | | |
| Based on Lee et al. (2003) [c] | Ocean anthropogenic $CO_2$ inventory | 1.9 to 2.4 |
| *1995* | | |
| Takahashi et al. (2002)[d] | Surface ocean $pCO_2$ | $2.8 \pm 1.5$ |
| *1990–1999* | | |
| Prentice et al. (2001) | Atmospheric $CO_2$ and $O_2/N_2$ | $1.7 \pm 0.5$ |
| Le Quéré et al. (2003) | Atmospheric $CO_2$ and $O_2/N_2$ [a] | $1.9 \pm 0.7$ |
| Keeling and Garcia (2002)[e] | Atmospheric $CO_2$ and $O_2/N_2$ [a] | $1.9 \pm 0.6$ |
| Plattner et al. (2002) | Atmospheric $CO_2$ and $O_2/N_2$ [a] | $2.4 \pm 0.6$ |
| Ciais et al. (1995); Tans et al. (1989); Trolier et al. (1996) | Atmospheric $CO_2$ and $\delta^{13}C$ [b] | 1.8 |
| Keeling and Piper (2000) | Atmospheric $CO_2$ and $\delta^{13}C$ | 2.4 |
| McNeil et al. (2003) | Ocean CFC inventory | $2.0 \pm 0.4$ |

[a]Gas fluxes corrected for variations in ocean $O_2$ content (see text for an explanation of the different estimates).

[b]Calculations updated by Prentice et al. (2001) based on the cited references.

[c]Assumes ocean and atmospheric $CO_2$ increases follow similar curves.

[d] Corrected by +0.6 $PgC\ y^{-1}$ to account for land-ocean river flux.

[e]1990–2000.

however, that the ocean $O_2$ flux is zero over several years, which would not be the case if the oceans were getting warmer. Budgets based on $O_2$ must be corrected to account for ocean $O_2$ fluxes. We can estimate this correction by assuming that $O_2$ and heat fluxes are proportional and using heat flux estimates. Keeling and Garcia (2002) estimated a proportionality factor of 4.9 nanomoles per joule (nmol/J), from geochemical observations, whereas Bopp et al. (2002) and Plattner et al. (2002) estimated values higher by ~20 percent from model results. The ocean heat content was directly estimated based on millions of temperature profiles for 1950–1996 (Levitus et al. 2000). For the 1980s, the ocean heat content decreased by $0.13 \times 10^{22}$ J $y^{-1}$; for 1990-1996, it increased by $1.0 \times 10^{22}$ J $y^{-1}$. The compilation of heat content measurements has not been made beyond 1996.

Corrections to the carbon budget for the 1990s depend a lot on the assumed changes in heat content for 1996–2000. Combining temperature and sea surface height observations, Le Quéré et al. (2003) estimated that the ocean heat content increased by $0.55 \times 10^{22}$ J $y^{-1}$, on average, over the 1990s. For comparison, Plattner et al. (2002) assumed that the rate of $1.24 \times 10^{22}$ J $y^{-1}$ observed during 1990–1995 was maintained throughout the 1990s, whereas Keeling and Garcia (2002) used a rate of 0.6 $\times 10^{22}$ J $y^{-1}$, based on model results including anthropogenic forcing but no natural variability. Bopp et al. (2002) did their budget only through 1996. The ocean $CO_2$ sink based on $O_2/N_2$ corrected with observational-based heat content is 1.8 ± 0.8 and 1.9 ± 0.7 PgC $y^{-1}$ for the 1980s and 1990s, respectively (Le Quéré et al. 2003).

It is much more difficult to estimate the regional fluxes of $CO_2$ than the global uptake (Colorplate 9 from Gloor et al. 2003). Atmospheric $CO_2$ measurements have small east-west gradients, which translate into large uncertainties in regional flux estimates from atmospheric inversions. Measurements of ocean $pCO_2$ have large uncertainties associated with the paucity of measurements relative to the regional and temporal variability, and the estimated fluxes, with our inability to accurately quantify gas exchange. Ocean inversions avoid uncertainties due to gas exchange but have uncertainties associated with their estimated ocean transport and data fields. Ocean inversions estimate regional $CO_2$ flux from measurements of DIC, $O_2$, alkalinity, and nutrients. Ocean models generally agree globally with one another but show differences in regional uptake by up to 100 percent of the mean flux (Orr et al. 2001). In spite of these uncertainties, all the methods show an outgassing of $CO_2$ in the tropics and an uptake at mid and high latitudes (Colorplate 9), consistent with our understanding of the underlying processes (Figure 12.1). The largest disagreement between the various estimates is in the Southern Ocean, where observations are sparser (Colorplate 9).

## Uncertainties and Future Research Needs

Since the first report of the Intergovernmental Panel on Climate Change in 1990, scientists' confidence in an ocean uptake of ~2 PgC $y^{-1}$ has increased tremendously. Early estimates, based on model results only, are now supported by a wide range of observations. The accuracy of this estimate, however, is hard to constrain. Estimates based on atmospheric measurements are hampered by the need to accurately quantify processes neglected so far (i.e., variations in land $\delta^{13}C$ fractionation, ocean $O_2/N_2$ outgassing); estimates based on ocean measurements, by the large spatial and temporal sampling frequency required; and ocean inversions and models, by the need to accurately quantify vertical transport and mixing processes in the ocean. On the other hand, it is possible to greatly reduce the uncertainty in regional flux estimates and in the global trend. Better constraints on regional ocean fluxes could substantially improve atmospheric inversions. The Southern Ocean is the worst-constrained region. The ocean south of 46°S covers 16 percent of the world oceans and ventilates about half of the deep ocean waters. Regional measurements suggest that this region is a sink of up to 1 PgC y$^{-1}$ (Takahashi et al. 2002), but atmospheric inversions see a sink of only 0.4 PgC y$^{-1}$ (Gurney et al. 2002). The discrepancies between the two methods may be due to the absence of local measurements during winter (Metzl et al. 1999, 2001; Figure 12.2), but this remains to be shown at the scale of the entire basin. The future response of Southern Ocean $CO_2$ uptake is one of the largest sources of uncertainty for predictions of future atmospheric $CO_2$ (Friedlingstein, Chapter 10, this volume).

Most of the uncertainties cited here apply to the mean state only. The difference between the 1980s and the 1990s appears to be more robust than the mean estimate for either decade (Le Quéré et al. 2003). Monitoring stations (HOT, BATS) (Winn et al. 1999; Bates et al. 2002) and repeated surveys (Inoue et al. 1995; Feely et al. 2002) have measured changes in $CO_2$ or DIC over the past decades thanks to the highly instrumental precisions and DIC standards (Dickson 2001). To understand and predict these trends, it is important to untangle variations in the processes that drive them, especially trends in ocean circulation and mixing, which have already been shown to affect DIC (Bates et al. 2002). For this we need large-scale monitoring of changes in ocean circulation and mixing and their effect on ocean biogeochemistry. The launching of drifting ARGO floats in the ocean will provide outstanding continuous measurements of temperature and salinity over the ocean. The attachment of $O_2$ sensors to these floats would provide extremely valuable information on changes in ocean biogeochemistry.

Scientists' understanding of the interactions between marine biology, biogeochemistry, and physics needs improvement. This inadequacy is best illustrated by the fact that we cannot explain what caused the variations of 80 parts per million (ppm) in atmospheric $CO_2$ during glaciations, although it is certain that the cause lies in ocean biogeochemistry (Joos and Prentice, Chapter 7, this volume). This problem must be tackled with well-validated ocean biogeochemistry models. Yet current models barely take into account ecosystem dynamics and the role of different plankton functional types. Gaps in experts' understanding of vertical mixing hinder development of models of whole biogeochemical cycles. Examination of observed interannual to decadal variations of the surface and intermediate carbon cycle is also an important task for detecting and

understanding trends or shifts and for verifying how present models are able to reproduce those variations. This examination can be achieved only if long-term time-series
stations, surface underway measurements, and basin-scale sections are maintained for
both dynamics and biogeochemistry in different ocean sectors and if data delivery and
synthesis of observations occur at an international level. Until such models can be
shown to reproduce glacial-interglacial variations in atmospheric $CO_2$ and modern
trends in biogeochemistry, our confidence in $CO_2$ flux predictions will be limited.

## Acknowledgments

We thank J. Greenblatt for providing Table 12.1; N. Gruber, C. Sabine, J. L.
Sarmiento, and two anonymous reviewers for constructive comments; people working
for the SeaWiFS Project (NASA Code 970.2) for providing data for Figure 12.1; people working on board the *R. V. Marion-Dufresne* for providing data for Figure 12.2
(OISO project supported by the French Institut National des Sciences de l'Univers,
Institut Pierre-Simon Laplace, and Institut Polaire Paul-Emile Victor); R. M. Key for
compiling global estimates of anthropogenic DIC for Figure 12.3; and M. Gloor for
providing Colorplate 9.

## Literature Cited

Archer, D., H. Kheshgi, and E. Maier-Reimer. 1997. Multiple timescales for
neutralization of fossil fuel $CO_2$. *Geophysical Research Letters* 24 (4): 405–408.
Bates, N. R., A. C. Pequignet, R. J. Johnson, and N. Gruber. 2002. A short-term sink for
atmospheric $CO_2$ in subtropical mode water of the North Atlantic Ocean. *Nature*
420:489–493.
Battle, M., M. Bender, T. Sowers, P. P. Tans, J. H. Butler, J. W. Elkins, J. T. Ellis, T. Conway, N. Zhang, P. Lang, and A. D. Clarke. 1996. Atmospheric gas concentrations over
the past century measured in air from firn at the South Pole. *Nature* 283:231–235.
Battle, M., M. Bender, P. P. Tans, J. W. C. White, J. T. Ellis, T. Conway, and R. J.
Francey. 2000. Global carbon sinks and their variability, inferred from atmospheric $O_2$
and $\delta^{13}C$. *Science* 287:2467–2470.
Beaugrand, G, P. C. Reid, F. Ibanez, J. A. Lindley, and M. Edwards. 2002. Reorganisation
of North Atlantic marine copepod biodiversity and climate. *Science* 296:1692–1694.
Behrenfeld, M. J., and P. G. Falkowski. 1997. A consumer's guide to phytoplankton pri-
mary productivity models. *Limnology and Oceanography* 42:1479–1491.
Bopp, L., P. Monfray, O. Aumont, J.-L. Dufresne, H. LeTreut, G. Madec, L. Terray, and
J. Orr. 2001. Potential impact of climate change on marine export production. *Global
Biogeochemical Cycles* 15:81–99.
Bopp, L., C. Le Quéré, M. Heimann, A. C. Manning, and P. Monfray. 2002. Climate-
induced oxygen oceanic fluxes: Implications for the contemporary carbon budget.
*Global Biogeochemical Cycles* 16, doi:10.1029/2001GB001445.
Boyd, P. W., and S. C. Doney. 2002. Modelling regional responses by marine pelagic

ecosystems to global climate change. *Geophysical Research Letters* 29, doi:10.1029/2001GL014130.

Chavez, F. P., J. R. Ryan, S. E. Lluch-Cota, and M. Ñiquen. 2003. From anchovies to sardines and back: Multidecadal change in the Pacific Ocean. *Science* 2999:217–221.

Ciais, P., P. Tans, J. W. White, M. Trolier, R. Francey, J. Berry, D. Randall, P. Sellers, J. Collatz and D. Schimel. 1995. Partitioning of ocean and land uptake of $CO_2$ as inferred by $\delta^{13}C$ measurements from the NOAA Climate Monitoring and Diagnostics Laboratory Global Air Sampling Network. *Journal of Geophysical Research* 100:5051–5070.

Dickson, A. 2001. Reference materials for oceanic measurements. *Oceanography* 14 (4): 21–22.

Feely, R., J. Boutin, C. Cosca, Y. Dandonneau, J. Etcheto, H. Inoue, M. Ishii, C. Le Quéré, D. Mackey, M. McPhaden, N. Metzl, A. Poisson, and R. Wanninkhov. 2002. Seasonal and interannual variability of $CO_2$ in the Equatorial Pacific. *Deep Sea Research* 49:2443–2469.

Gloor, M., N. Gruber, J. L. Sarmiento, C. L. Sabine, R. Feely, and C. Redenbeck. 2003. A first estimate of present and pre-industrial air-sea $CO_2$ flux patterns based on ocean interior carbon measurements and models. *Geophysical Research Letters* 30 (1), doi:10.1029/2002GL015594.

Gruber, N., and C. D. Keeling. 2001. An improved estimate of the isotopic air-sea disequilibrium of $CO_2$: Implications for the oceanic uptake of anthropogenic $CO_2$. *Geophysical Research Letters* 28:555–558.

Gurney, K. R., R. M. Law, A. S. Denning, P. J. Rayner, D. Baker, P. Bousquet, L. Bruhwiler, Y.-H. Chen, P. Ciais, S. Fan, I. Y. Fung, M. Gloor, M. Heimann, K. Higuchi, J. John, T. Maki, S. Maksyutov, K. Masarie, P. Peylin, M. Prather, B. C. Pak, J. Randerson, J. Sarmiento, S. Taguchi, T. Takahashi, and C.-W. Yuen. 2002. Towards robust regional estimates of $CO_2$ sources and sinks using atmospheric transport models. *Nature* 415:626–630.

Heimann, M., and E. Maier-Reimer. 1996. On the relations between the oceanic uptake of $CO_2$ and its carbon isotopes. *Global Biogeochemical Cycles* 10:89–110.

House, I. J., I. C. Prentice, and C. Le Quéré. 2002. Maximum impacts of future reforestation or deforestation on atmospheric $CO_2$. *Global Change Biology* 8:1–6.

Inoue, H. Y., H. Matsueda, H. Ishii, K. Fushimi, M. Hirota, I. Asanuma, and Y. Takasugi. 1995. Long-term trend of the partial pressure of carbon dioxide ($pCO_2$) in surface waters of the western North Pacific, 1984–1993. *Tellus* 47B:391–413.

Karl, D. M., R. Letelier, L. Lupas, J. Dore, J. Chrisitan, and D. Hebel. 1999. The role of nitrogen fixation in biogeochemical cycling in the subtropical North Pacific Ocean. *Nature* 388:533–538.

Keeling, R. F., and H. E. Garcia. 2002. The change in oceanic $O_2$ inventory associated with recent global warming. *Proceedings of the Natural Academy of Sciences* 99:7848–7853.

Keeling, C. D., and S. C. Piper. 2000. Interannual variations of exchanges of atmospheric $CO_2$ and $^{13}CO_2$ with the terrestrial biosphere and oceans from 1978 to 2000. III. Simulated sources and sinks. Scripps Institution of Oceanography Reference No. 00-14. San Diego: University of California, Scripps Institution of Oceanography.

Klaas, C., and D. Archer. 2002. Association of sinking organic matter with various types of mineral ballast in the deep sea: Implications for the rain ratio. *Global Biogeochemical Cycles,* doi:10.1029/2001GB0011765.

Laws, E. A., P. G. Falkowski, W. O. Smith Jr., H. Ducklow, and J. J. McCarthy. 2000. Temperature effects on export production in the open ocean. *Global Biogeochemical Cycles* 14:1231–1246.

Lee, K., S.-D. Choi, G.-H. Park, R. Wanninkhof, T.-H. Peng, R. M. Key, C. L. Sabine, R. A. Feely, J. L. Bullister, and F. J. Millero. 2003. An updated anthropogenic $CO_2$ inventory in the Atlantic Ocean. *Global Biogeochemical Cycles* (submitted).

Le Quéré, C., O. Aumont, L. Bopp, P. Bousquet, P. Ciais, R. Francey, M. Heimann, C. D. Keeling, R. F. Keeling, H. Kheshgi, P. Peylin, S. C. Piper, I. C. Prentice, and P. J. Rayner. 2003. Two decades of ocean $CO_2$ sink and variability. *Tellus* 55:649–656.

Levitus S., J. I. Antonov, T. P. Boyer, and C. Stephens. 2000. Warming of the world ocean. *Science* 287:2225–2229.

Levitus S., J. I. Antonov, J. Wang, T. L. Delworth, K. W. Dixon, and A. J. Broccoli. 2001. Anthropogenic warming of Earth's climate system. *Science* 292:267–270.

McNeil, B. I., R. J. Matear, R. M. Key, J. L. Bullister, and J. L. Sarmiento. 2003. Anthropogenic $CO_2$ uptake by the ocean based on the global chlorofluorocarbon data set. *Nature* 299:235–239.

Metzl, N., B. Tilbrook, and A. Poisson. 1999. The annual $fCO_2$ cycle and the air-sea $CO_2$ flux in the sub-Antarctic Ocean. *Tellus* 51B (4): 849–861.

Metzl, N., C. Brunet, A. Jabaud-Jan, A. Poisson, and B. Schauer. 2001. Summer and winter air-sea $CO_2$ fluxes in the Southern Ocean. Pp. 685–688 in *Proceedings of the 6th International Carbon Dioxide Conference,* Vol. 2. Tsukuba, Japan: Center for Global Environmental Research.

Murnane, R., J. L. Sarmiento, and C. Le Quéré. 1999. Spatial distribution of air-sea $CO_2$ fluxes and the interhemispheric transport of carbon by the oceans. *Global Biogeochemical Cycles* 13:287–305.

Orr, J. C., E. Maier-Reimer, U. Mikolajewicz, P. Monfray, J. L. Sarmiento, J. R. Toggweiler, N. K. Taylor, J. Palmer, N. Gruber, C. L. Sabine, C. Le Quéré, R. M. Key, and J. Boutin. 2001. Estimates of anthropogenic carbon uptake from four three-dimensional global ocean models. *Global Biogeochemical Cycles* 15:43–60.

Plattner, G.-K., F. Joos, and T. F. Stocker. 2002. Revision of the global carbon budget due to changing air-sea oxygen fluxes. *Global Biogeochemical Cycles,* doi:10.1029/2001GB001746.

Prentice, I. C., G. D. Farquhar, M. J. R. Fasham, M. L. Goulden, M. Heimann, and V. J. Jaramillo, H. S. Kheshgi, C. Le Quéré, R. J. Scholes, and D. W. R. Wallace. 2001. The carbon cycle and atmospheric $CO_2$. Pp. 183–237 in *Climate change 2001: The scientific basis (Contribution of Working Group I to the third assessment report of the Intergovernmental Panel on Climate Change),* edited by J. T. Houghton, Y. Ding, D. J. Griggs, M. Noguer, P. J. van der Linden, X. Dai, K. Maskell, and C. A. Johnson. Cambridge: Cambridge University Press.

Quay, P. D., B. Tilbrook, and C. S. Wong. 1992. Oceanic uptake of fossil fuel $CO_2$: Carbon-13 evidence. *Science* 256:74–79.

Riebesell, U., I. Zondervan, B. Rost, P. Tortell, R. Zeebe, and F. Morel. 2000. Reduced calcification of marine plankton in response to increased atmospheric $CO_2$. *Nature* 407 (6802): 364–367.

Sabine, C., L., R. M. Key, K. M. Johnson, F. J. Millero, J. L. Sarmiento, D. W. R.

Wallace, and C. D. Winn. 1999. Anthropogenic $CO_2$ inventory of the Indian ocean. *Global Biogeochemical Cycles* 13:179–198.

Sabine, C. L., R. A. Feely, R. M. Key, J. L. Bullister, F. J. Millero, K. Lee, T.-H. Peng, B. Tilbrook, T. Ono, and C. S. Wong. 2002. Distribution of anthropogenic $CO_2$ in the Pacific Ocean. *Global Biogeochemical Cycles,* doi:10.1029/2001GB001639.

Takahashi, T., S. C. Sutherland, C. Sweeney, A. Poisson, N. Metzl, B. Tilbrook, N. Bates, R. Wanninkhof, R. A. Feely, C. Sabine, J. Olafsson, and Y. Nojiri. 2002. Global sea-air $CO_2$ flux based on climatological surface ocean p$CO_2$, and seasonal biological and temperature effects. *Deep Sea Research II* 49:1601–1622.

Tans, P. P., T. J. Conway, and T. Nakazawa. 1989. Latitudinal distribution of the sources and sinks of atmospheric carbon dioxide derived from surface observations and atmospheric transport model. *Journal of Geophysical Research* 94:5151–5172.

Trolier, M., J. W. C. White, P. P. Tans, K. A. Masarie, and P. A. Gemery. 1996. Monitoring the isotopic composition of atmospheric $CO_2$-measurements from the NOAA Global Air Sampling Network. *Journal of Geophysical Research* 101:25897–25916.

Wanninkhof, R. S., S. Doney, T. Takahashi, and W. McGillis. 2001. The effect of using averaged winds on global air-sea $CO_2$ fluxes. Pp. 351–356 in *Gas transfer at water surfaces,* edited by M. A. Donelan, W. M. Drennan, E. S. Saltzman, and R. Wanninkhof. AGU Geophysical Monograph Series Vol. 127. Washington, DC: AGU Press.

Winn, C. D., Y.-H. Li, F. T. Mackenzie, and D. M. Karl. 1999. Rising surface ocean dissolved inorganic carbon at the Hawaii ocean time-series site. *Marine Chemistry* 60:33–47.

# 13

# Variability and Climate Feedback Mechanisms in Ocean Uptake of $CO_2$

Jeffery B. Greenblatt and Jorge L. Sarmiento

The major sinks of anthropogenic $CO_2$ are the oceans and terrestrial biosphere, which absorbed half the 6.3 ± 0.4 petagrams of carbon per year (PgC y$^{-1}$) emitted by fossil-fuel burning in the 1990s (Prentice et al. 2001). Because of the enormous size of these sinks, it is important to understand not only their mean size, but also their variations. Interannual variability appears to be significant for both the ocean and terrestrial sinks, though different methods currently yield conflicting results. In addition, global climate change resulting from emissions of $CO_2$ can induce changes in the sizes of the ocean and terrestrial sinks. Those changes may have already begun, as evidenced by ocean temperature increases (Levitus et al. 2000), and changes in ocean carbon uptake (Prentice et al. 2001). Based on deep oceanic chlorofluorocarbon (CFC) and radiocarbon ($\delta^{14}C$) measurements, Broecker et al. (1999) suggest that ocean circulation may have changed in the past century, and time-series measurements of atmospheric and oceanic $CO_2$ concentrations point to changes already taking place in the ocean's buffering chemistry (Sarmiento et al. 1995).

This chapter focuses on anthropogenic changes to the ocean carbon cycle, while Le Quéré and Metzl (Chapter 12) discuss natural processes regulating ocean $CO_2$ uptake, and Foley and Ramankutty (Chapter 14) treat the terrestrial carbon cycle. We begin with the current understanding of uptake variability, then explore the effects of climate change on the rates of $CO_2$ uptake, including long-term changes in uptake due to changing ocean chemistry. We also consider the potential effects of large-scale changes in ocean circulation, such as a collapse of the thermohaline circulation in the North Atlantic Ocean. We end with a discussion of future research needs.

## Variability in Ocean Uptake

There is considerable disagreement concerning the magnitude of interannual variability in ocean $CO_2$ uptake (Table 13.1). Studies that calculate uptake from ocean mod-

**Table 13.1.** Estimates of global ocean $CO_2$ uptake variability

| Study | Time period | Method or model | Variability |
|---|---|---|---|
| *Ocean models* | | | |
| Winguth et al. (1994) | 1981–1987 | LSG-HAMOCC3 | ±0.3[a] |
| Le Quéré et al. (2000) | 1979–1997 | OPA8-HAMOCC3 | ±0.4 |
| Le Quéré et al. (2002) | 1979–1999 | OPA8-HAMOCC3 | ±0.5 |
| *Ocean observations* | | | |
| Lee et al. (1998) | 1982–1995 | SST-$pCO_2$ regression | ±0.2[a] |
| Loukos et al. (2000) | 1982–1994 | SST-SSS-$pCO_2$ regression | ±0.2[b] |
| Gruber et al. (2002) | 1984–2002 | Box model driven by surface ocean $\delta^{13}C$, DIC and SST | ±0.3[c] |
| Bates (2002) | 1982–1992 | Calculated gas transfer using storm data from GTECCA | ±0.9 |
| *Atmospheric observations* | | | |
| Francey et al. (1995) | 1982–1992 | Atmospheric $CO_2$ and $\delta^{13}C$ | ±1.9[a] |
| Keeling et al. (1995) | 1978–1994 | Atmospheric $CO_2$ and $\delta^{13}C$ | ±2.0[a] |
| Joos et al. (1999b) | 1980–1996 | Atmospheric $CO_2$ and $\delta^{13}C$ | ±1.0 |
| Rayner et al. (1999) | 1980–1995 | Inversion of atmospheric $CO_2$, $\delta^{13}C$ and $O_2/N_2$ | ±1.0 |
| Bousquet et al. (2000) | 1980–1998 | Inversion of atmospheric $CO_2$ | ±1.3 |
| Baker (2001) | 1981–1996 | Inversion of atmospheric $CO_2$ | ±0.7 |

*Note:* The measure of variability is peak-to-peak amplitude (PgC $y^{-1}$), and not standard deviation, which is sometimes used.

[a]Annual average as reported in Le Quéré (2000).

[b]For equatorial Pacific Ocean (10°N–10°S and 170°W–96°W) only.

[c]Extrapolated from two sites near Bermuda to the entire North Atlantic Ocean; mean uptake of the North Atlantic is 0.7 ± 0.1 PgC $y^{-1}$.

els or from ocean observations report globally averaged interannual variability of up to ±0.5 PgC $y^{-1}$. Studies that calculate ocean uptake from atmospheric observations, however, show significantly higher interannual variability, on the order of ±1 PgC $y^{-1}$ or larger. This latter level of variability corresponds to ±50 percent or more of the long-term mean uptake of $CO_2$ (~2 PgC $y^{-1}$). While our understanding of the mechanisms responsible for interannual variability in $CO_2$ uptake is incomplete, Le Quéré and Metzl (Chapter 12, this volume) summarize what is known.

Differences among the ocean- and atmosphere-based studies are significant, but there are areas of agreement. Most studies show larger variability in the terrestrial biosphere than in the oceans (Joos et al. 1999b; Rayner et al. 1999; Battle et al. 2000 [$O_2$-based analysis]; Bousquet et al. 2000; Keeling and Piper 2000; Baker 2001; Manning 2001), such that the variability in atmospheric $CO_2$ accumulation (Figure 13.1) is dominated by the terrestrial component. There is also a strong correlation between uptake

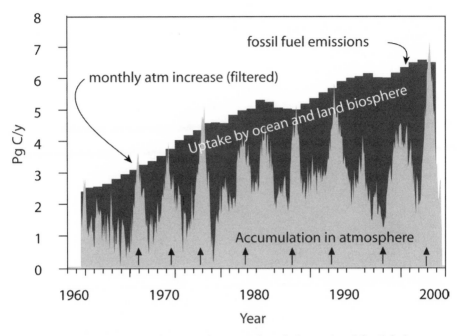

**Figure 13.1.** Accumulation of CO$_2$ in the atmosphere (light gray) and fossil-fuel emissions (dark gray). Uptake by ocean and terrestrial sinks is indicated by the difference in areas. El Niño years are indicated by vertical arrows. Note the large interannual variability in atmospheric accumulation, and hence in the total terrestrial plus ocean uptake (from Sarmiento and Gruber 2002).

and the El Niño–Southern Oscillation (ENSO) in most studies, with high oceanic uptake predominantly during the warm (El Niño) phase attributed to reduced upwelling of deep, CO$_2$-rich equatorial Pacific waters. The atmospheric growth rate also appears largest during the warm phase, implying reduced terrestrial uptake. A recent study by Russell and Wallace (2003), however, suggests that the oceans do play a role in the enhanced atmospheric growth rate through mechanisms that earlier studies missed.

Some studies that compare ocean- and atmosphere-based variability estimates find agreement among methods for some ocean regions. Bousquet et al. (2000) inverted atmospheric CO$_2$ data to obtain temporally and regionally resolved oceanic and terrestrial fluxes. In the equatorial region (20°S–20°N) and the Southern Ocean (50°S–90°S), where fluxes were relatively large, the inversion results agreed favorably with results of the ocean model of Le Quéré et al. (2000), as well as with ocean observations (Wong et al. 1993; Feely et al. 1997, 1999; Bousquet et al. 2000) (Figure 13.2). Outside these two regions, however, the inversion obtained much larger variability (±0.4 Pg C y$^{-1}$) than the

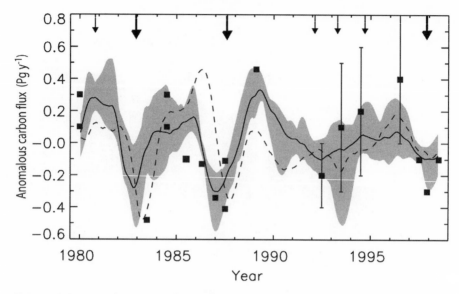

**Figure 13.2.** Anomalous sea-air fluxes of $CO_2$ in the equatorial Pacific ($20°S–20°N$) from the inversion study of Bousquet et al. (2000) (solid line; gray band indicates range of values obtained from sensitivity study), model results of Le Quéré et al. (2000) (dashed line), and oceanic observations from Wong et al. (1993), Feely et al. (1997, 1999), and Bousquet et al. (2000) (squares; vertical bars indicate uncertainties). Thick arrows indicate strong El Niño events characterized by minima in the Southern Oscillation index; thin arrows indicate weak El Niño conditions. Reprinted (abstracted/excerpted) with permission from Bousquet et al. (2000). ©2000 AASS.

ocean-based studies ($\pm0.1$ Pg C y$^{-1}$). For the global scale, Le Quéré et al. (2003) report significant interannual differences in global fluxes among inversion (Bousquet et al. 2000), ocean model (Le Quéré et al. 2000), and atmospheric synthesis (Francey et al. 2001; Keeling et al. 2001; Langenfelds et al. 2002) results (Heimann et al., Chapter 8, this volume, Colorplate 7, left center panel).

Shortcomings of the atmospheric inversion technique, such as an inability to cleanly separate ocean and terrestrial signals, may contribute to the large ocean variability such studies report. Also problematic is the exclusion of highly productive continental margins due to limited model resolution. These regions may be highly variable (Bousquet et al. 2000) and contribute large fluxes (Tsunogai et al. 1999). Strong restoring to climatological observations in semidiagnostic ocean biogeochemistry models (Le Quéré et al. 2000, 2002) may also dampen interannual variability in carbon uptake.

Gruber et al. (2002) analyzed long-term ocean observations from the Bermuda Atlantic Time Series and calculated significant interannual variability in $CO_2$ uptake. This result was then extrapolated over the North Atlantic, yielding an uptake variability of $\pm0.3$ PgC y$^{-1}$, nearly 50 percent of the mean North Atlantic Ocean uptake (0.7

± 0.1 PgC y$^{-1}$), a similar ratio to that from global inversion models. While extrapolation from one site to the entire ocean basin may be optimistic, the result highlights the need for model improvements to simulate the full range of observed conditions that affect CO$_2$ uptake. Along similar lines, Bates (2002) estimated the influence on CO$_2$ uptake of tropical and extra-tropical storms. In these storms, which mainly occur in regions of large air-sea CO$_2$ disequilibrium, strong winds increase air-sea gas exchange. Bates (2002) obtained total interannual variability of ±0.9 PgC y$^{-1}$, and claimed that current ocean-based studies neglect storm effects. This study used the cubic gas exchange relationship of Wanninkhof and McGillis (1999), however, which produced significantly higher uptake than the more established quadratic relationship (Wanninkhof 1992). Further work is needed to firmly establish the significance of tropical and extra-tropical storms in CO$_2$ uptake variability.

Effects of future climate change on variability are not well characterized. There is some evidence that El Niño may be increasing and will continue to do so as climate changes (Trenberth and Hoar 1996; Timmermann et al. 1999). Moreover, the mean climate may shift toward an El Niño-like state, punctuated by strong La Niña episodes with stronger interannual variability in equatorial sea surface temperatures (SSTs) (Timmermann et al. 1999). Although there is ample evidence for increased ocean uptake in El Niño years, estimating the effects of these large changes on interannual variability in CO$_2$ uptake is difficult.

## Climate Feedbacks on Ocean Uptake

As the oceans warm, changes projected due to increased radiative forcing and warmer atmospheric temperatures will include three main effects: changes in CO$_2$ solubility, changes in surface ocean stratification, and changes in marine biological productivity. Although there are some significant differences between existing models in the magnitude of the projected changes, all show decreased uptake due to changes in CO$_2$ solubility and surface ocean stratification and increased uptake due to changes in biological productivity. Overall, the net effect of increasing radiative forcing is a decrease in oceanic CO$_2$ uptake.

Carbon dioxide solubility is a strong function of surface ocean temperature, and increasing ocean temperatures (Levitus et al. 2000) are resulting in a decrease in CO$_2$ solubility and consequently less uptake. The magnitude of this term is fairly consistent across different models employing a variety of CO$_2$ emissions scenarios, resulting in cumulative changes in anthropogenic uptake of between −48 and −68 PgC (−9.4 percent to −14.0 percent) by 2065 or 2100 (Table 13.2).

Global warming also leads to increased surface ocean stratification, due to rising SSTs and increased precipitation at high latitudes, which results from an intensified hydrological cycle. This increased stratification decreases vertical exchange over all areas of the ocean, but particularly in the Southern Ocean and the North Atlantic. The decrease in

**Table 13.2.** Cumulative ocean uptake of $CO_2$ (PgC) due to different climate-induced feedback effects

| Study | Scenario and years | Climate baseline | Solubility effect | Stratification effect | Biological effect | Net effect |
|---|---|---|---|---|---|---|
| Sarmiento and Le Quéré (1996) | 1% $CO_2$ $y^{-1}$ for 100 years | 554 | −52 (−9.4%) | −117 (−21.1%) | +111 (+20.0%) | −58 (−10.5%) |
| Sarmiento et al. (1998) | IS92a-like, 1765−2065 | 401 | −56 (−14.0%) | −68 (−17.0%) | +108 (+26.9%) | −16 (−4.0%) |
| Matear and Hirst (1999) | IS92a, 1850−2100 | 376 | −48 (−12.8%) | −41 (−10.9%) | +33 (+8.8%) | −56 (−14.9%) |
| Joos et al. (1999a) | WRE550, 1765−2100 | 530 | −68 (−12.8%) | −15 (−2.8%) | +33 (+6.2%) | −50 (−9.4%) |
| Plattner et al. (2001) | WRE1000, 1765−2100 | 612 | −58 (−9.5%) | −27 (−4.4%) | +36 (+5.9%) | −48 (−7.8%) |

*Note:* "Climate baseline" refers to a simulation with anthropogenic $CO_2$ emissions but preindustrial ocean temperatures and circulation. "Effects" refer to uptake changes for various climate feedbacks and are expressed relative to the climate baseline. The "net effect" is the uptake change when all climate feedbacks are present (i.e., full climate change simulation).

vertical exchange is responsible for a decreased $CO_2$ uptake, because there is a reduced downward transport of surface water, which is enriched in $CO_2$ from the atmosphere, as well as reduced upward transport of deepwater deficient in anthropogenic $CO_2$ (Sarmiento et al. 1998; Sarmiento and Hughes 1999).

Models disagree over the magnitude of the stratification contribution. For example, Matear and Hirst (1999) obtained −41 PgC (−10.9 percent) during the period 1850–2100, whereas Sarmiento and Le Quéré (1996) obtained −117 PgC (−21.1 percent) during their 100-year simulation. While the time spans and emissions scenarios of the two studies differed, they produced similar net effects on $CO_2$ uptake, differing by only 2 PgC (Table 13.2). According to Matear and Hirst (1999), the large difference in stratification feedback in their study was caused by including the Gent et al. (1995) eddy parameterization in their model, which greatly increased Southern Ocean stratification in the absence of climate change and hence reduced the change in $CO_2$ uptake when stratification increased due to anthropogenic forcing. Joos et al. (1999a) and Plattner et al. (2001), by contrast, used simplified ocean models and obtained stratification effects that were smaller than those from Matear and Hirst (1999) (Table 13.2). The sensitivity of overall ocean uptake to Southern Ocean stratification is underscored by Friedlingstein et al. (2003), who found a factor of two difference in overall uptake between the models in their study (Cox et al. 2000; Dufresne et al. 2002), due mostly to differences in Southern Ocean convection. The Southern Ocean's role as a carbon sink is currently not very well understood and is an area of active research (Sabine et al. 1999; Caldeira and Duffy 2000; Matear et al. 2000; Orr et al. 2001).

Stratification also causes changes in marine biological productivity, which lead to indirect changes in $CO_2$ uptake. Decreased vertical exchange resulting from stratification reduces the supply to the surface of dissolved inorganic carbon (DIC) and nutrients such as nitrogen and phosphorus. This reduction in supply may result in a more efficient drawdown of DIC and surface nutrients by the biological pump and increased uptake of $CO_2$ from the atmosphere. The decreased supply of nutrients may also, however, lower biological productivity in areas where it is nutrient limited, and some modelers find that a decrease in surface alkalinity due to the decreased overturning reduces the buffering capacity of surface waters, reducing $CO_2$ uptake (Sarmiento and Le Quéré 1996; Joos et al. 1999a). In general, however, the decrease in supply of DIC to the surface is the predominating effect, resulting in a net increase in $CO_2$ uptake due to marine biology.

As Table 13.2 illustrates, studies disagree over the magnitude of the change in uptake due to biological productivity changes. These differences are based in part on what appears to be a strong coupling between stratification and the impact of the biological pump on the air-sea $CO_2$ distribution. As stratification increases, the biological pump tends to strip nutrients out of the surface more efficiently and thus reduce $CO_2$ to lower levels. This effect tends to compensate the slowdown in anthropogenic carbon uptake that occurs due to increased stratification. For instance, Sarmiento and Le Quéré

(1996) obtained a stratification effect of $-117$ PgC ($-21.1$ percent) and a biological effect of $+111$ PgC ($+20.0$ percent), whereas Matear and Hirst (1999) obtained much lower values for both effects, $-41$ Pg C ($-10.9$ percent) due to stratification and $+33$ Pg C ($+8.8$ percent) due to biology. Models do differ, however, in their assumptions about how biology will respond to nutrient availability. The studies by Sarmiento and Le Quéré (1996) and Sarmiento et al. (1998) assumed constant biological productivity, but the models of Matear and Hirst (1999), Joos et al. (1999a), and Plattner et al. (2001) were dependent on surface phosphate concentrations and other physical variables, so it is too early to draw hard conclusions about this.

The complexity of the marine ecosystem and our lack of a detailed understanding of many of the governing processes preclude definitive model projections for the significant changes expected to take place in the upper ocean by the late 21st century. As a result, most modelers have focused on sensitivity studies, which provide information about the importance of different mechanisms in controlling ocean $CO_2$ uptake. For example, Sarmiento et al. (1998) explored a number of biological parameterizations, including a low-uptake "constant phosphate" limit, and a high-uptake "super-biotic" limit where surface phosphate is completely consumed. The cumulative difference in carbon uptake from the atmosphere between 1990 and 2065 was $-70$ PgC ($-24.2$ percent) for the constant phosphate case, and $+120$ PgC ($+41.5$ percent) for the super-biotic case (Table 13.3). Joos et al. (1999a) also considered several cases, including a low uptake limit where the marine ecosystem composition shifts mainly toward $CaCO_3$ production (the "no-biota-DIC feedback" case). Cumulative uptake changes from 1765 to 2100 ranged from $-32$ PgC ($-6.0$ percent) for a constant productivity baseline to $-97$ PgC ($-18.3$ percent) for the no-biota-DIC feedback case. Plattner et al. (2001) improved upon the Joos et al. (1999a) study in several ways, but results were similar. Although some of the cases explored in these studies represent extreme conditions that would probably never be realized in nature, they illustrate the large sensitivity of uptake to different assumptions about biological mechanisms.

In summary, differences between model simulations of changes in carbon uptake are due primarily to differences in the parameterizations of vertical exchange and the biological carbon cycle. Further refinement in the understanding of ocean biology and the ability to model it, particularly responses of the marine biological system to significant changes in oceanic stratification and regional nutrient availability, is required before agreement can be reached on future changes in ocean uptake of $CO_2$.

## Long-Term Fate of Ocean Uptake

Carbon dioxide uptake is greatly enhanced by the buffering reaction with surface $CO_3^{2-}$, but because the reaction consumes $CO_3^{2-}$, the absorptive capacity of the ocean diminishes as more $CO_2$ is dissolved (Le Quéré and Metzl, Chapter 12). Simulations by Archer et al. (1997, 1998) project that for cumulative anthropogenic emissions of

**Table 13.3.** Biological effects on cumulative ocean uptake of $CO_2$ (PgC)

| Scenario | Cumulative ocean uptake of $CO_2$ (PgC) | |
|---|---|---|
| *Sarmiento et al. (1998),* | | |
| *IS92a-like, 1990–2065* | | |
| Climate baseline | 289 | |
| Super-biotic | +120 | (+41.5%) |
| Standard (constant productivity) | –27 | (–9.3%) |
| Constant phosphate | –70 | (–24.2%) |
| | | |
| *Joos et al. (1999a),* | | |
| *WRE550, 1765–2100* | | |
| Climate baseline | 530 | |
| Constant productivity | –32 | (–6.0%) |
| Standard (prognostic) | –50 | (–9.5%) |
| No biota feedback | –83 | (–15.6%) |
| No biota-DIC feedback | –97 | (–18.3%) |
| | | |
| *Plattner et al. (2001),* | | |
| *WRE1000, 1765–2100* | | |
| Climate baseline | 612 | |
| Constant productivity | –31 | (–5.1%) |
| No $CaCO_3$ feedback | –34 | (–5.6%) |
| Standard (prognostic) | –48 | (–7.9%) |
| No biota feedback | –84 | (–13.8%) |
| No OM feedback | –98 | (–16.1%) |

*Note:* "Climate baseline" refers to a simulation with anthropogenic $CO_2$ emissions but preindustrial ocean temperatures and circulation. "Standard" refers to uptake changes relative to the climate baseline for the full climate change simulation using the default biological model (indicated in parentheses). "Constant productivity" indicates fixed production, whereas "Prognostic" indicates production is determined from surface nutrient concentrations. "Super-biotic" indicates total surface phosphate consumption. "Constant phosphate" indicates that surface phosphate is forced to remain constant. "No biota-DIC feedback" refers to removing the changes in surface DIC concentration due to surface nutrient changes, whereas "No biota feedback" refers to removing the changes due to surface alkalinity as well. "No $CaCO_3$ feedback" indicates that changes in alkalinity due to the $CaCO_3$ cycle are removed. "No OM feedback" indicates that the organic matter contribution to alkalinity is removed.

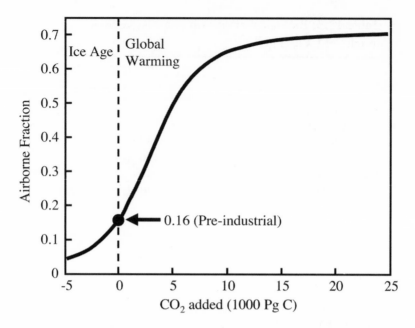

**Figure 13.3.** Marginal airborne fraction vs. $CO_2$ release. Results are for 5,000-year integrations, where the ocean-atmosphere system has reached equilibrium. Note that the effects of neither the terrestrial biosphere nor ocean sediments were included in these simulations (unpublished work by the authors, 2003).

~3000 PgC (a low estimate of total fossil fuel reserves; see Nakicenovic et al. 1996; Marland et al. 2000), deep Pacific $CO_3^{2-}$ concentrations are reduced by nearly 50 percent after 1,500 years, and the ocean is able to remove 19 percent less $CO_2$ from the atmosphere than the preindustrial ocean.

The long-term effectiveness of the ocean buffering system against additions of $CO_2$ can be described by the marginal airborne fraction (AF), which is defined as the fraction of a pulse of $CO_2$ added to the ocean that remains in the atmosphere after the ocean reaches equilibrium. An increase in the AF indicates a decrease in the ocean's buffering capacity. Figure 13.3 depicts changes to the AF over a wide range of $CO_2$ releases that we have calculated using a simplified ocean model with full carbon chemistry. Negative releases correspond to an ice age condition (−3,500 PgC or 180 ppm atmospheric $CO_2$), whereas large positive releases would be achievable if methane hydrates were exploited (Harvey and Huang 1995). Note that the AF, equal to 16 percent for the preindustrial ocean, plateaus near 70 percent above ~15,000 PgC, indicating near-total depletion of $CO_3^{2-}$, with ocean carbon accumulating primarily as dissolved $CO_2$.

Sea-floor $CaCO_3$ sediments provide additional neutralization of $CO_2$. Whereas Archer et al. (1997, 1998) suggest that sediment dissolution is slow, taking many thousands of years to exert an appreciable effect on atmospheric $CO_2$, recent observations in the Pacific and Indian Oceans indicate that dissolution of $CaCO_3$ rain may be taking place in areas where the saturation depth is shallow enough for anthropogenic $CO_2$ to reach (~1,000 m) (Feely et al. 2002; Sabine et al. 2002). Elevated levels of $CO_2$ may decrease calcification rates by lowering the $CaCO_3$ saturation state of surface waters (Iglesias-Rodriguez et al. 2002), which in turn may indirectly affect $CO_2$ levels by changing the organic to inorganic carbon export ratio or the ballasting of sinking particulate carbon (Armstrong et al. 2002). Changes in climate may also shift the distribution of marine calcifiers, which could impact the $CaCO_3$ budget and the ocean carbon cycle as a whole (Iglesias-Rodriguez et al. 2002). The magnitudes of these effects are still very uncertain, however, and further research is needed. Sediments are ultimately responsible for absorbing an additional 17 percent of anthropogenic $CO_2$ in the 3000 PgC release case, resulting in a total ocean uptake of 92 percent and a final atmospheric concentration of 425 ppm (Archer et al. 1998). Regeneration of sediments via river discharge requires more than 40,000 years for full recovery.

## Surprises

There exists the possibility of large, abrupt changes to the climate system, which may have significant impacts on the ocean carbon cycle. Some possible phenomena are the collapse of grounded ice sheets such as the West Antarctic Ice Sheet (Oppenheimer 1998; Church et al. 2001), the catastrophic release of methane from methane hydrates in ocean sediments (Harvey and Huang 1995), regime shifts in marine ecosystems resulting in changes in $CO_2$ uptake (Chavez et al. 2003), and the collapse of the North Atlantic thermohaline circulation (THC) (Manabe and Stouffer 1993).

Regime shifts in marine ecosystems observed during the past century encompass multidecadal, basinwide changes in temperature, thermocline depth, ecosystem structure, and $CO_2$ uptake. It is not currently understood, however, how the confluence of processes modulates ocean $CO_2$ uptake or whether these roughly 50-year oscillations are driven by natural or anthropogenic forcings (Chavez et al. 2003). Therefore, it is difficult to assess the role of ecosystem regime change under future climate change conditions. Among the other effects mentioned, only THC collapse has an appreciable likelihood during the 21st century. Therefore, we will restrict discussion in this section to that phenomenon.

The effect of a THC collapse on the ocean carbon cycle could be a large decrease in uptake, due to the breakdown of large-scale surface-to-deepwater transport (Sarmiento and Le Quéré 1996). Plattner et al. (2001) found that cumulative $CO_2$ uptake through 2500 is reduced by 16 percent for an emissions scenario that stabilizes at 550 ppm (WRE550), but the reduction jumps to 27 percent for a scenario (WRE1000) where the North Atlantic Deep Water formation collapses after 2100. The main contributor to this change in reduction is the circulation component, which is 2 percent in the WRE550 scenario and 21 percent in the WRE1000 scenario. An earlier study by Joos et al. (1999a) obtained similar results. As discussed, however, these studies both used simplified ocean models that obtained larger stratification effects than other studies, such as Matear and Hirst (1999). The paleoclimatic role of the THC in modulating atmospheric $CO_2$ is discussed by Joos and Prentice (Chapter 7).

In the early 1990s, Manabe and Stouffer demonstrated a long-term (> 500 years) reduction (>85 percent) of the THC maximum overturning under a $4 \times CO_2$ growth scenario (Manabe and Stouffer 1993, 1994; Dixon et al. 1999). Similar results have been obtained by a number of independent studies, using different models and a variety of atmospheric $CO_2$ growth trajectories (Table 13.4). Reductions in overturning vary over a wide range, from ~90 percent to as low as ~25 percent. In addition to the final level of atmospheric $CO_2$, the rate of atmospheric increase is also important in producing a THC collapse (Stocker and Schmittner 1997; Schmittner and Stocker 1999). Long-time integrations (> 1000 years) of $CO_2$ stabilization scenarios with simplified models (Fanning and Weaver 1996; Weaver et al. 1998; Wiebe and Weaver 2000; Ganopolski et al. 2001) show that the THC weakening is generally followed by full recovery over several hundred years (Cubasch et al. 2001), though there are some studies where the THC remains collapsed during a multi-thousand-year integration (Stocker et al. 1992; Manabe and Stouffer 1999).

Several studies, however, obtained essentially no reduction in overturning (Latif et al. 2000; Gent 2001; Sun and Bleck 2001). All these studies account for their result by showing that temperature increases are balanced by net salinization of North Atlantic waters, maintaining a density structure close to that of the preindustrial ocean. Gent (2001) and Sun and Bleck (2001) argue that the net salinization in their models is due to increased evaporation and reduced sea-ice melting in the North Atlantic. On the other hand, Latif et al. (2000) simulate strongly increased precipitation in the central equatorial Pacific, which is associated with reduced precipitation and warming over the Amazon rainforest and, therefore, reduced freshwater flux from the Amazon River basin into the tropical Atlantic. This result was also obtained by Cox et al. (2000), who employed a different coupled climate model. Understanding the relative roles of freshwater and heat is central to accurately modeling future changes in the THC, but two studies that specifically investigated these two factors obtained contradictory results (Dixon et al. 1999; Mikolajewicz and Voss 2000). Further investigations will be needed to fully elucidate the complex interactions involved.

## Future Research Needs

We have identified a number of important issues that remain to be resolved: (1) the disparity between estimates of ocean uptake variability obtained via ocean versus

**Table 13.4.** Studies of THC slowdown under global warming scenarios listed in order by the percentage reduction

| Study | Model | Scenario | Baseline | Pertur-bation | Year | Reduction |
|---|---|---|---|---|---|---|
| Stocker and Schmittner (1997) | Three-basin box model | $CO_2$ + 1% $y^{-1}$ to 750 ppm (99 yrs) | ~24 | ~2 | ~220 | ~90% |
| Stocker and Schmittner (1997) | Three-basin box model | $CO_2$ + 2% $y^{-1}$ to 650 ppm (42.5 yrs) | ~24 | ~2 | ~180 | ~90% |
| Joos et al. (1999a) | 2D ocean model | WRE1000 | ~24 | ~2 | ~2200 AD | ~90% |
| Manabe and Stouffer (1994) | GFDL R15 | $CO_2$ + 1% $y^{-1}$ to $4 \times CO_2$ (140 yrs) | 16–19 | ~2.5 | ~250 | ~85% |
| Manabe and Stouffer (1994) | GFDL R15 | $CO_2$ + 1% $y^{-1}$ to $2 \times CO_2$ (70 yrs) | 16–19 | ~8 | ~150 | ~50% |
| Stocker and Schmittner (1997) | Three-basin box model | $CO_2$ + 0.5% $y^{-1}$ to 750 ppm (198 yrs) | ~24 | ~13 | ~230 | ~45% |
| Mikolajewicz and Voss (2000) | MPI ECHAM3/LSG | IPCC scenario A to $4 \times CO_2$ (120 yrs) | 27–30 | ~16 | 150 | ~45% |
| Joos et al. (1999a) | 2D ocean model | WRE550 | ~24 | ~15 | ~2100 AD | ~40% |
| Bi et al. (2001) | CSIRO | IPCC IS92a to $3 \times CO_2$ (2082) | ~20 | ~13 | ~2100 AD | ~35% |
| Stocker and Schmittner (1997) | Three-basin box model | $CO_2$ + 1% $y^{-1}$ to $2 \times CO_2$ (70 yrs) | ~24 | ~16 | ~100 | ~35% |
| Wood et al. (1999); Thorpe et al. (2001 | HadCM3 | IPCC IS92a | ~23 | ~17 | 2100 AD | ~25% |
| Latif et al. (2000) | MPI ECHAM4/OPYC | IPCC IS92a | ~23 | ~23 | 2100 AD | ~0% |
| Sun and Bleck (2001) | GISS SI2000/HYCOM | $CO_2$ + 1% $y^{-1}$ to $2 \times CO_2$ (70 yrs) | ~23 | ~23 | 200 | ~0% |
| Gent (2001) | NCAR CSM | IPCC SRES A1 | ~29 | ~30 | 2100 AD | ~0% |

*Note:* "Baseline" refers to maximum overturning (Sv) in experiment with preindustrial $CO_2$ levels; "perturbation" refers to maximum overturning (Sv) for the $CO_2$ emissions scenario in the year of maximum THC slowdown. "Year" refers to time of maximum extent of THC slowdown atmosphere-based analyses; (2) the large possible contribution of the ocean thermohaline circulation to oceanic uptake and our poor understanding of how this will respond to global warming; and (3) the large potential impact of biological changes on the oceanic carbon sink. Addressing all of these will require improved ocean circulation and climate models, as well as more accurate and longer-term oceanic and atmospheric observations of relevant tracers such as those of the carbon system, oxygen, and carbon isotopes. An issue that we have not discussed in this chapter, and which may play a significant role, is the possibility that changes in ocean biology might lead to modifications in the stoichiometric ratio of carbon to nutrients in organic matter or $CaCO_3$ to organic matter, or other shifts in material exported from the surface of the ocean that could have a significant impact on the ocean carbon sink (e.g., Boyd and Doney 2002).

## Literature Cited

Archer, D., H. Kheshgi, and E. Maier-Reimer. 1997. Multiple timescales for neutralization of fossil fuel $CO_2$. *Geophysical Research Letters* 24:405–408.

———. 1998. Dynamics of fossil fuel $CO_2$ neutralization by marine $CaCO_3$. *Global Biogeochemical Cycles* 12:259–276.

Armstrong, R. A., C. Lee, J. I. Hedges, S. Honjo, and S. G. Wakeham. 2002. A new, mechanistic model for organic carbon fluxes in the ocean based on the quantitative association of POC with ballast minerals. *Deep-Sea Research II* 49:219–236.

Baker, D. F. 2001. Sources and sinks of atmospheric $CO_2$ estimated from batch least-squares inversions of $CO_2$ concentration measurements. Ph.D. thesis, Princeton University, Princeton, NJ.

Bates, N. R. 2002. Interannual variability in the global uptake of $CO_2$. *Geophysical Research Letters* 29 (5): 1059, doi:10.1029/2001GL013571.

Battle, M., M. L. Bender, P. P. Tans, J. W. C. White, J. T. Ellis, T. Conway, and R. J. Francey. 2000. Global carbon sinks and their variability inferred from atmospheric $O_2$ and $\delta^{13}C$. *Science* 287:2467–2470.

Bi, D., W. F. Budd, A. C. Hirst, and X. Wu. 2001. Collapse and reorganisation of the Southern Ocean overturning under global warming in a coupled model. *Geophysical Research Letters* 28:3927–3930.

Bousquet, P., P. Peylin, P. Ciais, C. Le Quéré, P. Friedlingstein, and P. P. Tans. 2000. Regional changes in carbon dioxide fluxes of land and oceans since 1980. *Science* 290:1342–1346.

Boyd, P. W., and S. C. Doney. 2002. Modelling regional responses by marine pelagic ecosystems to global climate change. *Geophysical Research Letters* 29, doi:10.1029/2001GL014130.

Broecker, W. S., S. Sutherland, and T.-H. Peng. 1999. A possible 20th-century slowdown of Southern Ocean deep water formation. *Science* 286:1132–1135.

Caldeira, K., and P. B. Duffy. 2000. The role of the Southern Ocean in uptake and storage of anthropogenic carbon dioxide. *Science* 287:620–622.

Chavez, F. P., J. Ryan, S. E. Lluch-Cota, and C. Miguel Ñiquen. 2003. From anchovies to sardines and back: Multidecadal change in the Pacific Ocean. *Science* 299:217–221.

Church, J. A., J. M. Gregory, P. Huybrechts, M. Kuhn, K. Lambeck, M. T. Nhuan, D.

Qin, and P. L. Woodworth. 2001. Changes in sea level. Pp. 639–693 in *Climate change 2001: The scientific basis (Contribution of Working Group I to the Third Assessment Report of the Intergovernmental Panel on Climate Change)*, edited by J. T. Houghton, Y. Ding, D. J. Griggs, M. Noguer, P. J. van der Linden, X. Dai, K. Maskell, and C. A. Johnson. Cambridge: Cambridge University Press.

Cox, P. M., R. A. Betts, C. D. Jones, S. A. Spall, and I. J. Totterdell. 2000. Acceleration of global warming due to carbon-cycle feedbacks in a coupled climate model. *Nature* 408:184–187.

Cubasch, U., G. A. Meehl, G. J. Boer, R. J. Stouffer, M. Dix, A. Noda, C. A. Senior, S. Raper, and K. S. Yap. 2001. Projections of future climate change. Pp. 525–582 in *Climate change 2001: The scientific basis (Contribution of Working Group I to the Third Assessment Report of the Intergovernmental Panel on Climate Change)*, edited by J. T. Houghton, Y. Ding, D. J. Griggs, M. Noguer, P. J. van der Linden, X. Dai, K. Maskell, and C. A. Johnson. Cambridge: Cambridge University Press.

Dixon, K. W., T. L. Delworth, M. J. Spelman, and R. J. Stouffer. 1999. The influence of transient surface fluxes on North Atlantic overturning in a coupled GCM climate change experiment. *Geophysical Research Letters* 26:2749–2752.

Dufresne, J.-L., P. Friedlingstein, M. Berthelot, L. Bopp, P. Ciais, L. Fairhead, H. Le Treut, and P. Monfray. 2002. On the magnitude of positive feedback between future climate change and the carbon cycle. *Geophysical Research Letters* 29:1405, doi:10.1029/2001GL013777.

Fanning, A. F., and A. J. Weaver. 1996. An atmospheric energy-moisture balance model: Climatology, interpentadal climate change, and coupling to an ocean general circulation model. *Journal of Geophysical Research* 101:15,111–15,128.

Feely, R. A., R. Wanninkhof, C. Goyet, D. E. Archer, and T. Takahashi. 1997. Variability of CO$_2$ distributions and sea-air fluxes in the central and eastern equatorial Pacific during the 1991–1994 El Niño. *Deep-Sea Research II* 44:1851–1867.

Feely, R. A., R. Wanninkhof, T. Takahashi, and P. Tans. 1999. Influence of El Niño on the equatorial Pacific contribution to atmospheric CO$_2$ accumulation. *Nature* 398:597–601.

Feely, R. A., C. L. Sabine, K. Lee, F. J. Millero, M. F. Lamb, D. Greeley, J. L. Bullister, R. M. Key, T.-H. Peng, A. Kozyr, T. Ono, and C. S. Wong. 2002. In situ calcium carbonate dissolution in the Pacific Ocean. *Global Biogeochemical Cycles* 16 (4): 1144, doi:10.1029/2002GB001866.

Francey, R. J., P. P. Tans, C. E. Allison, I. G. Enting, J. W. White, and M. Trolier. 1995. Changes in oceanic and terrestrial carbon uptake since 1982. *Nature* 373:326–330.

Francey, R. J., C. E. Allison, C. M. Trudinger, and P. J. Rayner. 2001. The interannual variation in global atmospheric $\delta^{13}$C and its link to net terrestrial exchange. Extended abstract of the 6th International Carbon Dioxide Conference, and in preparation for *Tellus*.

Friedlingstein, P., J.-L. Dufresne, P. M. Cox, and P. Rayner. 2003. How positive is the feedback between climate change and the carbon cycle? *Tellus* 55B (2): 692–700.

Ganopolski, A., V. Petoukhov, S. Rahmstorf, V. Brovkin, M. Claussen, A. Eliseev, and C. Kubatzki. 2001. CLIMBER-2: A climate system model of intermediate complexity. Part II. Model sensitivity. *Climate Dynamics* 17 (10): 735–751.

Gent, P. R. 2001. Will the North Atlantic Ocean thermohaline circulation weaken during the 21st century? *Geophysical Research Letters* 28:1023–1026.

Gent, P. R., J. Willebrand, T. J. McDougall, and J. C. McWilliams. 1995. Parameterizing eddy-induced tracer transports in ocean circulation models. *Journal of Physical Oceanography* 25:463–474.

Gruber, N., C. D. Keeling, and N. R. Bates. 2002. Interannual variability in the North Atlantic Ocean carbon sink. *Science* 298:2374–2378.

Harvey, L. D. D., and Z. Huang. 1995. Evaluation of the potential impact of methane clathrate destabilization on future global warming. *Journal of Geophysical Research* 100:2905–2926.

Iglesias-Rodriguez, M. D., R. Armstrong, R. Feely, R. Hood, J. Kleypas, J. D. Milliman, C. Sabine, and J. Sarmiento. 2002. Calcium carbonate in a changing ocean: Implications for the marine carbon cycle, carbonate producing organisms, and global climate. *Eos, Transactions, American Geophysical Union* 83:365–375.

Joos, F., G.-K. Plattner, T. F. Stocker, O. Marchal, and A. Schmittner. 1999a. Global warming and marine carbon cycle feedbacks on future atmospheric $CO_2$. *Science* 284:464–467.

Joos, F., R. Meyer, M. Bruno, and M. Leuenberger. 1999b. The variability in the carbon sinks as reconstructed for the last 1000 years. *Geophysical Research Letters* 26:1437–1440.

Keeling, C. D., and S. C. Piper. 2000. *Interannual variations of exchanges of atmospheric $CO_2$ and $^{13}CO_2$ with the terrestrial biosphere and oceans from 1978–2000. III. Simulated sources and sinks.* Scripps Institution of Oceanography Reference Series No. 00-14. San Diego: University of California, Scripps Institution of Oceanography.

Keeling, C. D., T. P. Whorf, M. Wahlen, and J. Vanderplicht. 1995. Interannual extremes in the rate of rise of atmospheric carbon dioxide since 1980. *Nature* 375:666–670.

Keeling, C. D., S. C. Piper, R. B. Bacastrow, M. Wahlen, T. P. Whorf, M. Heimann, and H. A. Meijer. 2001. *Exchanges of atmospheric $CO_2$ and $^{13}CO_2$ with the terrestrial biosphere and oceans from 1978 to 2000. I. Global aspects.* Scripps Institution of Oceanography Reference Series No. 00-06. San Diego: University of California, Scripps Institution of Oceanography.

Langenfelds, R., R. Francey, B. Pak, L. Steele, J. Lloyd, C. Trudinger, and C. Allison. 2002. Interannual growth rate variations of atmospheric $CO_2$ and its $\delta^{13}C$, $H_2$, $CH_4$, and CO between 1992 and 1999 linked to biomass burning. Art. no. 1048. *Global Biogeochemical Cycles* 16: 1048.

Latif, M., E. Roeckner, U. Mikolajewicz, and R. Voss. 2000. Tropical stabilization of the thermohaline circulation in a greenhouse warming simulation. *Journal of Climate* 13:1809–1813.

Le Quéré, C., J. C. Orr, P. Monfray, O. Aumont, and G. Madec. 2000. Interannual variability of the oceanic sink of $CO_2$ from 1979 through 1997. *Global Biogeochemical Cycles* 14:1247–1265.

Le Quéré, C., O. Aumont, P. Monfray, and J. Orr. 2002. Propagation of climatic events on ocean stratification, marine biology and $CO_2$: case studies over the 1979–1999 period. *Journal of Geophysical Research* (in press).

Le Quéré, C., O. Aumont, L. Bopp, P. Bousquet, P. Ciais, R. Francey, M. Heimann, D. C. Keeling, R. F. Keeling, H. Kheshgi, P. Peylin, S. C. Piper, I. C. Prentice, and P. J. Rayner. 2003. Two decades of ocean $CO_2$ sink and variability. *Tellus* 55B (2): 649–656.

Lee, K., R. Wanninkhof, T. Takohashi, S. C. Doney, and R. A. Feely. 1998. Low interannual variability in recent oceanic uptake of atmospheric carbon dioxide. *Nature* 396:155–159.

Levitus, S., J. I. Antonov, T. P. Boyer, and C. Stephens. 2000. Warming of the world ocean. *Science* 287:2225–2229.

Loukos, H., F. Vivier, P. P. Murphy, D. E. Harrison, and C. Le Quéré. 2000. Interannual variability of equatorial Pacific CO$_2$ fluxes estimated from temperature and salinity data. *Geophysical Research Letters* 27:1735–1738.

Manabe, S., and R. J. Stouffer. 1993. Century-scale effects of increased atmospheric CO$_2$ on the ocean-atmosphere system. *Nature* 364:215–218.

———. 1994. Multiple-century response of a coupled ocean-atmosphere model to an increase of atmospheric carbon-dioxide. *Journal of Climate* 7:5–23.

———. 1999. Are two modes of thermohaline circulation stable? *Tellus* 51A: 400–411.

Manning, A. C. 2001. Temporal variability of atmospheric oxygen from both continuous measurements and a flask sampling network: Tools for studying the global carbon cycle. Ph.D. thesis, University of California–San Diego, La Jolla, CA.

Marland, G., T. A. Boden, and R. J. Andres. 2000. Global, regional, and national CO$_2$ emissions. In *Trends: A compendium of data on global change*. Oak Ridge, TN: Carbon Dioxide Information Analysis Center, Oak Ridge National Laboratory, U.S. Department of Energy.

Matear, R. J., and A. C. Hirst. 1999. Climate change feedback on the future oceanic CO$_2$ uptake. *Tellus* 51B:722–733.

Matear, R. J., A. C. Hirst, and B. I. McNeil. 2000. Changes in dissolved oxygen in the Southern Ocean with climate change. *Geochemistry Geophysics Geosystems* 1: Paper number 2000GC000086.

Mikolajewicz, U., and R. Voss. 2000. The role of the individual air-sea flux components in CO$_2$-induced changes of the ocean's circulation and climate. *Climate Dynamics* 16:627–642.

Nakicenovic, N., A. Grübler, H. Ishitani, T. Johansson, G. Marland, J. R. Moreira, and H.-H. Rogner. 1996. Energy primer. Pp. 75–92 in *Climate change 1995: Impacts, adaptations and mitigation of climate change: Scientific-technical analysis*, edited by R. T. Watson, M. C. Zinyowera, R. H. Moss, and D. J. Dokken. Cambridge: Cambridge University Press.

Oppenheimer, M. 1998. Global warming and the stability of the West Antarctic Ice Sheet. *Nature* 393:325–332.

Orr, J. C., E. Maier-Reimer, U. Mikolajewicz, P. Monfray, J. L. Sarmiento, J. R. Toggweiler, N. K. Taylor, J. Palmer, N. Gruber, C. L. Sabine, C. Le Quéré, R. M. Key, and J. Boutin. 2001. Estimates of anthropogenic carbon uptake from four three-dimensional global ocean models. *Global Biogeochemical Cycles* 15:43–60.

Plattner, G.-K., F. Joos, T. F. Stocker, and O. Marchal. 2001. Feedback mechanisms and sensitivities of ocean carbon uptake under global warming. *Tellus* 53B:564–592.

Prentice, I. C., G. D. Farquhar, M. J. R. Fasham, M. L. Goulden, M. Heimann, V. J. Jaramillo, H. S. Kheshgi, C. Le Quéré, R. J. Scholes, and D. W. R. Wallace. 2001. The carbon cycle and atmospheric carbon dioxide. Pp. 183–238 in *Climate change 2001: The scientific basis (Contribution of Working Group I to the third assessment report of the Intergovernmental Panel on Climate Change)*, edited by J. T. Houghton, Y. Ding, D. J. Griggs, M. Noguer, P. J. van der Linden, X. Dai, K. Maskell, and C. A. Johnson. Cambridge: Cambridge University Press.

Rayner, P. J., I. G. Enting, R. J. Francey, and R. Langenfelds. 1999. Reconstructing the recent carbon cycle from atmospheric $CO_2$, $\delta^{13}C$ and $O_2/N_2$ observations. *Tellus* 51B:213–232.

Russell, J. L., and J. M. Wallace. 2003. ENSO-related variability in the rate of increase of atmospheric carbon dioxide. *Science* (submitted).

Sabine, C. L., R. M. Key, K. M. Johnson, F. J. Millero, A. Poisson, J. L. Sarmiento, D. W. R. Wallace, and C. D. Winn. 1999. Anthropogenic $CO_2$ inventory of the Indian Ocean. *Global Biogeochemical Cycles* 13:179–198.

Sabine, C. L., R. M. Key, R. A. Feely, and D. Greeley. 2002. Inorganic carbon in the Indian Ocean: Distribution and dissolution processes. *Global Biogeochemical Cycles* 16 (4): 1067, doi:10.1029/2002GB001869.

Sarmiento, J. L., and N. Gruber. 2002. Sinks for anthropogenic $CO_2$. *Physics Today* 55:30–36.

Sarmiento, J. L., and T. M. C. Hughes. 1999. Anthropogenic $CO_2$ uptake in a warming ocean. *Tellus* 51B:560–561.

Sarmiento, J. L., and C. Le Quéré. 1996. Oceanic carbon dioxide uptake in a model of century-scale global warming. *Science* 274:1346–1350.

Sarmiento, J. L., C. Le Quéré, and S. W. Pacala. 1995. Limiting future atmospheric carbon dioxide. *Global Biogeochemical Cycles* 9:121–137.

Sarmiento, J. L., T. M. C. Hughes, R. J. Stouffer, and S. Manabe. 1998. Simulated response of the ocean carbon cycle to anthropogenic climate warming. *Nature* 393:245–249.

Schmittner, A., and T. F. Stocker. 1999. The stability of the thermohaline circulation in global warming experiments. *Journal of Climate* 12:1117–1133.

Stocker, T. F., and A. Schmittner. 1997. Influence of $CO_2$ emission rates on the stability of the thermohaline circulation. *Nature* 388:862–865.

Stocker, T. F., D. G. Wright, and L. A. Mysak. 1992. A zonally averaged, coupled ocean-atmosphere model for paleoclimate studies. *Journal of Climate* 5:773–797.

Sun, S., and R. Bleck. 2001. Atlantic thermohaline circulation and its response to increasing $CO_2$ in a coupled atmosphere-ocean model. *Geophysical Research Letters* 28 (22): 4223–4226.

Thorpe, R. B., J. M. Gregory, T. C. Johns, R. A. Wood, and J. F. B. Mitchell. 2001. Mechanisms determining the Atlantic thermohaline circulation response to greenhouse gas forcing in a non-flux-adjusted coupled climate model. *Journal of Climate* 14:3102–3116.

Timmermann, A., J. Oberhuber, A. Bacher, M. Esch, M. Latif, and E. Roeckner. 1999. Increased El Niño frequency in a climate model forced by future greenhouse warming. *Nature* 398:694–697.

Trenberth, K. E., and T. J. Hoar. 1996. The 1990–1995 El Niño–Southern Oscillation event: Longest on record. *Geophysical Research Letters* 23:57–60.

Tsunogai, S., S. Watanabe, and T. Sato. 1999. Is there a "continental shelf pump" for the absorption of atmospheric $CO_2$? *Tellus* 51B:701–712.

Wanninkhof, R. 1992. Relationship between wind speed and gas exchange over the ocean. *Journal of Geophysical Research* 97:7373–7378.

Wanninkhof, R., and W. R. McGillis. 1999. A cubic relationship between air-sea $CO_2$ exchange and wind speed. *Geophysical Research Letters* 26:1889–1892.

Weaver, A. J., M. Eby, A. F. Fanning, and E. C. Wiebe. 1998. Simulated influence of carbon dioxide, orbital forcing and ice sheets on the climate of the last glacial maximum. *Nature* 394:847–853.

Wiebe, E. C., and A. J. Weaver. 2000. On the sensitivity of global warming experiments to the parameterisation of sub-grid scale ocean mixing. *Climate Dynamics* 15:875–893.

Winguth, A. M. E., M. Heimann, K. D. Kurz, E. Maier-Reimer, U. Mikolajewicz, and J. Segschneider. 1994. El Niño-Southern Oscillation related fluctuations of the marine carbon cycle. *Global Biogeochemical Cycles* 8:39–63.

Wong, C. S., Y. H. Chan, J. S. Page, G. E. Smith, and R. D. Bellegay. 1993. Changes in equatorial $CO_2$ flux and new production estimated from $CO_2$ and nutrient levels in Pacific surface waters during the 1986/87 El Niño. *Tellus* 45B:64–79.

Wood, R. A., A. B. Keen, J. F. B. Mitchell, and J. M. Gregory. 1999. Changing spatial structure of the thermohaline circulation in response to atmospheric $CO_2$ forcing in a climate model. *Nature* 399:572–575.

# PART IV
## The Carbon Cycle of the Land

# 14

# A Primer on the Terrestrial Carbon Cycle: What We Don't Know But Should

Jonathan A. Foley and Navin Ramankutty

The terrestrial biosphere is an integral component of the global carbon cycle. Carbon is brought into the terrestrial biosphere through photosynthesis, and it is released back to the atmosphere through plant respiration, microbial respiration (or "decomposition"), fires, and some human land use practices. At times when photosynthesis exceeds the sum of plant respiration, microbial respiration, fires, and land use releases, there is a net sink of carbon from the atmosphere into ecosystems. When photosynthesis is less than the sum of the other terms, there is a net source from ecosystems to the atmosphere.

On seasonal timescales, the gain and loss of carbon in terrestrial ecosystems is evident from subtle changes in atmospheric $CO_2$ concentrations—the seasonal "wiggles" in the Mauna Loa curve are a famous example of this phenomenon. But it is on longer timescales, from decades to centuries, that the carbon cycle of the terrestrial biosphere can significantly affect the $CO_2$ levels in the atmosphere and become important to the climate system. On these longer timescales, the amount of carbon stored in the terrestrial biosphere is the result of the balance between net primary productivity (NPP, the net accumulation of carbon through photosynthesis minus plant respiration over a year) and carbon losses through decomposition, land use, fires, and other disturbances (Figure 14.1).

The scientific community currently believes that the terrestrial biosphere has been acting, on average, as a net sink of atmospheric carbon. In fact, terrestrial ecosystems appear to be absorbing a significant fraction of anthropogenic $CO_2$ emissions (from fossil-fuel combustion and, to a lesser extent, cement production) during the past few decades—thereby keeping atmospheric $CO_2$ levels lower than they would otherwise be.

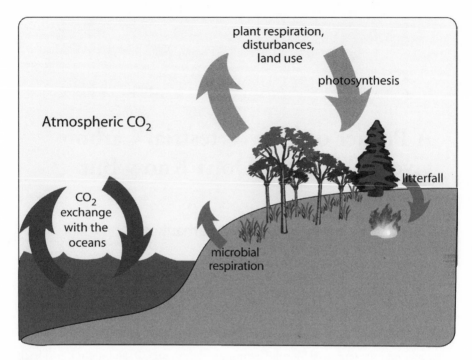

**Figure 14.1.** Relationships between terrestrial, oceanic, and atmospheric carbon pools. Adapted from Foley et al. (2003).

The terrestrial biosphere, in effect, has been "subsidizing" part of our fossil-fuel emissions—keeping a significant part of these $CO_2$ emissions out of the atmosphere.

Some basic numbers illustrate the point. During the decade of the 1990s, human activities were releasing roughly 6 petagrams of carbon per year (PgC y$^{-1}$) from fossil-fuel combustion and cement production (Prentice et al. 2001). During the same time period, humans may have been releasing another 1–2 PgC y$^{-1}$ through land use practices, primarily tropical deforestation (Houghton 2000). Taken together, these fossil-fuel and land use carbon emissions should lead to a 7–8 Pg increase in the carbon content of the atmosphere every year. Observations of atmospheric $CO_2$ levels, however, show that only about half of this anthropogenic carbon is accumulating in the atmosphere—the so-called airborne fraction. The other half of the anthropogenic carbon emissions is, for the time being, apparently absorbed in the oceans and terrestrial biosphere (Bousquet et al. 2000; Prentice et al. 2001; Schimel et al. 2001).

Recent analyses have attempted to determine the fate of anthropogenic $CO_2$ emissions in the atmosphere, oceans, and terrestrial biosphere. There is now overwhelming evidence to suggest that the oceans and terrestrial biosphere together absorb nearly 2 to

**Table 14.1.** Estimates of global budget of anthropogenic carbon emissions, as reported by the IPCC third assessment report (Prentice et al. 2001)

| Indicator | 1980s | 1990s |
|---|---|---|
| Measured increase in atmospheric $CO_2$ levels | $3.3 \pm 0.1$ | $3.2 \pm 0.1$ |
| Anthropogenic emissions from fossil fuels and cement production | $5.4 \pm 0.3$ | $6.4 \pm 0.4$ |
| Net ocean-atmosphere flux of carbon | $-1.9 \pm 0.6$ | $-1.7 \pm 0.5$ |
| Net land-atmosphere flux of carbon (including all sources from the land, such as tropical deforestation, as well as all land-based carbon sinks) | $-0.2 \pm 0.7$ | $-1.4 \pm 0.7$ |

*Note:* Units are in PgC ($10^{15}$g of carbon) per year.

3 billion tons of anthropogenic carbon emissions per year—and we can begin to estimate the approximate breakdown between oceanic and terrestrial uptake. For example, in the third assessment report of the Intergovernmental Panel on Climate Change (IPCC), Prentice et al. reported that the terrestrial biosphere absorbed roughly 1.4 PgC $y^{-1}$ (or about 22 percent of the anthropogenic emissions) during the 1990s and nearly 0.2 PgC $y^{-1}$ (or about 4 percent of the anthropogenic emissions) in the 1980s (Prentice et al. 2001; Table 14.1).

But how certain are these estimates? Even if we can roughly estimate the uptake of $CO_2$ by terrestrial ecosystems and the oceans, we do not know how these will change in the future. And as humans continue to pump more and more $CO_2$ into the atmosphere, we need to remember that the terrestrial biosphere may not always continue to absorb such a large share of our emissions. What if the terrestrial biosphere stopped absorbing so much of our emissions? Or what if the terrestrial biosphere started to release carbon dioxide instead of absorbing it? To address these questions, we need to understand what controls the uptake of carbon in the terrestrial biosphere.

## How Well Can We Estimate the Size of the Terrestrial Sources and Sinks?

In the IPCC third assessment report, Prentice et al. reported estimates of the net uptake of anthropogenic carbon emissions by the terrestrial biosphere: These estimates include the balance of all terrestrial carbon sources (from deforestation and other land use practices) and all terrestrial carbon sinks (the so-called residual terrestrial sink or missing sink).

The estimates of net terrestrial carbon uptake show considerable changes in the past two decades: from $-0.2$ ($\pm 0.7$) PgC $y^{-1}$ during the 1980s to $-1.4$ ($\pm 0.7$) PgC $y^{-1}$ during the 1990s (with a negative sign indicating a sink, or carbon uptake on land; Prentice et al. 2001). These figures were largely estimated using two factors: (1) overall constraints on the global carbon budget (by subtracting the measured atmospheric increase from the total anthropogenic emissions, which are both thought to be reliable estimates, to give the combined uptake of the oceans and terrestrial biosphere); and (2) the breakdown of oceanic versus terrestrial carbon uptake as indicated through simultaneous measurements of $CO_2$ and $O_2$ in the atmosphere, measurements of $\delta^{13}C$ in the atmosphere, and the results of ocean carbon cycle models (see Appendix 14.1). Le Quéré et al. (2003) recently updated the Prentice et al. estimates of oceanic and terrestrial carbon sinks and produce new figures for the net terrestrial carbon uptake during the 1980s ($-0.3 \pm 0.9$ PgC $y^{-1}$) and the 1990s ($-1.2 \pm 0.8$ PgC $y^{-1}$).

To determine why there are such large differences in net terrestrial uptake between the 1980s and 1990s and how this may change in the future, we must break down this net uptake value into the different sources and sinks of atmospheric carbon within terrestrial ecosystems.

## Estimating the Size of Terrestrial Carbon Sources

Land use practices, such as tropical deforestation and continued agricultural production in temperate latitudes, may lead to the loss of carbon from ecosystems. In particular, slash-and-burn agricultural practices in the tropics (where primary forests are cleared to make way for pastures and croplands) may lead to a substantial source of carbon to the atmosphere.

In a series of keystone papers, Houghton and his colleagues estimated the changes in terrestrial carbon balance associated with historical patterns of land use and land cover change (Houghton et al. 1983; Houghton 1999, 2003). These estimates are based on national land use inventories (using historical data collected for each region, including agricultural census records reported to the United Nations Food and Agriculture Organization [FAO]) and a simple "bookkeeping" model of ecosystem carbon cycling. In Houghton's bookkeeping model, changes in land use and land cover drive changes in terrestrial carbon storage through a series of heuristic rules about how ecosystem carbon pools (including vegetation and soil) respond to land use practices. Furthermore, the carbon removed from terrestrial ecosystems (e.g., wood products, agricultural products) is tracked and allowed to decompose or oxidize back to the atmosphere on timescales ranging from 1 to 100 years.

Houghton's estimates of carbon emissions from land use and land cover change have been used extensively in global budgets of anthropogenic carbon. For the 1980s Houghton estimated that land use and land cover change released roughly 2.0 PgC $y^{-1}$ into the atmosphere (Houghton 1999). This result implies that there is a large sink of carbon in the terrestrial biosphere in order to make the net land-atmosphere flux balance out to only $-0.2$ PgC $y^{-1}$. Accounting for Houghton's land use emissions estimates implies that there was a "residual" terrestrial sink of nearly $-2.2$ PgC $y^{-1}$ during the 1980s (see Table 14.2; Prentice et al. 2001; House et al. 2003).

**Table 14.2.** Global budget of anthropogenic carbon emissions for the 1980s, using Houghton (1999) land use emissions estimates

| 1980s carbon fluxes | Houghton budget |
| --- | --- |
| Atmospheric increase | 3.3 |
| Anthropogenic emissions | 5.4 |
| Net ocean-atmosphere flux | −1.9 |
| Net land-atmosphere flux | −0.2 |
| Modeled land use emissions | 2.0 |
| Implied residual terrestrial sink | −2.2 |

*Note:* Units are in PgC ($10^{15}$g of carbon) per year.

These estimates of terrestrial carbon sources (~2 PgC $y^{-1}$ from land use) and terrestrial carbon sinks (~2.2 PgC $^{-1}$ accumulating in ecosystems—the so-called missing sink) have formed the foundation for our understanding of the global carbon cycle during much of the past decade.

Several terms in this global carbon budget have a higher degree of certainty than others. For example, there are several different estimates of the atmospheric increase of $CO_2$ (~3.3 PgC $y^{-1}$ during the 1980s) and the fossil-fuel emissions rate (~5.4 PgC $y^{-1}$ during the 1980s): These are probably the "best-known" terms in the global carbon budget. Furthermore, the partitioning of the net flux of carbon into the oceans and terrestrial biosphere is well constrained by $CO_2$, $O_2$, and and $\delta^{13}C$ measurements (see Appendix 14.1). In addition, the uptake of carbon by the oceans has been estimated by a number of different ocean models, which generally agree with the observation-based budget estimates (Le Quéré and Metzl, Chapter 12, this volume). As a result, the most uncertain aspect of the anthropogenic global carbon budget is the breakdown of terrestrial sources (from land use) and terrestrial sinks (the so-called residual or missing sink).

Until very recently, Houghton's results were the only available estimates of land use emissions. But in the past few years, a few independent estimates of this part of the carbon cycle have been put forward. One of these new estimates was produced by the Carbon Cycle Model Linkages Project (CCMLP) (McGuire et al. 2001).

CCMLP aimed to evaluate the possible mechanisms behind terrestrial carbon sources and sinks, using a variety of new modeling tools and datasets (McGuire et al. 2001). In particular, CCMLP used four different process-based terrestrial ecosystem models driven by a combination of historical climate, $CO_2$, and land use data. One of the most novel aspects of this study was the use of the newly available historical land use data from Ramankutty and Foley (1999), which presented estimates of the extent of agricultural land on a 0.5° by 0.5° gridcell (~50 km on a side) for each year from 1700 to 1992.

**Table 14.3.** Comparison of global budget of anthropogenic carbon emissions for the 1980s*

| 1980s carbon fluxes | Houghton budget | CCMLP budget |
|---|---|---|
| Atmospheric increase | 3.3 | 3.3 |
| Anthropogenic emissions | 5.4 | 5.4 |
| Net ocean-atmosphere flux | −1.9 | −1.9 |
| Net land-atmosphere flux | −0.2 | −0.2 |
| Modeled land use emissions | 2.0 | 1.0 to 1.7 |
| Implied residual terrestrial sink | −2.2 | −1.2 to −1.9 |

*Using two different estimates of land use emissions—Houghton (1999) and CCMLP (see McGuire et al. 2001; Prentice et al. 2001; House et al. 2003).

*Note:* It should be noted that CCMLP estimated carbon emissions of $0.6-1.0$ PgC $y^{-1}$ due to cropland changes alone. We scale this by 60 percent (the ratio of Houghton's cropland emissions to total land use emissions), to estimate a *CCMLP* total land use flux of $1.0-1.7$ PgC $y^{-1}$. Units are in PgC ($10^{15}$g of carbon) per year.

CCMLP exercised the model in three different ways, each considering different combinations of drivers of terrestrial carbon dynamics: (1) $CO_2$ changes only; (2) $CO_2$ and climate changing together; and (3) land use, $CO_2$, and climate all changing together. These three simulations were conducted with four different models, each running from the mid-1800s to the early 1990s (McGuire et al. 2001).

Considering only the emissions of $CO_2$ from land use and land cover change during the 1980s, the CCMLP results were dramatically different from Houghton's. Houghton (1999) estimated that land use practices released approximately 2 PgC $y^{-1}$ during the 1980s, whereas CCMLP (using four different models) estimated a range of land use emissions from 1.0 to 1.7 PgC $y^{-1}$ (or 50 percent to 85 percent of Houghton's estimate).

Using these two different estimates for land use emissions in the global carbon budget, we see that there is a major difference in the two terrestrial terms. With *Houghton's* land use emissions estimate, the implied "residual" terrestrial carbon sink is roughly 2.2 PgC $y^{-1}$. However, using the CCMLP estimates of land use emissions, the implied "residual" sink ranges from 1.2 to 1.9 Pg C $y^{-1}$ (Table 14.3; McGuire et al. 2001; Prentice et al. 2001; House et al. 2003).

To date, there is no conclusive evidence for why the Houghton and CCMLP estimates of land use emissions (and the implied residual sink of terrestrial ecosystems) are so different. The differences could lie in the underlying land use data (for example, Ramankutty and Foley used the cropland area estimates from the U.N. FAO database, whereas Houghton estimated cropland changes based on other deforestation statistics) or in the way that carbon flows through ecosystems were simulated (Houghton used a bookkeeping model for several large regions of the world, whereas McGuire et al. used four process-based ecosystem models on each 0.5° by 0.5° gridcell of the world). These two groups are currently working together to understand the differences in the two estimates and how they can be reconciled in the future (e.g., House et al. 2003).

Two other recent studies (Archard et al. 2002; DeFries et al. 2002) also call into question the original land use emissions estimates of Houghton. DeFries et al. estimated that land use emissions were ~0.6 (±0.3–0.8) PgC $y^{-1}$ in the 1980s (compared with Houghton's estimate of 2 billion tons per year), and ~0.9 (±0.5–1.4) PgC $y^{-1}$ in the 1990s (compared with Houghton's estimate of 2.2 billion tons per year). A similar estimate was produced by Archard et al., who suggested land use emissions were 1.0 (±0.2) PgC $y^{-1}$ during the 1990s. Although the CCMLP, DeFries et al., and Archard et al. estimates all seem to point to significantly lower land use emissions than Houghton, it is still unclear which estimate is the most accurate.

At this point, it is most important to recognize this: We may have constrained the overall net flux of carbon between terrestrial ecosystems and the atmosphere (using atmospheric measurements of $CO_2$, $O_2$, and $^{13}C$; see Prentice et al. 2001), but the breakdown between terrestrial carbon sources (from land use) and terrestrial carbon sinks is still highly uncertain—and possibly in error by a factor of two.

## Estimating the Size of Terrestrial Carbon Sinks

In most early summaries of the global carbon budget, the size of the terrestrial sink was determined by difference. That is, estimates of land use emissions (typically from Houghton's studies) were subtracted from the net uptake of the terrestrial biosphere inferred from atmospheric measurements (e.g., Table 14.2, Table 14.3). That is, the terrestrial carbon sink was inferred from the difference of other fluxes—and was not described explicitly. This is why it was (until recently) usually referred to as the residual or missing sink—or the sink required to "balance the books" of the global carbon cycle.

But the CCMLP exercise simulated both terrestrial carbon sources (from land use) and potential terrestrial carbon sinks. The CCMLP models assumed that terrestrial sinks may have resulted from a combination of three factors: (1) ecosystem recovery land use/land cover change; (2) the effects of increasing $CO_2$ concentrations on plant productivity (the so-called $CO_2$ fertilization effect); and (3) the effects of recent climate variability and climatic change on ecosystems. The CCMLP study did not, however, include the potential effects of other ecological drivers, such as changing nitrogen deposition, forest management practices, or fire disturbances.

The CCMLP exercise was based on process-based terrestrial ecosystem models. These models use, to the extent possible, a "first principles" approach to simulating ecological and biogeochemical processes. The models generally include detailed calculations of ecosystem physiological processes (including photosynthesis, plant respiration, and microbial respiration), as well as heuristic representations of vegetation dynamics and disturbances. The models used in the CCMLP activity have also been tested against a wide variety of data, including in situ measurements of carbon and water fluxes, primary

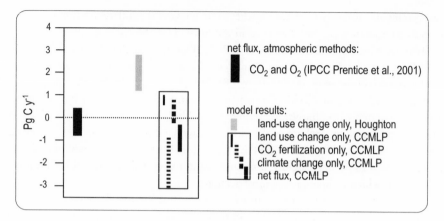

**Figure 14.2.** Estimates of the terrestrial carbon balance. Here the net carbon balance of the terrestrial biosphere (as inferred from atmospheric measurements) is presented from Prentice et al. (2001). Estimates of land use carbon emissions from Houghton (1999) are also reported. For comparison, recent modeling estimates of terrestrial carbon balance (arising from combinations of $CO_2$, climate, and land use change) are shown from the CCMLP exercise (McGuire et al. 2001). Unit is PgC y$^{-1}$. Figure adapted from House et al. (2003).

productivity, and soil moisture balance; satellite-based measurements of vegetation cover and plant phenology; and (by linking the output the ecosystem models to atmospheric transport models) measurements of seasonal atmospheric $CO_2$ concentrations gathered across the globe. Although the models still have many deficiencies, they represent our current understanding of how terrestrial ecosystems work and how they respond to climate, land use, and changing atmospheric chemistry.

As illustrated in Appendix 14.2, terrestrial carbon sinks can be generated through a variety of different mechanisms: (1) those that increase ecosystem productivity; (2) those that increase the residence time of carbon in vegetation, litter, or soil; and (3) those that change the loss of carbon through disturbances (both natural or anthropogenic). Although there is still a running debate on the relative importance of these different processes, the CCMLP simulations considered the possible carbon sinks associated with $CO_2$ fertilization (through enhancing NPP), climatic variability (through changes in NPP and residence time of carbon), and land use/land cover change (through changes in the average carbon residence time).

In the CCMLP results, the simulated net uptake of carbon for the 1980s was roughly –0.3 to –1.5 billion metric tons (McGuire et al. 2001). This range of estimates is in rough agreement (albeit on the high side) with net land-atmosphere flux estimates derived from atmospheric measurements and ocean modeling studies (Prentice et al. 2001; House et al. 2003)–see Figure 14.2.

**Table 14.4.** Breakdown of the terrestrial carbon budget for the 1980s, based on the CCMLP simulations

| CCMLP terrestrial carbon fluxes, 1980s | Flux estimates |
| --- | --- |
| Net land-atmosphere flux | −0.3 to −1.5 |
| Source from cropland change | +0.6 to +1.0 |
| Sink from $CO_2$ fertilization | −1.5 to −3.1 |
| Sink from climatic variability | −0.2 to +0.9 |

*Source:* See McGuire et al. 2001; Prentice et al. 2001; House et al. in press.

*Note:* Units are in PgC ($10^{15}$g of carbon) per year. Note that this reports only the changes in carbon fluxes from *cropland* change, not all land use changes.

The CCMLP results can be broken down into their individual components (Table 14.4). These results suggest that these two factors (ecosystem recovery from past land use, and $CO_2$ fertilization) could be responsible for the "missing" terrestrial carbon sink. Climatic variability actually appears to stimulate (on average) a net source of carbon from terrestrial ecosystems during the 1980s. Other processes not included in the CCMLP study—including increasing nitrogen deposition, shifting disturbances, and changing forest practices—may also be partly responsible for the sink. Interestingly, this "multifactor" explanation of the terrestrial carbon sink differs greatly from the explanation offered during the 1970s and 1980s—that the terrestrial carbon sink resulted only from the effects of $CO_2$ fertilization (Sabine et al., Chapter 2, this volume).

Here, it is important to remember two things. First, process-based ecosystem models are able to mimic the magnitude and timing (between the 1980s and 1990s) of the net land-atmosphere carbon flux, assuming that carbon sources are driven by land use and climatic variability and that carbon sinks are driven by ecosystem recovery from past land use and $CO_2$ fertilization. Second, this does not mean that these mechanisms have actually been "proven" to be responsible for the terrestrial carbon sink. The models are still largely based on local- to regional-scale representations of ecological processes and have had only limited testing against global data. Further work is needed to rigorously test different hypothesis regarding the generation of carbon sources and sinks in the terrestrial biosphere.

## Summary and Future Challenges

The observed increase in atmospheric $CO_2$ results from direct anthropogenic carbon emissions and small imbalances in the natural carbon fluxes between the atmosphere, terrestrial biosphere, and oceans (Prentice et al. 2001). Globally, it appears that the terrestrial biosphere is acting as a net sink of carbon (Prentice et al. 2001). But, on a region-by-region basis, terrestrial ecosystems may act as a net source or sink of atmospheric $CO_2$, depending on the place in question (e.g., Dixon et al. 1994; McGuire et al. 2001; Prentice et al. 2001).

Terrestrial ecosystems are both a source and a sink of atmospheric carbon. On one hand, humans are releasing $CO_2$ from the terrestrial biosphere through deforestation, cultivation, and other land use practices (Houghton 1999; McGuire et al. 2001). In addition, the effects of climate variability and climatic change on terrestrial ecosystems may be causing an additional source of carbon to the atmosphere (McGuire et al. 2001). On the other hand, a significant sink of atmospheric carbon in the terrestrial biosphere may result from several processes: $CO_2$ fertilization, land use recovery, changing patterns of disturbance or forest harvest, or nitrogen fertilization.

At this stage, the exact magnitude of terrestrial carbon sources and carbon sinks is not clear: Estimates vary as much as by a factor of two. Furthermore, we do not yet have a completely satisfactory explanation for the mechanisms of the terrestrial carbon sink. Although hypotheses have been put forward, there is still no "smoking gun" that clearly identifies one (or a combination) of these mechanisms as the culprit. A combination of large-scale observations, process-based modeling, ecosystem experimentation, and laboratory investigations is still needed to explain the "missing" terrestrial carbon sink.

Only when we understand the size of the terrestrial carbon sources and sinks, and the processes that generate them, can we accurately begin to forecast the future evolution of atmospheric $CO_2$ (Friedlingstein, Chapter 10, this volume). For example, different mechanisms for the terrestrial carbon sink will produce very different future behaviors. If $CO_2$ fertilization is largely responsible for the terrestrial sink, then the sink might be expected to increase as $CO_2$ levels continue to rise in the future. If the terrestrial sink is being primarily caused, however, by the recovery of past land use (e.g., ~50- to 100-year-old abandonment of agriculture in the temperate latitudes), then it is likely that the sink will eventually stop and disappear altogether in a few decades (Nabuurs, Chapter 16, this volume).

This is a fundamental question in understanding the future of the carbon cycle and the climate system: Will the terrestrial carbon sink continue to operate the same way in the future, as climatic changes become larger and larger? The bottom line is that the future evolution of atmospheric $CO_2$ will be determined not only by human activity, but also by the terrestrial biosphere and ocean. To accurately forecast changes in future climate, we must understand not only the possible paths of human emissions, but also the dynamics of carbon sources and sinks within the terrestrial biosphere and oceans.

## Acknowledgments

We would like to thank Chris Field for his leadership in pulling this important workshop and book together. In addition, we would like to thank Martin Heimann and Chris Field for their helpful comments on the manuscript. Finally, we thank Mary Sternitzky for preparing the illustrations for this chapter.

# Appendix 14.1. Methods of Estimating Terrestrial Carbon Fluxes

Here we review some of the methods used to determine the size and geographic locations of terrestrial carbon fluxes. This is a cursory overview of the topic. In a recent paper, House et al. (2003) provides a more complete review of the various methods of estimating terrestrial carbon sources and sinks.

Methods for estimating the size and geographic pattern of the terrestrial carbon sink first arose from the so-called atmospheric inversion technique of flux estimation. In an inversion method, terrestrial carbon fluxes are inferred by having to fulfill the requirement that the global carbon budget has to be balanced. Thus, knowing the fossil-fuel and land use sources of carbon and the amount of carbon stored in the atmosphere, one can estimate the terrestrial and oceanic sources or sinks by difference. Further, the ocean and terrestrial fluxes can be partitioned by one of three methods: (1) simultaneous measurements of atmospheric $CO_2$ and $O_2$; (2) observations of atmospheric $\delta^{13}C$; or (3) oceanic uptake as estimated by an ocean carbon cycle model. The inverse modeling approach has the advantage of being global in scale and of implicitly accounting for all the processes influencing the global carbon cycle. It has the disadvantage, however, of not being able to isolate the individual contributions of the various processes controlling the carbon cycle. Furthermore, while inversion methods are able to provide reasonably accurate estimates of global sources and sinks of carbon, and even sufficiently accurate estimates of the latitudinal north-south partitioning of the fluxes, they do not provide accurate longitudinal breakdown of the fluxes and of different regional fluxes.

Although inversion methods are useful, they are not sufficient to understand the functioning of the terrestrial carbon budget. More direct methods of observing the terrestrial sources and sinks of carbon have been developed. One such observational approach measures terrestrial carbon fluxes at the atmospheric boundary layer in flux towers using the "eddy covariance" technique. This technique takes advantage of the fact that transport in the boundary layer is dominated by turbulent eddies, and it uses turbulence theory and sophisticated instruments to measure vertical fluxes of carbon dioxide. Although flux measurements are useful to obtain terrestrial fluxes at local scales, they continue to be plagued by measurement errors when turbulence is low (such as at nighttime) and also have difficulty scaling up to regional levels and to decadal time scales.

Another method of estimating carbon fluxes directly is by using inventory methods. These methods are normally limited to observations of changes in aboveground biomass in forested ecosystems; from changes in biomass, sources or sinks of carbon can be inferred. The method has the advantage of comprehensively including all processes that affect an ecosystem but has the disadvantage of having limited consideration of belowground processes and nonforested ecosystems.

Finally, various numerical models have been used to estimate terrestrial sources and sinks of carbon. These models include representations of the processes that are thought to affect terrestrial carbon fluxes. In particular, the models include controls such as atmospheric $CO_2$ concentration, climate variability and change, atmospheric nitrogen deposition, and in a few cases anthropogenic land use and land cover change. The models have the advantage of being able to isolate the individual contributions of the various processes influencing the terrestrial carbon budget. The models are only as good, however, as our understanding of the processes, and moreover, they only include the processes that are currently hypothesized to influence the carbon budget.

In addition to all of these approaches to estimating present-day terrestrial carbon fluxes, many experimental approaches are in use to understand how terrestrial ecosystems might respond to changing atmospheric carbon dioxide concentrations and climate. In laboratories, greenhouses, and open-top chambers, plants are grown in conditions of increased (or decreased) ambient $CO_2$ concentrations to evaluate their response. This method has been further extended to the plot or stand scale in the Free Air $CO_2$ Enrichment (FACE) experiments, which aim to estimate the ecosystem-level response to increased $CO_2$. Furthermore, many soil-warming experiments around the world attempt to measure the response of microbial respiration to increased soil temperatures.

## Appendix 14.2. A Highly Simplified Model of Terrestrial Carbon Balance

In order to consider the mechanisms that may be producing a carbon sink in the terrestrial biosphere, it is useful to consider a highly simplified model of terrestrial carbon dynamics. Using just a few equations, we can describe the flow of carbon from the atmosphere into vegetation, litter, and soil carbon pools. Here we use a simplified model based on Foley (1995) (Figure 14.3).

In this highly simplified model, the carbon balance of vegetation can be written as:

$$\frac{\partial C_{v,total}}{\partial t} = \sum_{i=1}^{N} a_i NPP - \sum_{i=1}^{N} \frac{C_{v,i}}{\tau_{v,i}} - D$$

where $C_v$ is the carbon storage within vegetation biomass, $NPP$ is the total net primary productivity of the ecosystem, $a_i$ is the allocation coefficient for different vegetation pools (leaves, wood, and roots), $\tau_v$ is the average residence time of carbon in biomass pools, and $D$ is the removal of biomass through disturbances or land use.

When vegetation biomass is lost (through natural turnover or mortality), it is sent to litter—the amount of "dead" carbon that is beginning to decompose—forming $CO_2$ and soil organic matter. We can represent the carbon balance of the litter pool as:

$$\frac{\partial C_{v,total}}{\partial t} = \sum_{i=1}^{N} \frac{C_{v,i}}{\tau_{v,i}} - \sum_{i=1}^{N} \frac{C_{l,j}}{\tau_{l,j}}$$

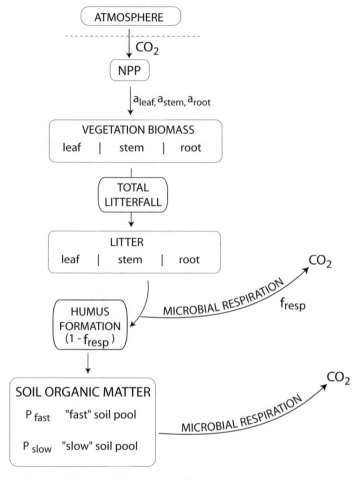

**Figure 14.3.** A highly simplified model of the terrestrial carbon cycle. Here carbon flows from the atmosphere into terrestrial vegetation, then into litter and soil organic matter. Along the way, carbon is lost back to the atmosphere through microbial respiration (or decomposition) of litter and soil organic materials. Adapted from Foley (1995).

where $C_l$ is the carbon content of the litter pools (dead leaves, dead roots, and dead wood) and $\tau_l$ is the average residence time of carbon in litter.

As litter decays, a fraction of this carbon is immediately lost to the atmosphere as $CO_2$, while the remaining carbon is converted into soil organic matter (as humus). The carbon balance of the soil carbon pools can be represented as:

$$\frac{\partial C_{s,total}}{\partial t} = (1-f) \sum_{i=1}^{M} \frac{C_{l,j}}{\tau_{l,j}} - \sum_{k=1}^{P} \frac{C_{s,k}}{\tau_{s,k}}$$

where $C_s$ is the carbon contained in soil organic material (or humus) pools, $(1-f)$ is the fraction of the litter that is converted to humus ($f$ is how much of the litter is respired to the atmosphere), and $\tau_s$ is the average residence time of carbon in soil pools. When evaluating the overall carbon balance of the terrestrial biosphere, we must consider the carbon budget of the vegetation, litter, and soil pools altogether. Looking at these pools together, we see that maintaining a carbon sink in terrestrial ecosystems requires that there must be a net increase in combined carbon mass in the vegetation, litter, and soil:

$$\frac{\partial (C_{v,total} + C_{l,total} + C_{s,total})}{\partial t} > 0$$

This simple terrestrial ecosystem model shows that we can produce a carbon sink only when carbon source terms are larger than the carbon loss terms. For example, there can be a sink in vegetation carbon if net primary productivity exceeds the loss of carbon (turnover and disturbances). Also, a sink can be produced if the inputs of litter or soil carbon are larger than the losses of carbon through microbial respiration.

In order for the sink to continue increasing over time and continue removing a sizable fraction of increasing $CO_2$ emissions in the future, the second-derivative of the carbon stocks must also be positive:

$$\frac{\partial^2 (C_{v,total} + C_{l,total} + C_{s,total})}{\partial t^2} > 0$$

In order to produce a long-term, sustained sink or source of carbon in terrestrial ecosystems, there must be a mechanism that maintains an imbalance between NPP, microbial respiration, and other losses of carbon.

One possible way to produce an increasing sink over time is the following:

$$\frac{\partial NPP}{\partial t} > 0$$

where *NPP* is increasing over time. Another way is the following:

$$\frac{\partial \tau_{v,l,s}}{\partial t} > 0$$

when the average residence time of carbon in vegetation, detritus, or soils is increasing.

## Literature Cited

Archard, F., H. D. Eva, H.-J. Stibig, P. Mayaux, J. Gallego, T. Richards, and J.-P. Malingreau. 2002. Determination of deforestation rates of the world's humid tropical forests. *Science* 297:999–1002.

Bousquet, P., P. Peylin, P. Ciais, C. Le Quéré, P. Friedlingstein, and P. P. Tans. 2000. Regional changes in carbon dioxide fluxes of land and oceans since 1980. *Science* 290:1342–1346.

DeFries, R. S., R. A. Houghton, M. C. Hansen, C. B. Field, D. Skole, and J. Townshend. 2002. Carbon emissions from tropical deforestation and regrowth based on satellite observations for the 1980s and 1990s. *Proceedings of the National Academy of Sciences* 99:14256–14261.

Dixon, R. K., S. Brown, R. A. Houghton, A. M. Solomon, M. C. Trexler, and J. Wisniewski. 1994. Carbon pools and flux of global forest ecosystems. *Science* 282:442–446.

Foley, J. A. 1995. An equilibrium model of the terrestrial carbon budget. *Tellus* 47B:310–319.

Foley, J. A., M. H. Costa, C. Delire, N. Ramankutty, and P. Snyder. 2003. Green surprise? How terrestrial ecosystems could affect Earth's climate. *Frontiers in Ecology and the Environment* 1 (1): 38–44.

Houghton, R. A. 1999. The annual net flux of carbon to the atmosphere from changes in land use 1850–1990. *Tellus* 51B:298–313.

———. 2000. A new estimate of global sources and sinks of carbon from land use change. *Eos* 81:S281.

———. 2003. Revised estimates of the annual net flux of carbon to the atmosphere from changes in land use 1850–2000. *Tellus* 55B:378–390.

Houghton, R. A., J. E. Hobbie, J. M. Melillo, B. Moore III, B. J. Peterson, G. R. Shaver, and G. M. Woodwell. 1983. Changes in the carbon content of terrestrial biota and soils between 1860 and 1980: A net release of $CO_2$ to the atmosphere. *Ecological Monographs* 53:235–262.

House, J. I., I. C. Prentice, N. Ramankutty, R. A. Houghton, and M. Heimann. 2003. Reducing apparent inconsistencies in estimates of terrestrial CO2 sources and sinks. *Tellus* (in press).

Le Quéré, C., O. Aumont, L. Bopp, P. Bousquet, P. Ciais, R. Francey, M. Heimann, C. D. Keeling, R. F. Keeling, H. Kheshgi, P. Peylin, S. C. Piper, I. C. Prentice, and P. J. Rayner. 2003. Two decades of ocean $CO_2$ sink and variability. *Tellus* (in press).

McGuire, A. D., S. Sitch, J. S. Clein, R. Dargaville, G. Esser, J. Foley, M. Heimann, F. Joos, J. Kaplan, D. W. Kicklighter, R. A. Meier, J. M. Melillo, B. Moore III, I. C. Prentice, N. Ramankutty, T. Reichenau, A. Schloss, H. Tian, L. J. Williams, and U. Wittenberg. 2001. Carbon balance of the terrestrial biosphere in the twentieth century: Analyses of $CO_2$, climate and land-use effects with four process-based ecosystem models. *Global Biogeochemical Cycles* 15:183–206.

Prentice, I. C., G. D. Farquhar, M. J. R. Fasham, M. L. Goulden, M. Heimann, V. J. Jaramillo, H. S. Kheshgi, C. Le Quéré, R. J. Scholes, and D. W. R. Wallace. 2001. The carbon cycle and atmospheric carbon dioxide. In *Climate change 2001: The scientific basis (Contribution of Working Group I to the third assessment report of the Intergovernmental Panel on Climate Change)*, edited by J. T. Houghton, Y. Ding, D. J. Griggs, M. Noguer, P. J. van der Linden, X. Dai, K. Maskell, and C. A. Johnson. Cambridge: Cambridge University Press.

Ramankutty, N., and J. A. Foley. 1999. Estimating historical changes in global land cover: croplands from 1700 to 1992. *Global Biogeochemical Cycles* 13 (4): 997–1027.

Schimel, D. S., J. I. House, K. A. Hibbard, P. Bousquet, P. Ciais, P. Peylin, B. H.

Braswell, M. J. Apps, D. Baker, A. Bondeau, J. Canadell, G. Churkina, W. Cramer, A. S. Denning, C. B. Field, P. Friedlingstein, C. Goodale, M. Heimann, R. A. Houghton, J. M. Melillo, B. Moore III, D. Murdiyarso, I. Noble, S. W. Pacala, I. C. Prentice, M. R. Raupach, P. J. Rayner, R. J. Scholes, W. L. Steffen, and C. Wirth. 2001. Recent patterns and mechanisms of carbon exchange by terrestrial ecosystems. *Nature* 414:169–172.

# 15

# Geographic and Temporal Variation of Carbon Exchange by Ecosystems and Their Sensitivity to Environmental Perturbations

## Dennis Baldocchi and Riccardo Valentini

Through the long-term monitoring of $CO_2$ concentration at Mauna Loa (Keeling and Whorf 1996) and an expanded network of $CO_2$ monitoring stations across the globe (Tans et al. 1996), we now possess an ability to observe the "breathing" of the biosphere. In general, the biosphere inhales $CO_2$ during its hemisphere's summer and exhales $CO_2$ during its winter. Superimposed upon this general pattern, however, is much complexity and variability due to temporal and spatial gradients in climate, available resources, plant structure and function, land use, and soil development.

Technically, the biosphere's breathing can be quantified by assessing net ecosystem productivity, the balance between natural sources and sinks of carbon dioxide. Net ecosystem productivity (NEP; grams of carbon per square meter per year [g C m$^{-2}$ y$^{-1}$]) is defined as the difference between gross primary productivity (GPP) and ecosystem respiration ($R_{eco}$):

$$NEP = GPP - R_{eco}$$

GPP is defined as the amount carbon dioxide that is assimilated by plants through photosynthesis. Alternatively, we can define GPP as the sum of net primary productivity, NPP, and autotrophic respiration, $R_{auto}$. As a rule of thumb, autotrophic respiration, or NPP, constitutes about half of GPP (Gifford 1994), and $R_{auto}$ is about two-thirds of ecosystem respiration (Falge et al. 2002).

Ecosystem respiration, $R_{eco}$, is more complex and consists of autotrophic respiration by the leaves, plant stems, and roots ($R_{auto}$) and heterotrophic respiration by the microbes and soil fauna ($R_{heteoro}$):

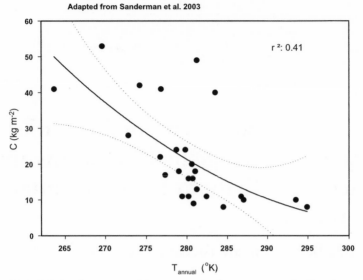

**Figure 15.1.** Relation between carbon content in the top meter of soil and mean annual temperature. The data are associated with the sites in the FLUXNET network (adapted from Sanderman et al. 2003).

$$R_{eco} = R_{plant} + R_{soil} = R_{auto} + R_{hetero}$$

Over many years, a more relevant measure of carbon balance is net biome productivity (NBP), which includes carbon losses due to fire, insect and pathogen damage, and other disturbances (Schulze et al. 2000).

Solar energy, temperature, and soil moisture are the main climate variables that modulate GPP on hourly, daily, and seasonal timescales (Leith 1975). Soil respiration is primarily a function of soil temperature, soil moisture, and whether or not the soil is waterlogged. On an annual timescale, soil respiration can be expressed as a function of the ratio between soil carbon content ($C$) and its turnover time ($\tau$). In principle, carbon stores in the soil increase with decreasing temperature ($T$) (Figure 15.1). This occurs, in part, because the soil turnover time decreases with increasing temperature. Whether or not soil respiration increases or decreases with changes in temperature, however, depends upon whether the temperature sensitivity of the numerator ($C$) outpaces that of the denominator ($\tau$) (Sanderman et al. 2003).

## Geographic Variability of the Response of NPP to Environmental Variables

To assess global carbon exchange of the biosphere, one must assess net carbon exchange for each biome on earth. This task is nontrivial, because each of the world's major biomes—tropical, temperate, and boreal forests, savannas, shrublands, grasslands, wetlands, tundra, and deserts—contribute differently to the global carbon budget. Factors to consider when constructing carbon budgets include their (1) respective land area; (2) physiological potential to assimilate and respire $CO_2$; (3) the size and perturbation status of carbon pools; and (4) sensitivity of GPP and $R_{eco}$ to environmental drivers (e.g., light, temperature, and soil moisture).

Biogeochemical models provide one tool for assessing the net exchange of carbon between the terrestrial biosphere and atmosphere at continental and global scales (Melillo et al. 1993; Cramer et al. 1999). These models account for the diversity and complexity of the natural world by dividing the terrestrial biosphere into broad vegetation classes that are defined by their function, structure, and climate (Bonan et al. 2002). The type and amount of vegetation at a particular location is evaluated either diagnostically, using remote sensing information derived from satellites (Running et al. 1999), or prognostically, using dynamic vegetation models (Foley et al. 1998). For a given plant class, photosynthesis and respiration are computed using algorithms that are a function of environmental variables, such as light, temperature, and soil moisture (Cramer et al. 1999). The implementation of these algorithms, however, requires weather or climate data for each unit of the model grid. Furthermore, the models require many assumptions about each unit, its representativeness in space, and its uniformity in time.

Nevertheless, with a validated biogeochemical model, one can quantify how spatial gradients in available sunlight, leaf area index, nitrogen content of leaves, and rainfall impose spatial patterns on annual photosynthesis (Churkina and Running 1998; Law et al. 2002) and how respiration scales with temperature, body size, soil moisture, and net primary productivity (Amthor 2000; Enquist 2002; Reichstein et al. 2002a). On a regional basis, NPP is light and temperature limited in northern climates and biomes, has a low degree of light and water limitation in tropical and humid temperate regions, and is limited by water availability in semi-arid climates (Colorplate 10).

## Geographic and Temporal Variability of the Response of NEP to Environmental Variables

Although the information in Colorplate 10 is instructive, it leaves many questions unanswered about ecosystem carbon cycling. For instance, how do geographic variations in climate and vegetation affect the responses of NEP to environmental forcings, and how do these forcing cause NEP to vary over the course of a year? Here, we attempt to answer these questions by distilling published data produced by the FLUXNET project (Baldocchi et al. 2001; Falge et al. 2002; Law et al. 2002). Specifically, we examine cases associated with major biomes: (1) temperate broad-leaved forests, (2) boreal and temperate coniferous forests, (3) grasslands, (4) mediterranean-type woodlands, (5) agricultural crops, and (6) northern wetlands. Highlighted in this analysis is the role of environmental switches—leaf on/leaf off; drought, frost (spring/fall), the presence or

Temperate Broadleaved
Deciduous Forest

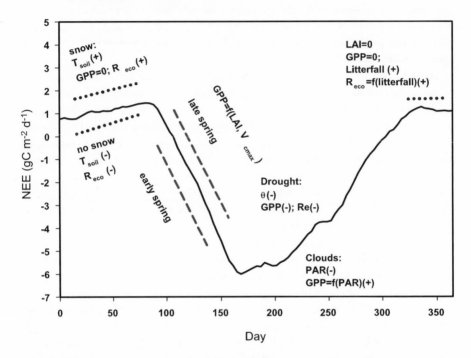

**Figure 15.2.** Seasonal variation in net $CO_2$ exchange of a temperate deciduous forest. Also shown are the environmental and biological factors that affect the seasonal dynamics of $CO_2$ exchange. $V_{cmax}$ represents photosynthetic capacity, LAI is leaf area index, and T is temperature. Data sources: Wofsy et al. (1993); Greco and Baldocchi (1996); Valentini et al. (1996); Goulden et al. (1996); Baldocchi et al. (2000); Granier et al. (2000, 2002); Schmid et al. (2000); Pilegaard et al. (2001); Saigusa et al. (2002). Data are from the Fluxnet database (http://www-eosdis.ornl.gov/FLUXNET/) and were gap filled and smoothed.

absence of snow, and, in wetlands, the height of the water table—on ecosystem carbon exchange. Data are presented as net ecosystem carbon exchange (NEE), which is the same quantity as, but opposite in sign from, NEP. NEE is calculated from an atmospheric perspective; it is negative when the atmosphere is losing carbon and the ecosystem is gaining carbon.

## Temperate Deciduous Broad-leaved Forests

Temperate deciduous broad-leaved forests exist across large regions of Asia, Europe, and North America, typically north of 30°. Globally, they occupy only 2 percent of the land area of the terrestrial biosphere (Melillo et al. 1993), but they contribute significantly to the global carbon budget because they occur in relatively wet and highly productive regions.

During winter, the trees are leafless and dormant, and the ecosystem is respiring (Figure 15.2). The presence or absence of snow has a major impact on soil temperatures and the rates of soil respiration from these forests (Goulden et al. 1996). Snow insulates the soil surface, so soil temperatures are warmer when snow is present. Consequently, greater rates of soil respiration occur from snow-covered ground than from bare soil, which is exposed to the cold winter weather.

During spring, a pronounced peak in ecosystem respiration reflects increased growth respiration as leaves emerge (Greco and Baldocchi 1996; Valentini et al. 1996; Granier et al. 2000, 2002). Daily respiration rates typically range between 5 and 7 g C m$^{-2}$ d$^{-1}$ during this period, which is double to triple respiration rates earlier in the spring. As leaf-out occurs, the ecosystem experiences a pronounced switch from a net source of carbon to a net sink. This switch can represent a net change in the magnitude of carbon exchange that approaches 10 g C m$^{-2}$ d$^{-1}$. Consequently, the date of leaf-out has a major impact on the net annual carbon exchange of this biome.

The date of leaf-out can vary by 30 days at a given site (Goulden et al. 1996; Wilson and Baldocchi 2000) and by more than 100 days across the deciduous forest biome (Figure 15.3). Consequently, differences in length of growing season can alter NEE by up to 600 g C m$^{-2}$ year$^{-1}$.

Once a forest has attained full canopy closure, changes in available sunlight explain most of the variability in hourly rates of carbon exchange, with a nonlinear and saturating response (Goulden et al. 1996; Valentini et al. 1996; Baldocchi 1997; Granier et al. 2000; Schmid et al. 2000; Pilegaard et al. 2001). The transparency of the atmosphere, however, complicates the sensitivity of NEE to changes in sunlight. At the canopy scale, the initial slope of the light response curve (also known as light use efficiency, LUE) increases as the fraction of diffuse radiation increases (Baldocchi 1997; Gu et al. 2002).

The net effect of increasing diffuse radiation, as a consequence of either clouds or atmospheric aerosols, is complex. In general, short-term NPP is twice as sensitive to changes in diffuse light as it is to changes in direct light (Gu et al. 2002). On an annual basis, increasing diffuse radiation by 20 percent without affecting total incident radiation (as can occur with aerosol loading in the atmosphere) can theoretically increase annual NEE of a temperate deciduous forest by 10–15 percent (Baldocchi et al. 2002), about 70 g C m$^{-2}$ year$^{-1}$. Increasing cloud fraction increases diffuse radiation but lowers total solar radiation. In this situation, an expected reduction in GPP (from the decrease in total solar radiation) is minimized at the canopy scale by an increase in light use efficiency (from the increased fraction of diffuse radiation).

Droughts tend to be episodic across the temperate deciduous forest biome. Summer drought and high vapor pressure deficits reduce daytime photosynthesis (Goulden et al. 1996; Greco and Baldocchi 1996; Baldocchi 1997). Drying of the soil also decreases soil

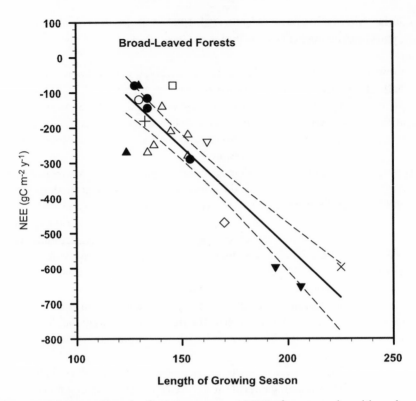

**Figure 15.3.** Impact of length of growing season on NEE of temperate broad-leaved forests. Length of growing season explains more than 80 percent of the variance of annual NEE across this biome (Baldocchi et al. 2001).

respiration (Hanson et al. 1993). Together, these features reduce NEE from values expected during normal, wet summer conditions (Figure 15.2).

In the autumn, photosynthesis ceases, and leaves senesce and drop. Soil respiration, on the other hand, experiences an enhancement due to the input of fresh litter that is readily decomposable, as soils are still warm and autumn rains stimulate litter decomposition (Goulden et al. 1996; Granier et al. 2000).

On a daily and monthly timescale, ecosystem respiration is mainly a function of soil temperature. The proportionality factor by which ecosystem respiration increases with a 10°C increase in temperature ($Q_{10}$) ranges between 1.6 and 5.4 (Schmid et al. 2000). Based on such data, one would expect that annual ecosystem respiration would increase as one moves south in the Northern Hemisphere to warmer climates. But across forests in Europe, ecosystem respiration was greater in northern regions (Valentini et al. 2000; Falge et al. 2002). This observation may be an artifact of relatively recent disturbance

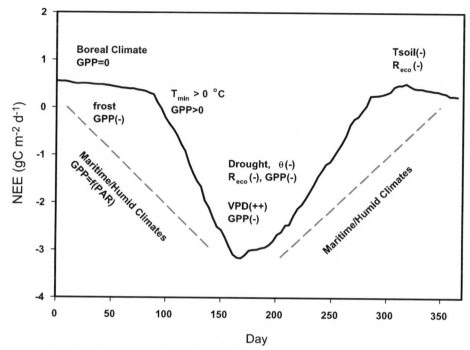

**Figure 15.4.** Seasonal variation of net $CO_2$ exchange of temperate, boreal, and alpine conifer forests. Also shown are the environmental and biological factors that affect the seasonal dynamics of $CO_2$ exchange. VPD is vapor pressure deficit. Data sources: Goulden et al. (1996); Jarvis et al. (1997); Lindroth et al. (1998); Anthoni et al. (2002); Hollinger et al. (1999); Goldstein et al. (2000); Berbigier et al. (2001); Markkanen et al. (2001); Dolman et al. (2002); Monson et al. (2002); Hui et al. (2003).

of carbon pools of northern European forests. Since carbon storage increases with decreasing temperature, any disturbance has the potential to release pools of previously stored carbon and produce relatively low values of NEE (Valentini et al. 2000).

## Evergreen Conifer Forests

Evergreen conifer forests occupy a diverse range of climates. They exist at the cold extremes of the boreal zone, in the mild, humid maritime regions, on alpine mountain slopes, and in the semi-arid interior regions of continents. Consequently, the seasonal cycle of carbon exchange of evergreen conifer forests and its sensitivity to environmental forcings is not universal. Boreal, maritime, and subtropical conifer forests experience different seasonal patterns of $CO_2$ exchange over a year's time (Figure 15.4).

Forests in maritime climates (Berbigier et al. 2001; Chen et al. 2002) and those growing close to the subtropics (Hui et al. 2003) have the potential to acquire carbon year-round. In contrast, conifer forests in cold boreal or alpine climates experience a restricted period for carbon uptake confined to the summer growing season (Goulden et al. 1996; Jarvis et al. 1997; Lindroth et al. 1998; Hollinger et al. 1999; Markkanen et al. 2001; Monson et al. 2002). During the short summer growing season in the extreme boreal and alpine climate zones, conifer forests frequently lose carbon on individual days. These situations either occur in low-productivity sites on hot, sunny days when high vapor pressure deficits promote stomatal closure (Baldocchi et al. 1997; Hollinger et al. 1999) or on cloudy days when canopy photosynthesis is reduced but the soil is warm and respiration is high (Monson et al. 2002).

Conifer forests in temperate, continental climates experience a longer growing season than forests in boreal and alpine climates. Because temperate conifer forests grow on more productive sites, they maintain higher rates of carbon uptake than do boreal forests (Aubinet et al. 2001; Dolman et al. 2002). Furthermore, they do not experience summer episodes of carbon loss, like boreal forests. Conifers growing in semi-arid regions, such as the western United States, acquire carbon best during the spring and fall and can lose carbon during the prolonged summer drought (Goldstein et al. 2000; Anthoni et al. 2002).

Whether or not conifers gain carbon on fine spring days before the true growing season starts remains an important question. Needles are present and have the potential to assimilate carbon. Physiological studies on *Picea abies* show that uptake rates are low but positive, on fine days with frost events, but that carbon assimilation is suspended when air temperature remains below freezing (Larcher 1975). At the canopy scale, $CO_2$ uptake for a boreal Scots pine forest is close to zero when the mean air temperature is above, but near, zero Celsius (Markkanen et al. 2001).

## Evergreen Broad-leaved Forests

Evergreen broad-leaved forests form dense, closed canopies in humid tropical regions of the Amazon, central Africa, and southeast Asia, as well as the temperate maritime climates of Japan and New Zealand. Evergreen, tropical forests occupy about 15 percent of the global land area. Broad-leaved evergreen trees also exist in a sparser woodlands in seasonally dry climates, such as the Mediterranean-type climates in Australia, California, Chile, Europe, and South Africa. Evergreen woodlands consist of a mix of herbaceous and woody plants, with mixtures ranging from 20/80 to 80/20 (Scholes and Archer 1997). Globally, these woodlands constitute more than 20 percent of the terrestrial biosphere (Melillo et al. 1993).

The aseasonal behavior of the tropical forests contrasts with the strongly seasonal behavior of subtropical and Mediterranean-type forests.

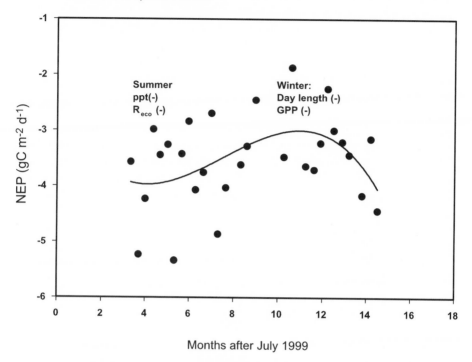

**Figure 15.5.** Seasonal variation of net ecosystem $CO_2$ exchange of a tropical forest growing in the Amazon near Manaus, Brazil. Adapted from Araujo et al. (2002).

### TROPICAL AND MARITIME CLIMATES

Published data on the seasonal patterns of NEE from evergreen tropical forests are relatively scarce (Grace et al. 1995; Malhi et al. 1998; Araujo et al. 2002). The emerging pattern is that the seasonality of daily-integrated NEE is weak, because of the small range in day length and temperature (Figure 15.5); day length is close to 12 hours long, year-round, and the amplitude of monthly mean air temperatures is less than 4°C. Furthermore, the nearly equal day and night length and warm temperatures experienced by evergreen tropical forests cause them to respire at greater rates and for a longer time than forests, during the growing season, at northern latitudes. As a consequence, daily net rates of carbon exchange are lower in evergreen tropical forests than in temperate broadleaved forests during the peak of their growing season (Malhi et al. 1998).

Day-to-day variations in GPP are mainly caused by changes in clouds and their modulation of sunlight. Ironically, the seasonal occurrence of drought, which reduces both GPP and $R_{eco}$, is a major factor modulating NEE of wet tropical forests.

In the temperate maritime climate of New Zealand, broad-leaved, evergreen *Nothofagus* forests experience rates of carbon exchange rates that resemble those of the tropical forests but with lower nocturnal respiration rates (Hollinger et al. 1994). The carbon exchange rates of these forests are also more seasonal than those of tropical forests, operating more like deciduous forests but without the dormant phase. Light-use efficiency of these forests is enhanced by diffuse radiation (Hollinger et al. 1994).

### Subtropical and Mediterranean Climates

In the subtropical regions of Africa (tiger bush), Australia (Eucalypt bush), and Brazil (cerrado), the timing of the wet and dry season is the major control on photosynthesis and respiration (Miranda et al. 1997; Eamus et al. 2001; Vourlitis et al. 2001). These woodlands are carbon sinks during the wet season, are carbon neutral during the dry season, and are net sources during the transition period from dry to wet. Changes in vapor pressure deficit and soil moisture are the major modifiers of the maximum value and initial slope of the relationship between sunlight and canopy photosynthesis (Eamus et al. 2001; Vourlitis et al. 2001).

Carbon dioxide exchange of Mediterranean-type ecosystems is out of phase with the seasonal pattern of carbon exchange in temperate forests (Figure 15.6). The prolonged summer drought limits gas exchange (photosynthesis and respiration) (Reichstein et al. 2002a), but rainfall is plentiful during the winter, so the plants are physiologically active when temperatures are cooler, days are shorter, and less sunlight is available than in summer. Consequently, Mediterranean-type ecosystems have less potential to acquire carbon than temperate systems that are active during the warmer and brighter summer.

Another prominent feature, relatively unique to savanna and Mediterranean-type ecosystems, is the effect of episodic rain on soil respiration during the summer (Reichstein et al. 2002b; Rey et al. 2002; Xu and Baldocchi 2003). Two mechanisms may produce enhanced respiration rates after rainfall. One is a physical displacement of soil air and $CO_2$ by the downward-moving front of water in the soil. But this effect is short-lived, and the volume of air in the soil profile is relatively small. The other mechanism is rapid activation of heterotrophic respiration (Birch 1958).

## Grasslands

Grasslands grow in regions with limited annual rainfall (Ehleringer et al. 1997) and possess several modes of function. In the Mediterranean-climate regions of the world, grasslands consist mainly of annual grasses that are functional during the winter and spring growing season and use the $C_3$ photosynthetic pathway. Grasslands in temperate continental regions are perennial and physiologically active during the summer growing season. Often, $C_3$ and $C_4$ grass species coexist, but the proportion of the mixture shifts toward $C_4$ species as one moves south. In the subtropics, most grasses photosynthesize with the $C_4$ pathway.

## Mediterranean Evergreen Forest

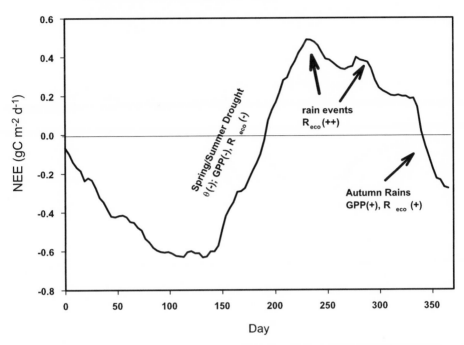

**Figure 15.6.** Seasonal variation of NEP of Mediterranean/subtropical woodland eco-systems and the environmental drivers that perturb the carbon fluxes. Data sources: Leuning and Attiwill (1978); Valentini et al. (1996); Eamus et al. (2001); Reichstein et al. (2002a); unpublished data from Joao Periera (Fluxnet data archive, http://www.daac.ornl.gov/FLUXNET/) and from Baldocchi and Xu (own data and data from AmeriFlux archive).

For Mediterranean-type grasslands, frost limits GPP during the winter growing season (Figure 15.7). Rapid growth occurs after the last frosts and ceases when the soil water profile is depleted and the plants die. On a year-to-year basis, timing of cessation of winter rains sets the end of the growing season. Rain pulses during the summer stimulate huge, short-term rates of respiration, as the plants are dead and GPP is zero (Xu and Baldocchi 2003). Seed germination and a new growing season commence with the arrival of the autumnal rains. In ecosystems with low NEE, such as Mediterranean grasslands, a few large pulses of soil respiration have the potential to change the ecosystem from being a sink of carbon to a source (Xu and Baldocchi 2003).

Perennial temperate grasslands are dormant during the winter, when they are often snow covered. During this period, material in the grass canopy is dead and the system

## Grasslands

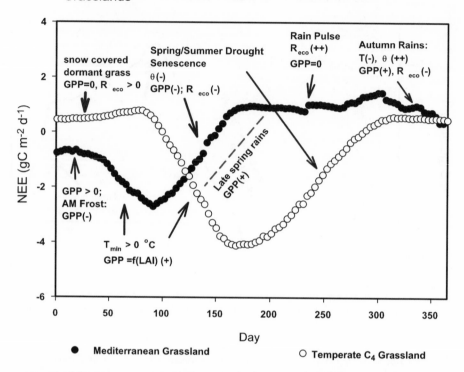

**Figure 15.7.** Seasonal variation in net $CO_2$ exchange of a Mediterranean annual grassland and a temperate continental $C_4$ grassland. Also shown are the environmental and biological factors that affect the seasonal dynamics of $CO_2$ exchange. Data sources: Valentini et al. (1996); Kim et al. (1992); Saigusa et al. (1998); Meyers (2001); Sukyer and Verma (2001); Flanagan et al. (2002); Xu and Baldocchi (2003).

is respiring. During spring, the grass begins to grow again. Initially, rates of photosynthesis are low, because photons are intercepted by living and dead material. Consequently, the slope of the light response curve changes markedly with leaf area index (LAI) (Suyker and Verma 2001; Flanagan et al. 2002; Xu and Baldocchi 2003). NEE saturates with high sunlight when LAI is low and becomes a quasi-linear function of sunlight as LAI increases and the canopy closes. Summer drought also has a marked effect on NEE, limiting photosynthesis and respiration (Kim et al. 1992).

In general, peak rates of $CO_2$ assimilation are greater for $C_4$ grasslands than for $C_3$ grasslands, and they are higher for grasslands with a summer growing season than those with a winter/spring growing season (Figure 15.7).

## Agricultural Crop

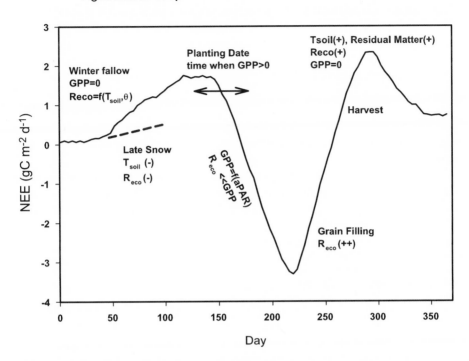

**Figure 15.8.** Seasonal variation of net $CO_2$ exchange of an agricultural crop. Also shown are the environmental and biological factors that affect the seasonal dynamics of $CO_2$ exchange. Data sources: Desjardins (1985); Baldocchi (1994); Rochette et al. (1995); Hanan et al. (2002); Gilmanov et al. (2003); Verma and Sukyer, AmeriFlux (AmeriFlux data archive http://public.ornl.gov/ameriflux/Participants/Sites/Map/index.cfm).

## *Agricultural Crops*

The seasonal course of NEE over agricultural regions is much different than that of the native ecosystems we have examined (Figure 15.8). During a long portion of the year, the surface is respiring because it is bare or covered with detritus. At the start of the growing season, the landscape experiences a long period when the canopy is sparse because crops must grow from seed. During this growth period, rates of net $CO_2$ uptake by crops are relatively low, compared with those of temperate and conifer forests. Maximum rates of carbon uptake occur after the crop achieves closure and are among the highest of all vegetation types, since crops are selected for high productivity, they grow on very fertile soils, they are fertilized, and they tend to have abundant soil moisture due to rainfall or irrigation (Ruimy et al. 1995). The period during which a crop canopy is closed is rather short, limiting the duration of maximum uptake rates. The occurrence of anthesis is another factor limiting carbon uptake rates of crops during the peak growing period. Respiration in wheat jumps distinctly after anthesis, as a substantial amount of assimilated energy is used to produce reproductive organs, which have high respiration costs (Baldocchi 1994; Rochette et al. 1995). During autumn and after harvest, ecosystem respiration peaks again, owing to the new input of fresh decomposable material and the warmth of the soils. Soil respiration during the fallow period is much lower than in native ecosystems, which have more decaying detritus on the ground or are maintaining the catabolism of a substantial overstory of trees and shrubs.

Because crops are fertilized and grow in regions with ample rainfall or are irrigated, sunlight is the main factor limiting NEE of crops. Like grasslands, the light response curve saturates when the canopy is sparse and becomes quasi-linear after full closure (Baldocchi 1994; Rochette et al. 1995; Ruimy et al. 1995). Photosynthetic pathway ($C_3$ or $C_4$) also affects maximum rates of $CO_2$ uptake.

## Northern Wetlands/Tundra

Northern wetlands and tundra ecosystems exist in the extremely cold desert climate of the Arctic. Yet, they often remain waterlogged during the summer, as the water balance is positive—precipitation exceeds evaporation. This combination of factors affects the seasonality and rates of carbon exchange experienced by northern wetlands and tundra (Figure 15.9). Despite extreme cold temperatures during the winter, these ecosystems continue to respire (Vourlitis et al. 2000). Respiration occurs because microbes can remain active and decompose organic matter deep in the soil in unfrozen zones.

Since the growing season is very short and the amount of vegetation is low, rates of net carbon uptake during the summer are low. Functionally, soil temperature and solar radiation control net $CO_2$ balance during the period from snow melt to early autumn. On a seasonal and interannual basis, these ecosystems are sensitive and vulnerable to changes in the water table, as large stores of carbon remain. Lowering the water table allows stored carbon pools to oxidize, causing the ecosystem to be a net source of carbon during the growing season (Vourlitis et al. 2000; Lloyd 2001).

## Conclusions

For dominant ecosystems of the world, NEP and its component fluxes, GPP and $R_{eco}$, exhibit diverse responses to seasonal changes in weather and climate. Ecosystems growing in a single environment and exposed to the same climate can show remarkable differences in the magnitude of carbon and energy fluxes, as well in their seasonal phasing. The implication of these patterns is that the complex mosaic of geographical

## Northern Wetlands/Tundra

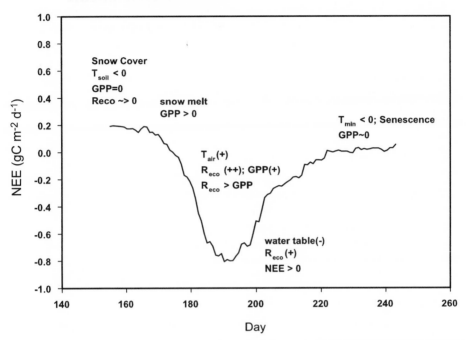

**Figure 15.9.** Seasonal variation in net $CO_2$ exchange of a northern wetland and tundra. Also shown are the environmental and biological factors that affect the seasonal dynamics of $CO_2$ exchange. Data are gap filled and smoothed. Data sources: Shurpali et al. (1995); Griffis et al. (2000); Soegaard et al. (2000); Vourlitis et al. (2000); Lloyd (2001).

variations in a given region cannot be simply represented and modeled as a single "green slab." A reasonably accurate land use/cover (forests, grasslands, crops, etc.) map is a prerequisite for correctly representing the aggregation of different functional responses and biospheric fluxes.

The examples in this chapter also highlight the vulnerability of the terrestrial carbon sink. Climate changes and disturbance regimes affect the rates and patterns of biospheric carbon exchanges. In boreal regions changes in seasonal dynamics and length of the growing season can have a profound impact on carbon oxidation, even when the annual mean temperature is not perturbed. And if warm and dry spells promote forest fires, the landscape can quickly change from a weak carbon sink to a large source of carbon to the atmosphere. In temperate regions water limitations on soil organic matter decomposition plays a significant role and can limit carbon oxidation, despite changes in temperature. In the tropics light-use efficiency and changes in the ratio of direct versus diffuse radiation can also play an important role in the future carbon exchanges.

These are only few examples of the complexity of the terrestrial ecosystem responses and the parameters that drive their carbon balance. More work is needed to understand these emergent properties, which can have important impacts on the future of the land carbon sink, and to improve our global models. For this reason the concept of "ecosystem physiology," the integrated response of a vegetation community to environmental factors including carbon, energy, and water, is becoming a useful new tool for understanding the dynamics of the terrestrial biota and its relation with Earth system processes.

## Acknowledgments

This research was supported by grants from the NASA Carbon Science Program (FLUXNET) and the U.S. Department of Energy's Terrestrial Carbon Program. We thank the many FLUXNET investigators for their contribution to this project. We are grateful to E.-Detlef Schulze and Paulo Artaxo for their expert reviews of this work. We also thank Chris Field and Mike Raupach for organizing the workshop on which this book is based and SCOPE for travel support.

## Literature Cited

Amthor, J. S. 2000. The McCree-de Wit-Penning de Vries-Thornley respiration paradigms: 30 years later. *Annals of Botany* 86:1–20.

Anthoni, P. M., J. Irvine, M. H. Unsworth, D. D. Baldocchi, D. Moore, and B. E. Law. 2002. Seasonal differences in carbon and water vapor exchange in young and old-growth ponderosa pine ecosystems. *Agricultural and Forest Meteorology* 111:203–222.

Araujo, A. C., A. D. Nobre, B. Kruijt, J. A. Elbers, R. Dallarosa, P. Stefani, C. V. Randow, A. O. Manzi, A. D. Culf, J. H. C. Gash, R. Valentini, and P. Kabat. 2002. Comparative measurements of carbon dioxide fluxes from two nearby towers in a central Amazonian rainforest: The Manaus LBA site. *Journal of Geophysical Research* 107, doi:10.1029/2001JD000676.

Aubinet, M., B. Chermanne, M. Vandenhaute, B. Longdoz, M. Yernaux, and E. Laitat. 2001. Long-term carbon dioxide exchange above a mixed forest in the Belgian Ardennes. *Agricultural and Forest Meteorology* 108:293–315.

Baldocchi, D. D. 1994. Comparative study of mass and energy exchange over a closed (wheat) and an open (corn) canopy. II. Canopy $CO_2$ exchange and water use efficiency. *Agricultural and Forest Meteorology* 67:291–322.

———. 1997. Measuring and modeling carbon dioxide and water vapor exchange over a temperate broad-leaved forest during the 1995 summer drought. *Plant, Cell and Environment* 20:1108–1122.

Baldocchi, D. D., and K. B.Wilson 2001. Modeling $CO_2$ and water vapor exchange of a temperate broadleaved forest across hourly to decadal time scales. *Ecological Modeling* 142:155–184.

Baldocchi, D. D., C. A. Vogel, and B. Hall. 1997. Seasonal variation of carbon dioxide exchange rates above and below a boreal jack pine forest. *Agricultural and Forest Meteorology* 83:147–170.

Baldocchi, D. D., J. J. Finnigan, K. W. Wilson, K. T. Paw U, and E. Falge. 2000. On measuring net ecosystem carbon exchange in complex terrain over tall vegetation. *Boundary Layer Meteorology* 96:257–291.

Baldocchi, D., E. Falge, L. H. Gu, R. Olson, D. Hollinger, S. Running, P. Anthoni, C. Bernhofer, K. Davis, R. Evans, J. Fuentes, A. Goldstein, G. Katul, B. Law, X. H. Lee, Y. Malhi, T. Meyers, W. Munger, W. Oechel, K. T. Paw U, K. Pilegaard, H. P. Schmid, R. Valentini, S. Verma, T. Vesala, K. Wilson, and S. Wofsy. 2001. FLUXNET: A new tool to study the temporal and spatial variability of ecosystem-scale carbon dioxide, water vapor, and energy flux densities. *Bulletin of the American Meteorological Society* 82:2415–2434.

Baldocchi, D. D., K. B. Wilson, and L.Gu. 2002. Influences of structural and functional complexity on carbon, water and energy fluxes of temperate broadleaved deciduous forest. *Tree Physiology* 22:1065–1077.

Berbigier, P., J.-M. Bonnefond, and P. Mellman. 2001. $CO_2$ and water vapour fluxes for 2 years above Euroflux forest site. *Agricultural and Forest Meteorology* 108:183–197.

Birch, H. F. 1958. The effect of soil drying on humus decomposition and nitrogen availability. *Plant and Soil* 10:9–31.

Black, T. A., G. den Hartog, H. Neumann, P. Blanken, P. Yang, Z. Nesic, S. Chen, C. Russel, P. Voroney, and R. Staebler. 1996. Annual cycles of $CO_2$ and water vapor fluxes above and within a Boreal aspen stand. *Global Change Biology* 2:219–230.

Bonan, G., S. Levis, L. Kergoat, and K. Oleson. 2002. Landscapes as patches of plant functional types: An integrating concept for climate and ecosystem models. *Global Biogeochemical Cycles* 16, doi:10.1029/2000GB001360.

Chen, J., M. Falk, E. Euskirchen, K. T. Paw U, T. H. Suchanek, S. L. Ustin, B. J. Bond, K. D. Brosofske, N. Phillips, and R. Bi. 2002. Biophysical controls of carbon flows in three successional Douglas-fir stands based on eddy-covariance measurements. *Tree Physiology* 22:169–177.

Churkina, G., and S. Running. 1998. Contrasting climatic controls on the estimated productivity of global terrestrial biomes. *Ecosystems* 1:206–215.

Cramer, W., D. W. Kicklighter, A. Bondeau, B. Moore III, G. Churkina, B. Nemry, A. Ruimy, and A. L. Schloss. 1999. Comparing global models of terrestrial net primary productivity (NPP): Overview and key results. *Global Change Biology* 5:1–15.

Desjardins, R. L. 1985. Carbon-dioxide budget of maize. *Agricultural and Forest Meteorology* 36:29–41.

Dolman, A. J., E. J. Moors, and J. A. Elbers. 2002. The carbon uptake of a mid latitude pine forest growing on sandy soil. *Agricultural and Forest Meteorology* 111:157–170.

Eamus, D., L. B. Hutley, and A. P. O'Grady. 2001. Daily and seasonal patterns of carbon and water fluxes above a north Australian savanna. *Tree Physiology* 21:977–988.

Ehleringer, J., T. Cerling, and B. Helliker. 1997. $C_4$ photosynthesis, atmospheric $CO_2$, and climate. *Oecologia* 112:285–299.

Enquist, B. J. 2002. Universal scaling in tree and vascular plant allometry: Toward a general quantitative theory linking plant form and function from cells to ecosystems. *Tree Physiology* 22:1045–1064.

Falge, E., D. Baldocchi, R. lson, P. Anthoni, M. Aubinet, Ch. Bernhofer, G. Burba, R. Ceulemans, R. Clement, H. Dolman, A. Granier, P. Gross, T. Grunwald, D. Hollinger,

N. O. Jenson, G. Katul, P. Keronen, A. Kowalski, C. T. Lai, B. Law, T. Meyers, J. Moncrief, E. J. Moors, W. Munger, K. Pilegaard, U. Rannik, C. Rebmann, A. Sukyer, J. Tenhunen, K. Tu, S. Verma, T. Vesala, K. Wilson, and S. Wofsy. 2001. Gap filling strategies for defensible annual sums of net ecosystem exchange. *Agricultural and Forest Meteorology* 107:43–69.

Falge, E., D. Baldocchi, J. Tenhunen, M. Aubinet, P. Bakwin, P. Berbigier, C. Bernhofer, G. Burba, R. Clement, and K. J. Davis. 2002. Seasonality of ecosystem respiration and gross primary production as derived from FLUXNET measurements. *Agricultural and Forest Meteorology* 113:53–74.

Flanagan, L. B., L. A. Wever, and P. J. Carlson. 2002. Seasonal and interannual variation in carbon dioxide exchange and carbon balance in a northern temperate grassland. *Global Change Biology* 8:599.

Foley, J. A., S. Levis, I. C. Prentice, D. Pollard, and S. L. Thompson. 1998. Coupling dynamic models of climate and vegetation. *Global Change Biology* 4: 561–579.

Gifford, R. M. 1994. The global carbon-cycle: A viewpoint on the missing sink. *Australian Journal of Plant Physiology* 21:1–15.

Gilmanov, T.G., S. B. Verma, P. L. Sims, T. P. Meyers, J. A. Bradford, G. G. Burba, and A. E. Suyker. 2003. Gross primary production and light response parameters of four Southern Plains ecosystems estimated using long-term $CO_2$-flux tower measurements. *Global Biogeochemical Cycles* 17, doi:10.1029/2002GB002023.

Goldstein, A. H., N. E. Hultman, J. M. Fracheboud, M. R. Bauer, J. A. Panek, M. Xu, Y. Qi, A. B Guenther, and W. Baugh. 2000. Effects of climate variability on the carbon dioxide, water, and sensible heat fluxes above a ponderosa pine plantation in the Sierra Nevada (CA). *Agricultural and Forest Meteorology* 101:113–129.

Goulden, M. L., J. W. Munger, S. M. Fan, B. C. Daube, and S. C. Wofsy. 1996. Measurements of carbon sequestration by long-term eddy covariance: Methods and a critical evaluation of accuracy. *Global Change Biology* 2:169–182.

Goulden, M. L., S. C. Wofsy, J. W. Harden, S. E. Trumbore, P. M. Crill, S. T. Gower, T. Fries, B. C. Daube, S.-M. Fan, D. J. Sutton, A. Bazzaz, and J. W. Munger. 1998. Sensitivity of boreal forest carbon balance to soil thaw. *Science* 279:214–217.

Grace, J., J. Lloyd, J. McIntyre, A. Miranda, P. Meir, H. Miranda, J. Moncrieff, J. Massheder, I. Wright, and J. Gash. 1995. Fluxes of carbon dioxide and water vapor over an undisturbed tropical forest in south-west Amazonia. *Global Change Biology* 1:1–13.

Granier, A., P. Biron, and D. Lemoine. 2000. Water balance, transpiration and canopy conductance in two beech stands. *Agricultural and Forest Meteorology* 100:291–308.

Granier, A., K. Pilegaard, and N. O. Jensen. 2002. Similar net ecosystem exchange of beech stands located in France and Denmark. *Agricultural and Forest Meteorology* 114:75–82.

Greco, S., and D. D. Baldocchi. 1996. Seasonal variations of $CO_2$ and water vapor exchange rates over a temperate deciduous forest. *Global Change Biology* 2:183–198.

Griffis, T., W. Rouse, and J. Waddington. 2000. Scaling net ecosystem $CO_2$ exchange from the community to landscape-level at a subarctic fen. *Global Change Biology* 6:459–473.

Gu, L., D. D. Baldocchi, S. B. Verma, T. A. Black, T. Vesala, E. M. Falge, and P. R. Dowty. 2002. Superiority of diffuse radiation for terrestrial ecosystem productivity. *Journal of Geophysical Research* 107, doi: 10.1029/2001JD001242.

Hanan, N., B. Burba, S. Verma, J. Berry, A. Suyker, and E. Walter-Shea. 2002. Inversion of net ecosystem CO$_2$ flux measurements for estimation of canopy PAR absorption. *Global Change Biology* 8:563–574.

Hanson, P. J., S. D. Wullschleger, S. A. Bohlman, and D. E. Todd. 1993. Seasonal and topographic patterns of forest floor CO$_2$ efflux from an upland oak forest. *Tree Physiology* 13:1–15.

Hollinger, D.Y., F. M. Kelliher, J. N. Byers, J. E. Hunt, T. M. Mcseveny, and P. L. Weir. 1994. Carbon-dioxide exchange between an undisturbed old-growth temperate forest and the atmosphere. *Ecology* 75:134–150.

Hollinger, D.Y., S. M. Goltz, E. A. Davidson, J. T. Lee, K. Tu, and H. T. Valentine. 1999. Seasonal patterns and environmental control of carbon dioxide and water vapour exchange in an ecotonal boreal forest. *Global Change Biology* 5:891.

Hui, D., Y. Luo, and G. Katul. 2003. Partitioning interannual variability in net ecosystem exchange between climatic variability and functional change. *Tree Physiology* 23:433–442.

Jarvis, P. G., J. M. Massheder, S. E. Hale, J. B. Moncrieff, M. Rayment, and S. L. Scott. 1997. Seasonal variation of carbon dioxide, water vapor, and energy exchange of a boreal black spruce forest. *Journal of Geophysical Research* 102:28953–28966.

Keeling, C. D., and T. P. Whorf. 1994. Atmospheric CO$_2$ records from sites in the SIO air sampling network. Pp. 16–26 in *Trends '93: A compendium of data on global change*. ORNL/CDIAC-65. Oak Ridge, TN: Oak Ridge National Laboratory (ORNL)/Carbon Dioxide Information Analysis Center (CDIAC).

Kim, J., S. B. Verma, and R. J. Clement. 1992. Carbon dioxide budget in a temperate grassland ecosystem. *Journal of Geophysical Research* 97:6057–6063.

Larcher, W. 1975. *Physiological plant ecology*. Berlin: Springer-Verlag.

Law, B. E., E. Falge, L. Gu, D. D. Baldocchi, P. Bakwin, P. Berbigier, K. Davis, A. J. Dolman, M. Falk, and J. D. Fuentes. 2002. Environmental controls over carbon dioxide and water vapor exchange of terrestrial vegetation. *Agricultural and Forest Meteorology* 113:97–120.

Lee, X., J. D. Fuentes, R. M. Staebler, and H. H. Neumann. 1999. Long-term observation of the atmospheric exchange of CO$_2$ with a temperate deciduous forest in southern Ontario, Canada. *Journal of Geophysical Research* 104:15975–15984.

Leuning, R., and P. M. Attiwill. 1978. Mass, heat and momentum exchange between a mature eucalyptus forest and atmosphere. *Agricultural Meteorology* 19:215–241.

Leith, H. 1975. Modeling primary productivity of the world. Pp. 237–263 in *Primary productivity of the biosphere*, edited by H. Leith and R. H. Whittaker. New York: Springer-Verlag.

Lindroth, A., A. Grelle, and A. S. Moren. 1998. Long-term measurements of boreal forest carbon balance reveal large temperature sensitivity. *Global Change Biology* 4:443–450.

Lloyd, C. R. 2001. The measurement and modelling of the carbon dioxide exchange at a high Arctic site in Svalbard. *Global Change Biology* 7:405–426.

Malhi, Y., A. D. Nobre, J. Grace, B. Kruijt, M. Pereira A. Culf, and S. Scott. 1998. Carbon dioxide transfer over a central Amazonian rain forest. *Journal of Geophysical Research* 31:593–612.

Markkanen T., U. Rannik, P. Keronen, T. Suni, and T. Vesala. 2001. Eddy covariance fluxes over a boreal Scots pine forest. *Boreal Environment Research* 6:65–78.

Melillo, J. M., A. D. McGuire, D. W. Kicklighter, B. Moore, C. J. Vorosmarty, and A. L.

Schloss. 1993. Global climate change and terrestrial net primary production. *Nature* 22:234–240.

Meyers, T. P. 2001. A comparison of summertime water and $CO_2$ fluxes over rangeland for well-watered and drought conditions. *Agricultural and Forest Meteorology* 106:205–214.

Miranda, A., H. Miranda, J. Lloyd, J. Grace, R. Francey, J. Mcintyre, P. Meir, P. Riggan, R. Lockwood, and J. Brass. 1997. Fluxes of carbon, water and energy over Brazilian cerrado: An analysis using eddy covariance and stable isotopes. *Plant, Cell and Environment* 20:315–328.

Monson, R. K., A. A. Turnipseed, J. P. Sparks, P. C. Harley, L. E. Scott-Denton, K. Sparks, and T. E. Huxman. 2002. Carbon sequestration in a high-elevation, subalpine forest. *Global Change Biology* 8:459.

Pilegaard, K., P. Hummelshoj, N. O. Jensen, and Z. Chen. 2001. Two years of continuous $CO_2$ eddy-flux measurements over a Danish beech forest. *Agricultural and Forest Meteorology* 107:29–41.

Reichstein, M., J. D. Tenhunen, O. Roupsard, J.-M. Ourcival, S. Rambal, F. Miglietta, A. Peressotti, M. Pecchiari, G. Tirone, and R. Valentini. 2002a. Severe drought effects on ecosystem $CO_2$ and $H_2O$ fluxes at three Mediterranean evergreen sites: Revision of current hypotheses? *Global Change Biology* 8:999.

Reichstein, M., J. D. Tenhunen, O. Roupsard, J.-M. Ourcival, S. Rambal, S. Dore, and R. Valentini. 2002b. Ecosystem respiration in two Mediterranean evergreen Holm Oak forests: Drought effects and decomposition dynamics. *Functional Ecology* 16:27–39.

Rey, A., E. Pegoraro, V. Tedeschi, I. De Parri, P. G. Jarvis, and R. Valentini. 2002. Annual variation in soil respiration and its components in a coppice oak forest in Central Italy. *Global Change Biology* 8:851–866.

Rochette, P., R. L. Desjardins, E. Pattey, and R. Lessard. 1995. Crop net carbon dioxide exchange rate and radiation use efficiency in soybean. *Agronomy Journal* 87:22–28.

Ruimy, A., P. G. Jarvis, D. D. Baldocchi, and B. Saugier. 1995. $CO_2$ fluxes over plant canopies and solar radiation: A review. *Advances in Ecological Research* 26:1–53.

Running, S. W., D. D. Baldocchi, D. Turner, S. T. Gower, P. Bakwin, and K. Hibbard. 1999. A global terrestrial monitoring network, scaling tower fluxes with ecosystem modeling and EOS satellite data. *Remote Sensing of the Environment* 70:108–127.

Saigusa, N, T. Oikawa, and S. Liu. 1998. Seasonal variations of the exchange of $CO_2$ and $H_2O$ between a grassland and the atmosphere: An experimental study. *Agricultural and Forest Meteorology* 89:131–139.

Saigusa, N., S. Yamamoto, S. Murayama, H. Kondo, and N. Nishimura. 2002. Gross primary production and net ecosystem exchange of a cool-temperate deciduous forest estimated by the eddy covariance method. *Agricultural and Forest Meteorology* 112:203–215.

Sanderman, J., R. Amundson, and D. D. Baldocchi. 2003. The effect of temperature on the residence time of soil organic carbon: A comparative analysis. *Global Biogeochemical Cycles* 17, doi:10.1029/2001GB001833.

Schmid, H. P., C. S. Grimmond, F. Cropley, B. Offerle, and H.-B. Su. 2000. Measurements of CO2 and energy fluxes over a mixed hardwood forest in the midwestern United States. *Agricultural and Forest Meteorology* 103:357–374.

Scholes, R. J., and S. R. Archer. 1997. Tree-grass interactions in savannas. *Annual Review of Ecology and Systematics* 28:517–544.

Schulze, E.-D., C. Wirth, and M. Heimann. 2000. Climate change: Managing forests after Kyoto. *Science* 289:2058–2059.

Shurpali, N. J., S. B. Verma, J. Kim, and T. J. Arkebauer. 1995. Carbon dioxide exchange in a peatland ecosystem. *Journal of Geophysical Research* 100:14319–14326.

Soegaard, H., C. Nordstroem, T. Friborg, B. U. Hansen, T. R. Christensen, and C. Bay. 2000. Trace gas exchange in a high-arctic valley. 3. Integrating and scaling $CO_2$ fluxes from canopy to landscape using flux data, footprint modeling, and remote sensing. *Global Biogeochemical Cycles* 14:725–744.

Suyker, A. E., and S. B. Verma. 2001. Year-round observations of the net ecosystem exchange of carbon dioxide in a native tallgrass prairie. *Global Change Biology* 7:279.

Tans, P., P. Bakwin, and D. Guenther. 1996. A feasible global carbon cycle observing system: A plan to decipher today's carbon cycle based on observations. *Global Change Biology* 2:309–318.

Valentini, R., P. DeAngelis, G. Matteucci, R. Monaco, S. Dore, and G. E. Scarascia-Mugnozza. 1996. Seasonal net carbon dioxide exchange of a beech forest with the atmosphere. *Global Change Biology* 2:199–207.

Valentini, R., G. Matteucci, A. J. Dolman, E. D. Schulze, C. Rebmann, E. J. Moors, A. Granier, P. Gross, N. O. Jensen, K. Pilegaard, A. Lindroth, A. Grelle, C. Bernhofer, T. Grunwald, M. Aubinet, R. Ceulemans, A. S. Kowalski, T. Vesala, U. Rannik, P. Berbigier, D. Loustau, J. Guomundsson, H. Thorgeirsson, A. Ibrom, K. Morgenstern, R. Clement, J. Moncrieff, L. Montagnani, S. Minerbi, and P. G. Jarvis. 2000. Respiration as the main determinant of carbon balance in European forests. *Nature* 404:861–865.

Verma, S. B., D. D. Baldocchi, D. E. Anderson, D. R. Matt, and R. E. Clement. 1986. Eddy fluxes of $CO_2$: Water vapor and sensible heat over a deciduous forest. *Boundary Layer Meteorology* 36:71–91.

Verma, S. B., J. Kim, and R. J. Clement. 1989. Carbon dioxide, water vapor and sensible heat fluxes over a tall grass prairie. *Boundary Layer Meteorology* 46:53–67.

Vourlitis, G. L., Y. Harazono, W. C. Oechel, M. Yoshimoto, and M. Mano. 2000. Spatial and temporal variations in hectare-scale net $CO_2$ flux, respiration and gross primary production of Arctic tundra ecosystems. *Functional Ecology* 14:203.

Vourlitis, G. L., N. Priante, M. M. S. Hayashi, J. D. Nogueira, F. T. Caseiro, and J. H. Campelo. 2001. Seasonal variations in the net ecosystem $CO_2$ exchange of a mature Amazonian transitional tropical forest (cerradao). *Functional Ecology* 15:388–395.

Wilson, K. B., and D. D. Baldocchi. 2000. Estimating annual net ecosystem exchange of carbon over five years at a deciduous forest in the southern United States. *Agricultural and Forest Meteorology* 100:1–18.

Wofsy, S., M. Goulden, J. Munger, S. Fan, P. Bakwin, B. Daube, S. Bassow, and F. Bazzaz. 1993. Net exchange of $CO_2$ in a midlatitude forest. *Science* 260:1314–1317.

Yamamoto, S., S. Murayama, N. Saigusa, and H. Kondo. 1999. Seasonal and interannual variation of $CO_2$ flux between a temperate forest and the atmosphere in Japan. *Tellus* 51B:402–413.

Xu, L., and D. Baldocchi. 2003. Seasonal variation in carbon dioxide exchange over a Mediterranean annual grassland in California. *Agricultural and Forest Meteorology* (in press).

# 16

# Current Consequences of Past Actions: How to Separate Direct from Indirect

Gert-Jan Nabuurs

In the 1820s a farmer owned a shrublike forest and grassland parcel in the foothills of the Alps. Because of changes in his agricultural regime, he was able to convert part of his land to high forest and decided to plant Norway spruce. In this chapter, I show how this pragmatic decision 180 years ago determined the carbon balance of that piece of land through the present day.

I show how past management of the biosphere and past socioeconomic factors have determined present land use and how the past thus determines the present carbon balance of large parts of the biosphere. This process is shown through a hectare-scale example, for which I follow three alternative management options, as well as published case studies from regions around the world.

Since this past management issue is closely related to "the dynamic effects of age structure" (UNFCCC 2002) mentioned in the Marrakesh Accords to the Kyoto Protocol, I touch upon the political discussions on separating direct and indirect effects as well.

## Characteristics of the Biosphere Carbon Balance

The terrestrial biosphere is a dynamic system with inherently long periods of slow buildup of biomass (where respiration is less than gross uptake) followed by short periods of large breakdown. The whole system thus fluctuates around an equilibrium, at which gross uptake of carbon is compensated by autotrophic and heterotrophic respiration. Rarely is an ecosystem ever in equilibrium, except at large scales and over long time periods. At a large scale, the system is composed of many local sources and sinks. Thus, the temporal variation in sinks and sources of a small ecosystem through time can be recognized at one point in time as the spatial distribution of the sinks and sources

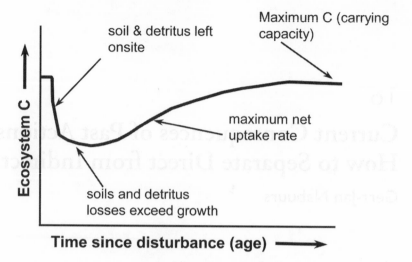

**Figure 16.1.** "An example of net changes in ecosystem carbon stocks over time. Changes in individual ecosystem components take place at different rates, but it is the net of the changes in all interconnected pools that determines the net flow to or from the atmosphere. In the example, the accumulation of biomass initially is at a lower rate than the decomposition of the dead organic matter stock so the stock of ecosystem C declines. Later in the cycle, dead organic matter stocks may increase, although other components have reached steady state. Maximum ecosystem stocks (highest value of Ecosystem C) occur at a later time than the maximum rate of net carbon uptake (steepest slope of the Ecosystem C line)" (quoted from Kauppi et al. 2001: 308).

through the landscape. These sources and sinks migrate through the landscape in the course of time.

Under stable environmental circumstances, an ecosystem can sequester carbon over long time frames only if it has lost significant amounts of carbon first. This is the direct causal link between the current dynamics of the biosphere and past (human or natural) actions (Figure 16.1). Therefore, it is very important to understand and assess past dynamics if we want to estimate current and future carbon dynamics. This requirement is valid from the local to the continental scale (Figure 16.2).

Forest dynamics on a single plot (Figure 16.1) emerge at continental scales in population structure. Figure 16.2, for example, displays the 1990 age class distribution of 140 million hectares (ha) of European forests. A major part of the area is in the age classes up to 70 years of age. This is partly a result of large afforestations in the early 20th century. The relatively small area with stands older than 70 years clearly shows the impact of conventional forest management in the past. This forest management

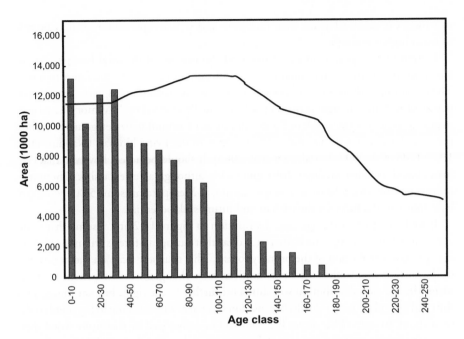

**Figure 16.2.** Age class distribution of the European forests (140 million ha in 30 countries) in 1990 (bars). This distribution is the result of afforestation in the early 20th century and a certain type of management following it. Purely hypothetically the age class distribution of the European primary forests may have resembled, at some stages in time, the drawn line (possibly some 275 million ha). Thousands of years of degradation completely diminished this primary forest, while 20th-century afforestation created a forest resource as given by the bars (Nabuurs 2001).

consisted of planting or reseeding after a clear-cut, followed by regular thinnings in a usually even-aged monospecies stand. Depending on the tree species, a clear-cut would follow again after some 60 to 150 years. This system made it possible for a forest owner to deliver a continuous stream of wood as a raw material to industry. This system led to the age class distribution in Figure 16.2 and has resulted in an average growing stock volume of stemwood of some 140 (cubic meters) $m^3$ $ha^{-1}$ (40 to 50 megagrams of carbon per hectare [Mg C $ha^{-1}$] in total tree living biomass) and an average age of 57 years. For comparison, the average age of forests in European Russia is in the range of 80 years, while in the United States it is 76 years (Pisarenko et al. 2000; Haynes 2002). Kurz and Apps (1999) reported an increasing average age of Canadian forests to 81.5 years in 1969, after which it leveled off and slightly declined owing to increases in natural disturbances. The leveling off is what causes the sink to go away, even though the decline is very small and in the range of annual fluctuations. This high peak in the average age is very interesting. If Canadian boreal forests with regular disturbances can reach 81 years, then European and U.S. temperate forests may be able to reach higher averages.

Currently, European countries harvest 50–60 percent of the total European stem-wood increment (the latter amounting to 750 million $m^3 y^{-1}$) for the forest industry. The remaining additions to the growing stock, together with changes in the wood products pool and the soil organic matter pool, are usually assessed by "inventory-based estimates" to yield a current sink in European forests of around 0.10 petagrams of carbon per year (Pg C $y^{-1}$). Of this, about 75 percent is in the living biomass (Nabuurs et al. 1997; Liski et al. 2002; Karjalainen et al. 2003). It should be noted, however, that estimates based on other methods differ quite a lot, although all assess a sink. Thousands of years of forest degradation in Europe, starting in Mesopotamia some 5,000 years ago, have thus laid the basis for the current and future sink potential.

If we look again at the age class distribution in Figure 16.2 and assume that management will not change much, then it is clear that European forests will not return to the age class distribution of the primary forests. This thus sets a cap on the persistence of the current sink (unless the site conditions have changed). Under current management, the age class distribution will shift a bit further to the right, but not much. This shift will take place rather slowly, because forest owners will continue to log stands older than about 80 years. This owner behavior will be influenced by the future wood market, owner goals, and the current stocking and age class distribution. This perspective on the present state of the forest provides a valuable starting point for assessing the future carbon sink of European forests as well as for other regions in the world.

The degree to which the current dynamics are fully controlled by past dynamics, however, in contrast to current influences, including nitrogen deposition, current weather, $CO_2$ fertilization, and current management (changes), is much less certain (Baldocchi and Valentini, Chapter 15, this volume).

## Hectare-Scale Impacts of Past Actions

I now return to the patch of shrub and grassland mentioned in the introduction to assess how the landowner's decision to convert to high forest 180 years ago affects the carbon balance today. I consider four alternative scenarios:

1. Maintain the land under the grazing regime and thus maintain the shrub-grass landscape (not plotted in Figure 16.3);
2. Plant beech and manage it in a 150-year rotation with four thinnings per rotation;
3. Plant Norway spruce in an 80-year rotation with four thinnings;
4. Abandon the land (i.e., stop grazing), allow natural regeneration, and do no further managing.

If we assume the shrub-grassland system with grazing was established in early Medieval times, then it was in equilibrium in the 1820s. If this system had continued until the year 2000, it would have a baseline carbon balance around 0 MgC $ha^{-1} y^{-1}$.

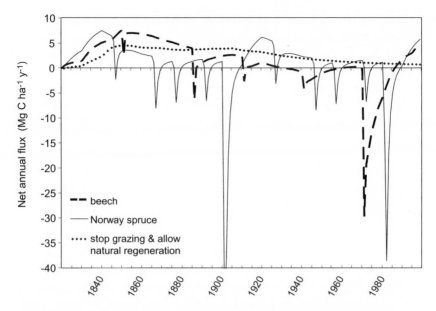

**Figure 16.3.** Comparison of the temporal development of the annual flux (positive = sink) in the total system of forest biomass, soils, and products for beech, Norway spruce, and the natural regeneration system. Peaks in sources are due to logging events. Results were obtained with the CO2FIX modeling framework (Masera et al. 2003).

The 80-year rotation of Norway spruce yields a rapid increase of the net sink after establishment, peaking at 7.5 MgC ha$^{-1}$ y$^{-1}$ after 21 years, in 1841 (Figure 16.3). Then, because of increasing density in the stand, the net sink starts to decrease. This decrease is counteracted by the thinnings at 25, 45, 55, and 70 years. These interventions produce peaks of carbon release sources at the level of the whole system. These peaks are the result of slash decomposing on the ground and wood products decaying. Immediately after the final felling in 1900 the system lost a lot of carbon, with a peak of the source of about 38 MgC ha$^{-1}$ y$^{-1}$. This cycle is repeated in the second rotation and leads in our reference year 2000 to a net sink of 5.75 MgC ha$^{-1}$ y$^{-1}$. The cumulative sum of all sources and sinks from 1820 until 2000 amounts to 132 MgC ha$^{-1}$ (mean net biome production, or NBP, is 0.73 MgC ha$^{-1}$ y$^{-1}$ over the whole period).

The pattern is similar for the beech system except that the accumulation is slower in the beginning but continues longer. The net sink also peaks at 7.5 MgC ha$^{-1}$ y$^{-1}$, but in 1850 instead of 1841. In 2000 the beech system was a net sink of 5.0 MgC ha$^{-1}$ y$^{-1}$. The overall net storage, however, is much larger than in the Norway spruce system, at 218 MgC ha$^{-1}$ (mean NBP is 1.2 MgC ha$^{-1}$ y$^{-1}$ over the whole period).

The third alternative of "stopping grazing + natural regeneration" shows very different dynamics. The growth is much slower in the beginning but continues much longer. In a managed forest, growth is stimulated by keeping the stand young, but in this case growth is more continuous and levels off from natural mortality. The overall net storage is much larger, with a final stocking of around 440 MgC ha$^{-1}$ (mean NBP is 2.4 MgC ha$^{-1}$ y$^{-1}$ over the whole period). In the reference year 2000, however, the remaining net sink is only 0.7 MgC ha$^{-1}$ y$^{-1}$.

Clearly the decision of a landowner in 1820 still has an impact on the carbon budget in 2000. This analysis, however, assumes "model" forest growth dynamics based on long-term measurements in permanent plots (the yield table information). It does not account for

• variability due to weather;
• N fertilization;
• CO$_2$ fertilization;
• other site amelioration from, for example, recovery from past grazing; or
• risks from natural disturbances.

As a consequence, this analysis probably overstates the impact of the landowner's decision in 1820. But the magnitude of the overstatement remains uncertain. We know that, at the local level, interannual variability can be very large (Kirschbaum 1999) and that CO$_2$ fertilization experiments have a short-term impact on NPP (King et al. 1997) but that longer-term impacts are not so easily detectable (Nadelhoffer et al. 1999; Caspersen et al. 2000). Furthermore we know that recovery from past land use change is apparent in Europe (Spiecker et al. 1996). From other analyses, it appears that the impact of current management on the carbon budget far outweighs the effects of CO$_2$ fertilization and N deposition (Houghton 2002).

## Patterns in Other Regions

The hectare-scale discussion so far reflects a range of socioeconomic factors that have determined land use and land use change in Europe over the past two centuries. Several published case studies for other regions highlight the impact of similar factors for temperate, boreal, and tropical regions (Figure 16.4). The recent and rapid exploration of the midwestern and western United States, between 1850 and 1900, led to a large efflux at that time (Houghton and Heckler 2000). During those decades, primary forests were harvested and native grasslands were cultivated. Later, abandonment of managed lands, due to changes in agriculture and job creation in industry, led to a rapid reduction of the net source from the biosphere of the United States. Around 1950 it was approximately in balance. Regrowing forests and low-till agriculture led to a small sink in the U.S. biosphere in the mid-1970s and early 1980s. This increasing sink was, however, already curbed by increased wildfire in the 1980s and early 1990s. The cumulative efflux from the United States between 1800 and 1950 presents a potential cap on future sink potential. Though this situation suggests a large potential, the recent increase in wildfires suggests that it may be difficult or impossible to achieve (Houghton and Hackler 2000).

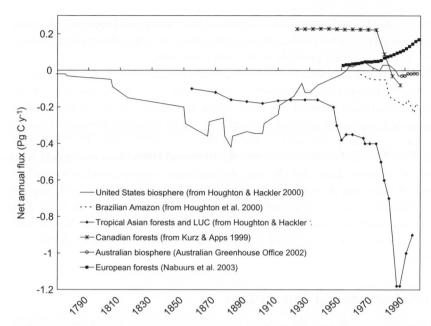

**Figure 16.4.** Temporal evolution of the net annual flux of carbon (sink = positive) for six global regional case studies. Note that for Canada the net flux is the result of changes in natural dynamics, whereas for the other continents the dynamics are the result of land use changes and vegetation rebound.

For European forests from 1950 to 1999, the sink is larger than in the United States, probably because the European data are for forests only. Further, the emission/degradation phase took place much earlier in Europe, starting perhaps 5,000 years ago. Europe experienced a very gradual process of overcutting and overgrazing that slowly expanded from the Mediterranean basin to Central Europe and only much later (16th and 17th centuries) into Scandinavia. The recent sink in Europe reflects an accumulation of biomass in existing forests rather than expansion of the forest area (Nabuurs et al. 2003). The sink increases over time because the increment increased over this time period, whereas logging has stayed approximately level since the 1950s. As in the United States, potential future carbon sequestration in Europe is strongly constrained by regular forest management.

The net flux is different for Australia (Australian Greenhouse Office 2002). Even though Australia's economic status resembles that in the United States and Europe, its biosphere still acts as a source, resulting from deforestation in some parts of the country.

In the tropics economies still focus primarily on natural resource extraction (Romero Lankao, Chapter 19, this volume), with sources from deforestation. In Asia the land use change rate seems to be declining (Houghton and Hackler 1999), whereas in the Amazon Basin deforestation started much more recently (Houghton et al. 2000). The largest potential for future carbon sequestration (as well as the optimal socioeconomic circumstances) is probably on the Asian continent.

In Canadian forests, with large areas of primary boreal forest, the net flux is mainly the result of changes in natural dynamics through time (Kurz and Apps 1999). This situation contrasts with the other continents, where the dynamics are the result of land use changes and vegetation rebound. Canadian forest accumulated biomass for several decades but gradually switched to a source in the early 1980s, owing to an increase in natural and human disturbance rates over the 1970s and 1980s (Kurz et al. 1995). More recent information shows that the natural disturbance rates in Canadian forests have decreased again since the late 1990s (see global burnt area at http://www.grid.unep.ch). As a whole, the 21st-century carbon budget of Canadian forests will be the result of the effects of disturbances (both fire and insect outbreaks are expected to increase with climate change) and the legacy of past disturbance history reflected in the age class structure.

## Policy Context and Conclusions

Current management may play only a minor role in the current sink of regions. It is the past that has determined the autonomous dynamics, with possible modification by current management, $CO_2$ fertilization, and other factors. To explicitly address the current effects of past action, the Marrakesh Accords state: "That accounting excludes removals resulting from: (i) elevated carbon dioxide concentrations above their pre-industrial level; (ii) indirect nitrogen deposition; and (iii) the dynamic effects of age structure resulting from activities and practices before the reference year" (UNFCCC 2002)— that is, direct and indirect human induced activities must be separated (Sanz et al., Chapter 24, this volume). Since this separation is extremely difficult, policy makers have a temporary solution for the first commitment period (2008–2012). Specifically, accounting rules or caps limit certain activities or the amount of credit that industrialized countries may obtain from measures involving land use, land cover, and forestry.

For subsequent commitment periods, however, direct effects will have to be separated from indirect effects in a more comprehensive way. A possible approach to this is given in Figure 16.5. The basic idea is that the direct impact of a measure slowly fades, relative to $CO_2$ fertilization and N deposition (Figure 16.5). This fading out is given by the curve in the B graph. For each management measure, the shape and angle of the transition from direct to indirect effects would have to be assessed. In case of organic matter addition to cropland, for example, most of the direct effect is gone after two to three years. With afforestation, the direct impact may be large for decades. This approach becomes complicated, however, when extended to every management and every region. Tackling this question raises large scientific challenges. Possibly, significant contributions can be made by combining age class distribution models with terrestrial ecosystem models.

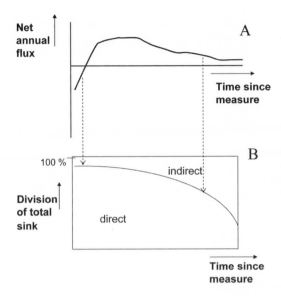

**Figure 16.5.** A possible approach for separating direct and indirect effects. A management measure is carried out at year 0 and has a sink impact as given in the top graph A, comparable to the trends given in Figure 16.3. The B graph then proposes how the sink in time can be assigned to direct impact of the human-induced measure versus the indirect impacts. It is likely that the direct impact declines through time, as the relative impact of weather variability, $CO_2$ fertilization, and N deposition increase.

## Acknowledgments

Most of this paper was written while the boys at home were asleep. I want to thank the organizers of the SCOPE meeting held in Ubatuba, for the very pleasant atmosphere at the meeting. Furthermore, I want to thank two anonymous reviewers for their comments and Skee Houghton and Mike Apps for helpful comments to the draft.

## Literature Cited

Australian Greenhouse Office. 2002. *Greenhouse gas emissions from land use change in Australia: Results of the national carbon accounting system.* Canberra, Australia.

Caspersen, J. P., S. W. Pacala, J. C. Jenkins, G. C. Hurtt, P. R. Moorcroft, and R. A. Birdsey. 2000. Contributions of land use history to carbon accumulation in US forests. Science 290:1148–1151.

Haynes, R., ed. 2002. *An analysis of the timber situation in the United States 1952–2050.* RPA Timber Assessment. Corvallis, OR: Pacific Northwest Research Station, Federal Forest Service, U.S. Department of Agriculture.

Houghton, R. A. 2000. Inter-annual variability in the global carbon cycle. *Journal of Geophysical Research* 105 (No. D15): 20, 121–120, 130.

———. Magnitude, distribution and causes of terrestrial carbon sinks and some implications for policy. *Climate Policy* 2:71–88.

Houghton, R. A., and J. L. Hackler. 1999. Emissions of carbon from forestry and land-use change in tropical Asia. *Global Change Biology* 5:481–492.

———. 2000. Changes in terrestrial carbon storage in the United States. 1. Role of agriculture and forestry. *Global Ecology and Biogeography* 9:125–144.

Houghton, R. A., D. L. Skole, C. A. Nobre, J. L. Hackler, K. T. Lawrence, and W. H.

Chomentowski. 2000. Annual fluxes or carbon from deforestation and regrowth in the Brazilian Amazon. *Nature* 403 (6767): 301–304.

Karjalainen, T., A. Pussinen, J. Liski, G. J. Nabuurs, T. Eggers, T. Lapveteläinen, and T. Kaipainen. 2003. Scenario analysis of the impact of forest management and climate change on the European forest sector carbon budget. *Forest Policy and Economics* 5:141–155.

Kauppi, P., R. J. Sedjo, M. Apps, C. Cerri, T. Fujimori, H. Janzen, O. Krankina, W. Makundi, G. Marland, O. Masera, G. J. Nabuurs, W. Razali, N. H. Ravindranath. 2001. Technical and economic potential of options to enhance, maintain and manage biological carbon reservoirs and geo-engineering. Pp. 302–343 in *Mitigation 2001*, edited by B. Metz, O. Davidson, R. Swart, and J. Pan. IPCC Third Assessment Report. Cambridge: Cambridge University Press.

King, A. W., W. M. Post, and S. D. Wullschleger. 1997. The potential response of terrestrial carbon storage to changes in climate and atmospheric $CO_2$. *Climatic Change* 35:199–227.

Kirschbaum, M. U. F. 1999. Modelling forest growth and carbon storage with increasing $CO_2$ and temperature. *Tellus* 51B:871–888.

Kurz, W. A., and M. J. Apps. 1999. A 70-year retrospective analysis of carbon fluxes in the Canadian forest sector. *Ecological Applications* 9:526–547.

Kurz, W. A., M. J. Apps, S. J. Beukema, and T. Lekstrum. 1995. 20th century carbon budget of Canadian forests. *Tellus* 47B:170–177.

Liski J., D. Perruchoud, and T. Karjalainen. 2002. Increasing carbon stocks in the forest soils of Western Europe. *Forest Ecology and Management* 169:159–172.

Masera, O. R., J. Garza-Caligaris, M. Kanninen, T. Karjalainen, J. Liski, G. J. Nabuurs, A. Pussinen, B. H. J. de Jong, and G. M. J. Mohren. 2003. Modelling carbon sequestration in afforestation, agroforestry, and forest management projects: The CO2FIX V2 approach. *Ecological Modelling* 3237:1–23.

Nabuurs, G. J. 2001. European forests in the 21st century: Impacts of nature-oriented forest management assessed with a large scale scenario model. Joensuu, Finland, and Wageningen, the Netherlands: Research Notes 121. Ph.D. thesis, University of Joensuu.

Nabuurs, G. J., R. Päivinen, R. Sikkema, and G. M. J. Mohren. 1997. The role of European forests in the global Carbon cycle: A review. *Biomass and Bioenergy* 13:345–358.

Nabuurs, G. J., M. J. Schelhaas, G. M. J. Mohren, and C. B. Field. 2003. Temporal evolution of the European forest sector carbon sink 1950–1999. *Global Change Biology* 9:152–160

Nadelhofer, K. J., B. A. Emmett, P. Gundersen, O. J. Kjonaas, C. J. Koopmans, P. Schleppi, A. Tietema, and R. F. Wright. 1999. Nitrogen deposition makes a minor contribution to carbon sequestration in temperate forests. *Nature* 398:145–148.

Pisarenko, A. I., V. V. Strakhov, R. Päivinen, K. Kuusela, F. A. Dyakun, and V. V. Sdobnova. 2000. *Development of forest resources in the European part of the Russian Federation*. Research Report 11. Joensuu, Finland: European Forest Institute.

Spiecker, H., K. Mielikainen, M. Köhl, and J. P. Skovsgaard. 1996. *Growth trends in European forests*. Research Report 5. Berlin: Springer for the European Forest Institute, Joensuu, Finland.

UNFCCC (United Nations Framework Convention on Climate Change). 2002. Report of the Conference of the Parties on its seventh session, held at Marrakech from 29 October to 10 November 2001. Addendum part two: Action taken by the Conference of the Parties, Volumes I to IV.

# PART V
# The Carbon Cycle
# of Land-Ocean Margins

# 17

# Pathways of Atmospheric CO$_2$ through Fluvial Systems

Jeffrey E. Richey

## Uncertainties in the Current "Global Fluvial Systems Model"

As the main pathway for the ultimate preservation of terrigenous production in modern environments, the transfer of organic matter from the land to the oceans via fluvial systems is a key link in the global carbon cycle (Ittekot and Haake 1990; Degens et al. 1991; Hedges et al 1992). Hence, the "role" of rivers in the global carbon cycle is most typically expressed as the fluvial export of total organic and dissolved inorganic carbon from land to the ocean (e.g., Likens et al. 1981). As will be discussed in more detail later in this chapter, the most common literature estimations of the magnitude of these fluxes are 0.4 petagrams of carbon per year (PgC y$^{-1}$) for total organic carbon (evenly divided between particulate and dissolved organic phases), and 0.4 PgC y$^{-1}$ for dissolved inorganic carbon. While these bulk fluxes are small components of the global C cycle, they are significant compared to the net oceanic uptake of anthropogenic CO$_2$ (Sarmiento and Sundquist 1992) and to the interhemispheric transport of carbon in the oceans (Aumont et al. 2001).

This chapter examines and expands on several inconsistencies in this conventional model. The first is in the estimates of the flux quantities and in the relative influence of natural and anthropogenic processes in determining these fluxes. While much river research has emphasized concentrations of carbon, sediments, and/or nutrients, with a focus on export to the oceans (Meybeck 1982, 1991; Degens et al 1991; Milliman and Syvitski 1992), considerable uncertainties remain. The second problem is that the role of fluvial systems may not be limited to fluvial exports to the coastal zone. Continental sedimentation may sequester large amounts of carbon in lower depressions and wetlands (Stallard 1998; Smith et al. 2001). More recent estimates indicate that CO$_2$ outgassing to the atmosphere from river systems may be an important pathway (Cole and Caraco 2001; Richey et al. 2002).

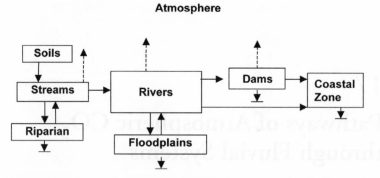

**Figure 17.1.** Major reservoirs and pathways of atmospheric $CO_2$ in fluvial systems. Atmospheric $CO_2$ fixed into streams and their riparian (near-stream) zones (as dissolved $CO_2$ or organic matter) is stored as alluvium, released back to the atmosphere, or transported down stream networks to larger river systems. This now-riverine carbon exchanges with floodplains, is outgassed, or is retained behind dams. Finally the "remaining" fluvial carbon is exported to the coastal zone.

## Dynamics of Fluvial Systems

Fluvial systems integrate hydrological and biogeochemical cycles, over scales from small streams to regional and ultimately to continental basins. The key issue here is the quantity of carbon removed from the atmosphere relative to the amount returned to the atmosphere, along the overall fluvial pathways. This sequence of processes can be summarized in the form of a box model (Figure 17.1). Briefly, three primary forms of carbon of atmospheric origin are transported through fluvial systems. Particulate organic carbon (POC) enters rivers from the erosion of soils (typically older materials) and as leaf litter (typically newly produced). Dissolved organic carbon (DOC) is produced through solubilization of soil organic carbon and enters streams via groundwater. Total dissolved inorganic carbon (DIC) is produced via weathering, as the dissolution of carbonate and silicate rocks. This process sequesters atmospheric $CO_2$, establishes the alkalinity, and influences the pH of water, which governs the subsequent partitioning of DIC between $pCO_2$, bicarbonate, and carbonate ions. The dynamics of carbon in fluvial systems are not defined solely by the export fluxes of bulk C. Rather, they are defined as a complex interplay of multiple C fractions; each exhibits distinct dynamics and compositional traits that hold over very broad ranges of geological, hydrological, and climatic conditions (Hedges et al. 1994).

Evaluating the model of Figure 17.1 is a challenge. The dynamics are complex, and multiple time constants are involved. Data are scarce, particularly in many of the most anthropogenically affected systems. The distribution of the constituent processes varies dramatically across the face of the globe (with some of the most important regions being the least measured).

## Mobilization from Land to Water and Riparian Zones

The fluxes from land to rivers are generally inferred directly from the fluxes out of a basin, especially at a global scale. Although there is considerable truth to this for dissolved species (especially conservative ones), it is less true for particulate species, especially with human intervention.

The modern terrestrial sediment cycle is not in equilibrium (Stallard 1998). Meade et al. (1990) estimated that agricultural land use typically accelerates erosion 10- to 100-fold, via both fluvial and Aeolian processes. Multiple other reports in the literature support this conclusion. With the maturation of farmlands worldwide, and with the development of better soil conservation practices, it is probable that the human-induced erosion is less than it was several decades ago. Overall, however, there has been a significant anthropogenic increase in the mobilization of sediments (and associated POC) through fluvial processes. The global estimates of the quantities, however, vary dramatically. Stallard (1998) poses a range of scenarios, from 24 to 64 Pg $y^{-1}$ of bulk sediments (from 0.4 to 1.2 Pg $y^{-1}$ of POC). Smith et al. (2001) estimate that as much as 200 Pg $y^{-1}$ of sediment is moving, resulting in about 1.4 Pg C $y^{-1}$ (using a lower percentage C than Stallard 1998).

Where does this material go? Does it all go downstream via big rivers, ultimately to the ocean, or is it stored inland? Stallard (1998) argues that between 0 and 40 Pg $y^{-1}$ of sediments (0 to 0.8 Pg C $y^{-1}$ as POC) is stored as colluvium and alluvium and never makes it downstream. Smith et al. (2001), using a different approach, estimate that about 1 Pg C $y^{-1}$ of POC is stored this way. If this movement is merely transferring POC from one reservoir to another, with the same residence times, there is no net change in the C cycle. Then the issue is, to what degree can the remaining soils sequester carbon by sorption to the newly exposed mineral soils? Both Stallard (1998) and Smith et al. (2001) argue that carbon is removed from the upper portion of the soil horizon, where turnover times are relatively rapid (decades or shorter) into either of two classes of environments with longer turnover times: wetlands and smaller, deeper depositional zones, coupled with new carbon accumulation at either erosional or depositional sites. Both assume that oxidation of organic C in transit is minimal, and both use quite conservative values for total suspended sediment (TSS) export. If true, this sequence of processes would result in a significant C sink, on the order of 1 PgC $y^{-1}$.

## Within-River Transport and Reaction Processes

Within-river transport processes carry these eroded materials downstream through the river network. Transport is not passive; significant transformations occur along the way. Rivers exchange with their floodplains (depending on how canalized and diked a river is). The movement of POC is, of course, directly linked to the movement of suspended sediments. Sediments are deposited and remobilized multiple times and over long timescales. In the Amazon, for example, Dunne et al. (1998) computed that as much sediment was being recycled within a reach as was leaving it. Presumably, a significant amount of the erosion-excess sediment discussed in the previous section makes it some distance downstream but is then slowed and retained within the alluvial floodplains.

An additional process within flowing water significantly affects organic matter (OM)—the mineralization to $pCO_2$. Most river and floodplain environments maintain $pCO_2$ levels that are supersaturated with respect to the atmosphere. High partial pressures of $CO_2$ translate to large gas evasion fluxes from water to atmosphere. Early measurements in the Amazon suggested that global $CO_2$ efflux (fluvial export plus respiration) from the world's rivers could be on the order of 1.0 PgC y$^{-1}$. Recent measurements of temperate rivers lead to estimates of global river-to-atmosphere (outgassing) fluxes of ~0.3 PgC y$^{-1}$, which is nearly equivalent to riverine total organic carbon (TOC) or dissolved inorganic carbon (DIC) export (Cole and Caraco 2001). Richey et al (2002) computed that outgassing from the Amazon alone was about 0.5 PgC y$^{-1}$. Assuming that the fluxes computed for the Amazon are representative of fluvial environments of lowland humid tropical forests in general, surface water $CO_2$ evasion in the tropics would be on the order of roughly 0.9 PgC y$^{-1}$ (three times larger than previous estimates of global evasion). Factoring in the recent Amazon results, a global flux of at least 1 PgC y$^{-1}$ directly from river systems to the atmosphere is likely.

What is the source of the organic matter being respired? Is it labile contemporary organic matter, recently fixed in the water by plankton or nearshore vegetation, or is it some fraction of the allochthonous (terrestrial) matter in transport? The prevailing wisdom is that riverborne organic matter is already very refractory and not subject to oxidation (after centuries on land). The "age" of riverine organic matter yields some important insights.

Measurements of the $^{14}$C ages of organic matter and $CO_2$ in river water are very few, but the results are intriguing (Hedges et al 1986). Cole and Caraco (2001) found that the POC (and to a lesser degree dissolved organic carbon [DOC]) entering the Hudson is greatly depleted in $^{14}$C, suggesting that the particulate material was originally formed on average ~5,000 years ago. Their analysis of $\delta^{13}$C suggests that this material is of terrestrial origin and unlikely to be ancient marine sedimentary rocks. Furthermore, they found that organic matter (OM) pools became selectively enriched in $^{14}$C downstream. Based on an inverse modeling approach, they hypothesized that this enrichment is due to utilization of old organic carbon, with dilution by recent primary production. That is, organic C that had resided in soils for centuries to millennia, without decomposing, is then decomposed in a matter of a few weeks in the riverine environment (how this happens is an intriguing question in its own right). Their results are not unique to the Hudson. Raymond and Bauer (2001) found that four rivers draining into the Atlantic are sources of old ($^{14}$C-depleted) and young ($^{14}$C-enriched) terrestrial dissolved organic carbon and of predominantly old terrestrial particulate organic carbon. Much of the younger (relatively speaking) DOC can be selectively degraded over the residence times of river and coastal waters, leaving an even older and more refractory component for oceanic export.

Thus, pre-aging and degradation may alter significantly the structure, distribution, and quantity of terrestrial organic matter before its delivery to the oceans. As noted by Ludwig (2001), the OM that runs from rivers into the sea is not necessarily identical to the OM upstream in river catchments. Cole and Caraco (2001) observe that the apparent high rate of decomposition of terrestrial organic matter in rivers may resolve the enigma of why OM that leaves the land does not accumulate in the ocean (sensu Hedges et al. 1997). Overall, this sequence of processes suggests that the OM that is being respired is translocated in space and time from its points of origin, such that, over long times and large spatial scales, the modern aquatic environment may be connected with terrestrial conditions of another time.

## Input to Reservoirs

Just because dissolved and particulate materials enter a river does not mean that they reach the ocean; modern reservoirs have had a tremendous impact on the hydrologic cycle. Starting about 50 years ago, large dams were seen as a solution to water resource issues, including flood control, hydroelectric power generation, and irrigation. Now, there are more than 40,000 large dams worldwide (World Commission on Dams 2000). This has resulted in a substantial distortion of freshwater runoff from the continents, raising the "age" of discharge through channels from a mean between 16–26 days and nearly 60 days (Vörösmarty et al. 1997). Whereas erosion has clearly increased the mobilization of sediment off the land, the proliferation of dams has acted to retain those sediments. Vorosmarty et al. (2003) estimates that the aggregate impact of all registered impoundments is on the order of 4–5 Pg y$^{-1}$ of suspended sediments (of the 15–20 Pg y$^{-1}$ total that he references). Stallard (1998) extrapolates from a more detailed analysis of the coterminous United States to an estimate of about 10 Pg y$^{-1}$ worldwide (versus 13 Pg y$^{-1}$ efflux to the oceans), for a storage of about 0.2 PgC y$^{-1}$ (which he includes as part of his overall calculation of continental sedimentation).

## Export to the Coastal Zone

The conventional wisdom is that the flux of POC and DOC are each about 0.2 PgC y$^{-1}$, and DIC is 0.4 PgC y$^{-1}$ (e.g., Schlesinger and Melack 1981; Degens 1982; Meybeck 1982, 1991; Ittekot 1988; Ittekkot and Laane 1991; Ludwig et al. 1996; Ver at al. 1999). That these analyses converge is not terribly surprising. They are all based on much of the same (very sparse) field data and use variations of the same statistically based interpolation schemes. Let us evaluate these numbers.

Because direct measurements are few, POC flux estimations are typically a product of the flux of total suspended sediments (TSS) and the estimated weight-percent organic C (w%C) associated with the sediment (because the bulk of POC is organic C sorbed to mineral grains). The first problem is an adequate resolution of the TSS flux. Data on TSS are frequently poor and of unknown quality. Many reported data are surface samples, and the depth integrations necessary to accurately characterize sediment flux are on the order of two to three times higher. Additionally, much sediment moves during episodic storm events, when measurements are almost never made.

As summarized by Vörösmarty et al. (2003), estimates of TSS transport to the oceans have ranged from 9 Pg y$^{-1}$ to more than 58 Pg y$^{-1}$, with more recent studies converging around 15 to 20 Pg y$^{-1}$. These estimates are generally based on extrapolations of existing data, which are weighted to the large rivers of passive margins and temperate regions. Milliman and Syvitski (1992) called attention to the much higher yield rates from steep mountainous environments (without directly computing a global total). More recently, Milliman et al. (1999) estimated that the total sediment flux from the East Indies alone (the islands of Borneo, Java, New Guinea, Sulawesi, Sumatra, and Timor), representing about 2 percent of the global land mass, is about 4 Pg y$^{-1}$, or 20–25 percent of the current global values. This type of environment (steep relief, draining directly to the oceans) is found elsewhere in the world, so the results are not likely to be unique. New data from Taiwan support these high levels, with isotopic analyses of the C showing that a significant part of the flux is human-driven (Kao and Liu 2002).

To obtain POC flux estimates, these values (and their uncertainties) must be multiplied by w%C. Meybeck (1991) divided particulate carbon into inorganic (PIC) and organic (POC) phases and assumed that high-sediment rivers have very low carbon fractions (0.5 w%C), representative of shale; he essentially does not consider the latter to be "atmospherically derived" and hence discounts it from estimates of fluxes to the ocean. More recent values for w%C tend to be in the 1–2 percent range, and higher for organic-rich systems (Richey et al. 1990; Stallard 1998; Gao et al. 2002).

To account for this range, POC flux can be computed as an ensemble based on different combinations of w%C and TSS fluxes, resulting in a range of 0.3 PgC y$^{-1}$ to 0.8 PgC y$^{-1}$, with a "more likely" level of about 0.5 PgC y$^{-1}$ (depending on assumptions used). Therefore, it is possible that the common estimate of 0.2 PgC y$^{-1}$ is low and that the overall value lies in the range of 0.2–0.5 PgC y$^{-1}$. The common estimate for PIC of 0.2 PgC y$^{-1}$ (Meybeck 1991; Ver et al. 1999) may also be underestimated, if sediment fluxes are higher.

The value of 0.2 PgC y$^{-1}$ for DOC export may also be low. DOC is also subject to sparse (and questionable) measurements, without the availability of a proxy like TSS for POC. Aitkenhead and McDowell (2000) developed a model of riverine DOC flux as a function of soil C:N. Using this model, they computed a global flux of 0.4 PgC y$^{-1}$, or twice the common estimate. That is, the total organic C output from fluvial systems may well be approximately double the original estimates, in the ~0.8 PgC y$^{-1}$ range.

## Marine Fate

Long-term preservation of terrestrially derived organic matter in the oceans occurs largely within sediments that accumulate along continental margins. Organic carbon within these sediments is thought to be preserved largely because it is adsorbed to mineral grains (Keil et al. 1994; Mayer 1994; Bishop et al. 1992). Hedges and Keil (1995) estimated that carbon preservation along continental margins over the Holocene was split roughly evenly between sediments accumulating within the delta or sedimentary plume of rivers and non-deltaic sediments accumulating outside the direct influence of major rivers (but within range of multiple smaller systems).

The efficiency of storage between deltaic and non-deltaic systems is different. The amount of organic carbon in non-deltaic continental shelf sediments falls in a narrow range (0.5–1.1 milligrams of carbon per square meter [mg C m$^{-2}$] of mineral surface), and typically > 90 percent of the preserved organic matter is adsorbed to mineral surfaces. Deltaic sediments are distinctly different, containing only a fraction of the organic carbon (by weight) found in other margin sediments. Suspended sediments from the Amazon River, for example, have loadings (~0.67 mg C m$^{-2}$) that are three times higher than the corresponding deltaic sediments (Keil et al. 1997), so more than two-thirds of the terrestrial particulate organic load delivered to the Amazon delta is lost from the mineral matrix and is not preserved. The Mississippi, Yellow, and other river/delta systems also show extensive loss of terrestrial organic matter. Thus many deltaic systems bury only a small fraction of the potential organic load normally sorbed to mineral particles, with the balance presumably desorbed or mineralized (and entering the DIC pool).

The organic matter lost by mineralization and not buried is one of the factors in maintaining the historical perspective that marginal seas are net heterotrophic (Chen, Chapter 18, this volume). But Chen (2003) reviews more recent evidence, based on direct measurements of pCO$_2$ (again, showing the critical importance of actual field measurements of key parameters!) and comes to the conclusion that these seas are net autotrophic, driven primarily by nutrients delivered via upwelling (with enhanced nutrients delivered by rivers leading to eutrophication constituting only a minor source), and net consumers of atmospheric CO$_2$. The overall implication of this sequence of processes is that much of the anthropogenically mobilized riverborne OM (and perhaps the naturally mobilized OM) is liable to remain in the marine environment over timescales longer than the current increase of atmospheric CO$_2$.

## Anthropogenic Transient

To what degree have these fluxes been influenced or impacted by human activities— that is, how much of this carbon is an anthropogenic transient? The consensus is that human-induced erosion has dramatically accelerated the movement of sediments (and

POC). While some of this material "hangs up" on land (sedimentation, reservoirs), some of it likely escapes to the sea (some of the regions with highest sediment yields have very few dams). There is evidence that anthropogenic processes have an affect on DIC. Raymond and Cole (2003) report an increase in the alkalinity of the Mississippi, which implies an increase in the consumption of atmospheric $CO_2$ through weathering. Jones et al. (2003), however, report a systematic decrease in $pCO_2$ in rivers across the United States, which they attribute to large-scale declines in terrestrial $CO_2$ production and import into aquatic ecosystems and not to terrestrial weathering or in-stream processes.

In the case of DOC, there is simply not enough information available to draw conclusions. Clair et al. (1999) suggested that DOC export from basins in Canada might increase by 14 percent with a doubling in atmospheric $CO_2$. An additional factor rarely addressed in rivers is direct loading from urban and industrial sources (Ver et al 1999; Abril et al. 2002). In evaluating the consequences of continental sedimentation and the potentially higher fluxes of POC, a net transient exported from land of roughly 1 PgC $y^{-1}$ is possible, with perhaps half of that going to the sea and the other half divided equally between outgassing and sedimentation.

## Overall Fluvial System–Atmosphere Exchange: Alternative Scenarios

Considerable uncertainty remains in the assessment of the carbon cycle of fluvial systems, including the magnitude of fluxes, how to include processes not previously considered, and delineation of anthropogenic and natural processes, all with an explicit recognition of geography. To assess the implications of each, the earlier discussions are summarized here via five "scenarios" (Figure 17.2).

The bulk transfer of atmospheric C through the land to fluvial systems (assuming a steady-state summation of downstream processes, in a non-steady-state environment) ranges from about 0.6 PgC $y^{-1}$ (conventional wisdom) to 2.6 PgC $y^{-1}$ (+ outgassing). Continental sedimentation results in a significant sink, but that sink is reduced with $CO_2$ outgassing (because of the way the sedimentation was computed). The inclusion of continental sedimentation, and then the larger export of OM to the sea (about twice conventional assumptions, under + POC, DOC), yields net sinks of atmospheric $CO_2$ of up to 1.6 PgC $y^{-1}$. If outgassing is included, however, then the fluvial net sink is reduced to 0.2 PgC $y^{-1}$. Although partitioning the total fluvial fluxes into natural conditions and anthropogenic transients is problematic at best, substantial evidence suggests that the mobilization of sediments has dramatically increased. While much of this material is captured in reservoirs, it is reasonable to expect that a considerable amount escapes to the sea (especially in non-deltaic regions with steep slopes and few dams).

Summarizing all the components of the riverine carbon cycle, several images emerge. As a global steady-state aggregate, there appears to be a sink (between continental sedimentation and marine sedimentation and dissolution) on the order of 1–1.5 PgC $y^{-1}$,

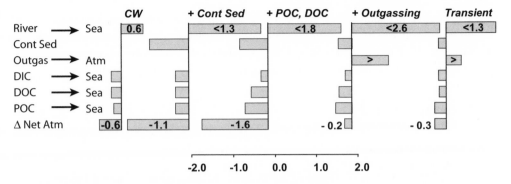

**Figure 17.2.** Scenarios of the fluxes of carbon through fluvial systems relative to atmospheric CO$_2$; including atmosphere to rivers (River → Atm, as sum of other fluxes), continental sedimentation (Cont Sed, alluvial, reservoirs), outgassing from rivers to atmosphere (Outgas → Atm), export of DIC to the sea (DIC → sea, as 50 percent of total DIC, to represent just atmospheric weathering component), DOC export to the sea (DOC → sea), export of POC to the sea (POC → sea), and by difference the net exchange with the atmosphere (→ Net Atm). For the purposes here, it is assumed that the export fluxes to the sea constitute a sink, with no return to the atmosphere, on an immediate decadal timescale. Particulate inorganic C is not included (as it is abiotic and does not interact with the other pools). Units are PgC y$^{-1}$ (scale at bottom lines up with the axis of each graph at 0). Scenarios are CW (conventional wisdom on DOC and POC export, no continental sedimentation or outgassing), + Cont Sed (adding continental sedimentation to CW), + POC, DOC (adding the higher values for POC and DOC fluxes discussed in the text to + Cont Sed), + Outgassing (adding outgassing to + POC, DOC), and Transient (inferring anthropogenic transients, based on the + Outgassing scenario).

with a significant anthropogenically enhanced component. A return flux to the atmosphere, on the order of 1 PgC y$^{-1}$, reduces the net sink to about 0.2 or 0.3 PgC y$^{-1}$. There are certainly disjunctures in space and time in this view, however. Because the organic matter in transport appears to be "old," the subsequent mobilization and oxidation/outgassing would essentially be mining old C. There are significant regional implications in this analysis. With its preponderance of land mass, extensive reservoirs, and agriculture, the bulk of continental sedimentation (and its implications for C sink), is focused in the Northern Hemisphere, between 30° and 50°N. C sequestration in paddy lands would be closer to the equator. The greatest amount of sediment flux to the ocean (and the greatest uncertainty) is in South and Southeast Asia and Oceania. Outgassing is function of both pCO$_2$ concentrations (driven by in situ oxidation) and surface area of water. It is likely most significant in the humid tropics, particularly during the peak of the wet seasons. The highly canalized temperate areas have less area available. The northern latitudes, particularly with warming, are liable to have significant fluxes.

Overall, more carbon is moving through the river system than previously assumed; however, the exact magnitudes remain uncertain. The degree to which fluvial systems constitute a net source/sink relative to the atmosphere is essentially governed by the interplay between mobilization of materials off the landscape and oxidation of those materials back to the atmosphere. Additional data will be necessary to constrain these magnitudes, and it is necessary to partition the model by geographic zone.

## Literature Cited

Abril, G., M. Nogueira, H. Etcheber, G. Cabecadas, E. Lemaire, and M. J. Brogueira. 2002. Behaviour of organic carbon in nine contrasting European estuaries. *Estuarine, Coastal and Shelf Science* 54:241.

Aitkenhead, J. A., and W. McDowell. 2000. Soil C:N as a predictor of annual riverine DOC flux at local and global scales. *Global Biogeochemical Cycles* 14:127–138.

Aumont, O., J. C. Orr, P. Monfray, W. Ludwig, P. Amiotte-Suchet, and J.-L. Probst. 2001. Riverine-driven interhemispheric transport of carbon. *Global Biogeochemical Cycles* 15:393–406.

Bishop, A. N., A. T. Kearsley, and R. L. Patience. 1992. Analysis of sedimentary organic materials by scanning electron microscopy: The application of backscattered electron imagery and light element X-ray microanalysis. *Organic Geochemistry* 18:431–446.

Clair, T. A., J. M. Ehrman, and K. Higuichi. 1999. Changes in freshwater carbon exports from Canadian terrestrial basins to lakes and estuaries under a 2xCO$_2$ atmosphere scenario. *Global Biogeochemical Cycles* 13:1091–1097.

Cole, J. J., and N. F. Caraco. 2001. Carbon in catchments: Connecting terrestrial carbon losses with aquatic metabolism. *Marine and Freshwater Research* 52 (1): 101–110.

Degens, E. T. 1982. Riverine carbon: An overview. In *Transport of carbon and minerals in major world rivers,* Part 1, edited by E. T. Degens. SCOPE/UNEP Sonderbd. Mitteilungen aus dem Geologisch-Paläontologischen Institut der Universität Hamburg 52, 1–12.

Degens, E. T., S. Kempe, and J. E. Richey. 1991. *Biogeochemistry of major world rivers.* New York: John Wiley and Sons.

Dunne, T., L. A. K. Mertes, R. H. Meade, J. E. Richey, and B. R. Forsberg. 1998. Exchanges of sediment between the flood plain and channel of the Amazon River in Brazil. *Geology Society of America Bulletin* 110 (4): 450–467.

Gao, Q., Z. Tao, C. Shen, Y. Sun, W. Yi, and C. Xing. 2002. Riverine organic carbon in the Xijiang River (South China): Seasonal variation in content and flux budget. *Environmental Geology* 41:826.

Hedges, J. I., and R. G. Keil. 1995. Sedimentary organic matter preservation: an assessment and speculative synthesis. *Marine Chemistry* 49:81–115.

Hedges, J. I., J. R. Ertel, P. D. Quay, P. M. Grootes, J. E. Richey, A. H. Devol, G. W. Farwell, F. W. Schmitt, and E. Salati. 1986. Organic carbon-14 in the Amazon River system. *Science* 231:1129–1131.

Hedges, J. I., P. G. Hatcher, J. R. Ertel, and K. J. Meyers-Schulte. 1992. A comparison of dissolved humic substances from seawater with Amazon River counterparts by [13]C-NMR spectrometry. *Geochimica et Cosmochimica Acta* 56:1753–1757.

Hedges, J. I., G. L. Cowie, J. E. Richey, P. D. Quay, R. Benner, and M. Strom. 1994. Origins and processing of organic matter in the Amazon River as indicated by carbohydrates and amino acids. *Limnology and Oceangraphy* 39 (4): 743–761.

Hedges, J. I., R. G. Keil, and R. Benner. 1997. What happens to land-derived organic matter in the ocean? *Organic Geochemistry* 27:195–212.

Ittekkot, V.1988. Global trends in the nature of organic matter in river suspensions. *Nature* 332:436–438.

Ittekkot, V., and B. Haake. 1990. The terrestrial link in the removal of organic carbon in the sea. Pp. 318–325 in *Facets of modern biogeochemistry,* edited by V. Ittekkot, S. Kempe, W. Michaelis, and A. Spitzy. Heidelberg: Springer-Verlag.

Ittekkot, V., and R. W. P. M. Laane. 1991. Fate of riverine particulate organic matter. Pp. 233–243 in *Biogeochemistry of major world rivers,* edited by E. T. Degens, S. Kempe, and J. E. Richey. New York: John Wiley and Sons.

Jones, J. B., E. H. Stanley, and P. J. Mulholland. 2003. Long-term decline in carbon dioxide supersaturation in rivers across the contiguous United States. *Geophysical Research Letters* 30:2-1–2-4.

Kao, S. J., and K.-K. Liu. 2002. Exacerbation of erosion induced by human perturbation in a typical Oceania watershed: Insight from 45 years of hydrological records from the Lanyang-Hsi River, northeastern Taiwan. *Global Biogeochemical Cycles* 16:1–7.

Keil, R.G., D. B. Montluçon, F. G. Prahl, and J. I. Hedges. 1994. Sorptive preservation of labile organic matter in marine sediments. *Nature* 370:549–552.

Keil, R. G., L. M. Mayer, P. D. Quay, J. E. Richey, and J. I. Hedges. 1997. Losses of organic matter from riverine particles in deltas. *Geochimica et Cosmochimica Acta* 61 (7): 1507–1511.

Likens, G., F. Mackenzie, J. Richey, J. Sedell, and K. Turekian, eds. 1981. *Flux of organic carbon by rivers to the oceans*. National Research Council/DOE Conf-8009140. Washington, DC: U.S. Department of Energy.

Ludwig, W. 2001. The age of river carbon. *Nature* 409:466–467.

Ludwig, W., J. L. Probst, and S. Kempe. 1996. Predicting the oceanic input of organic carbon by continental erosion. *Global Biogochemical Cycles* 10 (1): 23–41.

Mayer, L. M. 1994. Adsorptive control of organic carbon accumulation in continental shelf sediments. *Geochimica et Cosmochimica Acta* 58:1271–1284.

Meade, R. H., T. R. Yuzyk, and T. J. Day. 1990. Movement and storage of sediment in rivers of the United States and Canada. Pp. 255–280 in *Surface water hydrology,* edited by M. G. Wolman and H. C. Riggs. Boulder, CO: The Geology of North America, Geological Society of America.

Meybeck, M. 1982. Carbon, nitrogen and phosphorus transport by world rivers. *American Journal of Science* 282:401–425.

———. 1991. C, N, P, and S in rivers: From sources to global inputs. In *Interactions of C, N, P and S: Biogeochemical cycles,* edited by R. Wollast, F. T. MacKenzie, and L. Chou. Berlin: Springer Verlag.

Milliman, J. D., and J. P. M. Syvitski. 1992. Geomorphic tectonic control of sediment discharge to the ocean: The importance of small mountainous rivers. *Journal of Geology* 100 (5): 525–544.

Milliman, J. D., K. L. Farnswortha, and C. S. Albertin. 1999. Flux and fate of fluvial sediments leaving large islands in the East Indies. *Journal of Sea Research* 41:97–107.

Raymond, P. A., and J. E. Bauer. 2001. Riverine export of aged terrestrial organic matter to the North Atlantic Ocean. *Nature* 409:497–500.

Raymond, P. A., and J. J. Cole. 2003. Increase in the export of alkalinity from North America's largest river. *Science* 301:88–91.

Richey, J. E., J. I. Hedges, A. H. Devol, P. D. Quay, R. Victoria, L. Martinelli, and B. R. Forsberg. 1990. Biogeochemistry of carbon in the Amazon River. *Limnology and Oceanography* 35:352–371.

Richey, J. E., J. M. Melack, A. A. K. Aufdenkampe, V. M. Ballester, and L. Hess. 2002. Outgassing from Amazonian rivers and wetlands as a large tropical source of atmospheric $CO_2$. *Nature* 416:617–620.

Sarmiento, J. L., and E. T. Sundquist. 1992. Revised budget for the oceanic uptake of anthropogenic carbon-dioxide. *Nature* 356 (6370): 589–593.

Schlesinger, W. H., and J. M. Melack. 1981. Transport of organic carbon in the world's rivers. *Tellus* 33:172–187.

Smith, S. V., W. H. Renwick, R. W. Buddemeier, and C. J. Crossland. 2001. Budgets of soil erosion and deposition for sediments and sedimentary organic carbon across the conterminous United States. *Global Biogeochemical Cycles* 15 (3): 697–707.

Stallard, R. F. 1998. Terrestrial sedimentation and the carbon cycle: Coupling weathering and erosion to carbon burial. *Global Biogeochemical Cycles* 12:231–257.

Ver, L. M., F. T. Mackenzie, and A. Lerman. 1999. Biogeochemical responses of the carbon cycle to natural and human perturbations: Past, present and future. *American Journal of Science* 299:762–801.

Vörösmarty, C. J., K. Sharma, B. Fekete, A. H. Copeland, J. Holden, J. Marble, and J. A. Lough. 1997. The storage and aging of continental runoff in large reservoir systems of the world. *Ambio* 26:210–219.

Vörösmarty, C. J., M. Meybeck, B. Fekete, K. Sharma, P. Green, and J. Syvitski. 2003. Anthropogenic sediment retention: Major global impact from registered river impoundments. *Global and Planetary Change* 39:169–190.

World Commission on Dams. 2000. *Dams and development: A new framework for decision-making.* London: Earthscan.

# 18

# Exchanges of Carbon in the Coastal Seas

Chen-Tung Arthur Chen

In the natural carbon cycle, the time period for atmosphere-biosphere exchange ranges from only a few months to a few decades. The exchange of $CO_2$ between the atmosphere and the hydrosphere, by contrast, takes several hundred years if the interior of the oceans is taken into consideration. The exchange is much more rapid, however, on a time scale of a few years or even less, for only the terrestrial hydrosphere and the surface mixed layer of the oceans. The time for the atmosphere-lithosphere exchange is very long, requiring many thousands of years or more. The shallow sediments on the continental shelves interact relatively readily with the atmosphere. Some terrestrial material even crosses the shelves, which have a mean width of 70 kilometers (km) and a total area of $26 \times 10^6$ km$^2$, and efficiently reaches the slopes, which start at an average depth of 130 meters (m) (Gattuso et al. 1998). Dissolved organic matter may also be swiftly carried to the interior of the oceans through intermediate bodies of water in certain areas, including the Arctic, Okhotsk, Mediterranean, and Red Seas (Walsh 1995; Chen et al. 2003). The specific rates of productivity, biogeochemical cycling, and sequestration of $CO_2$ are higher in the continental margins than in the open oceans. The end result is that it may take only years, as opposed to hundreds of years, for the atmosphere, lithosphere, biosphere, and hydrosphere to interact in the continental margins. These zones may also act as major conveyor belts, transporting carbon to the interior of the oceans.

In general, the coastal oceans tend to absorb $CO_2$ in winter, when the water cools, and in spring, as a consequence of biological processes. In summer and fall, the processes of warming, respiration of marine organisms, and decomposition of organic matter release $CO_2$ back into the atmosphere. Bacterial processes involved in the production of $CH_4$ (methanogenesis) as well as in the biological production of dimethyl sulfide (DMS) on the shelves also release these important greenhouse or reactive gases into the atmosphere. Finally, direct and indirect human perturbations to the continental margins (e.g., pollution, eutrophication) are large and have dire consequences for marine ecosystems. Unfortunately, owing to the diversity and therefore complexity of the shelf systems, their precise roles in the carbon cycle have yet to be quantified with any degree of certainty. There is still, in fact, no consensus on the simple question posed by the Land-Ocean Interaction in the Coastal Zone project (LOICZ) in its first report: Are continental shelves carbon sources or sinks? (Kempe 1995).

Basing their argument on the imbalance between the total river transport of about 0.4 petagrams of carbon per year (PgC y$^{-1}$) and the oceanic organic carbon burial rate of around 0.14 PgC y$^{-1}$, Smith and Mackenzie (1987) and Smith and Hollibaugh (1993) noted that the ocean must be heterotrophic, releasing more $CO_2$ into the atmosphere than it takes up, in the absence of the anthropogenic perturbation of atmospheric $CO_2$. Over the long term, the difference of 0.26 PgC y$^{-1}$ is most likely returned to the atmosphere. Ver et al. (1999a,b) and Mackenzie et al. (2000) evaluated changes in the carbon cycle of the continental margins over the past three centuries. These three studies conclude that continental margin waters are still a source of $CO_2$ to the atmosphere in spite of increased invasion of $CO_2$ from the atmosphere to the continental margins driven by the rise in atmospheric $CO_2$. Fasham et al. (2001) adopts the same view and reports a net sea-to-air flux of 0.5 PgC y$^{-1}$ for continental margins.

## The Mass Balances

The most recent Joint Global Ocean Flux Study (JGOFS) synthesis, however, suggests that the marginal seas are sinks of $CO_2$ (Chen et al. 2003). This synthesis, with slight modifications to reflect some recent developments, indicates that the global average new production of phytoplankton on the shelf is 0.78 Pg y$^{-1}$ particulate organic carbon (POC) and 0.25 Pg y$^{-1}$ particulate inorganic carbon (PIC) (Figure 18.1, Table 18.1). This new production is only 13 percent of the average primary production rate and is mostly a result of upwelling (Chavez and Toggweiler 1995; Liu et al. 2000; Chen 2000, 2003a). Although net community productivity is rather high on global shelves (Lee 2001), only about 0.2 PgC y$^{-1}$ of PIC and the same amount of POC out of the total production (0.48 Pg y$^{-1}$ PIC and 6.2 Pg y$^{-1}$ POC) is buried and stored on the shelf. Downslope transport of modern particulate carbon is 0.5 PgC y$^{-1}$, 58 percent of which is organic (Chen et al. 2003). Results from major programs on the eastern United States, western European, East China Sea, and Mediterranean continental shelves all show that most of the biogenic particulate matter is remineralized over the shelves. Only a small proportion (< 8 percent) is exported to the adjacent slopes (Wollast and Chou 2001; de Haas et al. 2002; Chen 2003b), but higher export ratios have occasionally been reported.

The downslope transport of POC is only 23 percent of the offshore transport of dissolved organic carbon (DOC). Recent studies based on $^{13}C$, $^{14}C$, and $^{15}N$ show that a large portion of the off-shelf transport of POC may be old terrestrial or relic matter.

**Table 18.1.** Fluxes relevant to continental margins

| Category | Values |
|---|---|
| Rivers plus groundwater and ice | • DIC: 32 (1, 6)<br>• DOC: 30 (1), 27 (6)<br>• PIC: 15 (1, 6)<br>• POC: 20 (1), 18 (6)<br>• IC: 37 (7), 13 (9)<br>• OC: 34 (2), 30 (7), 31 (8) |
| Air-to-sea (gaseous) | • $CO_2$: 25 (1), 20 (2), 49 (3), 46–75 (4), 30 (6), 8.3 (7), 62 (10), 83 (11);<br>• $CH_4$: -0.1 (6);<br>• DMS: -0.07 (6) |
| Precipitation plus dust | • PIC: 0.3 (1, 6); OC: 0.2 (8) |
| Net burial plus fish catch | • PIC: 15 (1, 6, 9), 14.5 (9); POC: 15 (1, 6, 9), 14 (2)<br>• Total: 12.5 (7) |
| Gross upwelling plus surface inflow | • DIC: 2800 (1), 2827 (6);<br>• DOC: 80 (1), 70 (6);<br>• POC: 4 (1, 6) |
| Downslope export of particulates | • PIC: 20 (1, 6, 9);<br>• POC: 20 (1, 9), 27 (6);<br>• Total: 167 (7) |
| Gross surface water outflow | • DIC: 2800 (1,6);<br>• DOC: 120 (1, 6);<br>• PIC: 1.0 (1, 6);<br>• POC: 12 (1), 22 (6);<br>• Net: 58 (7) |
| Gross offshelf export (downslope + surface outflow) | • DIC: 2800 (1, 6);<br>• DOC 120 (1, 6);<br>• PIC: 21 (1, 6);<br>• POC 32 (1), 49 (6);<br>• Net: 225 (7) |
| Net offshore export (downslope + surface outflow-upwelling plus surface inflow) | • DIC: 0 (1), -27 (6);<br>• DOC: 40 (1), 50 (6), 33 (13);<br>• PIC: 21 (1, 6);<br>• POC: 28 (1), 45 (6);<br>• IC 4 (9);<br>• OC: 40 (2), 38 (12);<br>• Total: 58 (7) |

*(continued)*

**Table 18.1.** *(continued)*

| Category | Values |
|---|---|
| Primary productivity | • PIC: 40 (1, 6, 9), 24.5 (9);<br>• POC: 516 (1, 6,9);<br>• OC: 368 (2), 789 (5);<br>• Total: 830 (7) |
| New productivity | • DOC: 23 (6);<br>• PIC: 6 (1), 21 (6);<br>• POC: 75 (1), 42 (6);<br>• OC: 43 (2), 231 (5), 158 (13),<br>• Total: 167 (7) |
| f-ratio | • 0.15 (1), 0.12 (2), 0.29 (5), 0.2 (7), 0.13 (6) |

*Note:* All values except f-ratio are in $10^{12}$ moles C $y^{-1}$; numbers in parentheses are reference numbers. Traditionally these values are given in moles C and are not converted to PgC.

*Sources:* 1. Figure 17 and Table 8 of Chen et al. (2003) and the 27 references therein; 2. Rabouille et al. (2001); 3. Yool and Fasham (2001); 4. extrapolated from data on the European shelves by Frankignoulle and Borges (2001); 5. Gattuso et al. (1998); 6. this study; 7. Liu et al. (2000); 8. Smith et al. (2001); 9. Milliman (1993) and Wollast (1994) ; 10. Walsh and Dieterle (1994); 11. Tsunogai et al. (1999); 12. Alvarez-Salgado et al. (2001a) and 13. Hansell and Carlson (1998).

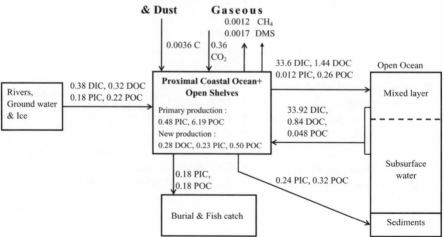

**Figure 18.1.** Schematic diagram for the annual carbon budget (in $10^{12}$ mol PgC $y^{-1}$) for the continental margins of the world (modified from Chen et al. 2003)

There have been reports of high PIC contents on the slopes (Chen 2002; Epping et al. 2002), but whether or not these are relic is unknown (Bauer et al. 2001). Net off-shelf transport of DOC is 0.60 PgC y$^{-1}$, about twice terrestrial input of 0.32 PgC y$^{-1}$ (Figure 18.1). The remaining 0.28 Pg y$^{-1}$ DOC is produced on the shelves and represents 35 percent of new organic carbon production, or 27 percent of total new carbon production, a finding that is consistent with the results of Hansell and Carlson (1998).

Marginal seas are also sources of $CH_4$ and DMS (Sharma et al. 1999; Marty et al. 2001). For $CH_4$, shelf sediments are likely to be the principal source, whereas for DMS, biological production in the water column is probably the main source. These fluxes are small (Figure 18.1) compared with the very large $CO_2$ fluxes, but they play an important role, based on their activity in absorbing solar energy and reactivity in the atmosphere.

## The pCO$_2$ of the Continental Margins

The discussion so far deals with carbon balances rather than the $CO_2$ partial pressure (pCO$_2$) of surface waters. Yet, for surface waters to be a source or sink for atmospheric $CO_2$, their pCO$_2$ must respectively be larger or smaller than atmospheric pCO$_2$. Estuaries are generally supersaturated with $CO_2$ largely as a result of the respiration of organic carbon input from rivers (Frankignoulle et al. 1998; Abril et al. 2002). The shelf systems have not been studied as thoroughly. The first comprehensive study of pCO$_2$ on a large shelf was conducted in the North Sea in May and June 1986 (Kempe and Pegler 1991). According to that study, the North Sea gained 1.4 mol C m$^{-2}$ y$^{-1}$. Since that study, many more pCO$_2$ data have become available. Frankignoulle and Borges (2001), for instance, measured pCO$_2$ during 18 cruises in the surface waters of many northwest European shelves. Their results show that these shelves are a sink of 0.09–0.17 PgC per year. This is an additional, appreciable fraction (45 percent) of the proposed flux for the open North Atlantic Ocean (Keir et al. 2001; Takahashi et al. 2002).

Data on surface water pCO$_2$, temperature, and salinity have been collected over all four seasons in the Yellow Sea and the East China Sea (ECS). These seas are a year-round $CO_2$ sink (Chen and Wang 1999; Tsunogai et al. 1999). Low pCO$_2$ has also been reported for the Bering and Mediterranean Seas, along the Californian coast, the Bay of Bengal, and many other locations (see tables compiled in Chen et al. 2003 and Chen 2003b). Gattuso et al. (1998) compiled all of the available data for coastal ecosystems and concluded that the proximal shelf regions that are directly influenced by the input of terrestrial organic matter are net heterotrophic. That is, they release $CO_2$ into the atmosphere because respiration is greater than biological production. The distal shelves are net autotrophic, owing to the smaller influence of terrestrial inputs and to the larger export of carbon to sediments and across the continental shelf break. The results compiled by Chen et al. (2003) and Chen (2003b) confirm the findings of Gattuso et al. (1998) and show that the global shelves are $CO_2$ sinks.

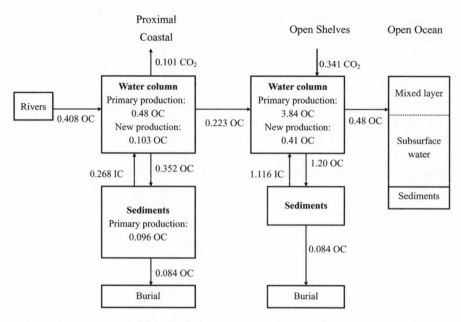

**Figure 18.2.** Organic carbon cycle in global coastal oceans in its preanthropogenic state. The boxes represent the reservoirs, and the arrows represent the fluxes between them. The air-sea fluxes do not include the net flux of $CO_2$ because the carbonate system is not included in the budget (data taken from Rabouille et al. 2001).

The temporal link between high winds and low sea-surface $pCO_2$ leads to a situation in which average sea surface $pCO_2$ can be supersaturated, and the net annual flux is still from the atmosphere to the sea. This situation occurs because the highest fluxes generally occur in winter and spring, during periods of undersaturation when the winds are strong. Exchange rates are lower in summer and fall when surface waters are generally more supersaturated and the winds are weaker (Memery et al. 2002). Many studies assume that an autotrophic system absorbs $CO_2$ from the atmosphere (e.g., Smith and Hollibaugh 1993), but intensive upwelling regions may be autotrophic and still release $CO_2$ to the atmosphere. This process can happen when the $pCO_2$ of shelf waters is reduced as a consequence of a decrease in the ratio of total $CO_2$:alkalinity due to dissolution of relic carbonate deposits and/or increased alkalinity due to sulfate reduction in the sediments (Chen 2002). In a recent study of the East China Sea continental shelf, Tsunogai et al. (1999) suggested that because the shallow seafloor restricts the convection of cooling water, cooling is greater for waters on the continental shelf than for waters in neighboring open oceans. This process leads to a "continental shelf pump," driven by the production of relatively cold and dense water, which, in combination with biological production, increases the absorption of $CO_2$ in the continental shelf zone. Based on a globally distributed sample of 33 shelves and marginal seas, Yool and Fashman (2001) found that the continental shelf pump accounts for a net oceanic uptake of 0.6 PgC y$^{-1}$. For the situation prior to human effects, Mackenzie and colleagues concluded that, "Before anthropogenic activities, the global coastal ocean was a net autotrophic system with a net export flux to sediments and the open ocean of 20 T mol organic C/yr (0.24 Pg C y$^{-1}$)" (Rabouille et al. 2001: 3615). These results support the conclusion that the proximal coastal oceans are $CO_2$ sources (0.10 PgC y$^{-1}$), whereas the distal coastal oceans are $CO_2$ sinks (0.34 PgC y$^{-1}$) (Figure 18.2; Gattuso et al. 1998).

## Response to Future Forcing

Though small in area, the continental margins are the focal point of land, sea, and atmosphere interactions. They have special significance, not only for biogeochemical cycling and processes, but also increasingly for human habitation. In comparison with the relatively uniform environment of the open oceans, or the rapidly mixing atmosphere, the spatial and temporal heterogeneity of the world's coastal zone is considerable. In addition, human impacts on these relatively small areas are disproportionately large (Pacyna et al. 2000).

Several recent studies show that the continental margins currently function as substantial carbon sinks. Ver et al. (1999a, b) and Mackenzie et al. (2000) estimated a source of 0.2 PgC y$^{-1}$ in the preindustrial era. They argue that the flux has decreased since 1800, a consequence of an ever-increasing nutrient discharge. Recently Mackenzie and coworkers (Rabouille et al. 2001) concluded that the continental margins were net sinks of 0.24 PgC y$^{-1}$ in the preanthropogenic state (Figure 18.2). Assuming that the estimates of Ver et al. (1999a, b) and Mackenzie et al. (2000) were consistently low by 0.44 PgC y$^{-1}$, the continental margins in 2000 would have been sinks of 0.34 PgC y$^{-1}$. Though this adjustment may be an oversimplification, it reconciles the air-sea flux of Ver et al. (1999a, b) and Mackenzie et al. (2000) (Figure 18.3) with my result of 0.36 PgC y$^{-1}$ (Figure 18.1). This is a significant fraction (16 percent) of the "global" air-to-sea flux of 2.2 PgC y$^{-1}$ (Takahashi et al. 2002).

The coastal oceans are also strongly influenced by river inputs, which are sensitive to numerous factors, including regional weather and climate, as well as dams and diversions. Rivers are the major conduits for the transfer of water, nutrients, organic material, and particulate matter from land to sea. Inputs of nutrients and organic matter provide critical support for fish breeding in the estuaries. Dams block the downstream transport of particulate matter, an important source of nutrients and food for aquatic biota. A large dam, such as the Nile River's Aswan Dam in Upper Egypt, can have a dramatic impact on fish stocks in connected estuaries. The major source of shelf nutrients, however, is onshore advection of subsurface waters (Chen 2000). Decreasing river out-

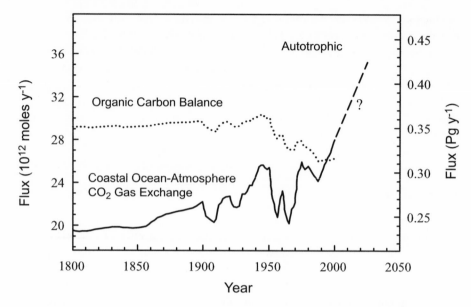

**Figure 18.3.** Organic carbon balance (dashed line) and net exchange flux of $CO_2$ across the air-seawater interface (solid line) for the coastal margin system, in units of $10^{12}$ moles C $y^{-1}$ and Pg $y^{-1}$. Positive values indicate the $CO_2$ flux is directed toward the surface waters (modified from Ver et al. 1999a by adding $36.7 \times 10^{12}$ mol $y^{-1}$ or 0.44 Pg $y^{-1}$ to their results).

flow will reduce the cross-shelf water exchange through a reduced buoyancy effect, and it will also diminish the onshore nutrient supply. Decreased primary production and fish catch on the shelf typically follow dam construction.

Globally, approximately 40 percent of the freshwater and particulate matter entering the oceans is transported by the 10 largest rivers. Buoyant plumes move much of the water and nutrients to the open shelves. Hence, these shelves also experience a diminished biological pump and fish production when damming reduces freshwater outflow. The fraction of riverine particulate carbon that is deposited in the deltas or beaches or converted to the ever-increasing DOC pool is not well known. On the other hand, the coastal oceans may be heterotrophic but nevertheless absorb $CO_2$.

## Conclusions

The first LOICZ report (Kempe 1995) concluded that coastal seas could be net sinks or sources of $CO_2$ for the atmosphere, with slim prospects for a quick resolution. Now, mass balance calculations, as well as direct $pCO_2$ measurements, indicate a consistent pattern. Taken together, continental shelves are significant sinks for atmospheric $CO_2$, absorbing 0.36 PgC a year. This flux is a composite over many estuaries, coastal waters, and intensive upwelling areas, which are typically supersaturated with respect to $CO_2$, and most open shelf areas, which are probably undersaturated. This "continental shelf pump" is primarily fueled by the cross-shelf transport of nutrients from nutrient-rich subsurface waters offshore. Though they are sinks for $CO_2$, the shelves release $0.1 \times 10^{12}$ mol $y^{-1}$ $CH_4$ and $0.07 \times 10^{12}$ mol $y^{-1}$ DMS into the atmosphere.

New production supported by the external sources of nutrients represents about 13 percent of primary production. The other 87 percent is respired and recycled on the shelf. Some of the organic material that is not recycled accumulates in the sediments, but most of the detrital organic matter, mainly in its dissolved form, is exported to the slopes and open oceans.

## Acknowledgments

I am indebted to the National Science Council of the Republic of China, which provided partial financial assistance (NSC 92-2611-M-110-003). N. Gruber made detailed and constructive criticisms, which strengthened the manuscript.

## Literature Cited

Abril, G., M. Nogueira, H. Etcheber, G. Cabecadas, E. Lemaire, and M. J. Brogueira. 2002. Behaviour of organic carbon in nine contrasting European Estuaries. *Estuarine, Coastal and Shelf Science* 54:241–262.

Alvarez-Salgado, X. A., D. M. Doval, A. V. Borges, I. Joint, M. Frankignoulle, E. M. S. Woodward, and F. G. Figueiras. 2001. Off-shelf fluxes of labile materials by an upwelling filament in the NW Iberian Upwelling System. *Progress in Oceanography* 51:321–337.

Bauer, J. E., E. R. M. Druffel, D. M. Wolgast, and S. Griffin. 2001. Sources and cycling of dissolved and particulate organic radiocarbon in the northwest Atlantic continental margin. *Global Biogeochemical Cycles* 15:615–636.

Chavez, F. P., and J. R. Toggweiler. 1995. Physical estimates of global new production: The upwelling contribution. Pp. 313–336 in *Upwelling in the ocean: modern processes and ancient records,* edited by C. P. Summerhayes, K.-C. Emeis, M.V. Angel, R. L. Smith, and B. Zeitzschel. Chichester, UK: John Wiley and Sons.

Chen, C. T. A. 2000. The Three Gorges Dam: Reducing the upwelling and thus productivity of the East China Sea. *Geophysical Research Letters* 27:381–383.

———. 2002. Shelf vs. dissolution generated alkalinity above the chemical lysocline in the North Pacific. *Deep-Sea Research II* 49:5365–5375.

———. 2003a. New vs. export productions on the continental shelf. *Deep-Sea Research II* 50:1327–1333.

Chen, C. T. A. 2003b. Air-sea exchanges of carbon and nitrogen in the marginal sea. In *Carbon and nutrient fluxes, I. Global continental margins,* edited by L. Atkinson, K. K. Liu, R. Quinones, and L. Talaue-McManus. New York: Springer-Verlag (submitted).

Chen, C. T. A., and S. L. Wang. 1999. Carbon, alkalinity and nutrient budget on the East China Sea continental shelf. *Journal of Geophysical Research* 104:20675–20686.

Chen, C. T. A., K. K. Liu, and R. MacDonald. 2003. Continental margin exchanges. Pp. 53–97 in *Ocean biogeochemistry: A JGOFS synthesis*, edited by M. J. R. Fasham. Berlin: Springer.

de Haas, H., T. C. E. van Weering, and H. de Stigter. 2002. Organic carbon in shelf seas: Sinks or sources, processes and products. *Continental Shelf Research* 22:691–717.

Epping, E., C. van der Zee, K. Soetaert, and W. Helder. 2002. On the oxidation and burial of organic carbon in sediments of the Iberian margin and Nazaré Canyon (NE Atlantic). *Progress in Oceanography* 52:399–431.

Fasham, M. J. R., B. M. Baliño, and C. Bowles, editors. 2001. Contributors: R. Anderson, D. Archer, U. Bathmann, P. Boyd, K. Buesseler, P. Burkill, A. Bychkov, C. Carlson, C. T. A. Chen, S. Doney, H. Ducklow, S. Emerson, R. Feely, G. Feldman, V. Garcon, D. Hansell, R. Hanson, P. Harrison, S. Honjo, C. Jeandel, D. Karl, R. Le Borgne, K. K. Liu, K. Lochte, F. Louanchi, R. Lowry, A. Michaels, P. Monfray, J. Murray, A. Oschlies, T. Platt, J. Priddle, R. Quiñones, D. Ruiz-Pino, T. Saino, E. Sakshaug, G. Shimmield, S. Smith, W. Smith, T. Takahashi, P. Tréguer, D. Wallace, R. Wanninkhof, A. Watson, J. Willebrand, and C. S. Wong. A new vision of ocean biogeochemistry after a decade of the Joint Global Ocean Flux Study (JGOFS). *Ambio Special Report* 10 (May): 4–31.

Frankignoulle, M., and A. V. Borges. 2001. European continental shelf as a significant sink for atmospheric carbon dioxide. *Global Biogeochemical Cycles* 15:569–576.

Frankignoulle, M., G. Abril, A. Borges, I. Bourge, C. Canon, B. DeLille, E. Libert, and5J.-M. Théate. 1998. Carbon dioxide emission from European estuaries. *Science* 282 :434–436.

Gattuso, J.-P., M. Frankignoulle, and R. Wollast. 1998. Carbon and carbonate metabolism in coastal aquatic ecosystems. *Annual Review of Ecology and Systematics* 29:405–434.

Hansell, D. A., and C. A. Carlson. 1998. Net community production of dissolved organic carbon. *Global Biogeochemical Cycles* 12:443–453.

Keir, R. S., G. Rehder, and M. Frankignoulle. 2001. Partial pressure and air-sea flux of $CO_2$ in the Northeast Atlantic during September 1995. *Deep-Sea Research II* 48:3179–3189.

Kempe, S. 1995. *Coastal seas: A net source or sink of atmospheric carbon dioxide?* LOICZ Reports and Studies No. 1. Texel, the Netherlands: Land Ocean Interaction in the Coastal Zone International Project Office.

Kempe, S., and K. Pegler. 1991. Sinks and sources of $CO_2$ in coastal seas: The North Sea. *Tellus* 43:224–235.

Lee, K. 2001. Global net community production estimated from the annual cycle of surface water total dissolved inorganic carbon. *Limnology and Oceanography* 46:1287–1297.

Liu, K. K., L. Atkinson, C. T. A. Chen, S. Gao, J. Hall, R. W. MacDonald, L. Talaue McManus, and R. Quinones. 2000. Exploring continental margin carbon fluxes on a global scale. *EOS* 81:641–642 plus 644.

Mackenzie, F. T., L. M. Ver, and A. Lerman. 2000. Coastal-zone biogeochemical dynamics under global warming. *International Geology Review* 42:193–206.

Marty, D., P. Bonin, V. Michotey, and M. Bianchi. 2001. Bacterial biogas production in coastal systems affected by freshwater inputs. *Continental Shelf Research* 21:2105–2115.

Mémery, L., M. Lévy, S. Vérant, and L. Merlivat. 2002. The relevant time scales in estimating the air-sea $CO_2$ exchange in a mid-latitude region. *Deep-Sea Research II* 49:2067–2092.

Milliman, J. D. 1993. Production and accumulation of calcium carbonate in the ocean: Budget of non-steady state. *Global Biogeochemical Cycles* 7:927–957.

Pacyna, J. M., H. Kremer, N. Pirrone, and K. G. Barthel, eds. 2000. Socioeconomic aspects of fluxes of chemicals into the marine environment. European Community Research Project report EUR 19089. Brussels: European Commission.

Rabouille, C., F. T. Mackenzie, and L. M. Ver. 2001. Influence of the human perturbation on carbon, nitrogen, and oxygen biogeochemical cycles in the global coastal ocean. *Geochimica et Cosmochimica Acta* 65:3615–3641.

Sharma, S., L. A. Barrie, D. Plummer, J. C. McConnell, P. C. Brickell, M. Levasseur, M. Gosseliln, and T. S. Bates. 1999. Flux estimation of oceanic dimethyl sulfide around North America. *Journal of Geophysical Research* 104 (D17): 21327–21342.

Smith, S. V., and J. T. Hollibaugh. 1993. Coastal metabolism and the oceanic carbon balance. *Reviews of Geophysics* 31:75–89.

Smith, S. V., and F. T. Mackenzie. 1987. The ocean as a net heterotrophic system: Implications from the carbon biogeochemical cycle. *Global Biogeochemical Cycles* 1:187–198.

Smith, S. V., W. H. Renwick, R. W. Buddemeier, and C. J. Crossland. 2001. Budgets of soil erosion and deposition for sediments and sedimentary organic carbon across the conterminous United States. *Global Biogeochemical Cycles* 15:697–707.

Takahashi, T., S. C. Sutherland, C. Sweeney, A. Poisson, N. Metzl, B. Tilbrook, N. Bates, R. Wanninkhof, R. A. Feely, C. Sabine, J. Olafsson, and Y. Nojiri. 2002. Global sea-air $CO_2$ flux based on climatological surface ocean $pCO_2$, and seasonal biological and temperature effects. *Deep-Sea Research II* 49:1601–1622.

Tsunogai, S., S. Watanabe, and T. Sato. 1999. Is there a "continental shelf pump" for the absorption of atmospheric $CO_2$? *Tellus* 51B:701–712.

Ver, L. M. B., F. T. Mackenzie, and A. Lerman. 1999a. Carbon cycle in the coastal zone: Effects of global perturbations and change in the past three centuries. *Chemical Geology* 159:283–304.

———. 1999b. Biogeochemical responses of the carbon cycle to natural and human perturbations: Past, present, and future. *American Journal of Science* 299:762–801.

Walsh, J. J. 1995. DOC storage in the Arctic seas: The role of continental shelves. *Coastal and Estuarine Studies* 49:203–230.

Walsh, J. J., and D. A. Dieterle. 1994. $CO_2$ cycling in the coastal ocean. I. A numerical analysis of the southeastern Bering Sea with applications to the Chuckchi Sea and the northern Gulf of Mexico. *Progress in Oceanography* 34:335–392.

Wollast, R. 1994. The relative importance of biomineralization and dissolution of $CaCO_3$ in the global carbon cycle. Pp. 13–34 in Past and present biomineralization processes: Considerations about the carbonate cycle, edited by F. Doumenge. *Bulletin de l'Institut Oceanographique, Monaco*, special issue 13.

Wollast, R., and L. Chou. 2001. Ocean margin exchange in the northern Gulf of Biscay: OMEX I. An introduction. *Deep-Sea Research II* 48:2971–2978.

Yool, A., and M. J. R. Fasham. 2001. An examination of the "continental shelf pump" in an open ocean general circulation model. *Global Biogeochemical Cycles* 15:831–844.

# PART VI
## Humans and the Carbon Cycle

# 19

# Pathways of Regional Development and the Carbon Cycle

## Patricia Romero Lankao

The distribution of carbon on the Earth has changed dramatically over the past 300 years (Houghton and Skole 1990; Ayres et al. 1994). Humans have exerted a strong influence on these transformations through diverse, interacting mechanisms, including agriculture, forestry, urbanization, and industrialization. These impacts on the carbon cycle occurred in the context of dynamic, demographic, and technological trends, institutional settings, and politics. Societies have been affected by these transformations (vulnerability), and they have also responded in ways that have the potential to feed back on the carbon cycle. These issues have been addressed in the literature in notions like adaptation and mitigation (Kelly and Adger 2000).

Different theories may be used to explain such transformations of the carbon cycle. Some, such as neoclassical or neoinstitutional, offer pathways of single-factor causation (Ostrom 1990). Others look for more complex explanations, such as direct causes, underlying forces (Turner et al. 1990; Geist and Lambin 2001), or the world economy approach (Braudel 1984). Here, I explore the key role of the history and world economy approaches for understanding patterns of change in the carbon cycle, and ways that carbon-relevant societal actions (land changes and energy use) manifest different regional and temporal dynamics. I focus on three regional development patterns with different carbon cycle consequences. These development patterns are core, rim, and peripheral.

Throughout history the interaction among urbanization, agriculture, and forestry has been complex and dynamic. Both regional and global perspectives can be used to identify carbon-relevant patterns of causation. The global perspective facilitates finding features common to all development patterns, while the regional view supports the exploration of key differences. The relationships among urbanization, agriculture, and forestry, which cannot be isolated from the energy sector, must be considered within the context of a world economy in which they are embedded. A world economy is "an economically autonomous section of the planet able to provide for most of its own needs, a section to which its internal link and exchanges give a certain organic unity" (Braudel 1984: 22).

## Key Historical Trends

Throughout their history, humans have transformed the carbon cycle through agriculture, forestry, trade, urbanization, and energy use in industry and transportation. In the past two centuries these processes have become so far-reaching that they have profoundly transformed the carbon cycle and related processes, including climate, water, and nutrient cycles. From 1700 to 1920, the area of croplands grew by $648 \times 10^6$ hectares (ha). In the next 60 years, it increased by $588 \times 10^6$ ha—an increase of 91 percent in 27 percent of the time. The world's forests and woodlands decreased by $537 \times 10^6$ ha between 1700 and 1920 and by $625 \times 10^6$ from 1920 to 1980 (Richards 1990: Table 10-1). These and related activities (biomass burning, intensification of livestock husbandry) are major contributors to the changes in carbon stored in vegetation and soils and the distribution of carbon between land and atmosphere (Houghton and Skole 1990).

Until the 1950s urbanization was too low and the number of large cities too small for cities to have other than local impacts on climate, the hydrological cycle, agricultural land use, and exploitation of natural resources. Since 1950 urbanization has become a major global factor, a phenomenon with important local, regional, national, and international impacts. Effects of urbanization on land cover are globally minor because urban areas comprise less than 2 percent of the Earth's surface (Lambin et al. 2001). On the other hand, urbanization's local effects and "ecological footprints"[1] on climate, air, land cover, economic activities, and quality of life now extend far into surrounding areas. In addition, large cities are now so numerous that their impact has become truly global. Before the middle of the 20th century, the largest and demographically most dynamic urban agglomerations like London, New York, and Tokyo were mostly in developed countries. Since 1990 this pattern has reversed, with most megacities now situated in developing countries (UNEP 2002).

The use of energy is an integral part of household activities, urban agglomerations, agriculture, and transportation. Biomass (plant matter) was the most important fuel until the middle of the 19th century. Since then, biomass has accounted for a decreasing share of the world's energy. It is now used mainly for domestic necessities in developing countries. In developed countries, coal and later oil and natural gas increased their share of international energy demand (Williams 1994). At least since the 15th century, forestry, agriculture, urbanization, and energy use have been interrelated with and driven by "world economies." It is useful to think of world economies as centered in a number of kinds of world cities, including centers of industrialization (London, New York, Berlin), gateway cities (Buenos Aires, Cairo, San Francisco, and Sydney), points of colonial penetration (La Paz, Lima), and centers of national power within developing countries (Manila, Mexico City). Such cities have experienced waves of carbon-relevant urbanization and industrialization. Among the waves with tangent impacts on the carbon cycle are the smokestack factories of the heavy-machinery phase in London (1850–1940) and the urban sprawl of Los Angeles and other big cities of the mass-production era (1930–1980).

World economies comprise three different regions: core, rim, and peripheral. In core regions within and surrounding world cities, land use changes and carbon releases to the atmosphere have been massive. Middle or rim areas are fairly developed regions that maintain tight trade interchanges with core areas and have similar carbon-relevant processes. There may also be a huge periphery. Sometimes these are located within a nation's boundary. In other cases, they are part of overseas empires or economic zones, representing "backwardness, archaism, and exploitation by others" (Braudel 1984: 39). Peripheries have functioned as resource-exporting hinterlands, linked to and environmentally affected by core regions through trade and political domination aimed at finding markets for industrialized goods, supplying heartlands with raw commodities (Galeano 1973; Ponting 1990), and, in recent decades, offering opportunities for relocation of activities perceived as environmentally destructive to the cores.

Linkages among core regions, rims, and peripheries are the essence of diverse regional pathways of development. Regional development has five key phases (Gruebel 1994; Lo 1994). These phases overlap because their technological paradigms need about 20 years to evolve from one to the other.

The first phase (1750–1820) was dominated by the textile, coal, and iron industries and led by England, plus some regions in Belgium and France. The processes leading to this phase included technical (factories, combinations of coke and stationary steam engines, wrought iron) and social (breakdown of medieval structures) aspects. Processes taking place in the peripheries driven by colonialism, namely, increasing extraction of timber and minerals, creation of croplands and plantations, and innovations in agriculture and animal husbandry, were also important drivers of this phase. Modifications in land cover were globally the main proximate cause of changes in the carbon cycle because plant matter was the main source of energy (Houghton and Skole 1990).

The second phase, also known as the steam era (1800–1870), saw the emergence of industrialization as a pervasive and global principle of economic growth, dominated by mobile steam power and inland navigation. England still led industrialization, but it soon spread to Belgium, France, Germany, and the United States. Key innovations appeared, including steel, railways, telegraphs, city gas- and coal-based chemical industry. Hinterland regions continued to supply raw materials, and they became widening markets for the products of core and middle regions. The bulk of the trade, however, remained within each realm (Gruebel 1994). Land use persisted as the main cause of changes in the carbon cycle (Houghton and Skole 1990).

The heavy engineering epoch (1850–1940), the most intensive period of industrialization for core regions, saw the emergence of oil, petrochemicals, synthetics, telephone, radio, electricity, and agricultural innovations. Economic expansion contrasted with the social and political inability to provide equitable standards of living, resulting in social conflicts. Production of steel, other primary commodities, and capital equipment were at the core of the international expansion of industrial and urban infrastructure (Ponting 1990). London and New York rose as world cities for key commodity markets, banking, and capital finance. Railway networks and ocean steamships linked distant regions through international trade dominated by industrialized core areas in Germany and the United States, and, to a lesser extent, by the emerging regions of Japan and Russia. Imperialism and colonialism were the political counterparts of these processes. Peripheries were granted access to markets in core regions in return for raw materials and food. Emissions of carbon from the combustion of fossil fuels became an important cause of changes in the carbon cycle, rivaling emissions from land use (Houghton and Skole 1990).

During the mass production and consumption era (1930–1980), the most intensive human transformation of the carbon cycle took place. Land use, land cover, and energy use increased rapidly with intensification of industrial activities, expansion of agricultural areas, and reduction of forests. Emissions from fossil fuels became by far the most important source of carbon releases. Transformations that might be called Fordism, Taylorism, and Welfare State led to consolidation of chemical industries, massive production of industrial commodities and consumer products such as plastics and petrochemicals, internal combustion engines, automobiles, agrochemical inputs, farm machinery, and consumer durables. New transport and communication systems facilitated cultural and information exchanges. They also facilitated the worldwide extension of industrialization. Nevertheless, the core regions continued to dominate most industrial output and trade. Only a few former rims (Austria, Canada, Japan, Scandinavia, and Switzerland) joined the heartlands. The rest remained in rim and periphery areas, depending heavily on exports of raw materials, and/or production of industrial commodities such as textiles and durables that had gradually shifted away from the core regions.

The current phase, not consolidated yet (1980–?), is characterized by the emergence of most industrialized countries, members of the Organisation for Economic Cooperation and Development (OECD), as core regions and new partners as rims. Regional economic blocs, such as the North American Free Trade Agreement, the European Union, and Asia-Pacific Economic Cooperation, are features of the current world economy led by the United States, Germany, and Japan, respectively. Computers, electronics, telecommunications, services, biotechnology, new materials, and to a lesser degree environmental technologies seem to be the main sources of transformation. These transformations are driven by new production systems (just-in-time, total quality control) and widely distributed production. Key trends include economic deregulation, liberalization, privatization, and new arenas of power with important differences between countries and regions. For example, between 1930 and 1980, many peripheries financed industrialization through export earnings from primary commodities, which allowed them to join the rims. Meanwhile, mass production was extended from core regions to rims, including parts of Brazil and Mexico (Marston et al. 2002). It is not yet clear which sources of energy will substitute for petroleum (Caldeira et al., Chapter 5, this volume). Genetic resources, biodiversity, water, soil, and "nature" may emerge as key "commodities."

## Current Relevant Pathways of Regional Development

The current world system has at least three patterns of regional development (Marston et al. 2002). The three share important characteristics, namely increased population in cities, fragmentation of surrounding landscapes, and agricultural intensification. On the other hand, some features of urbanization, agriculture, forestry, and energy are unique to each region.

The first pathway, centered on dominant world cities, takes place in core regions (Figures 19.1, 19.2, and 19.3). These core regions share increases in international trade, production, consumption, and carbon emissions.[2] World cities act as centers of technological innovation and international, managerial, and financial functions, as well as hubs of networks for less important urban centers (Lo 1994). Urban areas of the core regions experience lower population growth rates than those in rim or peripheral areas (Figure 19.4). Management policies and strategies have ensured investments in infrastructure, urban planning, and environmental issues. Local government expenditures tend to be between 15 and 40 times higher than in peripheral areas (Figure 19.5). The question is whether this might allow for development policies with positive carbon consequences reflecting capacity to promote mitigation and cope with the social implications of changes in the carbon cycle (Lo 1994; UNEP 2002).

Major cities in developed core and rim regions fall into two major categories with respect to carbon-relevant indicators such as urban land use, planning, and transport. One category is high-density agglomerations represented by Hamburg, London, Paris, Singapore, Stockholm, and Tokyo. Although recently experiencing some urban sprawl, they tend to have mixed land use, improved public transport, relatively low auto dependence, and enhanced viability of walking and cycling (Kenworthy and Laube 1996; UNEP 2002). The other category is low-density cities such as Canberra, Chicago, Detroit, and Los Angeles, which are sprawled, heavily zoned, and segregated. In these cities, dependence on autos is high and support for walking and cycling is low (Kenworthy and Laube 1996). Low-density cities tend to have more carbon-related local and global impacts than their high-density counterparts, including destruction of prime farming land and natural landscapes, photochemical smog, noise, and emissions of greenhouse gases.

North America, Western Europe, and Australia and New Zealand have 481, 473,

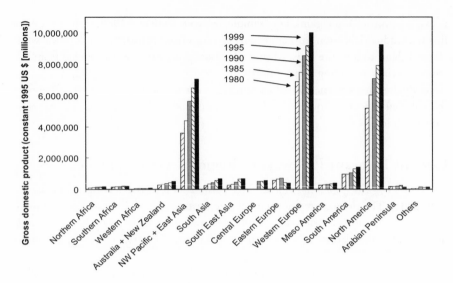

**Figure 19.1.** Share of world's GDP by region (1980–1999). (UNEP 2002)

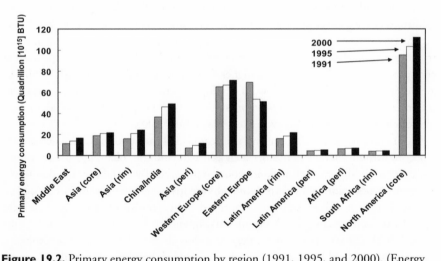

**Figure 19.2.** Primary energy consumption by region (1991, 1995, and 2000). (Energy Information Administration 2002)

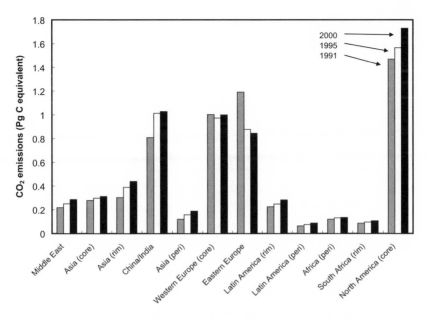

**Figure 19.3.** Carbon dioxide emissions by region (1991, 1995, and 2000). (Energy Information Administration 2002)

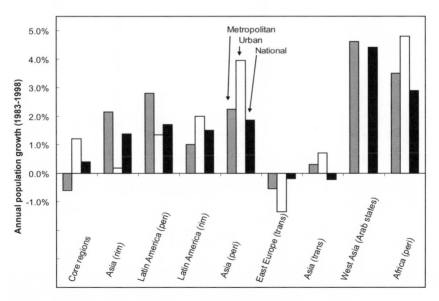

**Figure 19.4.** Demographic indicators (1993–1998) (UN-Habitat 2001).

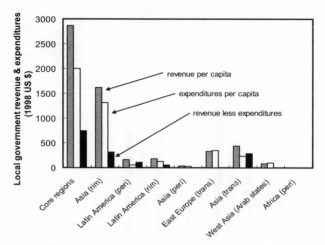

**Figure 19.5.** Local governmental revenue and expenditures (1998). (UN-Habitat 2001)

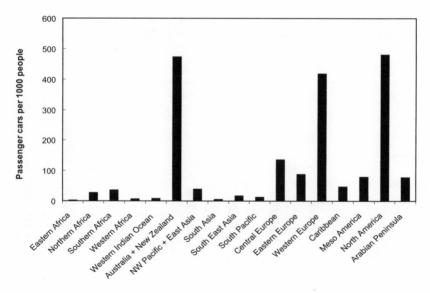

**Figure 19.6.** Number of passenger cars per thousand people (1995) (UNEP 2002).

and 419 cars per thousand people, respectively (Figure 19.6), well above the levels in the rest of the world. This could support the thesis that there is an inevitable relationship between rising incomes and increasing car use, or in more general terms that affluence, consumption, and environmental impact are intrinsically linked. But urbanization patterns in Western Europe and some Asian cities suggest that such a connection is not

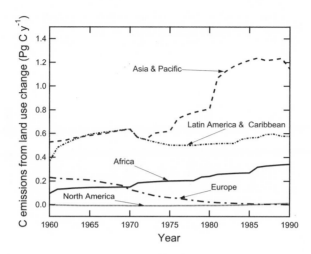

**Figure 19.7.** Annual net carbon flux to the atmosphere from land use change (1960–1990). Values are in millions of tons C (UNEP 2002).

inevitable. Other factors such as urban planning policies that favor alternatives to auto transport, good-quality transit systems, and environmental awareness and attitudes may be key drivers of these differences.

Notwithstanding these dissimilarities, both kinds of cities share a critical feature. They enjoy high standards of living, with ecological footprints in the air, land, water, and natural cycles of the earth that are much larger than in less-developed regions. In terms of land use, for instance, they run ecological deficits or footprints 8 to 15 times larger than their total national territories (Rees and Wackernagel 1996).

In core regions, forest areas have increased during recent decades while croplands have decreased (Richards 1990). Core regions' contribution to carbon emissions through land use change has therefore declined (Figure 19.7). Agriculture and forestry in core regions tend to be intensive in terms of yield, cultivation, products per unit area and time, and application of capital and inputs affecting the carbon and other global cycles.[3] Last but not least, primary activities tend to be more supported by the state through incentives and subsidies than they are in peripheries (Williams 1994).

A second development pathway occurs in rim or semiperipheral regions of Asia, Europe, and Latin America. One group of these, represented by Hong Kong, Korea, Singapore, and Taiwan (the "Asian tigers"), experienced rapid industrialization from 1970 to 1996. These cities expanded their share of world trade production and consumption of industrial goods through a shift from light manufacturing to durable consumer goods and machinery. The national states have been relatively able to support innovative research and development, social security, urban infrastructure, and local government revenues and expenditures (Figures 19.5 and 19.8).

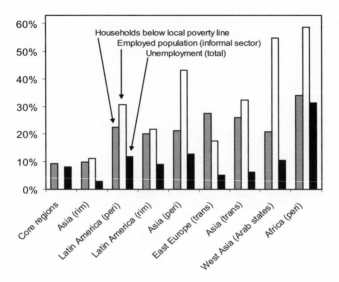

**Figure 19.8.** Urban social indicators (1998) (UN-Habitat 2001).

Brazil, Chile, and Mexico are examples of a second group of rim or semiperipheral cities. They developed industrial sectors (autos, assembly plants, synthetic fibers, and chemical fertilizers) with a strong transnational presence and exploited peripheral regions. In contrast to the first group, however, they still depend heavily on export of primary commodities as a source of international trade (Romero 2002). During the past two decades, under the influence of neoliberalism, their states promoted economic liberalization. Strongly impelled by international organizations, they weakened their ability to support domestic industries and invest in economic growth, infrastructure, and social security. The result has been recurrent economic and financial crises, increasing segregation, difficulty in financing social and urban infrastructural expenditure, and weakening of institutional settings (Romero 2002, and Figures 19.5, 19.8, and 19.9). High urbanization rates have caused increased segregation and sprawling informal settlements. Urban air pollution is a major problem in rim regions,[4] reflecting inadequate vehicle maintenance, poor fuel quality and road conditions, inefficient public transport, and long travel times (Lo 1994; UNEP 2002).

What is the carbon relevance of the differences in the two rim groups? Have they resulted in contrasting patterns of production, energy utilization, and $CO_2$ emissions? Although energy consumption and carbon emissions by rim regions have increased, they are still low compared with core regions (Figures 19.2 and 19.3). The Asian tigers' extended share of international manufacturing and export is linked to increased emissions, a feature of core regions during the first stage of industrialization (the heavy engineering phase). A relatively smaller percentage of carbon emissions by Latin American

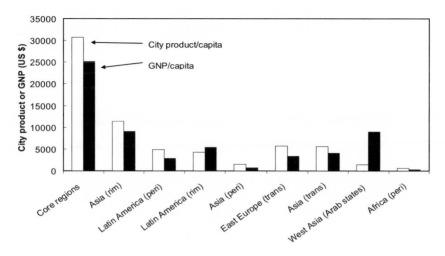

**Figure 19.9.** Economic indicators (1998) (UN-Habitat 2001)

rim areas is linked to changes in production structure (Luukanen and Kaivo-oja 2002). Mexican industry has become "lighter" and focused more on assembly industries. This is a controversial sector in development terms because it creates low-quality jobs and aggregate value, decreases government revenues, and lowers capacity to cope with environmental issues (Varady et al. 2002). Reduced carbon emissions in these countries could also be caused by deindustrialization related to recurrent economic, financial, and institutional crises.

One unsolved question is whether divergent development pathways might give Asian tigers increased capacity to adapt to changes in the carbon cycle. Although they are socially segregated, they enjoy relatively more wealth, technology, information, skills, infrastructure, access to resources, and management capabilities than their Latin American counterparts (Figures 19.5, 19.8, and 19.9). These advantages may provide a foundation for development policies with positive carbon-relevant consequences, including the ability both to promote mitigation and to cope with socioeconomic implications of changes in the carbon cycle.

The third development pathway taking place in the peripheries is represented by high primary commodity–export economies in Africa, Asia, and Latin America. These countries have recently experienced reductions in their already small share of world domestic product (Figure 19.1; Lo 1994). Their consumption of energy and $CO_2$ emissions has increased in recent years but is insignificant compared with that of the cores (Figures 19.2 and 19.3). Industrialized economies continue to be the main sources of trade and financing for these areas. Many of the peripheries maintain historical and structural characteristics that are vulnerable to the vagaries of international markets and are hit hard by recurrent weaknesses in the prices of their primary exports. For instance, in terms of 1990 U.S. dollars, the price index of all nonfuel primary commodities decreased from 118.1 to 89.3 from 1995 to 2000 (FAO 2001).

Falling prices and the embrace of economic liberalization in the peripheries both contribute to recurring financial crises, heavy external borrowing, inflation, stagnation of economic growth, high rates of urban and rural segregation, and weakening institutions. Besides hampering their ability to promote key carbon development policies, such factors might also hinder their ability to promote mitigation and cope with the social implications of changes in the carbon cycle.

Rim and peripheral regions share several aspects of land use changes caused by agriculture and forestry. Their forests have decreased and croplands have grown during recent decades (Richards 1990), leading to increased relative contributions to carbon emissions (Figure 19.7). Land use changes are caused by the interplay of several proximate and underlying factors. "Among the most frequently found factor combinations are the agriculture-wood-road connexus (mainly driven by economic, policy, institutional and cultural factors), the agricultural-wood connexus (mainly driven by technological factors), and population-driven agricultural expansion." There are regional variations, however, in causal connections, "with the agriculture-wood connexus featuring mainly Asian cases, and the road-agriculture connexus featuring mainly Latin American cases" (Geist and Lambin 2001, 95).

Rim and peripheral regions exhibit at least two carbon-relevant scenarios within the complex mosaic of agricultural and forest practices, namely intensification and diversification.[5] The first is common among commercial farmers, the second within the subsistence sector. Three sets of factors may drive intensification: land scarcity linked to population growth, markets, and intervention through projects sponsored by states, donors, or nongovernmental organizations (Lambin et al. 2001).

Diversification is often driven by interacting processes, which have adverse effects on agricultural-based livelihoods. This list of processes includes the embrace of liberalization, deregulation, and new regional trade agreements (Ellis 2000; Romero 2002). The impacts of intensification and diversification on the carbon cycle are diverse and potentially contradictory. For example, increased use of fertilizers may increase nitrous oxide emissions in most crops. But in rice the improved efficiency linked to intensification has led to decreased methane emissions (Milich 1999). Diversification strategies can lead to a range of results, including land degradation, abandoned land capital (terraces, irrigation), and regeneration through ecological succession (Romero 2002).

## Relevance of Common and Different Patterns of Development

The emergence of core, rim, and peripheral regions underlies diverse, carbon-relevant, regional pathways of development. Some carbon-relevant features are common to all regions, such as demographic concentration in urban areas and intensification of agriculture. These features can be managed through similar policy strategies, but only if they take into consideration the socioeconomic, political, and cultural configuration of each region.

Other key characteristics are differentiated in regional and historical terms. Core regions, for instance, have by far the highest share of international trade, production, energy consumption, and carbon emissions. Their responsibility in the design of mitigation and adaptation policies is consequently much higher. Some cores have become less intensive, while development patterns of some rims are leading to more intensive energy use and carbon emissions. Most of the peripheries are confronted with the basic issues of equity, governance, and the need for alternative models of economic growth.

Differences in development patterns demand a range of different carbon-related policies. Actions aimed at reducing $CO_2$ emissions are, for instance, most appropriate for cores. Policies aimed at reducing social disparities, increasing local financial capacities, and strengthening institutional settings are key for many rims and peripheries.

All regions are confronted by paradoxes. Industrialization, the key driver of development in core and rim regions, is at the heart of deep and irreversible transformations of the carbon cycle. Nevertheless, these industrialized regions are better endowed to promote development policies with positive consequences for the carbon cycle, such as urban and environmental planning. Might they also be better able than the peripheries to undertake mitigation and cope with the social implications of change? The latter bear a much smaller responsibility for altering the carbon cycle. In addition, they are confronted by pressing issues (social segregation, financial constraints), which may constrain their capacities for mitigation and adaptation.

## Notes

1. Ecological footprint, as a land-based surrogate measure of the population's demands on natural capital, "is the total area of productive land and water required continuously to produce all the resources consumed and to assimilate all the wastes produced, by a defined population" (Rees and Wackernagel 1996: 227–228).

2. Gruebel (1994), Lo (1994). As Figures 19.1, 19.2, and 19.3 show, core regions' gross domestic product (GDP) and share of energy consumption rose during recent years, even if there are differences in carbon dioxide emissions. Whereas North America's carbon dioxide emissions sharply increased, Asia's and Western Europe's tended to be stable.

3. The carbon relevance of intensive livestock in all regions is related to methanogenesis (Ayres et al. 1994: 130–131). Cattle from developed and developing countries produce 30 percent and 22.5 percent, respectively, of estimated annual $CH_4$ emissions by animals (Milich 1999: Figure 9).

4. Curitiba, Brazil, is an outstanding exception to this tendency (UNEP 2002).

5. Diversification in a livelihood context relates to the existence, at a point in time, of diverse income sources underpinned by circulation, migration, and off-farm employment (Ellis 2000).

# Literature Cited

Ayres, R., W. Schlesinger, and R. Socolow. 1994 . Human impacts on the carbon and nitrogen cycles. In *Industrial ecology and global change,* edited by R. Socolow, C. Andrews, F. Berkhout, and V. Thomas. Cambridge: Cambridge University Press.

Braudel, F. 1984. *Civilization and capitalism: 15th–18th century.* Vol. 3, *The perspective of the world.* New York: Harper and Row.

Ellis, F. 2000. *Rural livelihoods and diversity in developing countries.* New York: Oxford University Press.

Energy Information Administration. 2002. U.S. Department of Energy. http://www.eia.doe.gov.

FAO (Food and Agriculture Organization of the United Nations). 2001. *The state of food and agriculture 2001.* Rome. http://www.fao.org/ag.

Fernandez, L., and R. Carson, eds. 2002. *Both sides of the border: Transboundary environmental management issues facing Mexico and the United States.* Dordrecht, the Netherlands: Kluwer.

Galeano, E. 1973. *Open veins of Latin America: Five centuries of the pillage of a continent.* New York: Monthly Review Press.

Geist, H. J., and E. F. Lambin. 2001. *What drives tropical deforestation?* Report Number 4 of the LUCC (Land-Use and Land-Cover Change) Core Project of the International Geosphere Biosphere Programme (IGBP). Louvain-la-Neuve, Belgium: LUCC International Project Office. http://www.geo.ucl.ac.be/LUCC/lucc.html.

Gruebel, A. 1994. Industrialization as a historical phenomenon. In *Industrial ecology and global change,* edited by R. Socolow, C. Andrews, F. Berkhout, and V. Thomas. Cambridge: Cambridge University Press.

Houghton, R. A., and D. L. Skole. 1990. Carbon. In *The Earth as transformed by human action,* edited by B. L. Turner II, W. C. Clark, R. W. Kate, J. F. Richards, J. T. Mathews, and W. B. Meyer. Cambridge: Cambridge University Press.

Kelly, P. M., and W. N. Adger. 2000. Theory and practice in assessing vulnerability to climate change and facilitating adaptation. *Climate Change* 47:325–352.

Kennworthy, J., and F. Laube. 1996. Automobile dependence in cities: An international comparison of urban transport and land use patterns with implications for sustainability. *Environmental Impact Assessment Review* 16:279–308.

Lambin, E. F., B. L. Turner, H. J. Geist, S. B. Agbola, A. Angelsen, J. W. Bruce, O. T. Coomes, R. Dirzo, G. Fischer, C. Folke, P. S. George, K. Homewood, J. Imbernon, R. Leemans, X. B. Li, E. F. Moran, M. Mortimore, P. S. Ramakrishnan, J. F. Richards, H. Skanes, W. Steffen, G. D. Stone, U. Svedin, T. A. Veldkamp, C. Vogel, and J. C. Xu. 2001. The causes of land-use and land-cover change: Moving beyond the myths. *Global Environmental Change* 11:261–269.

Lo, F. 1994. The impacts of current global adjustment in shifting techno-economic paradigm on the world system. In *Mega city growth and the future,* edited by R. Fuchs, E. Brennan, J. Chamie, F. Lo, and J. I. Uitto. Tokyo: United Nations University Press.

Luukanen, J., and J. Kaivo-oja. 2002. Meaningful participation in global climate policy? Comparative analysis of the energy and $CO_2$ efficiency dynamics of key developing countries. *Global Environmental Change* 12:117–126.

Marston, S., P. Knox, and D. Liverman. 2002. *World regions in global context: Peoples, places and environments.* Upper Saddle River, NJ: Prentice-Hall.

Milich, L. 1999. The role of methane in global warming: Where might mitigation strategies be focused? *Global Environmental Change* 9:179–201.

Ostrom, E. 1990. *Governing the commons: The evolution of institutions for collective action.* Cambridge: Cambridge University Press.

Ponting, C. 1990. *A green history of the world.* London: Sinclair-Stevenson.

Rees, W., and M. Wackernagel. 1996. Urban ecological footprints: Why cities cannot be sustainable—and why they are a key to sustainability. *Environmental Impact Assessment Review* 16:223–248.

Richards, J. F. 1990. Land transformation. In *The Earth as transformed by human action,* edited by B. L. Turner II, W. C. Clark, R. W. Kate, J. F. Richards, J. T. Mathews, and W. B. Meyer. Cambridge: Cambridge University Press.

Romero, P. 2002. El peso de las políticas mexicanas en la 'sustentabilidad' de las recientes tendencias de desarrollo. In *La transición hacia el desarrollo sustentable: Perspectivas de América Latina y el Caribe,* edited by E. Leff, I. Pisanty, E. Ezcurra, and P. Romero. Mexico City: Instituto Nacional de Estadística (INE), Programa de las Naciones Unidas para el Medio Ambiente (PNUMA), and Universidad Autónoma Metropolitana–Unidad Xochimilco (UAM-X).

Turner, B. L., II, W. C. Clark, R. W. Kates, J. F. Richards, J. T. Mathews, and W. B. Meyer, eds. 1990. *The earth as transformed by human action: Global and regional changes in the biosphere over the past 300 years.* Cambridge: Cambridge University Press.

UNEP (United Nations Environmental Programme). 2002. *Global environmental outlook 3.* New York. http://www.unep.org/geo.

UN-Habitat (United Nations Human Settlements Programme). 2001. *GUO: Global Urban Indicators Database.* New York. http://www.unhabitat.org/programmes/guo/guo_indicators.asp

Varady, R., P. Romero-Lankao, and K. Hankins. 2002. Whither hazardous materials management in the US-Mexico Border? In *Both sides of the border: Transboundary environmental management issues facing Mexico and the United States,* edited by L. Fernandez and L. T. Carson. Dordrecht, the Netherlands: Kluwer.

Williams, R. 1994. Roles for biomass energy in sustainable development. In *Industrial ecology and global change,* edited by R. Socolow, C. Andrews, F. Berkhout, and V. Thomas. Cambridge: Cambridge University Press.

# 20

# Social Change and CO₂ Stabilization: Moving away from Carbon Cultures

## Louis Lebel

To achieve $CO_2$ stabilization in the atmosphere within the next century will require profound social changes in a world where most societies have already become, or aspire to be, carbon cultures. "Culture" is understood as a critical set of institutions that guide the ideas by which a society lives. Culture penetrates market, political, and other institutions and is far more than a "grab-bag" for all those factors that are not about economics or politics. In carbon cultures, much behavior and enterprise depends on carbon-rich fossil fuels.

Human activities affect the carbon cycle through a number of interacting pathways. Understanding is most advanced for the most direct factors treated one at a time and at a single scale. Especially important are the processes driving changes in land use toward lower carbon stocks and the incentives for continuing use of carbon-rich fuels. The underlying social structures and processes driving changes in the carbon cycle, however, are complex (e.g., Sayathe et al. 2001). Here, they are grouped into consumption, social organization, knowledge and values, technology, and institutions (Figure 20.1). This chapter considers each of these groups in turn.

## Consumption

Where and how people work, play, and move is critical for patterns in energy and land use. Although people do have some choice about how to meet particular needs and wants, social structures and processes greatly constrain these. Indeed some of these processes (historically) helped define those needs and wants. Consider the role of advertising, the size of marketing budgets, television programming, and the media in defining a desirable or normal "household." Corporations have a vested interest in widening the aspirational gap—the distance between what people have and what they feel they need. Ironically, the growth in the range of products does not necessarily mean

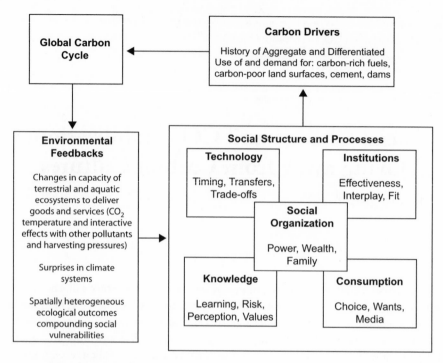

**Figure 20.1.** A simple conceptual framework for various social structures and processes influencing the global carbon cycle

more choice. Indeed, the choice to meet want without a purchase is being hidden from view. Control is not absolute, but the behavior of consumers is strongly channeled.

Everyday activities are, in their aggregate, extremely important for the future of the carbon cycle, because consumer goods and services, whether sports utility vehicles, household appliances, seafood delicacies, holiday homes, or exotic holidays in third world countries, often imply large direct and indirect energy use (Schipper 1997; Weber and Perrels 2000; Wier et al. 2001). Because much of the energy comes from burning fossil fuels, there is a strong relationship with carbon emissions. Although shifts from solid fuel (biomass and coal) to liquids (kerosene, fuel oil) to gas (LPG, natural gas, biogas) suggest a decarbonization of residential fuel use patterns, increases in overall consumption may outweigh these positive trends. For example, a detailed study of household energy consumption over 50 years in the Netherlands found no significant trend toward lower energy intensities or dematerialization in consumption patterns (Vinger and Blok 2000).

The skill of advertisers in driving consumption should not be underestimated. Often products that can no longer be functionally improved in a meaningful sense can still be differentiated by turning them into cultural symbols—or ways of making the purchases associate with things like freedom, sex, and feeling good about oneself (Sachs 1999). A watch does not just tell the time, but also identifies its wearer as a driver or an adventurer. Cars are full of gadgets that have nothing to do with driving performance. Of course, this satisfying of aspiration must be an empty promise so that the production-consumption machine can continue to move. A month later, a new model and set of symbols are launched to create a new round of demand. There is no saturation of demand or limit to expansion when commodities become cultural symbols (Sachs 1999).

Choices are not completely malleable by the media. Needs and wants also arise from the characteristics of a place. Climate, topography, and the accessibility and cost of land, energy, and other materials also help determine the range of choices available. In satisfying needs and wants, people often seek to maximize convenience and comfort as much as to minimize costs (Sayathe et al. 2001). Finally, various institutions, including property rights, taxation, and access to credit cards, provide additional incentives or disincentives for certain carbon-emitting behaviors.

The trend toward increased mobility is one of the areas that shows the least sign of slowing in growth, even in mature industrialized economies (Schipper 1997). Almost everywhere it is faster, further, and more. There are also big differences among cultures. Americans emit three to four times as much carbon per capita from personal vehicle use as do Europeans, in part because of greater driving distances, but also because U.S. cars use 25–30 percent more fuel (Schipper 1997). Historically, there have been major shifts in dominant modes of transport and infrastructure, from canals and rail to roads and air. What the future of transport in 100 years will be like is hard to predict and remains one of the larger sources of uncertainty for future carbon emissions.

Consumption of meat has a major impact on land use, because it is much less efficient than directly eating grain (Heilig 1995). Some trends in the developing world are staggering. Between 1990 and 1996 the number of McDonald's restaurants in Asia and Latin America quadrupled, and meat consumption there tripled over the past 25 years. Religious and cultural values that prohibit the use of certain meats, like pork or beef, ameliorate this trend. McDonald's restaurants in India, for example, serve many vegetarian meals. Demand for other export-oriented crops like coffee, cocoa, cotton, bananas, tea, and shrimp have also had major influences on forest cover in developing countries (Tucker 2002).

Countercurrents are present, but they cannot buy the airwaves, so their influence in defining culture is waning. Traditionally, many religious and other teachings have argued that an excess of material things is a distraction, wastes energy, reduces capacity to control one's life, or causes suffering. Buddhism often relates ideas of simplicity and sufficiency.

Underconsumption is also a problem. Per capita emissions in developing regions are very much lower than in the developed world (Romero Lankao, Chapter 19, this volume). Any just approach to reducing carbon emissions to close to zero will have to allow for major increases in some parts of the world over the next several decades. In these areas, additional consumption is important to improving well-being. For the more affluent parts of society, however, reducing over- and misconsumption is critical to local and global sustainability.

Globalization of trade and the liberalization of investment over the past several decades have resulted in many longer and more complex commodity chains (Conca 2002). Direct environmental feedback signals are easily distorted and lost (Princen 2002). Consumers now have almost no hope of reconstructing the environmental consequences of their purchase and use decisions. The challenge is acute for $CO_2$—a common substance that is only considered a pollutant in the aggregate and global sense. Input-output tables can capture most of the goods consumed in making a final product but still neglect services and goods obtained for free from ecosystems.

Finally, changing consumption of energy, goods, and services by firms, government agencies, and the military is at least as important as modifying household consumption patterns. Their purchases are large and may be distorted by special arrangements, subsidies, and incentives. Consumption by organizations is not independent of household consumption but may provide greater opportunities for directly reducing $CO_2$ emissions.

## Social Organization

Alternative patterns of consumption arise as a consequence of the choices provided (or constrained) by differences in wealth, entitlements, and power, as well as differences in things like family size, life-cycle stage, and geographic features. Wealth and power help determine allocation of buildings, equipment, and places to live. $CO_2$ emissions are positively correlated with household expenditure (e.g., Wier et al. 2001).

At the national level, a number of analyses suggest that, although emission intensities (per unit of gross domestic product) may peak at lower levels for countries at a given income level at a later date, even the lowest reasonable peaks may still be higher than many of the poorest countries can expect to reach in the next couple of decades (Dietz and Rosa 1997; Roberts and Grimes 1997). Moreover, this relationship may not continue for the poorest countries, especially if heavy manufacturing and other polluting industries tend to relocate to them (Bai 2002). In developed countries, education and employment status show little relationship with energy consumption after adjustment for wealth (Wier et al. 2001). The main reason some people consume so much whereas others consume so little is income, but around this basic relationship there is much variation.

The lifetime carbon emissions of a wealthy American differ by a huge amount from those of a poor slum dweller in Dhaka. The American has a correspondingly greater capacity to influence the aggregate outcome of world emission growth than does the resident of Bangladesh, even though Bangladesh is much more vulnerable to climate change. This asymmetry in interests, economic status, and capacity to change the global environment often coincide. They are, in part, a product of past distribution of power.

Family size and organization are important too. Per person, smaller families use more energy for housing and travel than large households. People living in houses use more energy than those living in apartments (Wier et al. 2001). These effects, however, are much smaller than those resulting from location, through the influence of climate on energy for heating and air-conditioning. Urbanization, if accompanied by good public transport infrastructure, may reduce carbon emissions through reduced use of personal vehicles and efficiencies in heating and cooling apartment buildings.

Falling rates of fertility are now producing patterns of rapid aging in much of the world. The effects of this pattern on energy use will depend on factors like whether or not older people are cared for by their children (as in many Asian cultures) or whether they remain economically independent and choose to travel overseas (as in some western countries).

Corporate organization is also likely to be very important for $CO_2$ emissions, above and beyond the effects of distancing emissions from consumer decisions. Vertical integration and special arrangements between states and energy producers can make emissions accounting difficult. In Vietnam and China, state enterprises have been very inefficient and polluting producers and users of energy.

Finally, the integration of carbon into development strategies of nations, regions, and even the world economic system will produce winners and losers. The winners will be mostly those that control the development of institutions at each scale, those with power and interests to pursue or protect.

## Knowledge and Values

Control of the production of knowledge (and the definition of legitimate sources of knowledge) is crucial to the development pathways a society follows. Engagement in modernization and globalization has greatly weakened the processes maintaining and building local and traditional knowledge. It has also undermined local institutions. Common property regimes for forest lands, for example, have been effective in maintaining tree cover in shifting cultivation systems for centuries, whereas modern practices that favor permanent agriculture and separate forest and conservation areas have failed to maintain forest cover and carbon stocks.

Perceptions of risk and vulnerabilities posed by elevated atmospheric $CO_2$ depend on the quality of assessments as well as how findings are represented. The Intergovernmental Panel on Climate Change (IPCC) process has been a largely successful attempt at independent assessment in an arena where policy stakes are very high. Attitudes on the environment may not, however, be as important in practice as other cultural values in which every day actions are embedded. Cultural values such as sharing, reciprocity, respect, humility, or patience can play a major role in moderating consumption and regulating land use. Religions like Buddhism, which place emphasis on breaking attachments with material things and people and controlling desires, argue against unnecessary consumption with the philosophy that material things do not, in themselves, produce long-term happiness or peace. This perspective is a counterpoint to today's emphasis on accumulation.

Culture fashions science and policy but is also changed by science. The success of industrialization, intensive agriculture, information technology, and medical science and genetics has given societies confidence in their ability to understand and control nature and how society interacts with it. This confidence underlies not only the idea that it is possible to integrate carbon management into development at local scales, but also the global institutions being formed to address global environmental change problems through "planetary management." The social construction of the $CO_2$ stabilization problem has been largely driven by the concerns of wealthy northern countries, which, though mostly to blame for current $CO_2$ levels, are calling for wider participation in "solutions" (Redclift and Sage 1998).

In contrast to this perspective, a growing body of researchers, practitioners, and policy makers have become more sceptical, or at least more humble, about the capacity of societies to forecast and manage complex systems (Gunderson and Holling 2002). A skeptical approach and a broad appreciation of uncertainties (and the prospects for reducing them) place an emphasis on learning from experience and safe-to-fail experiments in policy, land management, and energy technologies. Recognition of potential thresholds implies a commitment to avoid pushing systems to their limits (i.e., dangerous interference with the climate system).

Finally, crises may be essential for people to change their worldviews, even when other knowledge suggests that a change should have taken place long ago. Risks are abstract and not well understood; disasters and crises catch everybody's attention.

## Technology

Changes in energy technologies can greatly reduce emissions (Caldeira et al., Chapter 5, this volume). Achieving $CO_2$ stabilization will require massive changes in energy systems. The issue is one of timing, transfers, and trade-offs, each of which depends greatly on environmental politics and institutional arrangements between sectors and nations. History suggests that modal changes in energy systems take 50–100 years. Current trends in investments in energy research and development in both private and public sectors are not strongly oriented toward reducing carbon emissions, suggesting that a major shift in public policy is needed.

There is practically no limit to how much fossil fuels humans could transfer to the atmosphere: although conventional oil and gas is limited (Sabine et al., Chapter 2, this volume), the amount stored as coal and unconventional reserves is huge and will not "run out" during the 21st century (Edmonds et al. 2000). Emissions of CO$_2$ per unit of energy consumed are falling, but this trend toward decarbonization must be balanced against the continuing absolute increase in world consumption of energy (Nakicenovic 1997).

Nevertheless, a common expectation is that the elemental basis of the world economy will shift from carbon to hydrogen (Spearot 2000). There are important constraints. Some of the alternative sources of energy, including nuclear power and hydropower, have other environmental or health risks, which the public may resist. The same is also true for the wide range of carbon capture and storage technologies being envisaged now and likely to be technically feasible in the coming decades. Beyond health risks, other factors to consider include costs of storage, institutional issues related to monitoring for leakages, and risks of breakdown in the event of weak maintenance or major social disturbances like wars.

Changes to land management practices, such as reduced tillage and better fire management, can also reduce emissions. On the other hand, climate change feedbacks on ecological systems may result in the weakening of current sinks or the mobilization of new sources (Gruber et al., Chapter 3, this volume).

In developing countries where industrialization is just beginning, there are tremendous opportunities for making growth clean (Angel et al. 2000). Clean, renewable energy technologies could be accelerated in developing countries. One good mechanism for this involves encouraging developed countries to provide assistance, with the inducement that the emissions reductions in the developing countries that receive assistance can be used as credit toward national greenhouse gas emission reduction targets. This is the idea of joint implementation or activities implemented jointly under the United Nations Framework Convention on Climate Change (UNFCCC). Vertical transfers, where the technology remains with the investing company, can in some cases be desirable. This is especially true, for example, with advanced technologies, such as photovoltaics, where domestic producers are not ready to compete (Forsyth 1999). In many other cases, horizontal transfers involving local embedding and training of users and manufacturers will be much more desirable. Otherwise, there is a risk that industrialized-country exports subsidized by carbon credits from industrialized countries will decrease the competitiveness of developing-country producers, even in cases where local producers use technologies that are more appropriate for the settings. In the worst case, the transfers could result in no net increase in renewables, but an increased share of northern technologies in southern markets, leading to greater dependencies (Forsyth 1999). Local, national, and international interventions are required to ensure that private investments actually work toward carbon management goals.

## Institutions

Effective governance at a global scale that would help accelerate progress toward CO$_2$ stabilization still appears remote. Society is in the early stages of experimenting with international environmental institutions. So far, these institutions have had very minor effects, except in cases where change also made good business sense. For the Montreal Protocol on ozone, alternative technologies were available and only a few big chemical manufacturers needed to be brought on line. The situation for greenhouse gas emissions is much more complex because of the wide array of activities that result in emissions.

The Kyoto Protocol was significant in its attempt to include both emission reduction and sink enhancement. The details of accounting procedures and definitions of afforestation, reforestation, and deforestation under Article 3.3 can have a large effect on net carbon sequestration under the Protocol (Yamagata and Alexandrov 1999). It is easy to create perverse subsidies for forestry (Sanz et al., Chapter 24, this volume). An even greater challenge is striking the right balance between emission reduction and sink enhancement. If Annex I countries[1] are allowed to claim too much in carbon offsets, sink enhancement could act as a disincentive to emissions reductions. On the other hand, a too-small emphasis on sinks would discourage efforts to reevaluate forests, something likely to be beneficial beyond the potential for carbon sequestration potential. Yamagata and Alexandrov (2001) estimated, using the IMAGE land-use change model, that Annex I countries could claim a net carbon offset as high as 0.2 petagrams of carbon per year (Pg C $y^{-1}$) through afforestation, reforestation, and deforestation (ARD) activities. In practice, however, political and economic factors will be important determining in whether projects designed to enhance sinks will actually help store carbon and improve the sustainability of local livelihoods (Tomich et al. 2002).

The Kyoto Protocol includes several mechanisms designed to allow developing countries to help reduce emissions, while acknowledging that the basic responsibility at this time lies with the industrialized nations. All of these mechanisms, the Clean Development Mechanism, Joint Implementation, and Emissions Trading, pose many challenges for integrating international agreements with national policy, especially since the bulk of the commitments will have to be implemented by the private sector. Not surprisingly, the bargaining position adopted by country delegations has been strongly influenced by corporate sector interests. Politics will not disappear from future negotiations concerning carbon. Even less clear is how the interplay between the regime and other international and national institutions will unfold. Bottom-up approaches with a focus on things like environmental health, urban air pollution, and sustainable forest management, might, in the long run, be more effective than global efforts to control carbon emissions. Transnational networks of city bureaucrats, industry leaders, and civil society groups have begun launching voluntary programs with these goals, though results are still mixed (Gardiner and Jacobson 2002). Institutional arrangements not directly concerned with carbon or even the environment, such as those related to trade in agricultural commodities or foreign investment, may be even more important for carbon futures through their effects on consumption-production chains.

# Carbon Futures

Numerous technological and institutional trends have the potential to begin reducing the rates at which carbon dioxide is added to the atmosphere. The timing and speed at which these transformations are undertaken will determine how high $CO_2$ concentrations rise before stabilization. With higher concentrations, there is increased likelihood of undesirable surprises from the climate and ecological systems. Although scientists' understanding of the Earth system has increased impressively over the past few decades, our capacity to "manage" the planet is easily overstated. Uncertainties abound, and many surprises could be in store, from misunderstood feedbacks and unexpected interactions in a carbon cycle operating outside its historical range. The timeliness with which societies respond to these challenges depends a lot on their assessment of risks, the level of shared interests, and the power of various agents to effect changes in behavior, institutions, and technology.

New scenarios should explicitly address the topics discussed here. We need to consider how diverse institutions, including nations, and not just markets, interact with energy and land use. These scenarios should also address how societies learn (or do not learn) from environmental feedbacks and whether their current strategies for development are maintaining or destroying their capacity to adapt to surprises. The SRES scenarios prepared by the IPCC Working Group III (IPCC 2000) represent a substantial improvement over the original IPCC set, in terms of assumptions about carbon intensities of energy supply, sulfur emissions, and treatment of relative rates of economic growth in developing and developed countries. They still fall far short, however, of capturing key uncertainties in how societies, ecosystems, and the environment may interact over time as part of the global carbon system. For example, the scenarios do not provide much scope for exploring how development strategies may jointly influence emissions and sequestration. They intentionally neglect consideration of surprises or major discontinuities (IPCC 2000). Finally, whereas many economic processes are well represented, cultural and other social structures and processes are hardly mentioned.

Careful consideration of carbon also demands a broad discussion of regional energy security and consequently international relations (e.g., Stares 2000). For example, continuing rapid economic growth in China will require huge increases in energy imports, development of new domestic energy sources, and substantial efforts at maintaining and improving urban air quality (Gao 2000). China will likely displace the United States as the world's largest $CO_2$ emitter sometime in the 2020s, at which time per capita emissions will also exceed the world average.

The structure of society, including economic and political relations, will continue to drive growth in consumption of goods and services for at least several decades. It will not be easy to decouple this growth from its carbon consequences, but this is a prerequisite for $CO_2$ stabilization. It will be even more difficult to build the institutions and incentives for behavioral change in a manner that is just and ecologically sustainable. A move away from carbon cultures, not just through individual purchasing decisions but also through informal and formal collective action, will be essential for carbon management to succeed and global sustainability.

## Note

For the Kyoto Protocol, Annex I countries are those that are economically most developed, plus the countries of Eastern Europe and the European parts of the former Soviet Union. The list consists of Australia, Austria, Belarus, Belgium, Bulgaria, Canada, Croatia, Czech Republic, Denmark, Estonia, Finland, France, Germany, Greece, Hungary, Iceland, Ireland, Italy, Japan, Latvia, Liechtenstein, Lithuania, Monaco, Netherlands, New Zealand, Norway, Poland, Portugal, Romania, Russian Federation, Slovakia, Slovenia, Spain, Sweden, Switzerland, Turkey, Ukraine, United Kingdom of Great Britain and Northern Ireland, and United States of America.

## Literature Cited

Angel, D. P., M. T. Rock, and T. Feridhanusetyawan. 2000. Toward clean shared growth in Asia. Pp. 11–37 in *Asia's clean revolution industry, growth and the environment,* edited by D. P. Angel and M. T. Rock. Sheffield, UK: Greenleaf.

Bai, X. 2002. Industrial relocation in Asia: A sound environmental management strategy? *Environment* 44:8–21.

Conca, K. 2002. Consumption and environment in a global economy. Pp. 133–153 in *Confronting consumption,* edited by T. Princen, M. Maniates, and K. Conca. London: MIT Press.

Dietz, T., and E. A. Rosa. 1997. Effects of population and affluence on $CO_2$ emissions. *Proceedings of the National Academy of Sciences* 94:175–179.

Edmonds, J. A., J. F. Clarke, and J. J. Dooley. 2000. Carbon management: The challenge. Pp. 7–32 in *Carbon management: Implications for R&D in the chemical sciences and technology: A workshop report to the Chemical Sciences Roundtable.* Washington, DC: National Academies Press.

Forsyth, T. 1999. *International investment and climate change: Energy technologies for developing countries.* London: Earthscan in association with the Royal Institute of International Affairs.

Gao, S. 2000. China. Pp. 43–58 in *Rethinking energy security in East Asia,* edited by P. B. Stares. Tokyo: Japan Centre for International Exchange.

Gardiner, D., and L. Jacobson. 2002. Will voluntary programs be sufficient to reduce US greenhouse gas emissions? An analysis of the Bush administration's global climate change initiative. *Environment* 44 (8): 25–33.

Gunderson, L. H., and C. S. Holling. 2002. *Panarchy.* Washington, DC: Island Press.

Heilig, G. K. 1995. *Lifestyle and global land-use change: Data and theses.* Working Paper 95-91. Laxenburg, Austria: International Institute for Applied Systems Analysis.

IPCC (Intergovernmental Panel on Climate Change). 2000. *A special report of Working Group III of the Intergovernmental Panel on Climate Change: Emissions scenarios.* Geneva.

Nakicenovic, N. 1997. Freeing energy from carbon. Pp. 74–88 In *Technological trajecto-*

*ries and the human environment,* edited by J. H. Ausubel and H. D. Langford. Washington, DC: National Academies Press.

Princen, T. 2002. Distancing: Consumption and the severing of feedback. Pp. 103–131 in *Confronting consumption,* edited by T. Princen, M. Maniates, and K. Conca. London: MIT Press.

Redclift, M., and C. Sage. 1998. Global environmental change and global inequality: North/South perspectives. *International Sociology* 13:499–516.

Roberts, J. T., and P. E. Grimes. 1997. Carbon intensity and economic development 1962–91: A brief exploration of the environmental Kuznets curve. *World Development* 25:191–198.

Sachs, W. 1999. *Planet dialectics: Explorations in environment and development.* London: Zed Books.

Sathaye, J., D. Bouille, D. Biswas, P. Crabbe, L. Geng, D. Hall, H. Imura, A. Jaffe, L. Michealis, G. Pezko, A. Verbruggen, E. Worrell, and F. Yamba. 2001. Barriers, opportunities, and market potential of technologies and practices. Pp. 345–398 in *Climate change 2001: Mitigation (Contribution of Working Group III to the third assessment report of the Intergovernmental Panel on Climate Change),* edited by B. Metz, O. Davidson, R. Swart, and J. Pan. Cambridge: Cambridge University Press.

Schipper, L. T. 1997. Carbon emissions from travel in the OECD countries. Pp. 50–62 in *Environmentally significant consumption,* edited by P. Stern, T. Dietz, V. Ruttan, R. Socolow, J. Sweeney. Washington, DC: National Academies Press.

Spearot, J. A. 2000. Advanced engine and fuel systems development for minimizing carbon dioxide generation. Pp. 93–110 in *Carbon management: Implications for R&D in the chemical sciences and technology: A workshop report to the Chemical Sciences Roundtable.* Washington, DC: National Academies Press.

Stares, P. B., ed. 2000. *Rethinking energy security in East Asia.* Tokyo: Japan Centre for International Exchange.

Tomich, T. P., H. de Foresta, P. Dennis, Q. Ketterings, D. Murdiyarso, C. Palm, F. Stolle, Suyanto, and M. van Noordwijk. 2002. Carbon offsets for conservation and development in Indonesia. *American Journal of Alternative Agriculture* 17:125–137.

Tucker, R. 2002. Environmentally damaging consumption: The impact of American markets in tropical ecosystems in the twentieth century. Pp. 177–195 in *Confronting consumption,* edited by T. Princen, M. Maniates, and K. Conca. London: MIT Press.

Vinger, K., and K. Blok. 2000. Long-term trends in direct and indirect household energy intensities: a factor in de-materialization. *Energy Policy* 28:713–727.

Weber, C., and A. Perrels. 2000. Modelling lifestyle effects on energy demand and related emissions. *Energy Policy* 28:549–566.

Wier, M., M. Lanzen, J. Munksgaard, and S. Smed. 2001. Effects of household consumption patterns on CO$_2$ requirements. *Economic Systems Research* 13:260–273.

Yamagata, Y., and G. Alexandrov. 1999. Political implications of defining carbon sinks under the Kyoto Protocol. *World Resource Review* 11 (3): 346–359.

———. 2001. Would forestation alleviate the burden of emission reduction? An assessment of the future carbon sink from ARD activities. *Climate Policy* 1:27–40.

# 21

# Carbon Transport through International Commerce

Jeff Tschirley and Géraud Servin

In 1997 national governments agreed to adopt the Kyoto Protocol to the United Nations Framework Convention on Climate Change as a legally binding framework to reduce emissions of greenhouse gases (GHGs) (Sanz et al., Chapter 24, this volume). Although a number of GHGs are relevant, carbon dioxide is of overriding interest in part because of its sequestration in living organic matter that can be managed to regulate eventual fluxes to the atmosphere. The Kyoto Protocol has a predominant focus on the role of forests in carbon sequestration, but there is broad agreement in the science community on the need to understand the entire carbon cycle (Falkowski et al. 2000)—terrestrial, ocean, and atmosphere—and use these components in complementary and synergistic ways.

A comprehensive and systematic approach to carbon accounting would provide countries with the accuracy and consistency they need to monitor and manage their carbon stocks in the context of national, regional, and global efforts to mitigate and adapt to the effects of climate change. In addition to the focus on forestry, such an approach would require satellite and *in situ* observational data for land uses such as cropland and grazing land, wetlands, and coastal areas so that such information can be reflected in national, regional, and global carbon budgets (Cihlar 2002; Cihlar and Denning 2002; Cihlar et al. 2003).

With the exception of the service sector, all international commercial trade involves a transfer of carbon. The value of all global exports in 2000 was US$6,186 billion of which mining was US$813 billion, manufacturing US$4,613 billion, and agriculture US$558 billion (WTO 2003). This chapter focuses on transfers of carbon that take place between regions in the form of imports and exports of agricultural and forest products such as cereals and wood pulp (see Appendix 21.1). Although the economic dimension is significant, it is not addressed in this chapter. Furthermore, this chapter does not address the question of how credits (or debits) related to trade in products with carbon content could be applied in carbon accounting frameworks.

The intercomparison of a number of global ecosystem models indicates that approximately 55 petagrams of carbon (PgC) is generated each year in the form of net primary production (NPP) in terrestrial ecosystems (IGBP 1998). This production takes place on a global land area of 134 million square kilometers ($km^2$) (FAO 2003) consisting of forests and woodland, croplands, grasslands, and other ecosystems. Approximately 15 million $km^2$ is arable land in use at various levels of cropping intensity, ranging from rain-fed systems with low levels of external inputs to irrigated systems with high inputs.

An estimated 3.5 PgC are produced annually just by agricultural crops, accounting for about 7 percent of the total NPP in terrestrial ecosystems. Of this amount, approximately one-third is the harvested product. About half of the harvested product is used for direct human consumption and half for animal consumption (Goudriaan et al. 2001).

Only a few investigations have used global agriculture datasets to assess production relative to transfers of crop material across regional or national boundaries:

- Land use and cropland data were used to derive estimated long-term changes in global cropland areas that were then validated against a satellite-derived potential vegetation data set (Ramankutty and Foley 1999).
- Crop production data for 16 major food products were used to estimate horizontal carbon transfers (through trade) in agriculture. The data were converted to carbon equivalents to arrive at an estimated "displacement" in excess of 1 PgC per year ($y^{-1}$) (Ciais et al. 2001).
- Agricultural production statistics for 1991 were used to estimate the NPP of arable land by using fixed crop-specific coefficients that included dry matter content, harvest index, root production, and carbon and nitrogen content (Goudriaan et al. 2001).

The most relevant data on international trade of forest products appear in Michie and Kin (1999), which analyzes global trade flows of major forest products in six regions between 1983 and 1996, based on the UN ComTrade database. This comprehensive effort includes trade flow summaries and illustrated regional charts, maps, and spreadsheets.

## Data Availability and Quality

With the partial exception of forest products, little attention has been given to analyzing the quantities of carbon that flow annually between regions in the form of exports and imports. Potentially, a large number of items could be included. The UN Standard Industrial Trade Classification (SITC) scheme (United Nations 1986) uses a global framework with 10 categories for national export and import data (Table 21.1).

Each category contains numerous subcategories, most of them consisting of products that contain carbon. For example, category 0 contains 9 subcategories: meats, dairy products, fish, cereals, vegetables and fruit, sugars, harvested beverages, animal feeds, and miscellaneous edible products. These are further broken into 344 subgroups, with labels such as hides and skins, oil seeds and fruits, rubber, pulp and waste

**Table 21.1.** SITC framework for national export and import

| No. | Category |
| --- | --- |
| 0 | Food and live animals |
| 1 | Beverages and tobacco |
| 2 | Crude materials, inedible, except fuels |
| 3 | Mineral fuels, lubricants and related materials |
| 4 | Animal and vegetable oils, fats and waxes |
| 5 | Chemical and related products |
| 6 | Manufactured good classified chiefly by material |
| 7 | Machinery and transport equipment |
| 8 | Miscellaneous manufactured articles |
| 9 | Commodities and transactions not classified elsewhere |

*Source:* United Nations 1986

**Table 21.2.** Selected agriculture and forest product trade data availability

| Plant material | Woody material | Animals |
| --- | --- | --- |
| Cereals | Pulp | Live animals |
| Roots and tubers | Cellulosic material | Meat products |
| Vegetables | Paper | Milk products |
| Live trees | Paperboard | Fish and seafood |
| Coffee, tea, spices | Roundwood | Hides and skins |
| Oils, seeds, gums | Fibers and filaments | |
| Cotton | Wadding and weaving | |

*Source:* United Nations Statistical Office 2003

paper, fertilizers, and various kinds of agricultural machinery. Crude materials, fuels, chemicals, machinery, and manufactured goods are other relevant categories.

Carbon conversion coefficients have not been developed for most of the products in the SITC framework. Although in many cases the carbon content may not be sufficiently great to justify their inclusion in an accounting framework, more analysis is needed to ascertain the potential magnitude of the carbon content, the significance of trade in these items, and the viability of developing coefficients.

There is a wide variety of agricultural and forest products for which trade data are publicly available, measured in both quantity and value (Table 21.2). Of primary interest to the carbon science community are the volumes of carbon-based products that are transported through international commerce and their spatial and temporal distribution.

The primary sources for data related to commerce in agriculture and forestry are from the Food and Agriculture Organization of the United Nations (FAO 2003) and the UN Statistical Division database for commercial trade (United Nations Statistical Office 2003). The ComTrade database has several advantages. Foremost is the ability to quickly extract, manipulate, and display trade data in a variety of different ways. The agriculture and forest content is built primarily upon statistical data collected and managed by FAO. It is further supplemented with data from the Organisation for Economic Co-operation and Development (OECD) and selected other international sources, including the World Trade Organization (WTO) and the European Union (EU).

The most complete and reliable data at the global and continental levels are available only for the period since 1989. Longer time series could be helpful in projecting regional and global trends with better accuracy and temporal composition for individual crops and commodities. The ComTrade database does not aggregate trade quantities to higher levels, such as the two-digit group (represented by cereals, e.g., code 10), because of possible error arising from the use of different measurement units at the basic reporting level. Thus, estimates in this chapter have been derived manually from the four-digit groups (which includes wheat and rice, e.g., codes 1001 and 1006).

## Considerations in Data Analysis

Conversion rates for estimating carbon content of biomass rely primarily on the work of the Intergovernmental Panel on Climate Change (IPCC) in its assessment reports. A conversion factor of 45 percent was used to estimate the net dry amount of carbon in an agricultural crop that was made up of carbon. The figure of 48 percent may be more representative for the cereal group but would make little difference in the results (see Appendix 21.2). For forests and wood products, a conversion factor of 50 percent was used.

The accuracy of the carbon conversion factors is also assumed to be independent of production system. Specifically, countries that use high levels of external inputs such as fertilizers, pesticides, and conventional energy tend to produce crops with the same "embodied" carbon content as lower-income countries that use much lower inputs.

The embodied carbon contained in the cereal and wood products reflects only a portion of the total carbon consumed in producing, harvesting, and transporting those products. This additional carbon, consisting of fossil fuels arising from indirect sources, is (or should be) captured in National Greenhouse Gas Inventories (NGHGI). Although this additional carbon is probably not significant in calculating flows at the regional and global levels, the same may not necessarily be true at the national level, where the performance of individual countries in managing and accounting for their carbon stock changes takes on added economic and political importance (see Appendix 21.3).

The UN ComTrade database is country-based, and data are available only at the national level. For this study the country statistics were aggregated to produce estimates for six regions: Africa, Asia, Europe, North America, Oceania, and Other America (see Appendix 21.4 for region definitions). The aggregation is done by breaking down the

**Table 21.3.** Carbon exports and imports by continent for 2000 (PgC)

| Commodity group | Export (PgC) | Import (PgC) |
|---|---|---|
| *Cereals (10)* | | |
| Africa | 0.000 | 0.011 |
| Asia | 0.014 | 0.042 |
| Europe | 0.029 | 0.024 |
| North America | 0.049 | 0.003 |
| Oceania | 0.001 | 0.000 |
| Other America | 0.011 | 0.018 |
| Total cereals (10) | 0.105 | 0.098 |
| *Paper products (47,48,49)* | | |
| Africa | 0.001 | 0.001 |
| Asia | 0.009 | 0.023 |
| Europe | 0.046 | 0.044 |
| North America | 0.023 | 0.016 |
| Oceania | 0.001 | 0.001 |
| Other America | 0.007 | 0.034 |
| Total paper products (47, 48,49) | 0.087 | 0.118 |
| *Wood products (44,45,46)* | | |
| Africa | 0.060 | 0.011 |
| Asia | 0.011 | 0.039 |
| Europe | 0.071 | 0.055 |
| North America | 0.025 | 0.019 |
| Oceania | 0.004 | 0.000 |
| Other America | 0.006 | 0.004 |
| Total wood products (44,45,46) | 0.178 | 0.129 |
| *Grand total* | *0.370* | *0.345* |

region to its underlying countries and then summing the returned results. This approach means that the total trade from a region to a partner may be less than the total flow, as intraregional exchanges are included. For example, exports and imports from North America to North America mostly reflect trade between the United States and Canada.

Discrepancies exist between fluxes calculated from exports and those calculated from imports. At the global level the differences for all three commodity groups range from moderate (7 percent) for the cereals to large (26 and 38 percent respectively) for the paper and wood products (Table 21.3). At the regional level discrepancies reach several orders of magnitude. For example, the export of wood products from Africa to Asia is equivalent to 0.043 PgC whereas reported wood imports to Asia from Africa are only 0.002 PgC (Appendix 21.5). The reasons for such inconsistencies are not yet fully

**Table 21.4.** Volume and value of cereals, paper, and wood products exports by continent for 2000

| Commodity group | Net weight (billion tons) | % | Trade value (billion US$) | % | Unit price (US$/kg) |
|---|---|---|---|---|---|
| *Cereals (10)* | | | | | |
| Africa | 0.9 | 0 | 0.1 | 0 | 0.137 |
| Asia | 30.1 | 13 | 4.8 | 16 | 0.160 |
| Europe | 65.4 | 28 | 8.7 | 29 | 0.133 |
| North America | 108.6 | 47 | 12.6 | 43 | 0.116 |
| Oceania | 3.2 | 1 | 0.4 | 1 | 0.140 |
| Other America | 25.2 | 11 | 2.8 | 10 | 0.113 |
| Total cereals | 233.3 | 100 | 29.5 | 100 | 0.127 |
| *Paper products (47,48,49)* | | | | | |
| Africa | 1.9 | 1 | 1.2 | 1 | 0.630 |
| Asia | 18.7 | 11 | 18.2 | 12 | 0.970 |
| Europe | 92.5 | 53 | 79.7 | 54 | 0.861 |
| North America | 45.5 | 26 | 40.6 | 28 | 0.893 |
| Oceania | 1.9 | 1 | 1.1 | 1 | 0.560 |
| Other America | 13.7 | 8 | 6.8 | 5 | 0.498 |
| Total paper products | 174.2 | 100 | 147.6 | 100 | 0.847 |
| *Wood products (44,45,46)* | | | | | |
| Africa | 120.5 | 34 | 0.9 | 2 | 0.008 |
| Asia | 22.1 | 6 | 11.8 | 20 | 0.536 |
| Europe | 142.7 | 40 | 23.9 | 40 | 0.168 |
| North America | 50.5 | 14 | 18.3 | 31 | 0.362 |
| Oceania | 8.4 | 2 | 1.5 | 2 | 0.174 |
| Other America | 11.1 | 3 | 3.1 | 5 | 0.278 |
| Total wood products | 355.3 | 100 | 59.5 | 100 | 0.168 |

understood but may arise largely from differences among national reporting and accounting systems.

During data processing ComTrade estimates quantities (trade quantities and net weight in kilograms) at the six-digit level (e.g., durum wheat, code 100110) of the classification system and the four-digit level (e.g., wheat, code 1001). If, in the aggregation of quantities, there is a conflict in the quantity unit of measure, they follow the rule that "if the value associated with one quantity unit equals 75 percent or more of the total commodity value, then that quantity unit will be chosen as the unit of the given commodity and the remaining quantity will be estimated according to the proportion of its value." A "not reported" quantity is treated in the same sense and the quantity unit left blank.

**Table 21.5.** Carbon exports from region to region for 2000 (PgC)

| Commodity group | Africa | Asia | Europe | North America | Oceania | Other America | Grand Total |
|---|---|---|---|---|---|---|---|
| *Cereals (10)* | | | | | | | |
| Africa | 0.000 | 0.000 | 0.000 | 0.000 | 0.000 | 0.000 | 0.000 |
| Asia | 0.002 | 0.009 | 0.002 | 0.000 | 0.000 | 0.000 | 0.014 |
| Europe | 0.004 | 0.006 | 0.019 | 0.000 | 0.000 | 0.000 | 0.029 |
| North America | 0.008 | 0.023 | 0.002 | 0.003 | 0.000 | 0.013 | 0.049 |
| Oceania | 0.000 | 0.001 | 0.000 | 0.000 | 0.000 | 0.000 | 0.001 |
| Other America | 0.002 | 0.002 | 0.001 | 0.000 | 0.000 | 0.006 | 0.011 |
| Total cereals | 0.017 | 0.041 | 0.024 | 0.003 | 0.000 | 0.020 | 0.105 |

## Analysis and Main Findings

This chapter concentrates on the cereals, paper products, and wood products groups (see Appendix 21.1 for group definitions). Cereals include all coarse grains as well as rice and wheat. At the regional and global levels, this group is clearly the dominant one in both economic value and total trade volume. The group components tend to be homogenous in terms of the carbon content of the harvested product, which can range from 47 to 52 percent (Strehler and Stutzle 1987).

These three groups represented approximately 3.8 percent of total global trade value, and cereals represented 5.3 percent of the total trade value in agricultural products in 2000 (Table 21.4).

## Global Carbon Trade

Based on ComTrade data, exports trade in cereals, wood products, and paper products was approximately 0.370 PgC in 2000 (see Table 21.4). North America led in the export of cereals (0.049 PgC), and Europe led in the export of paper and wood products (0.046 PgC and 0.071 PgC).

If cereals (i.e., small and coarse grains) represent 60 percent (0.735 PgC) of the total NPP of harvested agricultural crops (1.225 PgC) then the carbon in cereal exports (0.105 PgC) represents 14.3 percent of the carbon content in global harvested cereal products (Table 21.3). If we consider the total NPP of agricultural harvested and nonharvested crops (3.5 PgC), then exported cereals represent 3 percent of the carbon content.

Globally, forests cover 39 million km$^2$ of which about 95 percent is natural forest and the remainder plantations (FAO 2001). In total, forests contain an estimated 340 PgC in live and dead material, representing more than 55 percent of the global carbon stored in vegetation (Niles et al. 2002), and have an estimated NPP of 45 PgC per year. From this base, exports of paper and wood products (0.265 PgC) represent a very small fraction (0.6 percent) of the carbon content in forest and woodland.

## Carbon Trade in Cereal Exports

Asia, Europe, North America, and Other America are the main exporters of cereals, representing 98 percent in volume and value (Table 21.4); Europe and North America account for three-quarters of that trade. Data consistency was cross-checked by calculating the price of a kilogram of cereals. The price ranges from 0.113US$ kg$^{-1}$ in Other America to 0.160US$ kg$^{-1}$ for Asia, from the world average of 0.127US$ kg$^{-1}$, broadly conform to the weight volume figures.

In the year 2000 North America exported 0.049 PgC of cereals followed by Europe, Other America, and Asia (Table 21.5). The three major flows are the exports from North America to Asia and Other America and intracontinental trade in Europe. These fluxes represent more than half (52 percent) of the global carbon flow in cereals. North America alone accounts for 40 percent of the total flow just in its exports to Asia and Other America.

## Global Carbon Net Flow (Exports minus Imports) for 2000

A central issue in agriculture and forest carbon commerce is the net position, the sum of total imports and exports. In 2000 North America was the largest net exporter of carbon for cereals (0.046 PgC) leading Europe and Oceania, which are small net exporters with 0.006 and 0.001 PgC, respectively. Asia and Other America are clearly net carbon importers. Africa is a net carbon importer for cereals but becomes a net exporter of carbon when wood products are considered (see Figure 21.1).

## Regional Carbon Net Flow (Exports minus Imports) from 1989 to 2000

North America and Europe have been net exporters of carbon since the first data sets in 1989, while Asia has been a net importer (Figures 21.2, 21.3, and 21.4). The trend since 1992 for North America (Figure 21.2) is stable with a net carbon export fluctuating between 0.040 and 0.050 PgC for cereals. Paper and wood products are also stable and less than 0.010 PgC.

In Europe (Figure 21.3) the net balances in cereals and paper products have been relatively stable since 1996. Wood product balances, however, increased steadily from 1997 (0.001 PgC) through 2000 (0.016 PgC).

Except for 1999, when Asia changed from being a net wood importer to a net

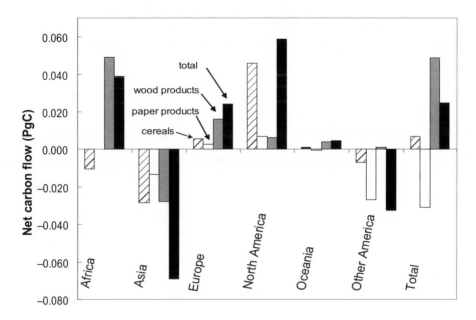

**Figure 21.1.** Net carbon flow (X – M) for all regions (2000)

exporter, the data are stable for the period 1989–2000. Asia is a net importer of carbon for all three commodity groups.

## Conclusions and Recommendations

FAO projects that cereal exports will continue to grow for the foreseeable future, as developing countries find it more efficient to import from lower-cost producers for a portion of their overall food needs (Bruinsma 2003). This trend implies (assuming constant climate change) that the present net exporting regions—Europe and North America—will further expand their present positions, possibly at growth rates exceeding 1 percent per annum. Global cereal production is projected to rise from $1.86 \times 10^9$ tons in 1999 to $2.83 \times 10^9$ tons in 2030 (Bruinsma 2003). At that rate, exports of cereals would represent only 7.6 percent of the 2.1 PgC produced yearly by cereals in 2030 but a fifth (21.7 percent) of the harvested cereal product (0.735 PgC), assuming that NPP does not change.

Even under conditions of a steadily growing trade in carbon commerce for agriculture and forestry, the total transfers among regions will remain almost insignificant (0.6 percent) in comparison to the total terrestrial carbon budget (55 PgC). From a national perspective, carbon commerce takes on an added importance, as a country's status (car-

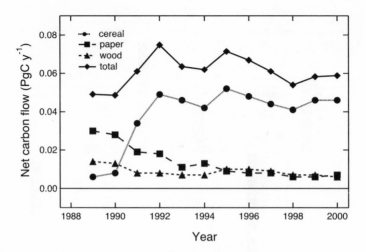

**Figure 21.2.** Net carbon flow (X – M) for North America (1989–2000)

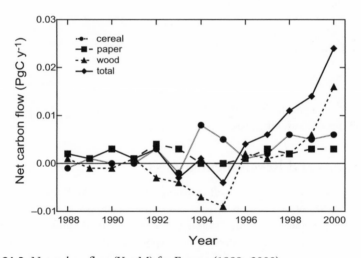

**Figure 21.3.** Net carbon flow (X – M) for Europe (1989–2000)

bon importer or carbon exporter) could have a significant bearing on its reporting to the Conference of the Parties of the UN Framework Convention on Climate Change (Sanz et al., Chapter 24, this volume). Under such conditions, it may be worthwhile to expand the analytical framework to include a wider variety of carbon-based products beyond the agriculture and forestry sectors, such as manufactured goods, chemicals, and other significant sectors.

Should a global emission trading system be implemented, the trade through commerce might be considered an element. An emission trading system could price carbon

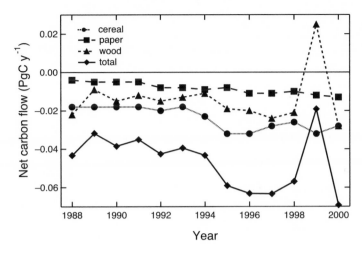

**Figure 21.4.** Net carbon flow (X – M) for Asia (1989–2000)

at US$14–US$23 per ton (Brown 1996), far above many of the prices (US$3–US$5 per ton) that have been used to date in various sequestration schemes. Higher prices could provide incentives for countries to assess potential carbon gains and losses from imports and exports.

This overview provides an initial analysis of the quantities of carbon that enter the trade stream from the agriculture and forest sectors. Future efforts could provide deeper analysis through more complete and consistent data, based on the following steps:

- Review possible differences in carbon content that may arise from technology, ecology, crop type, or variety.
- Assess algorithms for calculating carbon content in different types of export crops, in light of the technological package used, ecology, and crop variety.
- Further investigate the methods used to aggregate ComTrade data at the regional and global levels in order to undertake data assembly and develop reliable longer-term historical data for carbon commerce.
- Clean the ComTrade data to ensure that quantity and value data are consistent.
- Assess the feasibility and architecture required to develop a global carbon commerce database to analyze annual carbon commerce through queries using economic and tonnage data. Among the outputs could be calculations of the direction and velocity of flows, net carbon gains and losses, and country-to-country flows, as well as regional, continental, and global exchanges on and annual and decadal basis.
- Collaborate with the climate modeling community to achieve better spatial accounting for net carbon flows in commerce through improved data structure and definition of model requirements.

## Appendix 21.1. Selected definitions

*Cereals* (ComTrade code 10) are composed of wheat and meslin, rye, barley, oats, maize, rice, grain sorghum, and buckwheat, millet and canary seed, other cereals.

*Paper products* (ComTrade codes 47, 48, 49) are composed of pulp of wood, fibrous cellulosic material, waste, etc., paper and paperboard, articles of pulp, paper and board, printed books, newspapers, pictures, etc.

*Wood products* (ComTrade codes 44, 45, 46) include wood and articles of wood, wood charcoal, cork and articles of cork, manufactures of plaiting material, basketwork, etc.

*Primary crops (annual)* consist of cereals, pulses, roots and tubers, sugar crops, oil-bearing crops, fiber crops, vegetables, tobacco, fodder crops.

*Primary crops (perennial)* are composed of fruits and berries; nuts; permanent oil-bearing crops; spices, condiments, herbs; other permanents crops (e.g., cocoa, coffee, tea).

## Appendix 21.2. Crop carbon ratios (IPCC 1996)

| Product | Residue/ crop product | Dry matter content (%) | Carbon content (% dm) | Nitrogen- carbon (N/C) ratio |
|---------|------|--------|--------|--------|
| Wheat | 1.3 | 78–88 | 48.53 | 0.012 |
| Barley | 1.2 | 78–88 | 45.67 | |
| Maize | 1 | 30–50 | 47.09 | 0.02 |
| Oats | 1.3 | | | |
| Rye | 1.6 | | | |
| Rice | 1.4 | 78–88 | 41.44 | 0.014 |
| Millet | 1.4 | | | 0.016 |
| Sorghum | 1.4 | | | 0.02 |
| Pulse | | | | |
| Pea | 1.5 | | | |
| Bean | 2.1 | | | |
| Soya | 2.1 | | | 0.05 |
| Potatoes | 0.4 | 30–60 | 42.26 | |
| Feedbeet | 0.3 | 10–20[a] | 40.72[a] | |
| Sugarbeet | 0.2 | 10–20[a] | 40.72[a] | |
| Jerusalem artichoke | 0.8 | | | |
| Peanut | 1 | | | |

*Note:* Table mainly based on Strehler and Stulze (1987). Sugarbeet data from Ryan and Openshaw (1991). Nitrogen content from Barnard and Kristoferson (1985).

[a] Statistics are for beet leaves.

**Appendix 21.3.** Top importers and exports of cereals, paper products, and wood products, 2000

| Cereals (10) | | | Paper products (47,48,49) | | | Wood products (44,45,46) | | |
|---|---|---|---|---|---|---|---|---|
| Top exporters | Trade value (billion US$) | % of total | Top exporters | Trade value (billion US$) | % of total | Top exporters | Trade value (billion US$) | % of total |
| USA | 9.7 | 31 | Canada | 18.6 | 15 | Canada | 13.4 | 20 |
| France | 4.0 | 13 | USA | 16.3 | 13 | USA | 6.4 | 10 |
| Canada | 3.0 | 9 | Germany | 12.4 | 10 | Indonesia | 3.7 | 6 |
| Australia | 2.9 | 9 | Finland | 9.6 | 8 | Malaysia | 3.5 | 5 |
| Argentina | 2.4 | 8 | Sweden | 7.0 | 6 | Germany | 3.5 | 5 |
| Other countries | 9.3 | 30 | Other countries | 61.2 | 49 | Other countries | 35.6 | 54 |
| Total export | 31.3 | 100 | Total export | 125.1 | 100 | Total export | 66.2 | 100 |

| Cereals (10) | | | Paper products (47,48,49) | | | Wood products (44,45,46) | | |
|---|---|---|---|---|---|---|---|---|
| Top importers | Trade value (billion US$) | % of total | Top importers | Trade value (billion US$) | % of total | Top importers | Trade value (billion US$) | % of total |
| Japan | 3.7 | 12 | USA | 20.0 | 16 | USA | 17.0 | 23 |
| Mexico | 1.6 | 5 | Germany | 11.2 | 9 | Japan | 11.6 | 16 |
| Rep. of Korea | 1.5 | 5 | France | 8.8 | 7 | Germany | 4.7 | 6 |
| Iran | 1.5 | 5 | United Kingdom | 8.2 | 6 | United Kingdom | 3.8 | 5 |
| Brazil | 1.4 | 5 | China | 6.6 | 5 | China | 3.7 | 5 |
| Other countries | 21.1 | 68 | Other countries | 73.1 | 57 | Other countries | 33.5 | 45 |
| Total import | 30.9 | 100 | Total import | 127.9 | 100 | Total import | 74.2 | 100 |

**Appendix 21.4.** Country groupings (United Nations Statistical Office 2003)

| Africa | Africa (continued) | Asia (continued) |
|---|---|---|
| Algeria | Sao Tome and Principe | Lebanon |
| Angola | Senegal | Malaysia |
| Benin | Seychelles | Maldives |
| Botswana | Sierra Leone | Mongolia |
| Burkina Faso | Somalia | Myanmar |
| Burundi | South Africa | Nepal |
| Cameroon | Sudan | Pakistan |
| Cape Verde | Swaziland | Palestinian Terr. |
| Central African Republic | Tanzania | Oman |
| Chad | Togo | Philippines |
| Comoros | Tunisia | Qatar |
| Congo | Uganda | Korea Rep. |
| Côte d'Ivoire | Western Sahara | Korea Democratic People's |
| Djibouti | Zambia | Republic |
| Egypt | Zimbabwe | Ryukyu Island |
| Equatorial Guinea | | Saudi Arabia |
| Eritrea | *Asia* | Singapore |
| Ethiopia | Afghanistan | Sri Lanka |
| Gabon | Armenia | Syria |
| Gambia | Azerbaijan | Tajikistan |
| Ghana | Bahrain | Thailand |
| Guinea | Bangladesh | Turkey |
| Guinea-Bissau | Bhutan | Turkmenistan |
| Kenya | Brunei | United Arab Emirates |
| Lesotho | Cambodia | Uzbekistan |
| Liberia | China | Viet Nam |
| Libya | Cyprus | Yemen |
| Madagascar | East Timor | |
| Malawi | Georgia | *Europe* |
| Mali | India | Albania |
| Mauritania | Indonesia | Andorra |
| Mauritius | Iran | Austria |
| Morocco | Iraq | Belarus |
| Mozambique | Israel | Belgium |
| Namibia | Japan | Belgium-Luxembourg |
| Niger | Jordan | Bosnia Herzegovina |
| Nigeria | Kazakhstan | Bulgaria |
| Réunion | Kuwait | Croatia |
| Rwanda | Kyrgyzstan | Czech Rep. |
| Saint Helena | Laos | Denmark |

**Appendix 21.4.** *(continued)*

| *Europe (continued)* | *Oceania* | *Other America (continued)* |
|---|---|---|
| Estonia | Australia | Cayman Islands |
| Faeroe Islands | Christmas Island | Chile |
| Finland | Cocos Islands | Colombia |
| France | Cook Islands | Costa Rica |
| Germany | Fiji | Cuba |
| Gibraltar | French Polynesia | Dominica |
| Greece | Kiribati | Dominican Rep. |
| Hungary | Marshall Islands | Ecuador |
| Iceland | Micronesia | El Salvador |
| Ireland | Northern Mariana Islands | Falkland Islands (Malvinas) |
| Italy | Nauru | French Guiana |
| Latvia | New Caledonia | Grenada |
| Lithuania | New Zealand | Guadeloupe |
| Luxembourg | Niue | Guatemala |
| Macedonia | Norfolk Islands | Guyana |
| Malta | Oceania | Haiti |
| Moldova | Palau | Honduras |
| Netherlands | Papua New Guinea | Jamaica |
| Norway | Pitcairn | Martinique |
| Poland | Samoa | Mexico |
| Portugal | Solomon Islands | Montserrat |
| Romania | Tokelau | Netherlands Antilles and |
| Russian Federation | Tonga | Aruba |
| Slovakia | Tuvalu | Nicaragua |
| Slovenia | U.S. miscellaneous Pacific | Panama |
| Spain | islands | Paraguay |
| Sweden | Vanuatu | Peru |
| Switzerland | Wallis and Futuna Islands | Saint Kitts and Nevis |
| Ukraine | | Saint Lucia |
| United Kingdom | *Other America* | Saint Vincent and the |
| Yugoslavia (former) | Anguilla | Grenadines |
| | Antigua and Barbuda | Suriname |
| *North America* | Argentina | Trinidad and Tobago |
| | Bahamas | Turks and Caicos Islands |
| Bermuda | Barbados | Uruguay |
| Canada | Belize | U.S. Virgin Islands |
| Greenland | Bolivia | Venezuela |
| Saint Pierre and Miquelon | British Antarctic Territory | |
| United States | British Virgin Islands | |
| | Brazil | |

**Appendix 21.5.** Export and imports from reporter to partner, 2000 (United Nations Statistical Office 2003)

| Commodity group/exporter | Africa Export | Africa Import | Asia Export | Asia Import | Europe Export | Europe Import | North America Export | North America Import | Oceania Export | Oceania Import | Other America Export | Other America Import | Grand total |
|---|---|---|---|---|---|---|---|---|---|---|---|---|---|
| **Cereals (10)** | | | | | | | | | | | | | |
| Africa | 0.000 | 0.000 | 0.000 | 0.002 | 0.000 | 0.004 | 0.000 | 0.004 | 0.000 | 0.000 | 0.000 | 0.001 | 0.011 |
| Asia | 0.002 | 0.000 | 0.009 | 0.008 | 0.002 | 0.005 | 0.000 | 0.022 | 0.000 | 0.006 | 0.000 | 0.001 | 0.055 |
| Europe | 0.004 | 0.000 | 0.006 | 0.002 | 0.019 | 0.018 | 0.000 | 0.002 | 0.000 | 0.000 | 0.000 | 0.001 | 0.053 |
| North America | 0.008 | 0.000 | 0.023 | 0.000 | 0.002 | 0.000 | 0.003 | 0.003 | 0.000 | 0.000 | 0.013 | 0.000 | 0.052 |
| Oceania | 0.000 | 0.000 | 0.001 | 0.000 | 0.000 | 0.000 | 0.000 | 0.000 | 0.000 | 0.000 | 0.000 | 0.000 | 0.002 |
| Other America | 0.002 | 0.000 | 0.002 | 0.000 | 0.001 | 0.000 | 0.000 | 0.012 | 0.000 | 0.000 | 0.006 | 0.006 | 0.030 |
| *Total cereals (10)* | *0.017* | *0.000* | *0.041* | *0.012* | *0.024* | *0.027* | *0.003* | *0.042* | *0.000* | *0.006* | *0.020* | *0.010* | *0.203* |
| **Paper products (47,48,49)** | | | | | | | | | | | | | |
| Africa | 0.000 | 0.000 | 0.000 | 0.000 | 0.000 | 0.001 | 0.000 | 0.000 | 0.000 | 0.000 | 0.000 | 0.000 | 0.002 |
| Asia | 0.000 | 0.000 | 0.008 | 0.007 | 0.000 | 0.005 | 0.001 | 0.008 | 0.000 | 0.001 | 0.000 | 0.001 | 0.032 |
| Europe | 0.001 | 0.000 | 0.005 | 0.001 | 0.038 | 0.038 | 0.001 | 0.004 | 0.000 | 0.000 | 0.001 | 0.001 | 0.090 |
| North America | 0.000 | 0.000 | 0.006 | 0.001 | 0.003 | 0.001 | 0.010 | 0.013 | 0.000 | 0.000 | 0.004 | 0.001 | 0.039 |
| Oceania | 0.000 | 0.000 | 0.001 | 0.000 | 0.000 | 0.000 | 0.000 | 0.000 | 0.000 | 0.000 | 0.000 | 0.000 | 0.002 |
| Other America | 0.000 | 0.000 | 0.001 | 0.001 | 0.001 | 0.001 | 0.003 | 0.029 | 0.000 | 0.000 | 0.001 | 0.002 | 0.040 |
| *Total paper products (47,48,49)* | *0.001* | *0.001* | *0.020* | *0.009* | *0.043* | *0.046* | *0.016* | *0.055* | *0.001* | *0.001* | *0.006* | *0.005* | *0.205* |

## Wood products (44,45,46)

| | | | | | | | | | | | | |
|---|---|---|---|---|---|---|---|---|---|---|---|---|
| Africa | 0.003 | 0.010 | 0.043 | 0.000 | 0.014 | 0.001 | 0.000 | 0.000 | 0.000 | 0.000 | 0.000 | 0.000 | 0.071 |
| Asia | 0.000 | 0.002 | 0.010 | 0.011 | 0.001 | 0.008 | 0.001 | 0.007 | 0.000 | 0.008 | 0.000 | 0.002 | 0.050 |
| Europe | 0.001 | 0.002 | 0.010 | 0.001 | 0.059 | 0.050 | 0.001 | 0.001 | 0.000 | 0.000 | 0.000 | 0.002 | 0.127 |
| North America | 0.000 | 0.000 | 0.006 | 0.000 | 0.001 | 0.000 | 0.017 | 0.018 | 0.000 | 0.000 | 0.001 | 0.001 | 0.044 |
| Oceania | 0.000 | 0.000 | 0.003 | 0.000 | 0.000 | 0.000 | 0.000 | 0.000 | 0.001 | 0.000 | 0.000 | 0.000 | 0.005 |
| Other America | 0.000 | 0.000 | 0.002 | 0.000 | 0.001 | 0.000 | 0.002 | 0.004 | 0.000 | 0.000 | 0.001 | 0.001 | 0.010 |
| *Total wood products (44,45,46)* | *0.005* | *0.014* | *0.075* | *0.013* | *0.076* | *0.059* | *0.020* | *0.030* | *0.001* | *0.008* | *0.001* | *0.005* | *0.307* |
| *Grand total* | *0.023* | *0.015* | *0.136* | *0.035* | *0.143* | *0.133* | *0.039* | *0.127* | *0.002* | *0.016* | *0.027* | *0.020* | *0.715* |

# Literature Cited

Barnard, G., and L. Kristoferson. 1985. *Agricultural residues as fuel in the third world.* Produced by the Earthscan Energy Information Programme in collaboration with the Beijer Institute of the Royal Swedish Academy of Sciences. London: International Institute for Environment and Development (IIED).

Brown, S. 1996. Present and potential roles of forests in the global climate change debate. *Unasylva* 185 (47): 3–10.

Bruinsma, J., ed. 2003. *Agriculture towards 2015–2030.* Rome: Food and Agriculture Organization of the United Nations.

Ciais, P., T. Naegler, P. Peylin, A. Freibauer, and P. Bousquet. 2001. Horizontal displacement of carbon associated to agriculture and its impact on the atmospheric $CO_2$ distribution. Paper presented at sixth international carbon dioxide conference, Sendai, Japan, October 1–5.

Cihlar, J., ed. 2002. *Terrestrial carbon observations: The Ottawa assessment of requirements, status and next steps.* Environment and Natural Resources Report 2. Rome: Food and Agriculture Organization of the United Nations.

Cihlar, J., and S. Denning. 2002. *Terrestrial carbon observations: The Rio de Janeiro recommendations for terrestrial and atmospheric measurements.* Environment and Natural Resources Report 3. Rome: Food and Agriculture Organization of the United Nations.

Cihlar, J., M. Heimann, and R. Olson, eds. 2003. *Terrestrial carbon observations: The Frascati report on in situ carbon data and information.* Environment and Natural Resources Report 5. Rome: Food and Agriculture Organization of the United Nations.

Falkowski, P., R. J. Scholes, E. Boyle, J. Canadell, D. Canfield, J. Elser, N. Gruber, K. Hibbard, P. Högberg, S. Linder, F. T. Mackenzie, B. Moore III, T. Pedersen, Y. Rosenthal, S. Seitzinger, V. Smetacek, and W. Steffen. 2000. The global carbon cycle: A test of our knowledge of Earth as a system. *Science* 290: 290–296.

FAO (Food and Agriculture Organization of the United Nations). 1999. *Prevention of land degradation, enhancement of carbon sequestration and conservation of biodiversity through land use change and sustainable land management with a focus on Latin America and the Caribbean.* World Soil Resources Report 86. Rome.

———. 2001. *Global forest resources assessment 2000: Main report.* FAO Forestry Paper 140. Rome.

———. 2003. FAOSTAT (FAO statistical database). Rome. http://apps.fao.org/default.htm.

Goudriaan J., J. J. R. Groot, and P. W. J. Uithol. 2001. Productivity of agro-ecosystems. Pp. 301–313 in *Terrestrial global productivity*, edited by H. A. Mooney and B. Saugier. San Diego: Academic Press.

Heath, L., and K. Skogg. 2003. Contribution of forest products to the global carbon budget. In *2003 national report on sustainable forests.* Washington, DC: U.S. Forest Service.

IGBP (International Geosphere-Biosphere Programme). 1998. Global net primary productivity. In *GAIM, 1993–1997: The first five years: Setting the stage for synthesis,* edited by D. Sahagian and K. Hibbard. IGBP/GAIM (Global Analysis, Interpretation, and Modeling) Report No. 6. Durham, NH: IGBP, Institute for the Study of Earth, Oceans, and Space, University of New Hampshire. http://gaim.unh.edu/Products/Reports/Report_6/index.html.

IPCC (Intergovernmental Panel on Climate Change). 1996. *Greenhouse gas inventory reference manual,* Vol. 3, pp.4–75. Paris.

Michie, B., and S. Kin. 1999. *A global study of regional trade flows of five groups of forest products.* Helsinki: World Forest, Society and Environment Research Program.

Niles, J. O., S. Brown, J. Pretty, A. S. Ball, and J. Fay. 2002. Potential carbon mitigation and income in developing countries from changes in use and management of agricultural and forest lands. *Philosophical Transactions: Mathematical, Physical and Engineering Sciences* 360 (1797): 1621–1639.

Ramankutty, N., and J. A. Foley. 1999. Estimating historical changes in global land cover: Croplands from 1700 to 1992. *Global Biogeochemical Cycles* 13 (4): 997–1027.

Ryan, P., and K. Openshaw. 1991. *Assessment of bioenergy resources: A discussion of its needs and methodology.* Industry and Energy Development Working Paper, Energy Series Paper No. 48. Washington, DC: World Bank.

Strehler, A., and W. Stutzle. 1987. Biomass residues. In *Biomass regenerable energy,* edited by D. O. Hall and R. P. Overend. Chichester, UK: John Wiley and Sons.

United Nations. 1986. *Standard international trade classification.* United Nations Statistical Papers, Series M, no.34/rev.3. New York.

United Nations Statistical Office. 2003. ComTrade (Commercial trade database). New York. http://unstats.un.org/unsd/comtrade/.

WTO (World Trade Organization). 2003. *Trade statistics.* Geneva. http://www.wto.org/stats.

# PART VII
## Purposeful Carbon Management

# 22

# Near- and Long-Term Climate Change Mitigation Potential

Jayant A. Sathaye

Climate change is a critical global environmental challenge facing humanity with implications for food production, natural ecosystems, freshwater supply, health, and other conditions. According to the recent scientific assessment of the Intergovernmental Panel on Climate Change (IPCC), the Earth's climate system has demonstrably changed on both global and regional scales since the preindustrial era (Watson et al. 2001). Further, there is evidence to show that most of the warming (of 0.1°C per decade) observed over the past 50 years is attributable to human activities. The IPCC projects that the global mean temperature is likely to increase between 1.4 and 5.8°C by 2100. This unprecedented increase is expected to have severe impacts on the global hydrological system, ecosystems, sea level, crop production, and related processes.

Fossil-fuel combustion is the primary contributor to the formation of carbon dioxide, which accounts for about 72 percent of the temperature increase to 2100. Methane and nitrous oxide are the two other large contributors to the increase. Aerosols, ozone, and other gases tend to ameliorate the impact of the gases mentioned, but these effects are not as well understood. Fossil-fuel combustion released about 6.3 ± 0.3 petagrams of carbon per year (PgC y$^{-1}$), and emissions from land use change amounted to 1.6 ± 0.8 PgC y$^{-1}$ during 1989–98 (Watson et al. 2000). Terrestrial sinks captured 0.7 ± 1.0 PgC y$^{-1}$ over the same period. The impact of carbon and other greenhouse gases (GHGs) is cumulative, and hence it is important to examine both the current and past history of emissions and sinks in order to understand the full consequences of the contribution of these gases to climate change.

In this chapter I focus primarily on the potential for the mitigation of climate change, that is, the net reduction of GHG emissions, particularly from fossil-fuel combustion and land use change. Energy used to be viewed as an essential commodity for economic growth, but over the past 30 years it has become apparent that the link between energy use and economic growth can be decoupled. Energy intensity in the

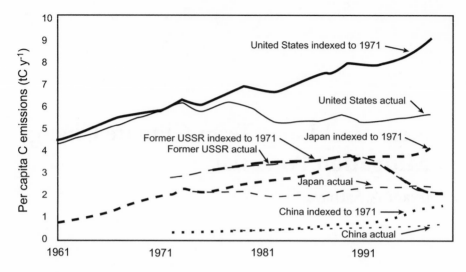

**Figure 22.1.** Actual per capita $CO_2$ emissions and per capita $CO_2$ emissions indexed to the ratio of 1971 GDP (1990 US$1,000 purchasing power parity) to 1971 $CO_2$ emissions for selected countries. t C = tons of carbon.

Japan, the United States, and other countries of the Organisation for Economic Co-operation and Development (OECD) has increased far slower than economic growth (Figure 22.1). China, too, has made significant gains, and its energy growth rate has been half that of its gross domestic product (GDP) growth rate since the early 1980s. Can similar trends continue in the future and not only slow but reverse the growth of energy use and consequent carbon emissions? To what extent do such trends need to continue in order to stabilize atmospheric GHG concentrations? Are technologies available to achieve near-term GHG emissions reduction and long-term stabilization? How much might these cost? I rely on the IPCC Third Assessment Report and other more recent reference materials to address these questions.

## Categories of Technical Mitigation Options

Mitigation options may be characterized by the sectors and GHGs they are intended to address. I focus below on carbon dioxide and methane, the two largest contributors to GHG concentrations. Many cost-effective technical strategies for greenhouse gas abatement have been identified for the energy sector. These options can be classified into two categories—improving energy efficiency and switching to less-carbon-intensive or carbon-neutral fuels. Improving energy efficiency reduces the energy used without reducing the level of service. Reduced energy use decreases associated environmental impacts, including emissions of greenhouse gases. Among fossil fuels, carbon content decreases from coal to oil to natural gas. Switching to less-carbon-intensive fuels is a viable strategy to reduce GHG emissions. Renewable energy sources such as wind, solar energy, and nuclear energy have no direct carbon content. Hydroelectric sources, however, may release $CO_2$ and methane, depending on the reservoir surface area and seasonal changes in the water depth, type of vegetation, and soils beneath their reservoirs (Fearnside 1997).

The forestry sector mitigation activities can be broadly categorized into three groups—forest carbon conservation and management measures; carbon storage management (expanding forest carbon sinks); and fossil-fuel substitution and management activities (Brown et al. 1996). Mitigation activities include conventional forest conservation, afforestation, and reforestation as well as modern activities such as substituting fossil fuels with bioenergy, fire protection techniques, reduced-impact logging practices, and recycling of forest products.

Improved crop and soil management can increase carbon levels in the currently cropped lands and reduce losses from newly cropped lands. Thus, altering management of agricultural soils, adopting new land use and management practices, and restoring degraded agricultural lands are potential measures to sequester atmospheric carbon and reduce greenhouse gas emissions.

Methane is expected to contribute about 24 percent of the total projected global warming over the next 100 years. The sources of methane in agriculture are enteric fermentation in ruminants, flooded rice fields, and anaerobic animal waste processing (Robertson, Chapter 29, this volume). Enteric fermentation in cattle and rice cultivation account for 31 and 12 percent of global methane emissions, respectively.

## Near-Term Mitigation Potential and Costs of Technical Options

Mitigation options to reduce carbon dioxide emissions from the energy sector have received the most attention in both technology characterization and global modeling studies to date. Only recently, options that address methane, nitrous oxide, chlorofluorocarbons, and other GHGs are being examined in models of the global economy.[1] The inclusion of biological and engineered sinks is yet to be explicitly considered in such models. One of the challenges to the inclusion of biological sinks in these models lies in their temporal dimension. Sinks accumulate carbon relatively slowly. Carbon releases from them can be either slow or rapid. The timing of monetary flows is very different from those of biological carbon. Biological sinks may release carbon in response to a warming climate or natural or other disturbances, and these disturbances would need to be carefully accounted.

Tables 22.1 and 22.2 show the potentials and costs estimated by the IPCC Third Assessment Report for the mitigation options in the energy and land use change sectors

**Table 22.1.** Estimates of potential global greenhouse gas emission reductions in 2010 and 2020

| Sector | Historic emissions in 1990 ($Pg\ C_{eq}.y^{-1}$) | Historic $C_{eq}$ annual growth rate in 1990–1995 (%) | Potential emission reductions ($Pg\ C_{eq}.y^{-1}$) in | | Net direct costs per ton of carbon avoided |
|---|---|---|---|---|---|
| | | | 2010 | 2020 | |
| Buildings[a] $CO_2$ only | 1.65 | 1.0 | 0.70–0.75 | 1.00–1.10 | Most reductions are available at negative net direct costs. |
| Transport $CO_2$ only | 1.08 | 2.4 | 0.10–0.30 | 0.30–0.70 | Most studies indicate net direct costs less than US$25/tC but two suggest net direct costs will exceed US$50/tC. |
| Industry, $CO_2$ only Energy efficiency | 2.300 | 0.4 | 0.30–0.50 | 0.70–0.90 | More than half available at net negative direct costs. |
| Material efficiency | | | ~0.20 | ~0.60 | Costs are uncertain. |
| Industry, non-$CO_2$ gases | 0.17 | | ~0.10 | ~0.10 | $N_2O$ emissions reduction costs are US$0–$10/$tC_{eq}$. |
| Agriculture[b] | | | | | Most reductions will cost between US$0 and US$100/$tC_{eq}$ with limited opportunities for negative net direct cost options. |
| $CO_2$ only | 0.21 | | | | |
| Non-$CO_2$ gases | 1.25–2.80 | n.a. | 0.15–0.30 | 0.35–0.7 | |

| | | | | | |
|---|---|---|---|---|---|
| Waste[b] $CH_4$ only | 0.24 | 1.0 | ~0.20 | ~0.20 | About 75% of the savings as methane recovery from landfills at net negative direct cost; 25% at a cost of $20/tC$_{eq}$. |
| Montreal Protocol replacement applications, non-$CO_2$ gases | 0 | n.a. | ~0.10 | n.a. | About half of reductions due to difference in study baseline and SRES baseline values. Remaining half of the reductions available at net direct costs below $200/tC$_{eq}$. |
| Energy supply and conversion,[c] $CO_2$ only | (1.62) | 1.5 | 0.05–0.15 | 0.35–0.70 | Limited net negative direct cost options exist; many options are available for less than $100/tC$^{eq}$. |
| Total | 6.90-8.40[d] | | 1.90–2.60[e] | 3.60–5.05[e] | |

[a]Buildings include appliances, buildings, and the building shell.

[b]The range for agriculture is mainly caused by large uncertainties about $CH_4$, $N_2O$, and soil-related emissions of $CO_2$. Waste is dominated by methane landfill, and the other sectors could be estimated with more precision as they are dominated by fossil $CO_2$.

[c]Included in sector values above. Reductions include electricity generation options only (fuel switching to gas/nuclear, $CO_2$ capture and storage, improved power station efficiencies, and renewables).

[d]Total includes all sectors reviewed in Chapter 3 for all six gases. It excludes non-energy-related sources of $CO_2$ (cement production, 160 MtC; gas flaring, 60 MtC; and land use change, 600–1,400 MtC) and energy used for conversion of fuels in the end-use sector totals (630 MtC). If petroleum refining and coke oven gas were added, global 1990 $CO_2$ emissions of 7,100 MtC would increase by 12 percent. Note that forestry emissions and their carbon sink mitigation options are not included.

[e]The baseline SRES scenarios (for six gases included in the Kyoto Protocol) project a range of emissions of 11,500–14,000 MtC$_{eq}$ for 2010 and of 12,000–16,000 MtC$_{eq}$ for 2020. The emissions reduction estimates are most compatible with baseline emissions trends in the SRES-B2 scenario. The potential reductions take into account regular turnover of capital stock. They are not limited to cost-effective options but exclude options with costs greater than US$100/tC$_{eq}$ (except for Montreal Protocol gases) or options that will not be adopted through the use of generally accepted policies.

**Table 22.2.** Estimates of potential global greenhouse gas emission reductions in 2010: Land use, land use change, and forestry

| Categories of mitigation options | Potential emission reductions in 2010 $(PgC_{eq.} \, Y^{-1})$ | Potential emission reductions ($PgC$) | Comments |
|---|---|---|---|
| Afforestation/ reforestation (AR)[a] | 0.197–0.584 | | Includes carbon in above- and belowground biomass. Excludes carbon in soils and in dead organic matter. |
| Reducing deforestation (D)[b] | | 1.788 | Potential for reducing deforestation is very uncertain for the tropics and could be in error by as much as ±50%. |
| Improved management within a land use (IM)[c] | 0.570 | | Assumed to be the best available suite of management practices for each land use and climatic zone. |
| Land use change (LC)[c] | 0.435 | | |
| Total | 1.202–1.589 | 1.788 | |

[a]*Source:* SRLUCF Table SPM-3 (Watson et al. 2000). Based on IPCC definitional scenario. Information is not available for other definitional scenarios. "Potential" refers to the estimated average stock change (MtC).

[b]*Source:* SRLUCF Table SPM-3 (Watson et al. 2000). Based on IPCC definitional scenario. Information is not available for other definitional scenarios. "Potential" refers to the estimated range of accounted average stock change 2008–2012 (MtC/y$^{-1}$).

[c]*Source:* SRLUCF Table SPM-4 (Watson et al. 2000). "Potential" refers to the estimated net change in carbon stocks in 2010 (MtC/y$^{-1}$). The list of activities is not exclusive or complete, and it is unlikely that all countries will apply all activities. Some of these estimates reflect considerable uncertainty.

(Watson et al. 2001). These estimates were based on bottom-up engineering approaches that examined the cost and potential of a set of technical options in each of the sectors identified in the tables. The potential reductions in Columns 4 and 5 of Table 22.1 can be compared with 1990 emissions in Column 2. Thus, for the buildings sector, the reduction potential is 0.70–0.75 $PgC_{eq}$ y$^{-1}$ in 2010, or 42–45 percent, of the historical emissions of 1.65 $PgC_{eq}$ y$^{-1}$ in 1990. The total reduction potential increases to 1.9–2.6 $PgC_{eq}$ y$^{-1}$ by 2020. In principle, this is more than enough to meet the Kyoto Protocol target of 5.2 percent. The cost of achieving this reduction is relatively low (Table 22.1, Column 6).

About half of the potential emissions reductions identified in the table can be achieved with direct benefits exceeding direct costs, and the other half at a net direct cost of up to US$100/t $C_{eq}$ (at 1998 prices).[2] Depending on the emissions scenario, this could allow global emissions to be reduced below 2000 levels in 2010–2020, at these net direct costs. The realized potential, however, is likely to be lower, owing to unaccounted implementation costs and barriers to project implementation.

Forests, agricultural lands, and other terrestrial ecosystems offer significant carbon mitigation potential as well. Conservation and sequestration of carbon, although not necessarily permanent, may allow time for other options to be further developed and implemented. The potential of biological mitigation options is on the order of 1.202–1.589 PgC y$^{-1}$ of emissions reductions by 2010 (Table 22.2), and 100 PgC (cumulative) by 2050, equivalent to about 10 to 20 percent of projected fossil-fuel emissions during that period. There are substantial uncertainties associated with this estimate (Watson et al. 2001). Cost estimates based on bottom-up analyses for biological mitigation vary significantly, from US$0.1/t C to about US$20/t C in several tropical countries and from US$20/t C to US $100/t C in nontropical countries. The cost calculations are subject to the same caveats that were noted for the energy sector.

In contrast to the relatively low estimates of costs and large opportunity for potential emissions reductions in the near term from bottom-up analyses, models of the global economy provide a generally more pessimistic picture of the costs. Costs in these models are reported in terms of the change (loss) of GDP in a future year, say 2010, or in terms of carbon taxes measured in U.S. dollars per ton of carbon that would be required to achieve a given reduction. Global modeling studies show national marginal costs to meet the Kyoto targets for each group of countries in Figure 22.2 to range from about US$20/tC up to US$600/tC without trading, and from about US$15/tC up to US$150/tC with Annex B trading.[3] In the absence of Annex B trading, losses range from 0.2 to 2 percent of GDP.[4] With Annex B trading, losses range from 0.1 to 1 percent of GDP. The range of results both within and across regions is caused, in large part, by varying assumptions about future GDP growth rates, carbon intensity, and energy intensity. The cost reductions from Annex B trading may depend on the details of implementation, including the compatibility of domestic and international mechanisms, and constraints on emissions trading.

How might one reconcile the bottom-up, technology-specific estimates that indicate low positive or even negative costs, and the higher and only positive costs from models of the global economy? One issue that would reduce the cost differential relates to the lack of a baseline in the technology estimates. Recall that the bottom-up estimates reported are not differences from a baseline scenario but from a 1990 base-year valuation. It is conceivable that some fraction of the negative or low-cost options would be absorbed in a bottom-up baseline scenario, thus leaving only the higher-cost options for a mitigation scenario. In global models, the low- or negative-cost options are assumed to be absorbed in the baseline and only positive costs remain. Using a common baseline would reduce the cost differential considerably.

Both types of models assume that all transactions are frictionless, and thus, transaction costs are not explicitly represented in the models. The technology cost estimates also do not account for the barriers that firms face to enter a carbon or energy efficiency

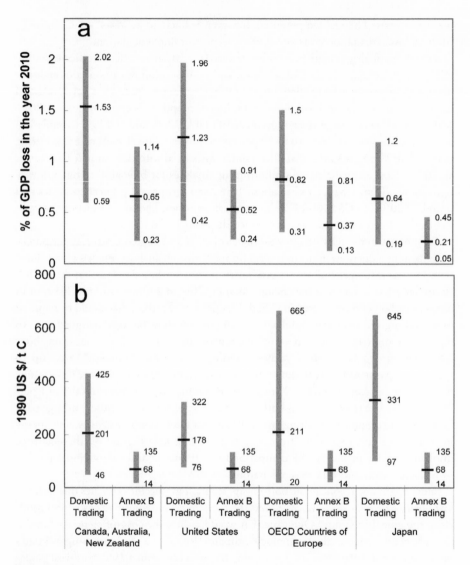

**Figure 22.2.** Projections of (a) GDP losses and (b) marginal cost in Annex B countries in 2010 from global models

market. Barriers such as high cost of capital, lack of information, misplaced incentives, network externalities, ill-defined property rights, market power, and others can prevent firms from entering the market or slow their entry (Sathaye et al. 2001). Overcoming such barriers adds to the cost of market entry and creates a disincentive to rapid and steady implementation of mitigation options. How much might it cost to overcome such barriers? What are the magnitudes of transaction costs? How do these change as a market evolves from nascent to more mature? Answering these questions requires further empirical research.

## Longer-Term (2100) Potential and Costs of Mitigation Options

The economic and other consequences of stabilizing climate change, as expressed by the concentration of carbon dioxide in the atmosphere, have been analyzed through several global economy models. The IPCC Third Assessment Report (Metz et al. 2001) noted that the technologies required to achieve stabilization are already commercialized or near commercialization. Stabilization will not require the development of exotic, unknown technologies.[5]

Figure 22.1 shows the historical changes in carbon emissions and GDP per capita. It is worth asking whether these historical changes in rates would be adequate to achieve stabilization of carbon dioxide concentration by 2100. Changes in carbon emissions of an economy may be expressed in terms of changes in two contributing factors: the energy and carbon intensities of the economy. Figure 22.3 shows how these two factors change from 2000 to 2100 in several IPCC mitigation scenarios. The figure also shows the range of historical changes. Historical rates of change in energy intensity are consistent with those needed for future stabilization of carbon dioxide concentrations, but those for carbon intensity are far slower. In other words, the mitigation scenarios include a more rapid shift away from carbon-intensive to less- or non-carbon-intensive forms of energy than has been achieved historically.

Would it be possible to achieve such a change in carbon intensity? This may be entirely possible, because the global energy mix will almost certainly change during the 21st century in response to limited proven conventional oil and gas reserves and conventional oil resources. The choice of energy mix and associated technologies and investments—toward more exploitation of unconventional oil and gas resources, exploitation of non-fossil energy sources, or use of fossil energy technology with carbon capture and storage—provides an opportunity for a shift toward the increased use of less-carbon-intensive fuels.

Stabilization levels depend more on cumulative emissions than on the emissions pathway. A gradual transition away from the world's present energy system toward a less carbon-emitting economy minimizes costs associated with premature retirement of existing capital stock, provides time for technology development, and avoids premature

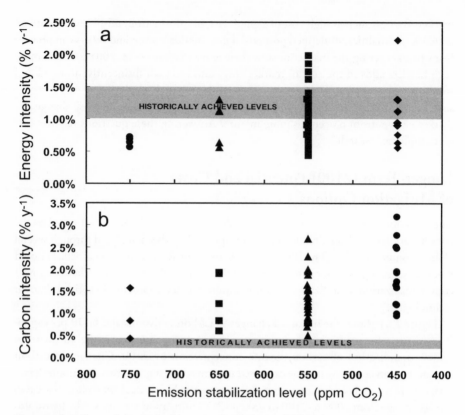

**Figure 22.3.** Rate of change in (a) energy intensity, and (b) carbon intensity, historically achieved levels (1860–1990). Based on data provided by Dr. Igor Bashmakov, Russian Center for Energy Efficiency, Moscow, in consultation with the late Prof. Tsuneyuki Morita, National Institute for Environmental Studies, Japan.

lock-in to early versions of rapidly developing low-emission technology. On the other hand, more rapid near-term action would decrease environmental and human risks and the associated costs of projected changes in climate, and may stimulate more rapid deployment of existing low-emission technologies and provide strong near-term incentives to future technological changes.

The costs of stabilizing carbon dioxide concentrations in the atmosphere increase as the concentration stabilization level declines (Watson et al. 2001). Different baselines can have a strong influence on absolute costs. While there is a moderate increase in the costs in shifting from a 750 ppm to a 550 ppm concentration stabilization level, there is a larger increase in shifting from 550 ppm to 450 ppm, unless the emissions in the baseline scenario are very low. In no case do the stabilization scenarios lead to significant declines in global GDP growth rates over this century. The losses, however, vary across regions and time. These studies do not incorporate carbon sequestration or non–carbon dioxide gases and did not examine the possible effect of more ambitious targets on induced technological change.

## Winners and Losers in Climate Change Mitigation

Plans for implementing climate change mitigation options will need to carefully consider the distributional impact on the affected stakeholders. As already noted, mitigation will aim at increasing the share of low- or non-carbon fuels and technologies, reduction of technologies that emit non-$CO_2$ GHGs, reduction of deforestation, and the increase of forest and other biomass cover and soil carbon. The obvious losers will be those that are associated with the supply and use of coal, and to a lesser extent oil, and deforesting countries whose forests would need to be conserved.

Within a country, not surprisingly, the impact of a carbon tax is likely to be felt primarily by carbon-intensive producers, such as those who use coal. The number of studies that have identified such sectoral impacts, however, is relatively small. One such study, Clean Energy Futures for the US, shows, for instance, that a US$100/t C tax results in a 6 percent decline in primary energy use in 2010, representing a 3 percent decline in oil consumption but a 37 percent decline in coal, compared with a business-as-usual scenario (IWG 2000). The consumption of other forms of energy increases from this carbon tax. These impacts also translate into reductions in employment, primarily in the coal but also in the oil industry. Carbon taxes could thus result in uneven distribution of impacts, and, without some form of compensation, the coal and oil industries have a strong disincentive to support a carbon reduction policy. The challenge for building broad support for a carbon tax is further exacerbated by the entrenched status of the losing industries, while the likely winners, industries that might benefit from a carbon tax like the energy efficiency and renewable energy industries, are dispersed and yet to be firmly established.

Across countries, the IPCC Third Assessment Report studies show that emission constraints on Annex I (industrialized) countries have varied spillover effects on non–Annex I (developing) countries. These spillovers include reductions in both projected GDP and oil revenues for oil-exporting, non–Annex I countries. The effects on these countries can be reduced by removal of subsidies for fossil fuels, energy tax restructuring according to carbon content, increased use of natural gas, and diversification of the economies of non–Annex I, oil-exporting countries.

Other non–Annex I countries (developing non-OPEC [Organization of Petroleum Exporting] countries) may be adversely affected by reductions in demand for their exports to OECD countries and by price increases of those carbon-intensive and other products they continue to import. These other non–Annex I countries may benefit from the reduction in fuel prices, increased exports of carbon-intensive products, and the transfer of environmentally sound technologies and know-how. The possible relocation of some carbon-intensive industries to non–Annex I countries and wider impacts on trade flows in response to changing prices may lead to carbon leakage[6] on the order of 5–20 percent.

## Climate Change Impacts and Mitigation

The impacts of adaptation options may be categorized as those caused by steady changes in conditions like temperature, precipitation, and sea level and those that are a result of extreme events, including sudden changes in storm surges and extreme weather, for example (Ravindranath and Sathaye 2002). In its Third Assessment Report the IPCC was unable to present much information about economic damage caused by climate change. Both the regional and temporal distribution of impacts continue to be difficult to predict, making it doubly difficult to translate these into economic damage.

Sea-level rise is one type of impact that may, however, be amenable to quantification. Increasing global temperatures cause thermal expansion of seawater and the melting of Antarctic and other ice caps, resulting in rising sea levels. The global sea level rose 10 to 20 centimeters (cm) during the 20th century, and IPCC scenarios project it to increase from 9 to 88 cm by 2100, depending on the projected increase of greenhouse gases (Watson et al. 2001). Populations that inhabit small islands and/or low-lying coastal areas are at particular risk of severe social and economic disruptions from sea-level rise and storm surges.

A 50-cm rise in global sea level could have significant impacts on U.S. coastal areas (Neumann et al. 2000). Estimates of land inundated are close to 24,000 km$^2$ (9,000 mi$^2$). Major coastal cities such as Miami, New Orleans, and New York will have to upgrade flood defenses and drainage systems or risk adverse consequences. Three options have been proposed for responding to coastal threats: planned retreat, accommodation, and protection. Impact and adaptation assessments evaluate where these responses might be implemented and then calculate the costs of implementation and the damage to resources that are not protected. Generally, property losses or the costs to protect property dominate the existing impact estimates for the United States. Based on a review of the existing literature, Neumann et al. (2000) estimate that cumulative impacts on coastal property of a 50-cm sea-level rise by 2100 range from about US$20 billion to US$150 billion.

Similar estimates could be derived for other vulnerable coastal areas to arrive at the cumulative global impact. Such a temporal and global estimate could be taken into consideration in deriving the timing and magnitude of mitigation options.

Adaptation options that also serve the purpose of mitigation will lower the cost of stabilizing climate change. Forestry options, such as mangrove forests, can help alleviate the impacts of sea-level rise while simultaneously mitigating climate change. Planting trees in strategic locations around homes or on rooftops in urban areas will reduce urban temperatures and thus the demand for electricity for space conditioning. The costs of these adaptation/mitigation options need to be examined from the perspective of their dual benefits.

## Notes

1. The Energy Modeling Forum–21 is currently assessing the impact of including non-$CO_2$ gases and sinks in models of the global economy.

2. These cost estimates are derived using discount rates in the range of 5 to 12 percent, consistent with public sector discount rates. Private internal rates of return vary greatly and are often significantly higher, affecting the rate of adoption of these technologies by private entities.

3. Annex B refers to industrialized countries that have signed on to an emissions target in the Kyoto Protocol. In the hypothetical case of full global emissions trading, marginal cost estimates drop to US\$5–90/t C, similar to the positive marginal cost estimates for potential emissions reductions from the bottom-up studies quoted earlier.

4. GDP is an oft-used but incomplete measure of welfare.

5. This fact does not imply that technologies based on as yet unknown materials and/or physical or chemical phenomena will not be developed. Indeed, technological progress is littered with many examples of discoveries that led to significant breakthroughs in improving energy and carbon efficiency. Such discoveries will accelerate the reduction in cost and the time frame over which stabilization may be achieved.

6. "Carbon leakage" refers to an unintended failure to reduce emissions, through shifting them from one location to another or from one sector to another.

## Literature Cited

Brown, S., J. Sathaye, M. Cannell, and P. Kauppi. 1996. Mitigation of carbon emissions to the atmosphere by forest management. *Commonwealth Forestry Review* 75 (1): 283–338.

Fearnside P. 1997. Greenhouse gas emissions from Amazonian hydroelectric reservoirs: The example of Brazil's Tucurui Dam as compared to fossil fuel alternatives. *Environmental Conservation* 24 (1): 64–75.

IWG (Interlaboratory Working Group). 2000. *Scenarios for a clean energy future.* ORNL/CON-476 and LBNL-44029. Oak Ridge, TN: Oak Ridge National Laboratory, and Berkeley, CA: Lawrence Berkeley National Laboratory.

Metz, B., O. Davidson, and R. Swart, eds. 2001. *Climate change 2001: Mitigation (Contribution of Working Group III to the Third Assessment Report of the Intergovernmental Panel on Climate Change).* Cambridge: Cambridge University Press.

Neumann, J. E., G. Yohe, R. Nicholls, and M. Manion. 2000. Sea-level rise and global climate change: A review of impacts to US coasts. Prepared for the Pew Center on Global Climate Change, Washington, DC.

Ravindranath, N. H., and J. Sathaye. 2002. *Climate change and developing countries.* Dordrecht, the Netherlands: Kluwer.

Sathaye, J., D. Bouille, D. Biswas, P. Crabbe, L. Geng, D. Hall, H. Imura, A. Jaffe, L.

Michaelis, G. Peszko, A. Verbruggen, E. Worrell, and F. Yamba. 2001. Barriers, opportunities, and market potential of technologies and practices. In *Climate change 2001: Mitigation (Contribution of Working Group III to the third assessment report of the Intergovernmental Panel on Climate Change),* edited by B. Metz, O. Davidson, and R. Swart. Cambridge: Cambridge University Press.

Watson, R., I. Noble, B. Bolin, N. Ravindranath, D. Verardo, and D. Dokken, eds. 2000. *Land use, land-use change, and forestry (Special report of the Intergovernmental Panel on Climate Change).* Cambridge: Cambridge University Press.

Watson, R. T., D. Albritton, T. Barker, I. Bashmakov, O. Canziani, R. Christ, U. Cubasch, O. Davidson, H. Gitay, D. Griggs, J. Houghton, J. House, Z. Kundzewicz, M. Lal, N. Leary, C. Magadza, J. McCarthy, J. Mitchell, J. Moreira, M. Munasinghe, I. Noble, R. Pachauri, B. Pittock, M. Prather, R. Richels, J. Robinson, J. Sathaye, S. Schneider, R. Scholes, R. Stocker, N. Sundararaman, R. Swart, T. Taniguchi, and D. Zhou. 2001. *Climate change 2001: Synthesis report.* Geneva: Intergovernmental Panel on Climate Change.

# 23

# Unanticipated Consequences: Thinking about Ancillary Benefits and Costs of Greenhouse Gas Emissions Mitigation

Jae Edmonds

The scale of human activities has grown to the point that they affect the Earth's global biogeochemical cycles. The concentrations of greenhouse gases such as carbon dioxide and methane have risen to levels that exceed any observations over very long periods. For $CO_2$ present concentrations exceed observed values over the past 400,000 to 23 million years (Houghton et al. 2001). Fossil-fuel carbon dioxide emissions are chiefly, but not exclusively, associated with the increase in concentration of $CO_2$.

The Framework Convention on Climate Change (FCCC; United Nations 1992) has as its goal the stabilization of the concentration of greenhouse gases in the atmosphere at a level that would prevent dangerous anthropogenic interference with the climate. Achieving that goal implies that a future in which emissions of greenhouse gases lead to higher-than-acceptable concentrations be superseded by one in which fossil-fuel $CO_2$ emissions peak and then decline and in which non-$CO_2$ greenhouse gas emissions are limited to acceptable levels. The implications of this goal are enormous for industry, infrastructure, mobility, and energy.

To the extent that policy intervention is necessary to achieve this end, costs will be incurred.[1] Measuring the cost of limiting greenhouse gas emissions is a nontrivial undertaking. In addition to the usual definitional questions of what cost is being measured, for example, average, marginal, or total, there are additional complexities. Methodological issues abound. Metz et al. (2001) addressed many of these issues, though quantification remains a matter of significant uncertainty.

One issue that is addressed in Hourcade et al. (2001) is the question of ancillary benefits of emissions mitigation policies. Interest in this issue dates at least to the mid-1990s

419

(Pitcher et al. 1995). Hourcade et al. (2001) paid particular attention to interactions between climate policy and local air quality. Here the imposition of a value on carbon leads to reductions in the use of fossil fuels, which are also associated with emissions of non-$CO_2$ gases such as CO, $SO_2$, volatile organic compounds (VOCs), and particulate matter (PM). In general, an ancillary policy benefit occurs whenever a policy established to address one problem improves, as an ancillary consequence, the state of another problem.

More recently, studies have begun to examine the converse question: What are the climate benefits of local air-quality policies? (See, for example, Joh 2002 or Shelby 2002.) These studies offer estimates of the implications for $CO_2$ emissions of nonclimate environmental policies. In fact, it is the pursuit of other goals that leads to the stabilization of greenhouse gases in the SRES B1, A1, and A1T scenarios (Nakicenovic and Swart 2000). In the B1 scenario, it is the aggressive pursuit of local and regional air-quality objectives that leads to a peak and subsequent decline in fossil-fuel $CO_2$ emissions. Similarly, it is the pursuit of economic growth through investments in technology that has similar consequences in the A1 and A1T scenarios.

The general proposition is that policies have impacts that extend beyond their initial focus. These impacts can be either beneficial or costly. This chapter will consider the broader problem of the ancillary consequences of climate policies and suggest that an appropriately formulated policy mix is impossible without explicit, simultaneous, dynamic consideration of the full suite of environmental problems.

## The Big Picture

In principle, the identification of an "optimal" policy for both climate and local air quality must be determined simultaneously. Operationally, this is never the case. The problem is complexity and scale. Temporal and spatial scales relevant to formulating local air-quality policy are fundamentally different from the temporal and spatial scales associated with formulating climate policy. Local air quality is associated with timescales that are short in comparison to the climate question. Local air-quality issues engage modeling systems that are of a similar complexity to climate systems. While temporal and special scales may overlap at the margin, modeling systems that span the full temporal and spatial scales relevant to both questions do not yet exist and are not likely to come into being in the near term.

The consistent treatment of abatement costs and benefits of greenhouse and nongreenhouse emissions presents important methodological challenges. Constructing an analysis from a set of heterogeneous and potentially inconsistent sources can compromise the integrity of the analysis.[2] Though the present state of the art relevant to formulating simultaneous climate and local air-quality policies provides grounds for pessimism, this state of affairs should not be assumed to be permanent. It was not long ago that the integration of models relevant to climate change seemed an impossible task.

## The Climate Policy Mix

Even in the realm of climate policy, models that address the full range of policy issues are at a relatively primitive state of development. Although most of this chapter focuses on the question of policies to address emissions mitigation, it is good to recall at the outset that a much broader set of policies affect climate change. These can be grouped into at least four major categories:

- emissions mitigation,
- technology development (for both mitigation and adaptation),
- adaptation, and
- reducing scientific uncertainty.

Though the primary focus of this chapter is on the control of greenhouse gas emissions, strategies within the other categories also have potential ancillary consequences. For example, the development of technologies for a hydrogen-based transportation system would have ancillary consequences for local and regional air quality, among many other things.[3]

The extent and nature of ancillary effects of a greenhouse gas emissions policy depend on a variety of factors, including (1) the nature of the pollutant, (2) the policy environment, (3) the technology associated with the emission and emission control options, and (4) system responses.

## The Nature of the Pollutant

The nature of the pollutant itself can affect the degree to which there are ancillary benefits to a policy intervention. Banning the production of chlorofluorocarbons (CFCs) had no direct effect on other pollutants. And yet CFCs have two natures. In their life as an atmospheric constituent, they act as a greenhouse gas. When they dissociate, they become an ozone-depleting substance. The motivation for banning the production of CFCs under the Montreal Protocol was primarily to protect stratospheric ozone. Yet there was an ancillary climate benefit.

From this perspective, sulfur emissions raise some thorny problems. They are a greenhouse-related emission in their own right, though their principal association is with acid deposition. They can be coproduced with $CO_2$ emissions. Unlike $CO_2$, their net effect in the atmosphere is likely to cool the surface. Their direct plus indirect effect is potentially large but highly uncertain. Their atmospheric residence time is extremely short compared with $CO_2$. And there is no established procedure for comparing their emission with that of $CO_2$. To the extent that control of $CO_2$ results in reductions of sulfur emissions, there is an ancillary benefit associated with reduced acid-precursor emissions and an ancillary cost associated with increased radiative forcing.

## The Policy Environment

Reductions in fossil-fuel carbon emissions may have different consequences under different technology and policy environments. For example, it is generally assumed that a policy that reduces carbon emissions reduces aerosol and particulate emissions. Yet if such a policy were applied to the utility sector, where sulfur emissions were controlled in a "cap and trade" regime, reductions in $CO_2$ emissions that resulted in lower fossil-fuel use would have little effect on aerosol production.

The policy background will vary over both time and place. Over time, controls over conventional pollutants have grown broader and tighter. Emissions that were once considered to be unavoidable or even benign by-products of the production process have come under controls. Power plant emissions of PM came under control early, followed by emissions of sulfur and nitrogen compounds. In general, controls on greenhouse gas emissions such as $CO_2$ will have a correspondingly smaller effect when local pollution is already tightly controlled. Improvements in local air quality will therefore likely be greater in those regimes in which relatively little control is exercised over emissions of local pollutants, though the precise interaction depends entirely on the nature of the policy regime.

This principle can be generalized to underscore the importance of the baseline. Results will be highly dependent on the circumstances defined by baseline assumptions. Reference emissions of both greenhouse and local air-quality pollutants will play a central role in determining the measured benefits and costs associated with any policy intervention.

## The Technology of Emissions and Emissions Control Options

Clearly, a climate mitigation or conventional pollutant abatement policy is affected by the very nature of the technology associated with the emission. Power plants have a different set of characteristics associated with their operation than a passenger automobile. Each sector, subsector, and technology can be expected to have its own unique profile. Policies that attempt to control emissions through control of the emission's activity level will have one profile of emissions mitigation and ancillary impacts. Consider the hypothetical example of a tax on cattle to reduce methane emissions. Reducing the herd size would result in, among other things, reducing nitrogen emissions also associated with cattle production. The entire premise of ancillary benefits is predicated on the observation that control technologies for one emission will result in the control of other emissions. For example, to the extent that carbon emissions from cars are controlled by higher-efficiency motors, generally lower local air-pollution emissions will also ensue.

## System Responses

Economic and energy systems are highly intertwined. It is virtually impossible to make a change in the operation of one sector and not see accommodating changes in other parts of the system. From the perspective of pollution, these changes can amplify the original change, dampen it, or even reverse it.

As the scale of the policy intervention increases, the value of changes in the ancillary activity also changes. Just as the marginal cost of emissions mitigation might be expected to rise as the most attractive options for mitigation are exhausted, leaving increasingly difficult reductions, the degree of ancillary benefit might decline as the ancillary activity is reduced in scale. Initial improvements in local air quality tend to have higher benefits than later reductions. The assumption of constant marginal benefits, often used in studies of ancillary benefits, is dubious for nonmarginal changes in local ancillary emissions.

The discussion of ancillary consequences of a policy to control greenhouse gas emissions usually focuses on pollutants. Still, one of the most important ancillary consequences of a policy to limit greenhouse gas emissions is the economic cost of the policy—that is, the aggregate value of the resources that must be obtained from other sectors of the economy to achieve the greenhouse gas emissions mitigation objective. Schelling (1996) has argued, for example, that for developing countries the benefits from economic growth are more valuable than the potential benefits from a moderation of climate change.

The implementation of environmental policies can have other effects as well. They require implementation, and for societies for which institutions are still being developed, the implementation of a policy affecting as pervasive a commodity as energy requires institutions capable of monitoring verification and enforcement. A potentially important ancillary benefit to the implementation of the policy could be institution building.

Institutions play another important role. They determine the efficiency with which a policy is implemented. To the extent that institutions are poorly developed and policy implementation is incomplete, costs will be higher and benefits lower. Even in a world in which institutions are well defined, the nature and role of institutions can affect both costs and benefits. For example, in a world in which clean development mechanism (CDM) credits are traded internationally, the value of ancillary benefits can affect both the supply of and demand for emissions mitigation.

## The Energy System

Kaya (1989) argued that emissions of carbon dioxide to the atmosphere could be usefully thought of as the product of population, per capita gross domestic product (GDP), energy intensity of GDP, and carbon intensity. Each of these four elements could, in principle, be the object of policy intervention to address $CO_2$ emissions.

It is difficult to imagine that policies to control either the number of persons in a society or the aggregate scale of economic activity would have climate change as their primary motivation. If such were the case, however, ancillary consequences would be pervasive. Some governments have argued that demographic policies, though not initially founded on the motivation to reduce greenhouse gas emissions, have nonetheless had a more profound effect than explicit climate policies in Annex I nations. The experience of the Russian Federation stands as a stark example of the implications of $CO_2$ emissions reductions by means of GDP decline.

A global energy system, consistent with increasing standards of living around the world, and an associated expansion in the demand for energy services will invariably be larger than the present system. Achieving the goal of the FCCC at concentrations of 750 parts per million (ppm) or lower means that the primary mechanisms by which the expanded provision of energy services is to be accomplished cannot simply be the expanded use of fossil fuels with freely vented carbon dioxide. While carbon can continue to be emitted throughout the century, stabilizing the concentration of $CO_2$ at 750 ppm or lower requires that emissions peak and then decline indefinitely thereafter. Such circumstances require the expanded deployment of some or all of the following classes of energy technology:

- energy intensity improvements
- low-carbon fossil fuels, such as methane or carbon capture and disposal
- renewables, such as wind and solar
- biomass
- nuclear power (fission and fusion)
- hydrogen

The control of greenhouse gas emissions and, in particular, $CO_2$ emissions will likely lead to the expanded scale of some or all of these technology classes.

A variety of ancillary consequences is likely, as these technologies expand their scale of activity. In many cases these potential consequences are complex, including both ancillary benefits and costs. The only technologies without ancillary consequences are those that have not yet been deployed at scale.

## Energy Intensity Improvements

Improvements in energy intensity, the ratio of energy consumption to some measure of product such as GDP, result from improvements in energy efficiency in end-use devices, as well as changes in the relative and absolute composition of energy-consuming activities. Improvement in energy efficiency occurs when the amount of energy required to provide a specific energy service declines.

Most improvements in energy efficiency bring with them improvements in all aspects of the energy service. They are driven by the forces of technology improvement and find their way into use via economic forces. Energy efficiency is generally not purchased at the cost of forgoing energy services. Improvements in energy intensity will have both proximate effects and pervasive effects. Direct effects through energy efficiency improvements will depend on the specific technology changes. Direct effects through changes in the composition of energy-using activities will have other and different effects. Indirect effects, other things being constant, would have pervasive ancillary consequences because they imply generally smaller energy supply, transformation, transport, and storage.

The direct consequences of changes in energy technology may be complex and difficult to measure. Consider, for example, improvements in energy efficiency in passenger transportation achieved via a reduction in the weight of the vehicle. The energy required to create the frame of an automobile may decline as the result of the substitution of newer, lighter materials for steel, and to the extent that the energy is associated with emissions of local and regional air pollutants, this change implies a net benefit. But the ancillary consequences can be still richer. For example, improving fuel economy in cars by reducing size and weight could result in either reduced or enhanced automotive service. For example, if the transportation service were provided by a smaller vehicle, the total transport service might be reduced because the smaller vehicle may provide less space and reduced structural resilience and safety. The development of advanced technologies could require exotic materials. The increased presence of exotic and heavy metals can themselves cause problems. For example, the development of more efficient transformers resulted in greater deployment of PCBs. On the other hand a smaller, lighter vehicle might be more maneuverable, provide extended range between fueling, and reduce overall vehicle cost, and the employment of new materials and designs could increase structural resilience. To the extent that the service provided by the non-emitting technology is different from the service provided by the technology it replaces, ancillary benefits and costs are created. A lower cost in the provision of transportation services, as measured by financial costs per passenger distance, can be enhanced or offset by nonfinancial costs associated with time commitments on the part of the vehicle driver.

## Low-Carbon Fossil Fuels

Not all fossil fuels release the same amount of carbon per unit energy. Natural gas releases less carbon per unit energy than oil, which in turn releases less than coal. A variety of options exist to reduce the emission of greenhouse gases by substituting one fossil fuel for another. Expanding the use of natural gas in substitution for other fossil fuels could have a variety of ancillary effects. First, because methane is a generally cleaner fuel than oil or coal, it can reduce the emission of local air pollutants, including PM, hydrocarbons, and sulfur. On the other hand, increased leakage from the natural gas transport and distribution system could introduce additional methane into the atmosphere, also a potent greenhouse gas. Expanding the use of natural gas significantly would require the construction of land-based transportation infrastructure. Natural gas pipelines are generally considered to convey a disadvantage to those through whose neighborhoods they pass. They bring some risk of blowouts and leakage and aboveground are visually unattractive. Furthermore, the transport of liquefied natural gas introduces a potential environmental and health risk through the vessels and storage systems. The high energy density of liquid natural gas could result in a spectacular discharge in the event of accident or sabotage.

Another potential technological response to incentives to reduce greenhouse gas emissions is the capture and disposal of carbon from the combustion of hydrocarbons.

The availability of such technology would enable fossil fuels to continue to be used while limiting greenhouse gas emissions. Preliminary analysis indicates that, depending on the cost of the systems, cumulative capture could be in the range of 100 to 300 PgC. Such quantities are wholly unprecedented. While carbon is presently captured for industrial purposes, the scale of current operations is three orders of magnitude smaller than these quantities.

The prospect of such scale raises a large number of concerns, including many questions of ancillary consequences. For example, it will be important to consider the health effects of shipments of $CO_2$ from capture to disposal sites. While $CO_2$ is presently shipped long distances, the scale of transport could become far greater in the future. Poor design of transport systems could cause ancillary damage. It will be important to monitor disposal sites to ensure that subterranean migration of $CO_2$ is monitored and that no unmonitored discharges occur. On the other hand, obtaining $CO_2$ from waste gas streams requires relatively clean exhaust gas, with the conventional pollutants removed. To the extent that the capture of $CO_2$ results in lower conventional pollutants, its deployment could have classic ancillary benefits.

## Renewable Energy

Renewable energy has long been attractive as a vehicle for energy production without concurrent greenhouse gas emissions. Furthermore, it involves no or limited risks from conventional pollutants, a clear ancillary benefit. Yet, as the scale of renewable technologies increases, issues may arise. Wind power, for example, provides electric power without either conventional pollution or greenhouse gas emissions. Reductions in cost have accelerated its deployment in the market. Yet with greater scale of deployment, a variety of unanticipated ancillary effects have come to the fore. Wind towers have been criticized for their noise, visual pollution, "ugliness," and avian impacts. Reliable mechanisms to prevent bird kills are still under development. Excellent wind sites have been abandoned in the United States for aesthetic reasons. Yet just as traditional fossil energy technologies have changed and adapted to address unintended consequences, especially pollution, renewable energy technology expansion can be expected to show a similar adaptive resilience.

Solar photovoltaic (PV) cells could meet similar reactions if deployed in large, stand-alone arrays. Deployed at a scale to deliver a significant fraction of an expanded global energy system, these arrays could simultaneously reduce greenhouse gas emissions and local air pollution by reducing the use of fossil fuels, a classic ancillary benefit. They also represent a significant land use with major consequences for local ecosystems, with potential for ancillary damage.

Just as for wind, this is unlikely to be the end of the story. Other deployment strategies are likely to evolve to mitigate undesirable effects. For example, deployed as part of building materials, PV arrays could provide power while utilizing the same land surface area as the infrastructure it serves. Yet the efficiency of collection would doubtless suffer, and the efficiency of power would be lower than power provided by a structure whose primary function was power generation. But if the cost of collection were sufficiently low, such arrays could, nevertheless, be deployed widely. The development of new technologies to be used at a large scale requires consideration of effects that are unimportant or trivial at small scales.

## Modern Commercial Biomass

Growing crops for their energy content is not a new idea. The first human use of energy employed the use of material of biological origin, such as sticks, dung, or straw. Traditional biomass fuels are not systematically produced. They are themselves by-products of some other activity. The supply either comes from nature (e.g., detritus or deforestation) or is a by-product of some other, controlling activity (animal waste, crop residues). Biomass fuels are generally employed with little or no processing. To the extent that population and natural production rates are in harmony, the employment of this energy form has only a modest direct effect on land use or ecosystems. It may, however, have a significant impact on indoor and local air quality. The environmental and health effects of traditional biomass are many, varied, and interconnected. To the extent that population and natural production rates are inconsistent, demands for traditional biomass fuels can imply significant pressure on local ecosystems or global deforestation, a major source of net carbon emissions to the atmosphere. Globally, the per capita use of traditional biomass continues to decline.

By contrast, modern commercial biomass is produced by cropping. It employs modern agricultural methods and could potentially expand to encompass a major fraction of managed lands. This eventuality could have profound consequences for land use. Several changes could follow in the wake of this development. Because the quality and extent of land resources are fixed, the introduction of biomass crops could imply a significant new farming activity. This in turn would imply an increased demand for managed lands. The increased demand for managed lands would, in turn, exert upward pressure on the value of land and the price of all land-utilizing products, including food, fiber, pastured cattle, and forest products. The increased value of land would invariably encourage the utilization of land not yet managed or, alternatively, slow the return of managed lands to unmanaged states. The significance of these effects will depend, to a substantial degree, on the rate of productivity growth in crops and other land-utilizing activities. Faster rates of productivity growth can be expected to lessen the adverse consequences, allowing continued reductions in the cost of food and fiber, continued improvements in human health and nutrition, and reduced pressure on unmanaged ecosystems. To the extent that historical rates of productivity growth are associated with the increased application of fertilizer to crops, the potential for productivity increases in the future may be limited, particularly in presently high-productivity settings, and may have implications for air and water resources. Advances in the biological sciences could prove useful in maintaining the rate of future productivity growth. But, genetically modified (GM) crops are not universally popular. Opposition to the deployment of GM technology cites the prospect for a variety of unintended and undesired consequences, ranging from irreversible modifications to unmanaged ecosystems to undesired health effects.[4]

## Nuclear Power

Nuclear power has the attractive property of not emitting greenhouse gases to the atmosphere in the process of producing electricity. Yet, the expanded deployment of nuclear power from fission reactors faces several challenges, most of which are associated with ancillary consequences. These include health and safety, weapons proliferation, and waste disposal. In addition, nuclear energy, like all energy technologies, must meet the test of the market. It must be able to provide power in a cost-competitive manner. In a greenhouse-constrained world, nuclear's lack of GHG emissions could be rewarded relative to competing technologies with net emissions, though not relative to nonemitting technologies. The magnitude and even existence of such a relative advantage is, however, completely dependent on the policy environment.

The ancillary issues facing nuclear power are nontrivial. Health and safety have always been matters of concern to the industry, and despite its comparatively favorable record in aggregate, major accidents at Three Mile Island and Chernobyl and other lesser events leave the question of health and safety as acute as ever. The potential for intentional misuse of nuclear fuels in the form of weapons is another issue. As the use of nuclear power grows, so too does the volume of nuclear materials to be protected and the prospect for a disastrous event. The deployment of fast-breeder reactor technology could magnify the problem by introducing potentially large volumes of weapons-grade fuels into circulation. The creation of nuclear waste is yet another ancillary consequence of the expanded deployment of nuclear technology. Technical strategies for addressing this issue exist, though "not in my back yard" (NIMBY) reactions to the establishment of long-term disposal sites persist.

## Hydrogen and Transport Systems

The attraction of hydrogen is obvious. It is an energy carrier whose by-product emissions are limited to water vapor.[5] Hydrogen can be employed in a variety of machines, including furnaces, engines, and fuel cells. Significant deployment raises a variety of questions about the nature of the production and delivery systems. Because hydrogen is not a primary energy form, it requires input of some other primary energy. Hydrocarbons—oil, gas, coal, or biomass—contain hydrogen and can be used as feed stocks to produce hydrogen for other applications. Hydrocarbons also contain carbon, which potentially could be released into the atmosphere unless that carbon can be collected and disposed in a way that permanently isolates it from the atmosphere. The exception is, of course, biomass, having derived its carbon from the atmosphere in the first place. If its carbon is captured and permanently isolated from the atmosphere, its use constitutes a negative net emission. But each of these primary energy forms has both ancillary benefits and costs. Hydrogen can also be produced by electrolysis, splitting water into hydrogen and oxygen. Again, however, the question arises as to the source of the electricity. Each of the potential producers of electricity has its own set of ancillary consequences. Furthermore, the hydrogen infrastructure question will doubtless have ancillary effects. Transport and storage will entail significant infrastructure requirements, and the creation of that infrastructure may bring its own set of issues. Health and safety concerns have been raised. With hydrogen as a major fuel, potentially significant water vapor will be created locally. If captured, this could be an ancillary benefit, particularly in places where water is scarce. On the other hand, the implication of large-scale consumption of $H_2$ for local atmospheric oxidization capacity is unknown.

## Final Comments

The simple conclusion of this essay is that the implementation of policies to mitigate the emission of greenhouse gases will have ancillary consequences—both benefits and costs. This is the inevitable implication of attempting to solve a multi-attribute control problem piecemeal. Piecemeal solutions work only with the simplest of problems or at the smallest of scales, and climate and local pollution problems are neither simple nor small in scale.

## Acknowledgments

The research reported in this chapter was made possible in part by support from the Integrated Assessment program in the Office of Science, U.S. Department of Energy, and by EPRI. I am further indebted to many people, whose comments and suggestions have served to improve the chapter, including Liz Malone, Charlette Geffen, John Clarke, Jim Dooley, Gerry Stokes, Rich Richels, and John Weyant. I retain responsibility for any remaining errors of opinion or fact.

## Notes

1. For concentrations ranging from 350 ppm to 750 ppm, stabilization of the concentration of $CO_2$ requires an emissions peak in the 21st century. Stabilization of the concentration of $CO_2$ in the atmosphere may or may not require policy intervention. Under some circumstances stabilization occurs as a consequence of the pursuit of other, nonclimate goals. For example, three of the six marker scenarios in Nakicenovic and Swart (2000), the SRES B1, A1, and A1T scenarios, are consistent with stabilization of $CO_2$ concentrations. For these scenarios, the pursuit of nonclimate goals has consequences for climate change. The converse can also occur. That is, the pursuit of climate change policies can have nonclimate consequences.

2. See the IPCC treatment of $SO_2$ benefits and costs, Chapter 8, Metz et al. (2001).

3. In addition, the mix of strategies would be expected to evolve over time. Strategies that play a supporting role in the near term may well move to center stage in the long term.

4. Similarly, dilemmas confront the development of rice strains that produce lower methane release, as they entail change in either the rice strain planted and/or cultural norms. Fewer concerns attend GM cattle, which could conceivably be introduced to reduce ruminant methane emissions.

5. While water vapor is a potent greenhouse gas in the atmosphere, the scale of emission associated with even large-scale deployment in the global energy system is presently thought to be of insufficient magnitude to affect global biogeochemical processes.

## Literature Cited

Houghton, J. T., Y. Ding, D. J. Griggs, M. Noguer, P. J. van der Linden, and D. Xiaosu, eds. 2001. *Climate change 2001: The scientific basis (Contribution of Working Group I to the third assessment report of the Intergovernmental Panel on Climate Change)*. Cambridge: Cambridge University Press.

Hourcade, J.-C., P. R. Shukla, L. Cifuentes, D. Davis, J. Edmonds, B. Fisher, E. Fortin, A. Golub, O. Hohmeyer, A. Krupnick, S. Kverndokk, R. Loulou, R. Richels, H. Segenovic, and K. Yamaji. 2001. Global, regional, and national costs and ancillary benefits of mitigation. Pp. 499–559 in *Climate change 2001: Mitigation (Contribution of Working Group III to the third assessment report of the Intergovernmental Panel on Climate Change)*, edited by B. Metz, O. Davidson, R. Swart, and J. Pan. Cambridge: Cambridge University Press.

Joh, S. 2002. Hybrid top-down and bottom-up modeling approach to integrated climate change and air quality strategies in Korea. Paper presented to Sino-Korea-U.S. Economic and Environmental Modeling Workshop, Beijing, November 7–8.

Kaya, Y. 1989. Impact of carbon dioxide emission control on GNP growth: Interpretation of proposed scenarios. Presentation to the Energy and Industry Subgroup, Response Strategies Working Group, Intergovernmental Panel on Climate Change, Paris, France.

Metz, B., O. Davidson, R. Swart, and J. Pan, eds. 2001. *Climate change 2001: Mitigation (Contribution of Working Group III to the third assessment report of the Intergovernmental Panel on Climate Change)*. Cambridge: Cambridge University Press.

Nakicenovic, N., and R. Swart, eds. 2000. *Emissions scenarios (Special report of the Intergovernmental Panel on Climate Change)*. Cambridge: Cambridge University Press.

Pitcher, H. M., C. MacCracken, S. Kim, M. Wise, R. Sands, E. Malone, K. Fisher-Vanden, and J. Edmonds. 1995. *Ancillary benefits of stabilizing $CO_2$ emissions: Conventional air pollutants*. DE-AC06-76L001831, DE-AC06-76L001830. Prepared by Pacific Northwest Laboratory for the U.S. Environmental Protection Agency and for the U.S. Department of Energy.

Schelling, T. C. 1996. The economic diplomacy of geoengineering. *Climatic Change* 33:303–307.

Shelby, M. 2002. The ancillary carbon benefits of $SO_2$ reductions from a small-boiler policy in Taiyuan, China. Paper presented to Sino-Korea-U.S. Economic and Environmental Modeling Workshop, Beijing, November 7–8.

United Nations. 1992. *United Nations Framework Convention on Climate Change*. New York.

# 24

# International Policy Framework on Climate Change: Sinks in Recent International Agreements

Maria José Sanz, Ernst-Detlef Schulze, and Riccardo Valentini

Climate change is one of the most significant sustainable development challenges facing the international community. It has implications not only for the health and well-being of the Earth's ecosystems, but also for the economic enterprises and social livelihoods that we have built upon this base. Creative responses based on solid research, shared knowledge, and the engagement of people at all levels are required to meet the challenge posed by climate change. International agreements are one of the fundamental tools for effective action. To be effective, international agreements depend on scientific inputs built on a foundation of the best technology available.

## International Policy Framework to Address Global Climate Change

Climate change was first recognized as a serious problem by a major intergovernmental meeting in February 1979. The First World Climate Conference, held in Geneva, was an important scientific event (WMO 1979). It issued a declaration calling on the world's governments "to foresee and prevent potential man-made changes in climate that might be adverse to the well-being of humanity."

A large number of international conferences on climate change have been convened since then: Attended by government policy makers, scientists, and environmental groups, they have addressed both scientific and policy issues (Table 24.1). The Second World Climate Conference, held in 1990 in Geneva, was a particularly crucial step toward a binding global convention on climate change. Some of these meetings have taken place under the auspices of the United Nations and its specialized agencies. Others have been held within regional and global forums such as the European Union, the Commonwealth, and the South Pacific Forum or have been convened by individual governments.

At the 1992 Earth Summit in Rio de Janeiro, the international community adopted Agenda 21, an unprecedented global plan of action for sustainable development. The summit also agreed on the Rio Declaration on Environment and Development, a set of principles defining the rights and obligations of nations, and on a Statement of Forest Principles to guide more sustainable management of the world's forests. Agenda 21 was a landmark achievement in integrating environmental, economic, and social concerns into a single policy framework. It contains over 2,500 wide-ranging recommendations for action, including detailed proposals for reducing wasteful consumption patterns, combating poverty, protecting the atmosphere, oceans, and biodiversity, and promoting sustainable agriculture. The proposals presented in Agenda 21 remain sound, and they have since been expanded and strengthened at several major United Nations conferences on population, social development, women, cities, and food security.

The 1992 UN Framework Convention on Climate Change (UNFCCC) is the first binding international legal instrument to specifically address climate change. After 15 months of intensive negotiations, it was adopted in May 1992 within the Intergovernmental Negotiating Committee for a Framework Convention on Climate Change (INC/FCCC). In June 1992 it was opened for signature in Rio de Janeiro at the UN Conference on Environment and Development (UNCED). The INC negotiators drew on the First Assessment Report (FAR) of the Intergovernmental Panel on Climate Change (IPCC), a body established jointly by the United Nations Environment Programme and the World Meteorological Organization (WMO). They were also influenced by the Ministerial Declaration issued by the Second World Climate Conference and by policy statements adopted by numerous other climate conferences.

The Convention provided a general framework for addressing the climate change issue. The Convention was signed by 154 states (including the United States and the European Union) during UNCED. Other states have signed since then, and some national legislatures have ratified. The FCCC entered into force after it was ratified by 50 states.

Even before the Convention was adopted, some countries had already taken unilateral action at the national level. Most member states of the Organisation for Economic Co-operation and Development (OECD) have set national targets for stabilizing or reducing their emissions of greenhouse gases. In 1990 the Council of the European Communities (EC) adopted a policy that provides for stabilizing the emissions of carbon dioxide—the most significant greenhouse gas—at 1990 levels by the year 2000. A strategy to limit carbon dioxide emissions and to improve energy efficiency is currently being elaborated by the EC Commission.

In addition, two other international environmental treaties address climate change indirectly. The amended 1987 Montreal Protocol on Trace Gases that Deplete the Ozone Layer legally obliges its parties to phase out chlorofluorocarbons (CFCs) by the year 1996. Although inspired by concern over the destruction of the ozone layer, this

**Table 24.1.** Major conferences on global climate change

| Period | When | What | Where |
|---|---|---|---|
| Pre-UNCED | June 1988 | The Changing Atmosphere (UNEP) | Toronto, Canada |
| | May 1989 | Forum on Global Change | Washington, DC |
| | July 1989 | Summit of the Arch (G-7) | Paris, France |
| | Nov. 1989 | Ministerial Conference | Noordwijk, Netherlands |
| | May 1990 | Interparliamentary Conference | Washington, DC |
| | July 1990 | Economic Summit (G-7) | Houston, TX |
| | Nov. 1990 | Second World Climate Conference | Geneva, Switzerland |
| | 1989–1992 | INC negotiations: U.N. FCCC adopted | New York, NY |
| UNCED | June 1992 | UNCED (Earth Summit) | Rio de Janeiro, Brazil |
| (FCCC opened for signature) | March 1994 | FCCC enters into force | |
| Post-UNCED | April 1995 | COP-1, Berlin Mandate | Berlin, Germany |
| | July 1996 | COP-2, Ministerial Declaration | Geneva, Switzerland |
| Pre-Kyoto | Oct. 1997 | "Challenge of Global Warming" Conference | Washington, DC |
| Kyoto | Dec. 1997 | COP-3, U.N. Kyoto Protocol text adopted | Kyoto, Japan |
| Post-Kyoto | | COP-4 | |
| | | COP-5 | |
| | Oct. 2000 | COP-6 | The Hague, Netherlands |
| | June 2001 | COP-6bis, Bonn Agreement | Bonn, Germany |
| | Nov. 2001 | COP-7, Marrakech Accords | Marrakech, Morocco |
| | Aug.–Sept. 2002 | World Submit on Sustainable Development (WSSD) | Johannesburg, South Africa |
| | Oct. 2002 | COP-8 | New Delhi, India |

protocol is also significant for climate change, since CFCs are potent greenhouse gases (Prinn, Chapter 9, this volume). Similarly, the 1997 Geneva Convention on Long-Range Trans-boundary Air Pollution and its protocols regulate the emission of noxious gases, some of which are precursors of greenhouse gases. These treaties, however, do not address the complex set of interrelated climate issues.

While the UNFCCC focused on voluntary actions to be taken by the year 2000 for long-term control of the total concentration of atmospheric greenhouse gases, subsequent attention focused on regulating and reducing greenhouse gas emissions after the year 2000. As early as 1995, when UNFCCC parties were advised that it was unlikely that the voluntary goals of that treaty would be met, a Conference of Parties (COP) authorized under the FCCC began to consider legally binding measures to reduce greenhouse gas emissions and proposed to craft a protocol or some other legal instrument that would be binding on the industrialized and developing countries. The Berlin Mandate, which was adopted at the first meeting of COP (COP-1) in 1995, proposed dealing with future climate change by strengthening existing commitments under the FCCC. It also, however, continued to exempt developing countries, which are parties, from any new binding commitments related to controls on greenhouse gas emissions. Shortly thereafter, in December 1995, the IPCC released its second assessment report on climate. Despite debate on its scientific findings, the United Nations endorsed the second assessment report as the basis and scientific guidelines for negotiations on further action to limit possible human alteration of the climate system. One of the main findings of the second assessment report—and one that has attracted considerable debate—was that "the balance of evidence suggests a discernible human influence on the climate system" (Houghton et al. 1996: 4). At the conclusion of COP-1, a two-year analysis and assessment phase was undertaken to consider the possible elements of a regulatory instrument to limit greenhouse gas emissions.

## The Kyoto Protocol

Successive negotiations by the Conference of Parties (COP) of the UNFCCC helped to forge a December 1997 accord, the UNFCCC Kyoto Protocol, an international treaty that, if it enters into force, would implement the first legally binding reduction of greenhouse gas emissions with the aim of stabilizing (if not reducing) atmospheric concentrations of these pollutants at some point in the future.[1] Different countries would be bound by different levels of responsibility and compliance under the Protocol, but the combined efforts required of industrialized countries alone would be expected to reduce global emissions of greenhouse gases by approximately 5 percent from 1990 levels by the year 2012. The protocol (as yet unratified) tries to achieve a balance between technically and economically feasible ways to reduce the anthropogenic emissions and to increase storage of carbon in terrestrial ecosystems through legally binding commitments. The final aim is to reduce the concentration of $CO_2$ and other greenhouse gases (GHGs) in the atmosphere. To enter into force, the Kyoto Protocol requires at least 55 parties of the UNFCCC to ratify, and it requires that ratifying countries account for at least 55 percent of the base year emissions (1990).

Fossil-fuel emissions were identified as the main source of greenhouse gases—especially $CO_2$. The focus on $CO_2$ immediately opened a debate on energy use in the industrialized world, and it initiated major efforts to invest in research and development in the field of alternative and renewable energy sources. The emission of fossil carbon as $CO_2$ is part of a much larger, predominantly natural cycle of C assimilation (photosynthesis) and respiration. This natural cycle has been disturbed, however, not only by fossil-fuel emission, but also by land use changes (mainly deforestation and harvest of primary forest) (Prentice et al. 2001). The carbon cycle links biology with industrial and agricultural production, with the consequence that practically no group in society is unaffected by changes in the carbon cycle or its management. This is presumably why it has proved so difficult to turn an apparently simple request—to begin to adjust fossil-fuel emissions downward toward a rate that can be reabsorbed by natural processes—into action.

The Kyoto Protocol core elements can be summarized as follows (Schulze et al. 2002):

1. The industrialized countries and those with economies in transition (the 45 "Annex B" countries) commit themselves to a reduction in fossil-fuel emissions in the first commitment period (2008–2012) compared with emissions in the base year 1990 (Article 3.1). The reduction should be at least 5 percent when averaged across all Annex B countries. The specific commitment, however, varies among countries. This results in country-specific "assigned amounts" for fossil-fuel emissions in 2008. Some countries are allowed to increase emissions, such as Australia (+8 percent). Others have to make larger reductions. The EU (–8 percent) internally has a "burden sharing" agreement that requires some parties to carry out reductions whereas others can increase emissions. The extremes are 21 percent reductions for Denmark and Germany and a 30 percent increase for Greece. The accounting will take place according to specific rules in the first commitment period between 2008 and 2012. The year 1990 (or 1997) will be the baseline year for several definitions and actions.

2. Countries are allowed to create carbon sinks in order to offset emissions. This can be done by planting new forests (afforestation and reforestation), decreasing deforestation (Article 3.3: ARD), or applying new management approaches (Art. 3.4: Additional activities[2]) in forestry, agriculture, and grazing that can create new carbon pools. These new C pools can become tradable resources in their own right.

3. The protocol can be interpreted in ways that allow countries and the private sector to trade $CO_2$ equivalent units from technological developments (e.g., improving power station efficiency) and from additional activities in forestry and agriculture. Carbon units can be traded between Annex B countries (Art 6: Joint implementations, JI) as well as between industrialized Annex B countries and developing non–Annex B countries (Art. 12: Clean development mechanism, CDM).

Some possible ramifications and implications of the treaty were not fully considered at the time the protocol was written (1997). Only after the Kyoto Protocol was cast into legal text did it become apparent that many terms in the text were not clearly defined and that the language was sometimes ambiguous, allowing counterproductive interpretations (i.e., underaccounting of sources). These "unwanted" effects might, at worst, reverse the intention of the protocol (see IGBP 1998; WBGU 1998).

Sinks from ecosystem management have been among the most contentious features of the Kyoto Protocol. Objections to carbon sinks are based primarily on two arguments. First, sinks may allow developed nations to delay or avoid technological adjustments—the "loophole" argument. Second, technical and operational difficulties would reduce the value of sinks, allowing for inflated claims of carbon offsets—the "floodgates" argument. Objections to carbon sink projects in non–Annex I countries concern "permanence" (will the stored carbon remain in storage?), "additionality" (is the carbon storage above and beyond that in the absence of direct management?), "leakage" (does carbon storage in one location increase release from another?), measurement, verification, and lack of technology transfer. Although contentious, well-designed forestry projects for mitigating global carbon emissions can provide significant environmental and socioeconomic benefits to host countries and local communities (Edmonds, Chapter 23, this volume). A well-designed regulatory framework, including adherence to international agreements on biodiversity, desertification, and wetlands, would strengthen sustainable development while also enhancing efforts to address climate change.

The Subsidiary Body on Scientific and Technological Advice (SBSTA, a committee that gives advice to the member states) responded to the arguments on carbon sinks with a request that the IPCC clarify and define the terminology and make suggestions on procedural issues. This advice was compiled in the special report *Land Use, Land Use Change and Forestry* (SR-LULUCF) (Watson et al. 2000). The IPCC was asked to be politically relevant but not prescriptive. Decisions are to be made by policy makers on the basis of options to be provided by the IPCC. Based on this IPCC report and driven by the timing of the Kyoto Protocol itself, the sixth Conference of the Parties (COP 6) met in The Hague in 2000 to settle the unresolved issues and to prepare a legal document as a basis for ratification of the protocol by the signatory nations.

## The Kyoto Protocol Concerns and Implications

A number of major ecological concerns still complicate the Kyoto Protocol and the following negotiations.

### Definitions

Even defining a "forest" is a problem because of different national circumstances. During the negotiations, it was agreed, for the first commitment period, that forests have a minimum area of 0.05 hectare (ha, i.e., 500 square meters [$m^2$]), >10 percent crown cover, and plants more than 2 m high. One problem with this definition is that deforestation may (according to the Kyoto definitions) be preceded by degradation (with associated emissions), until 10 percent of the initial cover and the corresponding amount of C is reached. If the area is then deforested, only 10 percent of the initial C pool in biomass would be counted as emission. Similarly, a partial forest harvest before deforestation would reduce the accountable loss. In contrast to the activities that create a sink (afforestation, reforestation, revegetation), many land use activities that cause emissions remained undefined (degradation, conversion of forest types, harvest) or neglected by the protocol. The inclusion of the harvest cycle still remains a matter of discussion, although any change in C stocks is supposed to appear in national reports.

## *"Additional Human-Induced Activities since 1990" in Reference to Article 3.4*

The wording of this article opens two issues: What is additional, and what is meant by "human-induced since 1990"? The basic idea of the Kyoto Protocol was to stimulate additional activities in natural carbon sequestration that exceed the pre-1990 activities. But the wording can be interpreted in many other ways and has led to major differences in philosophy and strategy in the subsequent negotiations.

## *Role of Old Forests and Plantations*

Exploitation of primary forest results in large C losses, because of the harvesting of a very large C pool that will never again be reached in a future plantation (Nabuurs, Chapter 16, this volume). Thus, this is a permanent loss of C to the atmosphere. The average lifetime of wood products, currently estimated at 15 to 20 years (Harmon et al. 1996), is much shorter than the lifetime of wood in living trees. If only "human-induced" plantations can be accounted and if natural forests are assumed to be C-neutral, it becomes attractive in the context of the Kyoto Protocol to convert non-human-induced, unaccountable old-growth forest into plantations (the harvest may remain unaccounted as normal forest practice) and claim credit for stem growth of these plantations—even though the real effect of such action is to increase the amount of $CO_2$ released to the atmosphere. In this context, primary forests are at maximum risk, and the Kyoto Protocol offers an incentive to accelerate their demise.

The negotiations started at a Conference of the Parties at The Hague (COP6) and continued in Bonn (COP6bis) and Marrakesh (COP7). The Bonn Agreement and the Marrakesh Accords represent major breakthroughs in the international efforts concerning climate change. Nevertheless, major ecological issues that emerged from the beginning of the negotiations remained untouched (Schulze et al. 2002). Afforestation and deforestation projects under the clean development mechanism are especially critical. These provisions are potentially susceptible to interpretations that create perverse incentives that lead to a loss of carbon and/or pristine forests in exchange for plantations (Schulze et al. 2003).

Many issues, it is hoped, will be resolved in the second commitment period. It is quite clear (Houghton et al. 2001) that $CO_2$ stabilization in the atmosphere requires GHG emissions ultimately to be reduced far more drastically than anticipated in the Bonn Agreement. This might require additional tools that are not yet available in the toolbox of the Kyoto Protocol. Protecting actual stocks, that is, primary existing forest, has a potential even larger than the cumulative increase in emissions expected until 2050. Further, protecting existing stocks has important associated ancillary benefits (i.e., biodiversity conservation). Though there remains a long way to go to reach the modest goals of Kyoto, the protocol can become an important step in a critical process.

# Notes

1. The legal text of the Kyoto Protocol is available at www.unfccc.org/resource/docs/cop3/01a01.pdf.

2. Forest management, revegetation, cropland management, and grassland management.

# Literature Cited

Harmon, M. E., J. M. Harmon, W. K. Ferrell, and D. Brooks. 1996. Modeling carbon stores in Oregon and Washington forest products: 1900–1992. *Climate Change* 33:521–550.

Houghton, J. T., L. G. Meira Filho, B. A. Callender, N. Harris, A. Kattenberg, and K. Maskell, eds. 1996. *Climate change 1995: The science of climate change (Contribution of Working Group I to the second assessment report of the Intergovernmental Panel on Climate Change)*. Cambridge: Cambridge University Press.

Houghton, J. T., Y. Ding, D. J. Griggs, M. Noguer, P. J. van der Linden, and D. Xiaosu, eds. 2001. *Climate change 2001: The scientific basis (Contribution of Working Group I to the third assessment report of the Intergovernmental Panel on Climate Change)*. Cambridge: Cambridge University Press.

IGBP (International Geosphere-Biosphere Programme). 1998. The terrestrial carbon cycle: Implications for the Kyoto Protocol. *Science* 280:1393–1394.

Prentice, I. C., G. D. Farquhar, M. J. R. Fasham, M. L. Goulden, M. Heimann, V. J. Jaramillo, H. S. Kheshgi, C. Le Quéré, R. J. Scholes, and D. W. R. Wallace. 2001. The carbon cycle and atmospheric carbon dioxide. Pp. 183–237 in *Climate change 2001: The scientific basis (Contribution of Working Group I to the third assessment report of the Intergovernmental Panel on Climate Change)*, edited by J. T. Houghton, Y. Ding, D. J. Griggs, M. Noguer, P. J. van der Linden, X. Dai, K. Maskell, and C. A. Johnson. Cambridge: Cambridge University Press.

Schulze, E. D., R. Valentini, and M. J. Sanz. 2002. The long way from Kyoto to Marrakesh: Implications of the Kyoto Protocol negotiations for global ecology. *Global Change Biology* 8:505–518.

Schulze, E. D., D. Mollicone, F. Achard, G. Matteucci, S. Federici, H. D. Eva, and R. Valentini. 2003. Making deforestation pay under the Kyoto Protocol? *Science* 299:1669.

Watson, R. T., I. R. Noble, B. Bolin, N. H. Ravindranath, D. J. Verardo, and D. J. Dokken, eds. 2000. *Land use, land-use change, and forestry (Special report of the Intergovernmental Panel on Climate Change)*. Cambridge: Cambridge University Press.

WBGU (German Advisory Council on Global Change). 1998. *Accounting of biological sources and sinks in the Kyoto Protocol*. Special report to the government by the German Advisory Council on Global Change. Bremerhaven.

WMO (World Meteorological Organization). 1979. *Proceedings of the World Climate Conference: A conference of experts on climate and mankind*. WMO-No.537. Geneva.

# 25

# A Multi-Gas Approach
# to Climate Policy

Alan S. Manne and Richard G. Richels

Although the Kyoto Protocol (Conference of the Parties 1997) encompasses a number of radiatively active gases, assessments of compliance costs have focused almost exclusively on the costs of reducing carbon dioxide ($CO_2$) emissions.[1] There are a number of reasons why this is the case. Carbon dioxide is by far the most important anthropogenic greenhouse gas (Houghton et al. 1996); until recently, few economic models have had the capability to conduct comprehensive multi-gas analyses;[2] and the quality of data pertaining to other greenhouse gases (GHGs) is poor (both spatially and temporally). Nevertheless, focusing exclusively on $CO_2$ may bias mitigation cost estimates and lead to policies that are unnecessarily costly. In this chapter, we examine the implications of a multi-gas approach for both short- and long-term climate policy.

A number of anthropogenic gases have a positive effect on radiative forcing (Houghton et al. 1996). We consider the three thought to be the most important: $CO_2$, methane ($CH_4$), and nitrous oxide ($N_2O$). We also consider the cooling effect of sulphate aerosols. We, however, exclude the so-called second basket of greenhouse gases included in the Kyoto Protocol. These are the hydrofluorocarbons (HFCs), the perfluorocarbons (PFCs), and sulphur hexafluoride ($SF_6$). This omission is not believed to alter the major insights of the analysis.

When dealing with multiple gases, it is necessary to find some way to establish equivalence among gases. The problem arises because the gases are not comparable. Each gas has its own lifetime and specific radiative forcing. Houghton et al. (1996) suggested the use of global warming potentials (GWPs) to represent the relative contribution of different greenhouse gases to the radiative forcing of the atmosphere. A number of studies have, however, pointed out the limitations of this approach (Prinn, Chapter 9, this volume), noting that in order to derive optimal control policies, it is important to consider both the impacts of climate change and the costs of emissions abatement.[3] GWPs do neither. Also problematic is the arbitrary choice of time horizon for calculating

439

GWPs.[4] In this chapter, we examine alternatives to GWPs that may provide a more logical basis for action.

## The Model

The analysis is based on the MERGE model (a model for evaluating the regional and global effects of greenhouse gas reduction policies). MERGE is an intertemporal general equilibrium model. Like its predecessors, the version used for the present analysis (MERGE 4.0) is designed to be sufficiently transparent so that one can explore the implications of alternative viewpoints in the greenhouse debate. It integrates submodels that provide a reduced-form description of the energy sector, the economy, emissions, concentrations, temperature change, and damage assessment.

MERGE combines a bottom-up representation of the energy supply sector with a top-down perspective on the remainder of the economy. For a particular scenario, a choice is made among specific activities for the generation of electricity and for the production of non-electric energy. Oil, gas, and coal are viewed as exhaustible resources. There are introduction constraints on new technologies and decline constraints on existing technologies. MERGE also provides for endogenous technology diffusion. That is, the near-term adoption of high-cost carbon-free technologies in the electricity sector leads to accelerated future introduction of lower-cost versions of these technologies.

Outside the energy sector, the economy is modeled through nested constant elasticity production functions. The production functions determine how aggregate economic output depends upon the inputs of capital, labor and electric and non-electric energy. In this way, the model allows for both price-induced and autonomous (non-price) energy conservation and for interfuel substitution. It also allows for macroeconomic feedbacks. Higher energy and/or environmental costs will lead to fewer resources available for current consumption and for investment in the accumulation of capital stocks. Economic values are reported in U.S. dollars of constant 1990 purchasing power.

The world is divided into nine regions: (1) the USA; (2) OECDE (Western Europe); (3) Japan; (4) CANZ (Canada, Australia, and New Zealand); (5) EEFSU (Eastern Europe and the Former Soviet Union); (6) China; (7) India; (8) MOPEC (Mexico and OPEC); and (9) ROW (the rest of world). Note that the countries belonging to the Organisation for Economic Co-operation and Development (OECD) (Regions 1 through 4) together with the economies in transition (Region 5) constitute Annex B of the Kyoto Protocol.

Each of the model's regions maximizes the discounted utility of its consumption subject to an intertemporal budget constraint. Each region's wealth includes not only capital, labor, and exhaustible resources, but also its negotiated international share of emission rights. Particularly relevant for the present calculations, MERGE provides a general equilibrium formulation of the global economy. We model international trade in emission rights, allowing regions with high marginal abatement costs to purchase emission rights from regions with low marginal abatement costs. There is also trade in oil, gas, and energy-intensive goods. International capital flows are endogenous.

MERGE can be used for either cost-effectiveness or cost-benefit analysis. For the latter purpose, the model translates global warming into its market and nonmarket impacts. Market effects are intended to measure direct impacts on gross domestic product (GDP), such as agriculture, timber, and fisheries. Nonmarket effects refer to those not traditionally included in the national income accounts, such as impacts on biodiversity, environmental quality, and human health. These effects are even more difficult to measure than market effects.

For Pareto-optimal outcomes—that is, those scenarios in which the costs of abatement are balanced against the impacts of global climate change—each region evaluates its future welfare by adjusting the value of its consumption for both the market and nonmarket impacts of climate change. The market impacts represent a direct claim on gross economic output—along with energy costs, aggregate consumption, and investment. Nonmarket impacts enter into each region's intertemporal utility function and are viewed as an adjustment to the conventional value of macroeconomic consumption. For more on the model, see our web site: http://www.stanford.edu/group/MERGE/.

## The Treatment of Greenhouse Gases and Carbon Sinks

MERGE requires information on the sources of the gases under consideration, their geographical distribution, how they are likely to change over time, and the marginal costs of emissions abatement. Unfortunately, the quality of the data is uneven, particularly for the non-$CO_2$ greenhouse gases. In many instances, we have had to rely on a great deal of judgment to arrive at globally disaggregated time series. Similarly, there is a paucity of data related to the potential for carbon sinks. In this section, we identify the main sources for our estimates. In many instances, however, the cited data required some interpretation to meet the demands of the present analysis. Again, for details, see the computer program shown on our web site.

For purposes of the present analysis, greenhouse gas emissions are divided into two categories: energy related and non-energy related. MERGE tracks energy-related releases of both $CO_2$ and $CH_4$. For the reference case, the model is calibrated so that global $CO_2$ emissions approximate the Houghton et al. (1995) central case no-policy scenario (IS92a). This has been done through the adjustment of several key supply- and demand-side parameters in the energy-economy submodel. Table 25.1 presents estimates of energy- and non-energy-related $CH_4$ emissions for 1990. When a constraint is placed on GHG emissions, the choice of technologies for the energy sector is influenced by their emission characteristics.

We next turn to non-energy-related emissions. In the case of $CO_2$, we must account for other industrial releases (primarily cement production) and the net changes associated with land use. According to Houghton et al. (1995), other industrial emissions are

**Table 25.1.** Methane emissions, 1990 (millions of tons)

| Emissions source | USA | OECDE | Japan | CANZ | EEFSU | China | India | MOPEC | ROW | World |
|---|---|---|---|---|---|---|---|---|---|---|
| *Non-energy-related emissions* | | | | | | | | | | |
| Enteric fermentation | 7.8 | | | | | 5.2 | 12.0 | | 60.0 | 85.0 |
| Rice paddies | 0.0 | | | | | 18.0 | 18.0 | | 24.0 | 60.0 |
| Biomass burning | 0.0 | | | | | 10.0 | 10.0 | | 20.0 | 40.0 |
| Landfills | 8.0 | 8.0 | | 2.0 | 4.0 | 4.0 | 2.0 | | 12.0 | 40.0 |
| Animal waste | 5.0 | 2.0 | | 2.0 | 2.0 | 5.0 | 3.0 | | 6.0 | 25.0 |
| Domestic sewage | 5.0 | 5.0 | 1.0 | 1.0 | 1.0 | 1.0 | 1.0 | | 10.0 | 25.0 |
| Subtotal | 25.8 | 15.0 | 1.0 | 5.0 | 7.0 | 43.2 | 46.0 | 0.0 | 132.0 | 275.0 |
| *Energy-related emissions* | | | | | | | | | | |
| Gas | 3.5 | 3.3 | 0.4 | 1.2 | 29.2 | 0.6 | 0.2 | 12.0 | 4.7 | 55.0 |
| Coal | 7.1 | 2.6 | 0.0 | 1.0 | 10.0 | 20.8 | 0.5 | 0.0 | 2.9 | 45.0 |
| Subtotal | 10.7 | 5.9 | 0.4 | 2.2 | 39.2 | 21.4 | 0.7 | 12.0 | 7.6 | 100.0 |
| Anthropogenic | 36.5 | 20.9 | 1.4 | 7.2 | 46.2 | 64.6 | 46.7 | 12.0 | 139.6 | 375.0 |
| Natural | | | | | | | | | | 160.0 |
| Total | | | | | | | | | | 535.0 |

*Sources:* Houghton et al. (1995) and IEA (1998).

**Table 25.2.** Potential sink enhancement in 2010 at a marginal cost of US$100 per ton of carbon (million tons of carbon)

| Region | Potential sink enhancement |
|--------|:--------------------------:|
| USA | 50 |
| OECDE | 17 |
| Japan | 0 |
| CANZ | 50 |
| EEFSU | 34 |
| China | 25 |
| India | 13 |
| MOPEC | 25 |
| ROW | 250 |
| World | 464 |

*Source:* Houghton et al. (1996).

relatively small. These are exogenous inputs into MERGE. With regard to land use, we assume that, in the absence of policy, the mass of carbon in the terrestrial biosphere remains constant.

This raises the issue of carbon sink enhancement. The protocol states that Annex B commitments can be met by the net changes in greenhouse gas emissions from sources and removal by sinks resulting from direct human-induced land use change and forestry activities limited to aforestation, reforestation, and deforestation since 1990, measured as verifiable changes in stocks in each commitment period (Conference of the Parties 1997). There is some confusion, however, regarding the treatment of soil carbon. This issue has been flagged for further study in the protocol. For the present analysis, we have adopted the values shown in Table 25.2 for 2010. We suppose that marginal sink enhancement costs are proportional to the quantity of enhancement. We also assume that the potential for sink enhancement increases over time.

Table 25.1 also includes non-energy-related $CH_4$ emissions. Reductions from the reference path are determined by a set of time-dependent marginal abatement cost curves. In 2010 the curve is calibrated based on Reilly et al. (1999).[5] For later years, the marginal cost of emissions abatement declines as a result of technical progress.

Nitrous oxide emissions are treated in a manner similar to non-energy-sector $CH_4$ emissions. Table 25.3 reports estimates for 1990. A marginal abatement cost curve for each region is constructed for each commitment period. For 2010 we again rely on the work of Reilly et al. (1999). Similarly, for later years, we assume that the marginal cost of emission abatement declines with technical progress.

**Table 25.3.** Anthropogenic nitrous oxide emissions, 1990 (millions of tons)

| Region | Anthropogenic nitrous oxide emissions |
|--------|---------------------------------------|
| USA | 1.1 |
| OECDE | 0.8 |
| Japan | 0.1 |
| CANZ | 0.3 |
| EEFSU | 0.3 |
| China | 0.7 |
| India | 0.5 |
| MOPEC | 0.2 |
| ROW | 1.7 |
| World | 5.7 |

*Source:* Houghton et al. (1995).

## Alternative Approaches to GWPs

A multi-gas approach to climate policy raises the issue of trade-offs among gases. In this chapter we explore two alternatives to GWPs—one based on cost-effectiveness and the other based on the balancing of costs and benefits. In each case the relative contribution of each gas to achieving the goal is an endogenous output rather than an exogenous input. That is, we make an endogenous calculation of the incremental value of emission rights for $CH_4$ and $N_2O$ relative to $CO_2$ and examine how the relationships might change over time.

In a cost-effectiveness analysis, the goal is to minimize the cost of achieving a particular objective. In the area of climate policy, objectives have included limits on emissions; cumulative emissions; atmospheric concentrations; the rate of temperature change; absolute temperature change; and damage.

For purposes of illustration, we begin by assuming that the goal of climate policy is to limit the increase in mean global temperature over the next two centuries. Figure 25.1 shows the price of emission rights for ceilings of 2° and 3°C. Not surprisingly, the rate of increase of these values is greater with the more stringent target. The calculations are made under the assumption of full where and when flexibility. With the first, reductions take place where it is cheapest, regardless of geographical location. With the second, they take place when it is cheapest.[6]

Figure 25.2 shows the prices of $CH_4$ and $N_2O$ relative to that of carbon. It also shows the 100-year GWPs for each gas.[7] Notice that the relative prices vary over time. This is particularly so for $CH_4$. With a relatively short lifetime, a ton emitted in the

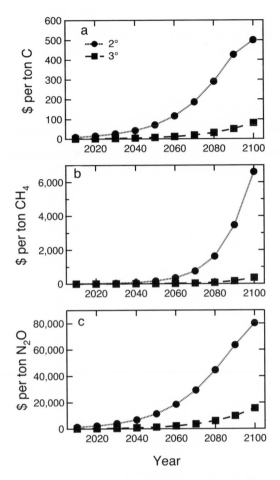

**Figure 25.1.** Incremental value of emission rights
for (a) carbon, (b) $CH_4$, and (c) $N_2O$, 2010–2100
(alternative temperature ceilings).

early decades of the 21st century will have a negligible effect on temperature in the late 21st century. As we approach the temperature ceiling, however, emitting $CH_4$ becomes increasingly problematic. By contrast, $N_2O$ has a lifetime more commensurate with that of $CO_2$. Hence, its price ratio is less volatile. The price ratios are sensitive not only to the proximity to the ceiling, but also to the ceiling itself. Limiting the temperature increase to 2° rather than 3°C produces an entirely different set of weights.

Some have suggested that damage may be sensitive to both absolute temperature

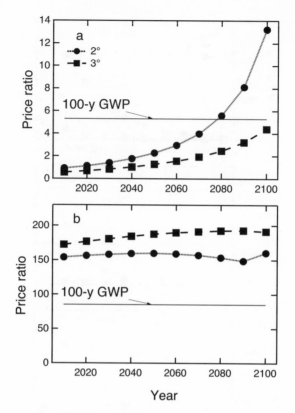

**Figure 25.2.** Prices of a ton of (a) $CH_4$ and (b) $N_2O$ relative to carbon (alternative temperature ceilings).

change and the *rate* of temperature change (Peck and Teisberg 1994; Alcamo and Kreileman 1996; Toth et al. 1997; Petschel-Held et al. 1999). To explore this possibility, we impose an additional constraint on the two temperature scenarios by limiting the allowable increase during a single decade to 10 percent of the total allowable increase. That is, decadal temperature change is limited to 0.2° and 0.3°C, respectively.

With a 2°C ceiling on absolute temperature change, there are decades during the 21st century where the limit on the rate of temperature change would be binding. From Figure 25.3, note that the price ratios for $CH_4$ reflect what we observed earlier—the closer we are to the temperature constraint, the more valuable $CH_4$ becomes. This appears to be true whether the constraint is on absolute temperature change or on the decadal rate of temperature change.

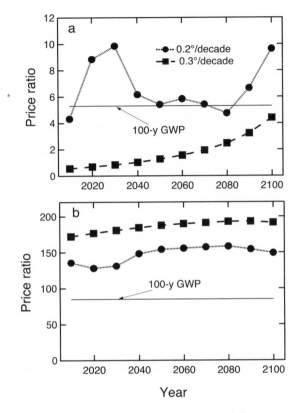

**Figure 25.3.** Prices of a ton of (a) CH₄ and (b) N₂O relative to carbon (constraint on absolute decadal temperature change).

## When the Objective Is Balancing Costs and Benefits

Thus far, we have assumed that the objective is cost-effectiveness. That is, the objective is to minimize the cost of meeting a specified target. In this section, we illustrate how the model can be used to identify relative prices when the goal is to balance the costs of abatement against possible achievements in terms of reducing environmental damage. Before proceeding, however, a major caveat is in order. Given the rudimentary state of the existing knowledge base, it would be unwise to place too much weight on any particular set of damage estimates. The purpose of this section is to illustrate how such an analysis might proceed. Nevertheless, even an illustrative analysis can yield some useful insights into the nature of the price ratios.

We begin by dividing impacts into two categories—market and nonmarket. Smith

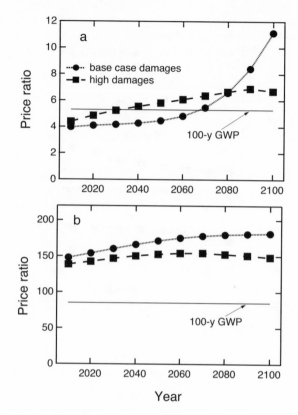

**Figure 25.4.** Prices of a ton of (a) $CH_4$ and (b) $N_2O$ relative to carbon when the objective is balancing costs and benefits.

(1996) reviewed several studies of the potential impacts of climate change on the United States. With regard to market impacts, these studies showed considerable disagreement at the sectoral level. For the present analysis, we assume that a 2.5°C increase in temperature (beyond current levels) would result in market damage on the order of 0.25 percent of GDP to Annex B countries.

Unfortunately, there has been relatively little analysis of the potential impacts of climate change on developing countries. Because they tend to have a larger share of their economies concentrated in climate-sensitive sectors, we assume that market damage is apt to be a higher percentage of GDP. Specifically, we assume that a 2.5°C increase in temperature would result in market damage of 0.50 percent of GDP in developing countries. For all regions, market damage is projected as proportional to the amount of temperature change.

For our base case estimates of nonmarket damage, we assume that when per capita incomes approach US$40,000 (about twice the 1990 level in the OECD countries), consumers would be willing to pay 2 percent of their income to avoid a 2.5°C increase in temperature. (As a point of reference, the United States today spends about 2 percent of its GDP on all forms of environmental protection.) For high-income countries, we assume that damage would increase quadratically with increases in the mean global temperature. For example, consumers would be willing to pay 8 percent of their income to avoid a 5°C increase. Of all the parameters in our model, this is perhaps the most uncertain. For purposes of sensitivity analysis, we explore a much more pessimistic scenario, one in which consumers are willing to pay 2 percent of their income to avoid just a 1°C increase in temperature and 8 percent of their income to avoid a 2°C increase.

Figure 25.4 compares the prices of $CH_4$ and $N_2O$ relative to that of carbon. These results are similar in many respects to that of the cost-effectiveness analysis with a constraint on absolute temperature change. With a temperature ceiling, we are in essence adopting an L-shaped damage function. That is, we are assuming that damage is zero below the ceiling and infinite thereafter. By contrast, with the aggregate damage function employed in our benefit-cost analysis, we have damage rising slowly in the early years and more rapidly later on. While not exactly an L-shaped damage function, it nevertheless produces qualitatively similar results. That is, we again find the value of the non-$CO_2$ gases relative to $CO_2$ to vary over time. We also find the value of $CH_4$ to increase significantly as we approach the bend in the damage function. Finally, we find that the price ratios are sensitive to the damage functions adopted for the analysis.

## Some Concluding Thoughts

Until recently, climate policy analyses have focused almost exclusively on $CO_2$ emissions abatement. This is not surprising, given the importance of $CO_2$ relative to other greenhouse gases, the capabilities of existing models, and the paucity of data related to the non-$CO_2$ gases. Nevertheless, a focus on $CO_2$ emissions only can lead to significant biases in the estimation of compliance costs. Accordingly, existing models are being modified to account for a broader array of greenhouse gases.

This chapter documents one such modeling effort. Before summarizing our conclusions, we again stress the preliminary nature of the results. Calculating the long-term implications of $CO_2$ emissions abatement has always been a daunting task. The problem is compounded with the addition of other gases. Still, we believe that the present analysis, although illustrative in nature, highlights some important considerations for policy makers.

Given the lack of a rationale for choosing one set of GWPs over another, we examined two alternatives for establishing trade-offs between gases. The first was based on cost-effectiveness analysis, the second on benefit-cost analysis. In the former, we imposed a ceiling on the mean global temperature increase. We then identified the least-cost strategy for not exceeding the ceiling. We found that the price of the non-$CO_2$ gases relative to that of carbon tended to vary over time. This was particularly so in the case of $CH_4$. With its relatively short lifetime, its value increased as we approached the temperature ceiling. The tighter the ceiling, the more rapid the increase in value. We observed a similar phenomenon when we added a constraint on the rate of temperature change. The incremental value of $CH_4$ was particularly high when the rate of temperature change approached the prescribed limit.

A number of studies have noted that in order to derive optimal control policies, we must consider the discounted damage of emissions from each gas. Here, the goal is to balance the cost of abatement with what such reductions might achieve in reducing environmental damage. Although this approach provides a more rational basis for climate policy, we currently lack the necessary knowledge base for specifying the shape of the damage functions and for valuing undesirable impacts. Nevertheless, through sensitivity analysis we were able to develop some insights about the relative weights of the various gases.

Both the cost-effectiveness analysis and the benefit-cost analysis highlight the shortcomings of GWPs for establishing equivalence among gases. Not only do the efficiency price ratios vary over time, but they also are sensitive to the ultimate goal. Ideally, the price ratios would be the product of an analysis that minimized the discounted present value of damage and mitigation costs. Unfortunately, given the current state of knowledge regarding potential damage, such an approach may be premature. If indeed this is the case, focusing on temperature change may have distinct advantages over GWPs. It could serve as a useful surrogate for benefit-cost analysis—at least for the time being.

## Acknowledgments

We have benefited from discussions with Joe Aldy, Jae Edmonds, Larry Goulder, John Reilly, John Weyant, Tom Wigley, Larry Williams, and Tom Wilson. We are indebted to Christopher Gerlach and Robert Parkin for research assistance. Funding was provided by EPRI (Electric Power Research Institute). The views presented here are solely those of the individual authors. Comments should be sent to rrichels@epri.com.

## Notes

1. See the 1999 special issue of *The Energy Journal* on the Costs of the Kyoto Protocol: A Multi Model Evaluation for numerous examples.

2. See Reilly et al. (1999) for an example of a multi-gas analysis of the costs of complying with the Kyoto protocol.

3. See, for example, Reilly and Richards (1993), Schmalensee (1993), Fankhauser (1995), Kandlikar (1995), Tol (1999), and Reilly et al. (1999). Also see Wigley (1998) for a discussion of GWPs and their shortcomings.

4. See Schmalensee (1993).

5. In contrast to Reilly et al. (1999), we use linear marginal abatement cost curves. Both coincide at US$100 per ton. For discussions of the potential for $CH_4$ abatement, see Kruger (1998), USEPA (1999), and Houghton et al. (1996). The latter also contains a discussion of the potential for $N_2O$ abatement.

6. Equity need not be sacrificed in order to achieve efficiency. There are alternative ways to make side-payments between regions.

7. Note that we have converted the GWPs from carbon dioxide equivalents to carbon equivalents.

## Literature Cited

Alcamo, J., and G. J. J. Kreileman. 1996. Emission scenarios and global climate protection. *Global Environmental Change* 6 (4): 305–334.

Conference of the Parties. 1997. Kyoto Protocol to the United Nations Framework Convention on Climate Change, Third Session, Kyoto, December 1–10.

Fankhauser, S. 1995. *Valuing climate change: The economics of the greenhouse*. London: Earthscan.

Hammitt, J. K., A. K. Jain, J. L. Adams, and D. J. Wuebbles. 1996. A welfare-based index for assessing environmental effects of greenhouse-gas emissions. *Nature* 381: 301–303.

Houghton, J. T., L. G. Meira Filho, J. P. Bruce, H. Lee, B. A. Callander, and E. F. Haites, eds. 1995. *Climate change 1994: Radiative forcing of climate change and an evaluation of the IPCC 1992 IS92 emission scenarios*. Cambridge: Cambridge University Press.

Houghton, J. T., L. G. Meira Filho, B. A. Callender, N. Harris, A. Kattenberg, and K. Maskell, eds. 1996. *Climate change 1995: The science of climate change (Contribution of Working Group I to the second assessment report of the Intergovernmental Panel on Climate Change)*. Cambridge: Cambridge University Press.

IEA (International Energy Agency). 1998. *Abatement of methane emissions*. IEA Greenhouse Gas R&D Program technical report. Paris.

Kandlikar, M. 1995. The relative role of trace gas emissions in greenhouse abatement. *Energy Policy* 23 (10): 879–883.

Kruger, D. 1998. Integrated assessment of global climate change: Modeling of non-$CO_2$ gases. *Energy Modeling Forum No. 16*. May 15. Washington, DC: Methane and Utilities Branch, Atmospheric Pollution Prevention Division, Office of Air and Radiation, U.S. Environmental Protection Agency.

Nordhaus, W. 1991. To slow or not to slow: The economics of the greenhouse effect. *Energy Journal* 101 (407): 920–937.

Peck, S., and T. Teisberg. 1994. Optimal carbon emission trajectories when damages depend on the rate or level of global warming. *Climatic Change* 28:289–314.

Petschel-Held, G., H.-J. Schellnhuber, T. Bruckner, F. L. Toth, and K. Hasselmann. 1999. The tolerable windows approach: Theoretical and methodological foundations. *Climatic Change* 41 (3–4): 303–331.

Reilly, J., and K. Richards. 1993. Climate change damage and trace gas index issue. *Environmental and Resource Economics* 3:41–61.

Reilly, J., R. Prinn, J. Harnisch, J. Fitzmaurice, H. Jacoby, D. Kicklighter, J. Melillo, P. Stone, A. Sokolov, and C. Wang. 1999. Multi-gas assessment of the Kyoto Protocol. *Nature* 401:549–555.

Schmalensee, R. 1993. Comparing greenhouse gases for policy purposes. *Energy Journal* 14 (1): 245–255.

Smith, J. 1996. Standardized estimates of climate change damages for the United States. *Climatic Change* 32 (3): 313–326.

Tol, R. S. J. 1999. Methane emission reduction: An application of FUND. IVM-D99/3. Amsterdam: Institute for Environmental Studies, Vrije Universiteit.

Toth, F. L., T. Bruckner, H.-M. Füssel, M. Leimbach, G. Petschel-Held, and H.-J. Schellnhuber. 1997. The tolerable windows approach to integrated assessments. Pp. 403–430 in *Climate change and integrated assessment models: Bridging the gaps—Proceedings of the IPCC Asia Pacific Workshop on Integrated Assessment Models, United Nations University, Tokyo, Japan, March 10-12, 1997,* edited by O. K. Cameron, K. Fukuwatari, and T. Morita. Tsukuba, Japan: Center for Global Environmental Research, National Institute for Environmental Studies.

USEPA (U.S. Environmental Protection Agency). 1999. *Costs of reducing methane emissions in the United States.* Washington, DC: Methane Branch, USEPA.

Wigley, T. 1998. The Kyoto Protocol: $CO_2$, $CH_4$ and climate implications. *Geophysical Research Letters* 25 (13): 2285–2288.

# 26

# Storage of Carbon Dioxide by Greening the Oceans?

Dorothee C. E. Bakker

The widespread use of fossil energy results in the release of the greenhouse gas carbon dioxide ($CO_2$) to the atmosphere at an average rate of 5.9 petagrams of carbon per year (PgC yr$^{-1}$; 1 Pg = $10^{15}$ g) (Sabine et al., Chapter 2, this volume). The resulting rapid increase of atmospheric $CO_2$ is a major factor in global warming (Houghton et al. 2001). Because of the potential risks of climate change, it is crucial to consider how to curb fossil-fuel $CO_2$ emissions and whether techniques are available to absorb atmospheric $CO_2$. Such a consideration requires weighing the economic, societal, and environmental costs of climate change relative to those of reducing $CO_2$ emissions and implementing $CO_2$ sequestration.

The oceans absorb roughly 30 percent of contemporary $CO_2$ emissions caused by human activity (Sabine et al., Chapter 2, this volume). Sluggish exchange between surface waters and the deep ocean sets the maximum rate for net oceanic $CO_2$ uptake and makes it a slow process relative to the rapid anthropogenic $CO_2$ release. If anthropogenic $CO_2$ emissions were stopped now, the oceans would eventually take up 70–80 percent of the $CO_2$ emissions, over a period of several centuries (Maier-Reimer and Hasselmann 1987; Archer et al. 1997).

Past and current oceanic $CO_2$ uptake reduces the capacity for future oceanic $CO_2$ absorption (Matear and Hirst 1999). It also influences oceanic pH (Archer et al. 1998), calcification (Gattuso et al. 1998; Riebesell et al. 2000), and the marine ecosystem (Houghton et al. 2001). Warming of surface waters and possible changes in oceanic circulation caused by climate change will have indirect effects on the physical and biological carbon pumps in the oceans (Sarmiento et al. 1998; Matear and Hirst 1999). These changes will provide positive, as well as negative, feedbacks on oceanic $CO_2$ uptake, the atmospheric $CO_2$ content, and climate change.

Because the oceans absorb so much of the $CO_2$ emitted by human activity, it is not surprising that scientists, entrepreneurs, and politicians are investigating whether it is possible to accelerate net oceanic $CO_2$ absorption. One possible technique is fertilization of ocean regions where low nutrient availability currently limits phytoplankton growth. Addition of the macronutrients nitrogen and phosphorus and the trace element iron has been suggested. In this chapter I consider the principle, efficiency, duration, costs, and environmental consequences of $CO_2$ storage by fertilization.

## Iron Limitation of Algal Growth

The idea of iron fertilization has received wide attention since John Martin (1990a: 11) commented, "With half a shipload of iron, I could give you an ice age." Large areas of the oceans, notably the Southern Ocean, the equatorial Pacific Ocean, and the subarctic Pacific Ocean, have low algal growth, despite high year-round nutrient concentrations in surface water (Figure 26.1). Hart (1934) suggested that low iron levels might limit phytoplankton growth in Antarctic waters.

Dust is a major iron source for the open oceans. Dust input to the oceans strongly depends on the proximity of dust sources on land and prevailing wind patterns. Ocean regions with high nutrient concentrations and low algal growth receive relatively little atmospheric dust (Duce and Tindale 1991; Watson 2001). An increase of atmospheric $CO_2$ upon a reduction in dust deposition at the end of glacial periods suggests a feedback between dust-derived iron, algal growth in the oceans, atmospheric $CO_2$, and climate (Martin 1990b; Watson et al. 2000).

The arrival of accurate methods for iron analysis allowed tests and confirmation of the iron hypothesis (De Baar et al. 1990; Hudson and Morel 1990; Martin et al. 1990). In situ iron enrichment experiments have been conducted in the equatorial Pacific Ocean (IronEx I, II), the Southern Ocean (SOIREE, EisenEx, SOFeX), and the subarctic Pacific Ocean (SEEDS, SERIES) (Table 26.1). These mesoscale experiments were started only after extensive site surveys. The rapid subduction of the fertilized waters in IronEx I demonstrates the critical importance of site selection for these experiments. Iron was added once in some experiments and repeatedly in others. Inert tracer, drifting buoys, and ADCP (Acoustic Doppler Current Profiler) measurements were used to track the iron-enriched waters.

The in situ iron additions promoted development of an algal bloom (Figure 26.2a–d) (Martin et al. 1994; Coale et al. 1996; Boyd et al. 2000; Smetacek 2001; Johnson et al. 2002; Tsuda et al. 2003). Diatoms, large algae with siliceous shells, thrived upon iron addition (Coale et al. 1996; Boyd et al. 2000), whereas no changes in calcifying algae have been recorded. Algal growth decreased the concentrations of nutrients and $CO_2$ in the mixed layer (Table 26.1, Figure 26.2e–f). The algal $CO_2$ uptake either reduced $CO_2$ release to the atmosphere or increased the uptake of atmospheric $CO_2$ by surface waters (Cooper et al. 1996; Watson et al. 2000), depending on their $CO_2$ saturation level (Figure 26.2e).

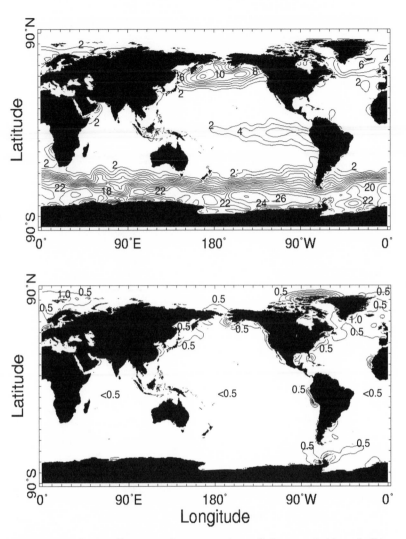

**Figure 26.1.** Seasonally averaged concentrations of nitrate and chlorophyll in oceanic surface waters with contours at 2 µmol l⁻¹ and 0.5 µmol l⁻¹ intervals, respectively. Data are from Conkright et al. (1998a, b, c).

Iron-induced algal growth resulted in the production of particulate organic carbon (POC) (Bidigare et al. 1999; Nodder and Waite 2001) and presumably also of dissolved organic carbon (DOC). A carbon budget for the SOIREE bloom suggests significant production of DOC (Bakker et al. 2003). The quantity and quality of DOC production following iron fertilization remain uncertain.

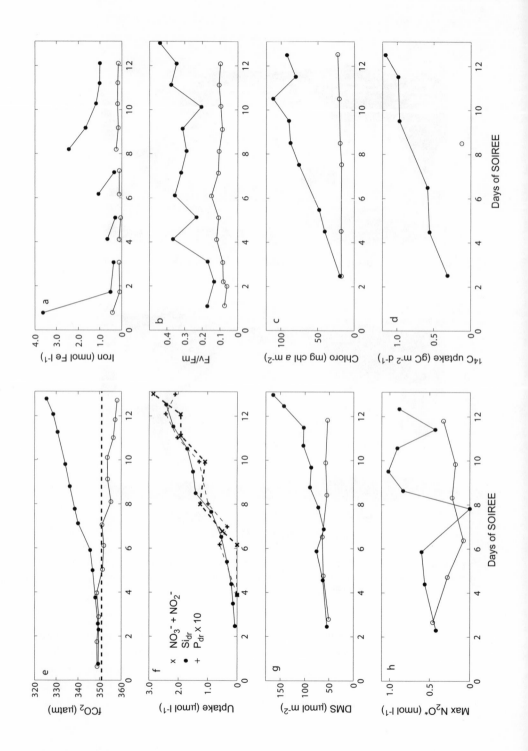

**Figure 26.2.** Parameters inside (closed symbols) and outside (open symbols) the iron-fertilized waters (or "patch") in the Southern Ocean Iron RElease Experiment (after Boyd et al. 2000; Watson et al. 2000; Law and Ling 2001). Shown are (a) dissolved iron in surface water, (b) photosynthetic competency (Fv/Fm) in surface water, (c) depth-integrated chlorophyll, (d) depth-integrated [14]C uptake, a measure for primary production, (e) the fugacity of $CO_2$ ($fCO_2$) in surface water and air (dashed line, mean value of $fCO_2$ in air for SOIREE), (f) macronutrient uptake in surface water, (g) depth-integrated dissolved DMS (dimethyl sulphide) and (h) maximum $N_2O^*$ (nitrous oxide) in the pycnocline. Samples for dissolved iron were taken at roughly 3 m depth, samples for Fv/Fm, $fCO_2$ and nutrients at $-5$ m depth. Underway data for dissolved iron, photosynthetic competency, and nutrients are expressed as the daily mean inside the patch (defined at >50 percent of peak levels of the tracer sulphur hexafluoride [$SF_6$] on that day) and outside the patch (defined at an $SF_6$ concentration <10 f mol l[-1]). The previously reported data for [14]C uptake have been corrected by multiplication with a factor 12/14. Vertically integrated data have been integrated for the upper 65 m, the depth of the mixed layer. $Si_{dr}$ and $P_{dr}$ are dissolved reactive silica and phosphorus. Nitrate and phosphate depletion were obtained by subtraction of levels in the patch from values in ambient waters, which remained constant during SOIREE. Underway data for $fCO_2$ are indicated as a daily mean of $fCO_2$ measurements inside and outside the patch (obtained from $fCO_2$ at the 10 percent highest and 10 percent lowest $SF_6$ values, respectively). Note the reversed $y$ axis for $fCO_2$. Surface water $fCO_2$ outside the patch increased during SOIREE, partly as the result of surface water warming. Maximum $N_2O^*$ is the difference between the measured $N_2O$ concentration (inside or outside the patch) and that predicted from the $N_2O$-density relationship outside the patch (Law and Ling 2001).

**Table 26.1.** The amount of added iron (Fe) and the observed drawdown of dissolved inorganic carbon (DIC) in the mixed layer during three iron fertilization experiments

| Experiment | IronEx I | IronEx II | SOIREE |
|---|---|---|---|
| Area | Equatorial Pacific | Equatorial Pacific | Southern Ocean |
| Time | 10/1993 | 05–06/1995 | 02/1999 |
| Fe added ($10^3$ mol) | 7.8[a] | 8.0[c] | 31[g] |
| $-\Delta$DIC ($\mu$mol kg$^{-1}$) | 6–7[b] | 27[d] | 18[h] |
| $-\Delta$DIC patch ($10^8$ mol) | n.d. | 2.4[e] (13–16 d) | 0.3–1.1[h, i] (13 days), 0.5–2.5[j] (42 days) |
| $-\Delta$C:$\Delta$Fe ($10^4$) | n.d. | 3.0 | 0.11–0.35 (13 days), 0.16–0.81 (42 days) |
| Downward C transport | Subduction after 2–3 days | 7-fold increase in export[f] | Constant export[k, l] |

*Note:* Note that DIC drawdown in the mixed layer strongly overestimates the short-term (<1 year) storage of atmospheric $CO_2$, as significant remineralization of organic carbon is likely to occur in the mixed layer before $CO_2$ has been taken up from the atmosphere by slow air-sea exchange. $-\Delta$C:$\Delta$Fe is the ratio of observed DIC drawdown to added iron. N.d. indicates parameters, which were not determined.

*Sources:* [a] Martin et al. (1994), [b] Watson et al. (1994), [c] Coale et al. (1996), [d] Steinberg et al. (1998), [e] Cooper et al. (1996), [f] Bidigare et al. (1999), [g] Bowie et al. (2001), [h] Bakker et al. (2001), [i] Boyd et al. (2000), [j] Abraham et al. (2000), [k] Charette and Buesseler (2000), [l] Nodder and Waite (2001).

## Carbon Storage by Iron Addition

Iron-induced algal growth, the composition of organic matter, downward carbon transport, remineralization, and future ventilation of the water jointly determine the efficiency and duration of carbon storage following iron addition. Unfortunately, the last three terms and the interactions between them are difficult to quantify.

Downward carbon transport occurs by sinking of particles, deep mixing, and subduction of carbon-rich surface water. Export of organic particles increased in the IronEx II and SOFeX experiments (Bidigare et al. 1999; Buesseler et al. 2003) but remained constant in SOIREE (Table 26.1) (Charette and Buesseler 2000; Nodder and Waite 2001). Downward carbon transport by deep mixing was not observed in any of the iron enrichment experiments. Subduction of the young IronEx I bloom transported some organic carbon downward (Watson et al. 1994), presumably both as POC and DOC.

Remineralization, the breakdown of organic matter, strongly decreases the downward flux of organic carbon. Most organic carbon that leaves the mixed layer is remineralized at shallow and intermediate depths, from which the $CO_2$, released by remineralization, may be ventilated to the atmosphere by deep mixing and upwelling. Only a fraction of the organic matter reaches the deep ocean and ocean sediments, where the carbon can be stored away from the atmosphere for centuries or more.

Carbon storage by subduction in IronEx I was expected to be short-lived because the subsurface waters of the equatorial Pacific Ocean are part of a confined circulation system (Broecker and Peng 1982; Reverdin 1995) and resurface roughly every 15 years (for a thermocline depth of 250 meters [m] and an upwelling rate of 16–40 m y$^{-1}$). Thus, any organic carbon stored by subduction would be rapidly remineralized and released as $CO_2$ back into the atmosphere. Continuous iron fertilization of the equatorial Pacific Ocean between 18°N and 18°S would therefore provide little carbon storage (Sarmiento and Orr 1991).

By contrast, some Southern Ocean waters leave the surface for decades to centuries by deep mixing and subduction. Near the Subantarctic Front, intermediate water sinks below more saline water. Any extra carbon in the "new" intermediate water would be stored away from the atmosphere for long periods. The production of slowly degradable DOC in algal blooms at or south of the Subantarctic Front could favor long-term carbon storage. Sarmiento and Orr (1991) modeled continuous iron fertilization of surface waters south of 31°S. Iron addition for a 100-year period would reduce the atmospheric $CO_2$ increase by 17 percent in an IPCC business-as-usual scenario (this scenario is discussed in Chapter 4 of this volume). Breakdown of organic matter would rapidly lower the initial carbon storage (12 PgC yr$^{-1}$) to 2 PgC yr$^{-1}$ after 20 years and 1 PgC yr$^{-1}$ after 100 years of continuous iron fertilization. It is important to note that much of the stored carbon would be rapidly returned to the atmosphere as soon as the iron addition was stopped, highlighting the limited duration of carbon storage. This study probably overestimates the potential for carbon storage by its assumption of complete nutrient depletion of the fertilized waters, which has not been observed in the iron fertilization experiments. Other modeling studies vary in the amount of carbon stored but present a similar overall pattern (Broecker and Peng 1991; Joos et al. 1991a, 1991b). These calculations probably represent an upper limit for the potential of $CO_2$ sequestration by iron fertilization.

Local, rather than basinwide, iron fertilization could sequester small amounts of $CO_2$. The controls on and consequences of basinwide iron fertilization equally apply to small-scale programs, albeit at a more limited scale. Site selection and monitoring of the amount of carbon stored over a range of timescales will be major challenges to iron fertilization schemes.

## Drawbacks of Iron Fertilization

In the iron enrichment experiments, the biomass of bacteria and small algae remained fairly constant (Coale et al. 1996; Hall and Safi 2001; Gervais et al. 2002), probably as a consequence of increases in grazing by small zooplankton (Coale et al. 1996). The biomass of larger phytoplankton, notably diatoms, increased. By promoting a shift toward larger algal species, iron addition thoroughly changes the structure of the marine food web. It is conceivable that continuous iron fertilization would decrease biodiversity by consequently favoring a small number of species. Long-term, widespread iron fertilization could potentially allow the spread of harmful algal blooms, as has been observed in coastal waters, which are "fertilized" by high nutrient inputs from rivers (eutrophication).

Iron fertilization and iron-induced algal growth reduce nutrient availability to downstream ecosystems (Sarmiento and Orr 1991), which decreases the downstream productivity and changes the structure of the marine food web. One may speculate that the biomass of large algae will be affected more than that of small algae in downstream waters.

The increases in algal growth, carbon export, and remineralization upon iron addition reduce oxygen levels in subsurface waters. Long-term or large-scale fertilization could create conditions with zero oxygen concentrations ("anoxic conditions") at intermediate depths (Fuhrman and Capone 1991; Sarmiento and Orr 1991). In the modern ocean the greenhouse gases nitrous oxide ($N_2O$) and methane ($CH_4$), which are 275 and 62 times more potent than $CO_2$ over a 20-year time scale (Houghton et al. 2001), are produced in anoxic conditions. For example, nitrous oxide is produced below seasonal algal blooms in the northwestern Indian Ocean (Law and Owens 1990). As some nitrous oxide escapes to the atmosphere, these waters are important natural sources for this potent greenhouse gas. By analogy, low oxygen levels and anoxic conditions upon iron fertilization are likely to promote production of nitrous oxide and methane. In SOIREE, nitrous oxide concentrations increased at the bottom of the mixed layer (Figure 26.2h), but this did not result in extra release of $N_2O$ to the atmosphere (Law and Ling 2001). Methane concentrations did not change in SOIREE (C. S. Law, pers. comm.) nor have changes been reported for other experiments. The negative effects of $N_2O$ and $CH_4$ release to the atmosphere would counteract and possibly outweigh the benefits of the atmospheric $CO_2$ sequestration from iron addition (Jin and Gruber 2002; Jin et al. 2002). Low oxygen levels and anoxic conditions following iron fertilization would also negatively affect marine biota that depend on oxygen, such as fish.

Iron fertilization may change the marine production of important gases, such as dimethylsulphide (DMS), halocarbons, and alkyl nitrates. Dimethylsulphide is a major source of atmospheric sulphate and is thought to indirectly affect aerosol formation, albedo, and climate (Charlson et al. 1987). Volatile halocarbons and alkyl nitrates strongly influence atmospheric chemistry. Marine biological processes promote the production of DMS (Turner et al. 1996) and certain halocarbons (Chuck 2002). They may also play an important role in the production of alkyl nitrates (Chuck 2002; Chuck et al. 2002). The production and release of DMS to the atmosphere increased in the IronEx II and SOIREE iron enrichment experiments (Figure 26.2g; Turner et al. 1996; Turner et al. 2003). The concentration of some halocarbons and an alkyl nitrate changed in the EisenEx experiment (Chuck 2002). Changes in the marine source of these gases following iron fertilization could open a Pandora's box with feedbacks on atmospheric chemistry and global climate (Fuhrman and Capone 1991; Lawrence 2002). Iron fertilization has a wide range of observed and potential side effects. These require further study before any large-scale scheme is undertaken.

## Iron Fertilizer and Costs

Ratios of algal carbon uptake to added iron were $0.1 \times 10^4 : 1$ to $3.0 \times 10^4 : 1$ in the IronEx II and SOIREE experiments (Table 26.1). A ratio of $0.7 \times 10^4 : 1$ was calculated between the POC export at 100 m depth and the added iron in SOFeX (Buesseler and Boyd 2003; Buesseler et al. 2003). These ratios are one to two orders of magnitude lower than the carbon-to-iron ratio of $3 \times 10^5 : 1$ in severely iron-stressed diatoms (Sunda and Huntsman 1995). Iron losses (Bowie et al. 2001) and a higher carbon-to-iron content in iron-replete than in iron-stressed algae (Sunda and Huntsman 1995) may explain the difference in the ratios. Continuous iron fertilization of waters south of $31°S$ would require $6 \times 10^6$ tons Fe $y^{-1}$ (1 ton = $10^6$ g) for the maximum possible algal growth at a carbon-to-iron ratio of $1 \times 10^4 : 1$ (after Sarmiento and Orr 1991), which is a 10-fold lower ratio than Sarmiento and Orr used in their calculations.

This iron demand corresponds to 1 percent of the global crude steel output of $700 \times 10^6$ tons Fe $y^{-1}$ (Nagasawa 1995). Purposeful iron mining to meet such a large demand would strongly raise the economic and energy cost of iron fertilization. Iron for small-scale additions might be acquired at relatively low cost as a by-product from industry. Any iron fertilizer used for ocean fertilization should be of good quality, with only traces of other constituents. Use of high-quality ferrous scrap could compete with iron recycling for other purposes (Nagasawa 1995).

The cost of $CO_2$ storage by iron fertilization has been estimated as US\$5–US\$100 $ton^{-1}$ of carbon stored (Ritschard 1992). It is unclear how Ritschard accounted for the efficiency and duration of carbon storage and whether these estimates include the cost of iron mineral. The values do not include the cost of monitoring the efficiency, duration, and side effects of iron fertilization. The $CO_2$ production by acquisition of iron fertilizer and by iron fertilization itself should also be taken into account in calculating the net amount of carbon stored.

## Carbon Storage by Adding Nitrogen and Phosphorus Fertilizer

Enhancement of algal growth by fertilization with nitrogen and phosphorus has also been suggested as an option for carbon storage (Ritschard 1992). In a modeling study, nitrogen and phosphorus were added to the equatorial oceans between $18°N$ and $18°S$ (Orr and Sarmiento 1992). Continuous nutrient addition for 100 years would increase oceanic $CO_2$ uptake by 0.7 PgC $yr^{-1}$ on average, if the algal remains were removed and stored away from the atmosphere. This scenario would require a nutrient input higher than the current global nitrogen consumption for fertilizer and close to that for phosphorus. This fertilizer demand would be very costly, both economically and in energy terms. If the algae were not harvested, continuous, widespread fertilization could store $0.4$ $PgC$ $yr^{-1}$. Degradation of algal material, however, would result in large-scale anoxic conditions in subsurface waters.

Stimulation of macro-algal growth by the addition of macronutrients seems to be an inefficient tactic for carbon storage (Orr and Sarmiento 1992). Any such scheme would drastically change the local marine ecosystem in ways that might resemble the effects of high nutrient inputs in coastal seas (eutrophication), including damage to coral reefs, toxic algal blooms, and fish kills.

## Legal Issues

The 1972 London Convention bans "dumping," which it defines as " any deliberate disposal at sea of wastes or other matter from vessels, aircraft, platforms, or other manmade structures at sea" (Article 3.1.a (i) in IMO 1972). A 1991 amendment explicitly prohibits dumping of industrial waste. The 1996 Protocol to the Convention, which awaits ratification, removes some of the exemptions with respect to dumping platforms in the 1972 Convention (IMO 1972, 1996). The London Convention, however, does allow the placement of matter for a purpose other than its mere disposal, provided that such placement does not cause pollution or harm marine life. Because fertilizer, originating from recycling or industrial processes, and even anthropogenic $CO_2$ could be regarded as industrial waste (Johnston et al. 1999), it is uncertain whether these provisions would allow $CO_2$ storage by ocean fertilization.

The Protocol on Environmental Protection to the Antarctic Treaty (Anonymous 1991) states that activities in the treaty area shall limit changes of and adverse impacts on the Antarctic environment and associated ecosystems. This provision appears to rule out widespread fertilization of waters south of $60°S$, once the protocol has been ratified.

Thus, it is unclear whether fertilization of international waters for the purpose of $CO_2$ storage would be allowed under international law. Of course, laws are designed to promote public policy goals. If iron fertilization were considered an appropriate public policy, then international agreements could be modified.

## Conclusion

Major uncertainties are attached to the efficiency and duration of $CO_2$ storage from ocean fertilization. Modeling studies indicate largely short-lived carbon storage upon basinwide, continuous iron fertilization of the Southern Ocean, the oceanic region with the highest potential for carbon storage from iron addition. It is unknown how climate change, changes in oceanic circulation, and increases in the atmospheric $CO_2$ content will affect future oceanic carbon storage.

The addition of nitrogen and phosphorus fertilizer can be discarded as a serious option for carbon storage based on the small amount of carbon stored and the large supply of fertilizer required. The advantages of carbon storage by iron addition (mostly temporary carbon storage with some durable carbon storage) may not justify the negative aspects of the technique, which would include a reduction of subsurface oxygen concentrations, other major changes of marine chemistry, a transformation of the marine ecosystem, possible losses in marine biodiversity, and substantial financial and energy costs. The production of dimethylsulphide, halocarbons, and alkyl nitrates could adversely affect atmospheric chemistry and global climate. Increased emission of the greenhouse gases nitrous oxide and methane would counteract, or even outweigh, the benefits of atmospheric $CO_2$ reduction by iron addition. The potential negative side effects of iron fertilization require serious investigation before a balanced judgment can be made on the applicability of the technique as a strategy for sequestering $CO_2$ and reducing climate change.

Oceanic fertilization does not provide a full solution to the impacts of the fossil-fuel-driven economy on global climate. At best, it might delay the atmospheric $CO_2$ increase and buy time. The international community should carefully weigh the expected benefits and adverse effects of temporary $CO_2$ storage, reduction of $CO_2$ emissions, and climate change and act accordingly.

## Acknowledgments

The work was supported by the European Community ORFOIS grant (EVK-CT-2001-00100). Constructive comments by Adrian Webb, Peter Brewer, Ken Caldeira, and Walker Smith helped to improve the manuscript.

## Literature Cited

Abraham, E. R., C. S. Law, P. W. Boyd, S. J. Lavender, M. T. Maldonado, and A. R. Bowie. 2000. Importance of stirring in the development of an iron-fertilized phytoplankton bloom. *Nature* 407:727–730.

Anonymous. 1991. *Protocol on environmental protection to the Antarctic Treaty.* http://www .polarlaw.org/treaties.htm.

Archer, D., H. Kheshgi, and E. Maier-Reimer. 1997. Multiple timescales for neutralization of fossil fuel $CO_2$. *Geophysical Research Letters* 24:405–408.

———. 1998. Dynamics of fossil fuel $CO_2$ neutralization by marine $CaCO_3$. *Global Biogeochemical Cycles* 12:259–276.

Bakker, D. C. E., A. J. Watson, and C. S. Law. 2001. Southern Ocean iron enrichment promotes inorganic carbon drawdown. *Deep-Sea Research II* 48:2483–2507.

Bakker, D. C. E., P. W. Boyd, M. A. Charette, M. P. Gall, J. A. Hall, S. D. Nodder, K. Safi, R. J. Singleton, T. W. Trull, A. M. Waite, A. J. Watson, J. Zeldis, E. R. Abraham, C. S. Law, and K. Tanneberger. 2003. A pelagic carbon budget for the Southern Ocean Iron RElease Experiment (SOIREE), February 1999 (in revision).

Bidigare, R. R., K. L. Hanson, K. O. Buesseler, S. G., Wakeham, K. H. Freeman, R. D. Pancost, F. J. Millero, P. Steinberg, B. N. Popp, M. Latasa, M. R. Landry, and E. A. Laws. 1999. Iron-stimulated changes in C-13 fractionation and export by equatorial Pacific phytoplankton: Toward a paleogrowth rate proxy. *Paleoceanography* 14:589–595.

Bowie, A. R., M. T. Maldonado, R. D. Frew, P. L. Croot, E. P. Achterberg, R. F. C. Mantoura, P. J. Worsfold, C. S. Law, and P. W. Boyd. 2001. The fate of added iron during a mesoscale fertilization experiment in the Southern Ocean. *Deep-Sea Research II* 48:2703–2743.

Boyd, P. W., A. J. Watson, C. S. Law, E. R. Abraham, T. Trull, R. Murdoch, D. C. E. Bakker, A. R. Bowie, K. O. Buesseler, H. Chang, M. A. Charette, P. Croot, K. Downing, R. D. Frew, M. Gall, M. Hadfield, J. A. Hall, M. Harvey, G. Jameson, J. La Roche, M. I. Liddicoat, R. Ling, M. Maldonado, R. M. McKay, S. D. Nodder, S. Pickmere, R. Pridmore, S. Rintoul, K. Safi, P. Sutton, R. Strzepek, K. Tanneberger, S. M. Turner, A. Waite, and J. Zeldis. 2000. A mesoscale phytoplankton bloom in the polar Southern Ocean stimulated by iron fertilization. *Nature* 407:695–702.

Broecker, W. S., and T. H. Peng. 1982. *Tracers in the sea.* Palisades, NY: Eldigio Press.

———. 1991. Factors limiting the reduction of atmospheric $CO_2$ by iron fertilization. *Limnology and Oceanography* 36:1919–1927.

Buesseler, K. O., and P. W. Boyd. 2003. Will ocean fertilization work? *Science* 300:67–68.

Buesseler, K. O., J. Andrew, S. Pike, and M. Charette. 2003. Does iron fertilization enhance carbon sequestration in the Southern Ocean? (in revision).

Charette, M. A., and K. O. Buesseler, 2000. Does iron fertilization lead to rapid carbon export in the Southern Ocean? *Geochemistry, Geophysics, Geosystems* 1, 2000GC000069.

Charlson, R. J., J. E. Lovelock, M. O. Andreae, and S. G. Warren. 1987. Oceanic phytoplankton, atmospheric sulphur, cloud albedo and climate. *Nature* 326:655–661.

Chuck, A. L. 2002. Biogenic halocarbons and light alkyl nitrates in the marine environment. Ph.D. thesis, University of East Anglia, Norwich, UK.

Chuck, A. L., S. M. Turner, and P. S. Liss. 2002. Direct evidence for a marine source of $C_1$ and $C_2$ alkyl nitrates. *Science* 297:1151–1154.

Coale, K. H., K. S. Johnson, S. E. Fitzwater, R. M. Gordon, S. Tanner, S. Tanner, F. P. Chavez, L. Ferioli, C. Sakamoto, P. Rogers, F. Millero, P. Steinberg, P. Nightingale, D. Cooper, W. P. Cochlan, M. R. Landry, J. Constantinou, G. Rollwagen, A. Trasvina, and R. Kudela. 1996. A massive phytoplankton bloom induced by an ecosystem-scale iron fertilization experiment in the equatorial Pacific Ocean. *Nature* 383:495–511.

Conkright, M. E., S. Levitus, T. P. Boyer, and T. O'Brien. 1998a. *World ocean atlas 1998.* Vol. 10, *Nutrients and chlorophyll of the Atlantic Ocean.* NOAA (National Oceanic and Atmospheric Administration) Atlas, NESDIS (National Environmental Satellite, Data, and Information Service) 36. Washington, DC: U.S. Government Printing Office.

———. 1998b. *World ocean atlas 1998.* Vol. 11, *Nutrients and chlorophyll of the Pacific Ocean.* NOAA (National Oceanic and Atmospheric Administration) Atlas, NESDIS (National Environmental Satellite, Data, and Information Service) 37. Washington, DC: U.S. Government Printing Office.

———. 1998c. *World ocean atlas 1998.* Vol. 12, *Nutrients and chlorophyll of the Indian Ocean.* NOAA (National Oceanic and Atmospheric Administration) Atlas, NESDIS (National Environmental Satellite, Data, and Information Service) 38. Washington, DC: U.S. Government Printing Office.

Cooper, D. J., A. J. Watson, and P. D. Nightingale. 1996. Large decrease in ocean-surface $CO_2$ fugacity in response to in situ iron fertilization. *Nature* 383:511–513.

De Baar, H. J. W., A. G. J. Buma, R. F. Nolting, G. C. Cadée, G. Jacques, and P. J. Tréguer. 1990. On iron limitation of the Southern Ocean: Experimental observations in the Weddell and Scotia Seas. *Marine Ecology Progress Series* 65:105–122.

Duce, R. A., and N. W. Tindale. 1991. Atmospheric transport of iron and its deposition in the ocean. *Limnology and Oceanography* 36:1715–1726.

Fuhrman, J. A., and D. G. Capone. 1991. Possible biogeochemical consequences of ocean fertilization. *Limnology and Oceanography* 36:1951–1959.

Gattuso, J.-P., M. Frankignoulle, I. Bourge, S. Romaine, and R. W. Buddemeier. 1998. Effect of calcium carbonate saturation of seawater on coral calcification. *Global Planetary Change* 18:37–46.

Gervais, F., U. Riebesell, and M. Y. Gorbunov. 2002. Changes in the size-fractionated primary productivity and chlorophyll a in response to iron fertilization in the southern Polar Frontal Zone. *Limnology and Oceanography* 47:1324–1335.

Hall, J. A., and K. Safi. 2001. The impact of Fe fertilisation on the microbial food web in the Southern Ocean. *Deep-Sea Research II* 48:2591–2613.

Hart, T. J. 1934. On the phytoplankton of the south-west Atlantic and the Bellingshausen Sea, 1929–31. *Discovery Reports* 8:1–268.

Houghton, J. T., Y. Ding, D. J. Griggs, M. Noguer, P. J. van der Linden, X. Dai, K. Maskell, and C. A. Johnson, eds. 2001. *Climate change 2001: The scientific basis. (Contribution of Working Group I to the third assessment report of the Intergovernmental Panel on Climate Change).* Cambridge: Cambridge University Press.

Hudson, R. J. M., and F. M. M. Morel. 1990. Iron transport in marine phytoplankton: Kinetics of cellular and medium coordination reactions. *Limnology and Oceanography* 35:1002–1020.

IMO (International Maritime Organization). 1972. *Convention on the prevention of marine pollution by dumping of wastes and other matter.* London. http://www .londonconvention.org.

———. 1996. Protocol to the convention on the prevention of marine pollution by dumping of wastes and other matter, 1972, and resolutions adopted by the special meeting. London. http://www.londonconvention.org.

Jin, X., and N. Gruber. 2002. Offsetting the radiative benefit of ocean iron fertilization by enhancing ocean $N_2O$ emissions. *Eos, Transactions, American Geophysical Union* 83 (AGU Fall Meeting Supplement), Abstract U21A-0006.

Jin, X, C. Deutsch, N. Gruber, and K. Keller. 2002. Assessing the consequences of iron fertilization on oceanic $N_2O$ emissions and net radiative forcing. *Eos, Transactions, American Geophysical Union* 83 (Ocean Sciences Supplement), Abstract OS51F-10.

Johnson, K. S., J. K. Moore, and W. O. Smith Jr. 2002. Workshop investigates role of iron dynamics in ocean carbon cycle. *US JGOFS Newsletter* 12 (1): 8–9.

Johnston, P., D. Santillo, and R. Stringer. 1999. Ocean disposal/sequestration of carbon dioxide from fossil fuel production and use: An overview of rationale, techniques and implications. Technical Note 01/99. Exeter, UK: Greenpeace Research Laboratories.

Joos, F., J. L. Sarmiento, and U. Siegenthaler. 1991a. Estimates of the effect of Southern Ocean iron fertilization on atmospheric $CO_2$ concentrations. *Nature* 349:772–774.

Joos, F., U. Siegenthaler, and J. L. Sarmiento. 1991b. Possible effects of iron fertilization in the Southern Ocean atmospheric $CO_2$ concentration. *Global Biogeochemical Cycles* 5:135–150.

Law, C. S., and R. Ling. 2001. Nitrous oxide flux and response to increased iron availability in the Antarctic Circumpolar Current. *Deep-Sea Research II* 48:2509–2527.

Law, C. S., and N. J. P. Owens. 1990. Significant flux of atmospheric nitrous oxide from the North-west Indian Ocean. *Nature* 346:826–828.

Lawrence, M. G. 2002. Side effect of oceanic iron fertilization. *Science* 297:1993.

Maier-Reimer, E., and K. Hasselmann. 1987. Transport and storage of $CO_2$ in the ocean: An inorganic ocean-circulation carbon cycle model. *Climate Dynamics* 2:63–90.

Martin, J. M. 1990a. A new iron age, or a ferric fantasy? *US JGOFS News* 1:5 and 11.

———. 1990b. Glacial to interglacial $CO_2$ change: The iron hypothesis. *Paleoceanography* 5:1–13.

Martin, J. M., S. E. Fitzwater, and R. M. Gordon. 1990. Iron deficiency limits phytoplankton growth in Antarctic waters. *Global Biogeochemical Cycles* 4:5–12.

Martin, J. M., K. H. Coale, K. S. Johnson, S. E. Fitzwater, R. M. Gordon, S. J. Tanner, C. N. Hunter, V. A. Elrod, J. L. Nowicki, T. L. Coley, R. T. Barber, S. Lindley, A. J. Watson, K. Van Scoy, C. S. Law, M. I. Liddicoat, R. Ling, T. Stanton, J. Stockel, C. Collins, A. Anderson, R. Bidigare, M. Ondrusek, M. Lasata, F. J. Millero, K. Lee, W. Yao, J. Z. Zhang, G. Friederich, C. Sakamoto, F. Chavez, K. Buck, Z. Kolber, R. Greene, P. Falkowski, S. W. Chisholm, F. Hoge, R. Swift, J. Yungel, S. Turner, P. Nightingale, A. Hatton, P. Liss, and N. W. Tindale. 1994. Testing the iron hypothesis in ecosystems of the equatorial Pacific Ocean. *Nature* 371:123–129.

Matear, R. J., and A. C. Hirst. 1999. Climate change feedback on the future oceanic $CO_2$ uptake. *Tellus* 51B:722–733.

Nagasawa, T. 1995. How the Japanese steel industry is addressing global environmental problems: Focusing on measures to control carbon dioxide emissions. *Energy Conservation and Management* 36:839–844.

Nodder, S. D., and A. M. Waite. 2001. Is Southern Ocean organic carbon and biogenic silica export enhanced by iron-stimulated increases in biological production? Sediment trap results from SOIREE. *Deep-Sea Research II* 48: 2681–2701.

Orr, J. C., and J. L. Sarmiento. 1992. Potential of marine macroalgae as a sink for $CO_2$: Constraints from a 3-D general circulation model of the global ocean. *Water, Air and Soil Pollution* 64:405–421.

Reverdin, G. 1995. The physical processes of open ocean upwelling. Pp. 123–148 in *Upwelling in the ocean: modern and ancient records,* edited by C. P. Summerhayes, K. C. Emeis, M. V. Angel, R. L. Smith, and B. Zeitschel. Chichester, UK: John Wiley and Sons.

Riebesell, U., I. Zondervan, B. Rost, P. D. Tortell, R. E. Zeebe, F. M. M. Morel. 2000. Reduced calcification of marine plankton in response to increased atmospheric $CO_2$. *Nature* 407:364–367.

Ritschard, R. L. 1992. Marine algae as a $CO_2$ sink. *Water, Air and Soil Pollution* 64:289–303.

Sarmiento, J. L., and J. C. Orr. 1991. Three-dimensional simulations of the impact of Southern Ocean nutrient depletion on atmospheric $CO_2$ and ocean chemistry. *Limnology and Oceanography* 36:1928–1950.

Sarmiento, J. L., T. M. C. Hughes, R. J. Stouffer, and S. Manabe. 1998. Simulated response of the ocean carbon cycle to anthropogenic climate warming. *Nature* 393:245–249.

Smetacek, V. 2001. EisenEx: International team conducts iron experiment in Southern Ocean. *US JGOFS News* 11:11–14.

Steinberg, P. A., F. J. Millero, and X. Zhu. 1998. Carbonate system response to iron enrichment. *Marine Chemistry* 62:31–43.

Sunda, W. G., and S. A. Huntsman. 1995. Iron uptake and growth limitation in oceanic and coastal phytoplankton. *Marine Chemistry* 50:189–206.

Tsuda A., S. Takeda, H. Saito, J. Nishioka, Y. Nojiri, I. Kudo, H. Kiyosawa, A. Shiomoto, K. Imai, T. Ono, A. Shimamoto, D. Tsumune, T. Yoshimura, T. Aono, A. Hinuma, M. Kinugasa, K. Suzuki, Y. Sohrin, Y. Noiri, H. Tani, Y. Deguchi, N. Tsurushima, H. Ogawa, K. Fukami, K. Kuma, and T. Saino. 2003. A mesoscale iron enrichment in the western Subarctic Pacific induces a large centric diatom bloom. *Science* 300:958–961.

Turner, S. M., P. D. Nightingale, L. J. Spokes, M. I. Liddicoat, and P. S. Liss. 1996. Increased dimethyl sulphide concentrations in seawater from in situ iron enrichment. *Nature* 383:513–517.

Turner, S. M., M. J. Harvey, C. S. Law, and P. S. Liss. 2003. Iron-induced changes in oceanic sulphur biogeochemistry (in preparation).

Watson, A. J. 2001. Iron limitation in the oceans. Pp. 9–39 in *The biogeochemistry of iron in seawater,* edited by D. R. Turner and K. A. Hunter. Chichester, UK: John Wiley and Sons.

Watson, A. J., C. S. Law, K. A. van Scoy, F. J. Millero, W. Yao, G. E. Friederich, M. I. Liddicoat, R. H. Wanninkhof, R. T. Barber, and K. H. Coale. 1994. Minimal effect of iron fertilization on sea surface carbon dioxide concentrations. *Nature* 371:143–145.

Watson, A. J., D. C. E. Bakker, A. J. Ridgwell, P. W. Boyd, and C. S. Law. 2000. Effect of iron supply on Southern Ocean $CO_2$ uptake and implications for glacial $CO_2$. *Nature* 407:730–733.

# 27

# Direct Injection of $CO_2$ in the Ocean

## Peter G. Brewer

It is now 25 years since Marchetti (1977) first suggested bypassing atmospheric disposal of some fraction of industrial $CO_2$ emissions and using direct deep-ocean disposal as one means of ameliorating climate change. After all, the alkalinity of the ocean already provides the dominant long-term sink for atmospheric $CO_2$, and deep-ocean injection may logically be seen as simply accelerating a "natural" process. Behind this apparently simple suggestion lies great complexity, fascinating science, and strongly held opinions. The scale of modern $CO_2$ fluxes, and the interest in finding safe and economically viable ways to avoid "dangerous anthropogenic interference with climate," has led to new efforts to investigate this possible solution. Until very recently, only cartoon sketches were available to describe the possibilities of this field (Hanisch 1998). Now, important and challenging small-scale field experiments (Brewer et al. 1999) and new ocean modeling studies (Drange et al. 2001; Caldeira et al. 2002) have been conducted, and they illuminate both the possibilities and the challenges facing this strategy. A significant international scientific literature now exists (Handa and Ohsumi 1995), and the field has become established as a growing part of ocean science.

There are many variants of proposed ocean $CO_2$ disposal strategies. For example, Caldeira and Rau (2000) proposed using limestone to neutralize saturated solutions of captured $CO_2$, followed by shallow ocean disposal of the resulting solution. All of these concepts will have to be subjected to experimental test, and much of this work remains to be done. Early experiments on small-scale $CO_2$ injections form the basis of this background chapter.

The scale of surface ocean $CO_2$ uptake from the atmosphere is now so large that any distinction between a "natural" process and an industrial policy of only slightly indirect surface ocean disposal is hard to maintain. The average rate of surface ocean $CO_2$ uptake for the 1980s and 1990s was approximately 2.0 petagrams of carbon per year ($PgC\,y^{-1}$). This is 21 million tons of $CO_2$ per day, and fluxes of such scale will have biogeochemical impacts. Surface ocean waters are already >0.1 pH units lower than in preindustrial times (Brewer 1997), and if the IS92a "business-as-usual" scenario of the Intergovernmental Panel on Climate Change (IPCC) is followed, by the end of this century surface ocean carbonate ion concentrations will drop by 55 percent, with anticipated major effects on coral reef systems (Langdon et al. 2000) and calcareous plankton (Riebesell et al. 2000).

With this as background, it is clear that the enormous engineering demands would limit any industrial effort to dispose of $CO_2$ by direct injection in the deep sea to a very small fraction of the massive current and future surface fluxes. Nonetheless, such techniques could be very useful as part of a portfolio of atmospheric stabilization strategies, if they are shown to be safe and cost effective. The criticisms leveled at this approach are essentially that it would be harmful to marine life (Seibel and Walsh 2001) and that it would inevitably increase the already large oceanic burden of fossil-fuel $CO_2$.

Such comments at once expose scientific unknowns, for there are simply no standards to refer to here, and standards are urgently needed. How sensitive are marine animals to pH changes? If we permit atmospheric $CO_2$ levels to rise to ~600 parts per million (ppm) and remain there, then surface ocean waters, and eventually all ocean waters, will experience an ~0.3 pH change from preindustrial levels. Is that an acceptable limit, and why? What burden of fossil fuel $CO_2$ is "acceptable" for climate or geochemical or biological reasons? How might we view the trade-off between the direct effects of $CO_2$ and the induced effects of climate and temperature? The focus of the ocean science community has been on observing the evolving fossil-fuel $CO_2$ tracer signal, and these questions have simply (and astonishingly) not yet been posed. It is clear that the atmosphere, and therefore also the ocean, has experienced very large changes in $CO_2$ over Phanerozoic time (Berner 1990). And in today's ocean there are large pH gradients between the Atlantic and Indo-Pacific oceans and within ocean basins. Many marine animals migrate vertically through large pH gradients each day. How these observations relate to the modern fossil-fuel signal remains to be tested.

Given these questions, the current series of deep-ocean $CO_2$ injection experiments can be viewed not simply as disposal strategy tests, but also as controlled enrichment experiments that may allow us to mimic the elevated $CO_2$ levels of a future ocean in much the same way that ecosystems are manipulated on land (DeLucia et al. 1999; Shaw et al. 2002).

The cost of $CO_2$ capture dominates the economics of both geologic and oceanic disposal schemes. An excellent introduction to this problem is provided by the International Energy Agency's Greenhouse Gas Program (http://www.ieagreen.org.uk/). The most widely used technology is scrubbing the gas stream with an amine solvent and regenerating the pure $CO_2$ by heating the amine. This technique is routinely used in the food and beverage industries and for capturing $CO_2$ for geologic disposal in the Sleipner Field off Norway. The costs of the capture process presently exert an ~25 percent energy penalty, and there are now strenuous efforts to improve this.

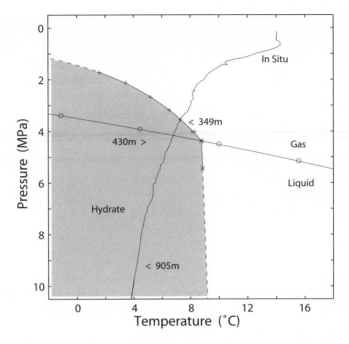

**Figure 27.1.** The phase behavior of $CO_2$ in seawater, showing the gas-liquid and hydrate phase boundaries with a typical in situ P-T profile (from Brewer et al. 1999)

## The Near-Field Fate of $CO_2$ in Seawater

Ocean scientists are well versed in the thermodynamics and ocean distributions of the extraordinarily dilute (~2.2 millimolar oceanic $CO_2$ system (Wallace 2001). But such concepts must be radically extended to investigate the science of deep-ocean $CO_2$ injection. Figure 27.1 shows the phase diagram for pure $CO_2$ superimposed on an oceanic pressure-temperature (P-T) scale (Brewer et al. 1999), with a temperature profile from a station off northern California overlaid.

The shaded area in Figure 27.1 indicates the zone in which $CO_2$ will react with seawater to from a solid hydrate ($CO_2.6H_2O$), with profound changes in physical behavior. For the P-T profile indicated, $CO_2$ gas will first form a hydrate at a depth of about 350 meters (m). The transition from gas to liquid occurs at about 400 m deep. For warm water regions (such as the Sargasso and Mediterranean Seas), these profiles will be shifted, but Figure 27.1 is generally applicable over much of the world's oceans. Liquid $CO_2$ is highly compressible, and at depths < 2,750 m, it is less dense than seawater; thus $CO_2$ released will form a buoyant plume that will dissolve in the surrounding ocean (Alendal and Drange 2001). The dissolution rate of $CO_2$ droplets is close to 3

Copyright 2001 Monterey Bay Aquarium Research Institute     Dive# 330
Tiburon/2001/177/01_28_55_15.rgb (MAIN)                     Lat= 36.70801163
Tue Jun 26 17:08:49 2001 GMT (local +7)  esecs=993575329   Lon= -123.52588654
co2-corral: identity-reference 2, perspective close-up, contains gas-hydrate

Depth= 3602.2 m  Temp= 1.546 C  Sal= 34.671 PSU  Oxy= 2.92 ml/l  Xmiss= 74.8%

**Figure 27.2.** A "frost heave" of $CO_2$ hydrate on the sea floor at 3,600 m depth resulting from massive hydrate formation.

micromoles per square centimeter per second ($\mu mol\ cm^{-2}\ sec^{-1}$) (Brewer et al. 2002a), with the result that, for a release of small droplets, 90 percent of the plume is dissolved within 30 minutes and within about 200 m above the release point.

Below a depth of about 3,000 m, the high compressibility of liquid $CO_2$ results in formation of a fluid of greater density than seawater, and a gravitationally stable release is possible. This finding has led to suggestions of storing liquid $CO_2$ as a "lake" on the deep ocean floor. Here considerable complexity exists. The solubility of $CO_2$ in seawater is so high ($\approx 0.8$ molar at 1(C and 30 megapascals [MPa]) and the partial molal volume so low ($\approx 31$ milliliters per mole [ml mol$^{-1}$]) that a dense boundary layer can form (Aya et al. 1997), inhibiting mixing with the ocean above. Such a system would have much in common with naturally occurring pools of dense brines on the seafloor. The formation of hydrate in such a system, however, can occur spontaneously, with complex self-generated, fluid dynamic instabilities and resulting large volume changes from the incorporation of six molecules of water for every molecule of $CO_2$ (Brewer et al. 1999). A spectacular example of this is shown in Figure 27.2, where an experimental pool of $CO_2$ placed on the seafloor at 3,600 m depth has penetrated the upper sediments and formed a massive hydrate "frost heave."

It was earlier thought that the formation of a hydrate would possibly result in "permanent" storage of $CO_2$ on the seafloor as a hydrate because the P-T conditions are so strongly favorable. The essential condition for hydrate stability, however, is equality of the chemical potential in all phases. Because seawater is so strongly undersaturated with respect to $CO_2$, the hydrate will dissolve. The oceanic dissolution rates of both hydrate-coated $CO_2$ droplets (Brewer et al. 2002b) and of the solid hydrates (Rehder et al. 2002) have been directly measured.

Possible technologies for future applications may include either pipeline or tanker injection techniques (Aya et al. 2003). With either approach, injection depth, physical state, or local pH perturbation can be optimized.

## The Far-Field Fate of $CO_2$ in Seawater

Once fossil-fuel $CO_2$ injected into deep-ocean water becomes dissolved in the background ocean, it yields a tracer signal that can be modeled. Aumont et al. (2001) compared seven ocean models to explore the long-term fate of purposefully injected $CO_2$. Essentially all show a strong correlation between depth of injection and efficiency of retention. For injection at 3,000 m ocean depth, overall retention efficiencies are ~ 85 percent over a 1,000-year timescale. Some surface ocean reexposure and reabsorption from the atmosphere is included in this estimate. For a 3,000 m injection, this fraction is about 25 percent of the total.

## Experimental Activities

For many years, only model concepts and sketches were available to describe this field. Plans for a medium-scale experiment (several tons $CO_2$) off the coast of Hawaii, and later off the coast of Norway, were frustrated by environmental permitting issues. This situation changed in 1996, with the successful adaptation of remotely operated vehicle (ROV) technology to contain and release small, controlled quantities of $CO_2$ in the deep ocean for scientific study (Brewer et al. 1998). Rapid advances in technique then led to the ability to conduct experiments at great depth (Brewer et al. 1999). These experiments have now been extended to elegant biological response studies (Barry et al. 2002) and sophisticated chemical measurements (Brewer et al. 2002b).

The most recent experimental system used is a 56-liter (l), carbon-fiber, composite piston-accumulator (Hydratech, Fresno, CA), with a 23-centimeter (cm) outside diameter and 194-cm length, rated at 3,000 pounds per square inch gauge (psig), for ROV operation (Figure 27.3). Two tandem cylinder pumps, with capacities of 128 milliliters (ml) and 970 ml, provide power for accurate delivery of the contained $CO_2$. The $CO_2$ cools and compresses during descent to ocean depth, so that under typical conditions (900 psig on deck at 16°C; 1.6°C at 3,600 m), 45 l are available per dive for experimental purposes.

The availability of these experimental quantities permits tests of ecosystem responses

**Figure 27.3.** A 56-l carbon-fiber-wound $CO_2$ delivery system installed on ROV Tiburon, showing end cap with gauges, delivery pumps on top, and valves to the left. The dispensing valve is attached to the robotic arm in front of the vehicle and is not shown (from Brewer et al. 2003).

to elevated deep-ocean $CO_2$ levels. One recent experimental arrangement is shown in Figure 27.4. Here, a set of small $CO_2$ "corrals" has been placed on the seafloor at a depth of 3,600 m, where the liquid $CO_2$ is sufficiently dense to be gravitationally stable. Around these $CO_2$ sources are arranged animal cages, and cores are taken to investigate the benthic infauna. Current meters, a time-lapse camera, and recording pH sensors complete the observing system.

Such experiments are at a very early stage, and rapid progress is expected. The challenges of carrying out field experiments in the deep sea should, however, not be underestimated. For example, the well-known tidal flows force the low pH plume from the $CO_2$ corrals, so that the experimental cages are exposed to varying pH signals (Figure 27.5). It should be possible to advance experimental design to improve this.

## Conclusions

Deep-ocean $CO_2$ sequestration is clearly technically possible, although there are many variants in the details, and few of these have yet been explored thoroughly. The primary costs of any $CO_2$ capture and disposal technology lie in the capture step, and this situation is the same for either geologic or oceanic disposal. Early ideas of "permanent"

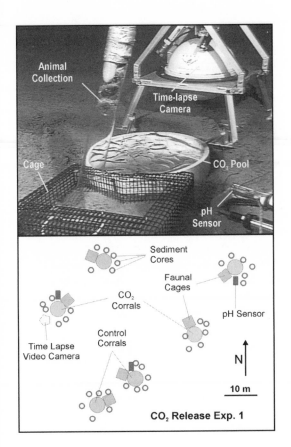

**Figure 27.4.** A sketch of a deep-sea $CO_2$ enrichment experimental site designed to investigate the response of marine organisms to locally elevated $CO_2$ levels. Note that deep ocean $CO_2$ levels will rise even in the absence of any active carbon sequestration program (from Barry et al. 2002).

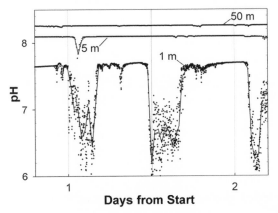

**Figure 27.5.** The pH signals recorded at distances of 1 m, 5 m, and 50 m from a $CO_2$ source placed on the seafloor. The effects of electrode drift have been removed, and the data sets are offset for visual clarity. The background oceanic pH is about 7.6. The effect of the tidal velocity ellipse is to advect the $CO_2$ plume past each sensor with about a 12.4-hour period (from Barry et al. 2002).

disposal as a hydrate on the seafloor are not realistic because the hydrate will readily dissolve in the unsaturated ocean water. Concepts of a "lake" of $CO_2$ on the seafloor, with a stable, dense boundary layer above, remain to be tested, but these would be locally specific solutions where seafloor topography permits. In most locations, $CO_2$ would quickly become dissolved in seawater and transported as a tracer plume by the abyssal circulation. The mean ventilation time of the Atlantic Ocean is ~250 years, and that of the Pacific Ocean ~550 years (Stuiver et al. 1983). Models show the expected Antarctic atmospheric reexposure of $CO_2$ on approximately these timescales. Much of the $CO_2$ is reabsorbed, with the result that, over a timescale of about 1,000 years, the injection is some 85 percent efficient.

The primary concerns over oceanic injection are the possible effects on marine life, and the fact that it will add to the already substantial ocean fossil-fuel $CO_2$ burden. There are as yet no standards to refer to here, and these are urgently needed. The current invasion rate of fossil-fuel $CO_2$ from the atmosphere to the surface ocean now approximates 20 million tons per day. This flux has already changed surface ocean pH, and a 0.3 pH reduction in ocean waters can be expected if atmospheric $CO_2$ is held at about 600 ppm. Thus, deep-sea ecosystems will inevitably experience change, and it is likely that any disposal strategy will be of significantly smaller scale than the surface invasion signal.

Release of $CO_2$ directly into the ocean may be less harmful than release of $CO_2$ into the atmosphere. In this case, if injected $CO_2$ results in diminished atmospheric peak concentrations, then direct injection of $CO^2$ in the ocean could reduce the overall adverse consequences of fossil-fuel burning. If oceanic injection simply adds to atmospheric releases, however, overall adverse consequences could increase. Thus, the ability of oceanic injection to contribute to diminished adverse consequences of fossil-fuel use depends both on the science and technology of oceanic $CO_2$ injection and on the roles that that science and technology might play in our energy economy.

## Acknowledgments

I acknowledge the David and Lucile Packard Foundation, the U.S. Department of Energy Ocean Carbon Sequestration Program, and an international research grant from the New Energy and Industrial Technology Organization for support of this research.

## Literature Cited

Alendal, G., and H. Drange. 2001. Two-phase, near field modeling of purposefully released $CO_2$ in the ocean. *Journal of Geophysical Research* 106:1085–1096.
Aumont, O., J. C. Orr, A. Yool, K. Plattner, F. Joos, E. Maier-Reimer, M.-F. Weirig, R. Schlitzer, K. Caldeira, M. Wickett, and R. Matear. 2001. Efficiency of purposeful $CO_2$

injection in the deep ocean: Comparison of seven ocean models. *IGBP Open Science Conference 2001, Amsterdam, The Netherlands.* http://www.ipsl.jussieu.fr/OCMIP /phase2/poster/aumont.pdf.

Aya, I., K. Yamane, and H. Nariai. 1997. Solubility of $CO_2$ and density of $CO_2$ hydrate at 30 MPa. *Energy* 22:263–271.

Aya, I., R. Kojima, K. Yamane, P. G. Brewer, and E. T. Peltzer. 2003. In situ experiments of cold $CO_2$ release in mid-depth. Pp. 739–744 in *Greenhouse gas control technologies*, edited by J. Gale and Y. Kaya. Amsterdam: Pergamon.

Barry, J., B. A. Seibel, J. Drazen, M. Tamburri, C. Lovera, and P. G. Brewer. 2002. Field experiments on direct ocean $CO_2$ sequestration: The response of deep-sea faunal assemblages to $CO_2$ injection at 3200m off Central California. *Eos, Transactions, American Geophysical Union* 83:OS51F-02.

Berner, R. A. 1990. Atmospheric carbon dioxide levels over Phanerozoic time. *Science* 249:1382–1386.

Brewer, P. G. 1997. Ocean chemistry of the fossil fuel $CO_2$ signal: The haline signature of "business as usual." *Geophysical Research Letters* 24:1367–1369.

Brewer, P. G., F. M. Orr Jr., G. Friederich, K. A. Kvenvolden, and D. L. Orange. 1998. Gas hydrate formation in the deep sea: In situ experiments with controlled release of methane, natural gas, and carbon dioxide. *Energy and Fuels* 12:183–188.

Brewer, P. G., G. Friederich, E. T. Peltzer, and F. M. Orr Jr. 1999. Direct experiments on the ocean disposal of fossil fuel $CO_2$. *Science* 284:943–945.

Brewer, P. G., E. T. Peltzer, G. Friederich, and G. Rehder. 2002a. Experimental determination of the fate of rising $CO_2$ droplets in seawater. *Environmental Science and Technology* 36:5441–5446.

Brewer, P. G., J. Pasteris, G. Malby, E. T. Peltzer, S. White, J. Freeman, B. Wopenka, M. Brown, and D. Cline. 2002b. Laser Raman spectroscopy used to study the ocean at 3600-m depth. *Eos, Transactions, American Geophysical Union* 42:469–470.

Brewer, P. G., E. T. Peltzer, G. Rehder, R. Dunk. 2003. Advances in deep-ocean $CO_2$ sequestration experiments. Pp.1667–1670 in *Greenhouse gas control technologies*, edited by J. Gale and Y. Kaya. Amsterdam: Pergamon.

Caldeira, K., and G. H. Rau. 2000. Acclerating carbonate dissolution to sequester carbon dioxide in the ocean: Geochemical implications. *Geophysical Research Letters* 27:225–228.

Caldeira, K., M. E. Wickett, and P. B. Duffy. 2002. Depth, radiocarbon and the effectiveness of direct ocean injection as an ocean carbon sequestration strategy. *Geophysical Research Letters,* doi:10.1029/2001GL014234.

DeLucia, E. H., J. G. Hamilton, S. L. Naidu, R. B. Thomas, J. A. Andrews, A. Finzi, M. Lavine, R. Matalama, J. E. Mohan, G. R. Hendrey, and W. H. Schlesinger. 1999. Net primary production of a forest ecosystem with experimental $CO_2$ enrichment. *Science* 284:1177– 1179.

Drange, H., G. Alendal, and O. M. Johannessen. 2001. Ocean release of fossil fuel $CO_2$: A case study. *Geophysical Research Letters* 13: 2637–2640.

Handa, N., and T. Ohsumi. 1995. *Direct ocean disposal of carbon dioxide.* Tokyo: Terra.

Hanisch, C. 1998. The pros and cons of carbon dioxide dumping. *Environmental Science and Technology* 32:20–24A.

Langdon, C., T. Takahashi, C. Sweeney, D. Chipman, J. Goddard, F. Marubini, H. Aceves, H. Barnett, and M. J. Atkinson. 2000. Effect of calcium carbonate saturation state on the calcification rate of an experimental coral reef. *Global Biogeochemical Cycles* 14:639–654.

Marchetti, C. 1977. On geoengineering and the $CO_2$ problem. *Climatic Change* 1:59–68.

McNeil, B., R. J. Matear, R. M. Key, J. L. Bullister, and J. L. Sarmiento. 2003. Anthropogenic $CO_2$ uptake by the ocean based on the global chlorofluorocarbon data set. *Science* 299:235–239.

Rehder, G., S. H. Kirby, W. B. Durham, L. Stern, E. T. Peltzer, J. Pinkston, and P. G. Brewer. 2002. Dissolution rates of pure methane hydrate and carbon dioxide hydrate in undersaturated sea water at 1000m depth. *Geochimica et Cosmochimica Acta* (submitted).

Riebesell, U., I. Zondervan, B. Rost, P. D. Tortell, R. E. Zeebe, and F. Morel. 2000. Reduced calcification of marine plankton in response to increased atmospheric $CO_2$. *Nature* 407:364–367.

Seibel, B., and P. J. Walsh. 2001. Potential impacts of $CO_2$ injection on deep-sea biota. *Science* 294:319–320.

Shaw, H. R., E. S. Zavaleta, N. R. Chiariello, E. L. Cleland, H. A. Mooney, and C. B. Field. 2002. Grassland responses to global environmental changes suppressed by elevated $CO_2$. *Science* 298:1987–1990.

Stuiver, M., P. D. Quay, and H. G. Ostlund. 1983. Abyssal water carbon-14 distribution and the age of the world oceans. *Science* 219:849–851.

Wallace, D. W. R. 2001. Storage and transport of excess $CO_2$ in the oceans: The JGOFS/WOCE Global $CO_2$ survey. In *Ocean circulation and climate*, edited by G. Siedler, J. Church, and J. Gould. San Diego: Academic Press.

# 28

# Engineered Biological Sinks on Land

## Pete Smith

Terrestrial sinks are not permanent. If a land management or land use change is reversed, the carbon accumulated will be lost. New C stocks, once established, need to be preserved in perpetuity. A number of current biosphere C stocks are vulnerable to current and future climatic change and human activity. Because losses from these will militate against activities to increase sink strength in the future, potential carbon losses from land need to be considered when estimating the potential for increasing land carbon sinks.

Under the Kyoto Protocol, sinks include those arising from afforestation and reforestation (deforestation also needs to be accounted for), forest management, cropland management, grazing land management, and revegetation. Estimates for the sequestration potential of these activities range from about 30 to 80 grams of carbon per square meter per year (g C m$^{-2}$ y$^{-1}$). The best options for enhancing sinks occur where there are co-benefits associated with a sink-enhancing activity—"win-win" management options have the greatest potential for implementation. In many cases the techniques to enhance carbon sinks are well known, but economic and educational constraints present the greatest barriers to their implementation. Terrestrial C sink enhancement needs to be tackled hand-in-hand with related environmental, political, and socioeconomic problems.

## Engineered Biological Carbon Sinks on Land— Their Place in the Global C Cycle

During the 1980s and 1990s fossil fuel-combustion and cement production emitted 5.9 petagrams (Pg) C y$^{-1}$ to the atmosphere, while land use change emitted 12 PgC y$^{-1}$ (Sabine et al., Chapter 2, this volume). Atmospheric C increased at a rate of 3.2 ± 0.1 PgC y$^{-1}$, the oceans absorbed ~2.3 PgC y$^{-1}$, and the estimated terrestrial sink was 1.8 PgC y$^{-1}$.

The size of the biological terrestrial carbon pools are 2,300 PgC for soil organic carbon and 650 PgC for vegetation (Sabine et al., Chapter 2). The annual fluxes between land and atmosphere (global net primary production [NPP], respiration, and fire) are of the order of 60 PgC $y^{-1}$ (Sabine et al., Chapter 2). The pool sizes are therefore large compared with both the gross and net annual fluxes of carbon to and from the terrestrial biosphere. Manipulating the size of these terrestrial carbon pools is at the heart of efforts to engineer carbon sinks on land.

Engineered sinks on land are typically designed to replace stocks that were present before human intervention. Historically, for example, soils have lost between 40 and 90 PgC globally through cultivation and disturbance (Schimel 1995; Houghton 1999; Lal 1999) with total system losses closer to 180 PgC (DeFries et al. 1999; Houghton et al. 1999). The engineered soil C sink is therefore replacing the stock lost through earlier human activities. The goal of engineering biological sinks on land is to increase the C stock relative to the current C stock.

Biological carbon sinks on land can be engineered by increasing the size of the soil organic carbon or vegetation pools. This is achieved by increasing the net flux of carbon from the atmosphere to the terrestrial biosphere by increasing global NPP, by storing more of the carbon from NPP in the longer-term carbon pools in the soil, or by slowing decomposition. In essence, engineered carbon pools on land rely on either increasing the standing stock of carbon in vegetation or sequestering carbon in soil.

For vegetation carbon sinks, the best (but not the only) option for carbon storage is via tree planting (afforestation, reforestation). For soil carbon sinks, the best options are to increase C stocks in soils that have been depleted in carbon, that is, agricultural soils and degraded soils. Other options are available, and these will be listed later.

Estimates for soil carbon sequestration potential vary widely. Based on studies in European cropland (Smith et al. 2000a), U.S. cropland (Lal et al. 1998), global degraded lands (Lal 2003), and global estimates (Cole et al. 1996; Watson et al. 2000), global soil carbon sequestration potential is estimated to be $0.9 \pm 0.3$ PgC $y^{-1}$ (Lal 2004), between one-third and one-quarter of the annual increase in atmospheric carbon levels. Over 50 years, the level of C sequestration suggested by Lal (2004) would restore a large part of the carbon lost from soils historically. Estimates based on Europe and North America by Smith et al. (2000b) suggest a lower potential (one-third to one-half of historical soil C loss over 100 years).

Estimates for the carbon sequestration potential for forests range between about 1 PgC $y^{-1}$ (the lower figure of IPCC 1997) and 2 PgC $y^{-1}$ (Trexler 1988 [cited in Metting et al. 1999]), between one-third and two-thirds of the annual increase in atmospheric carbon levels. As well as absorbing $CO_2$ from the atmosphere, a change in the balance between deforestation and afforestation/reforestation would have the added benefit of reducing the net efflux of $CO_2$ to the atmosphere from land use change, that is, it would decrease the size of a current C source as well as increase the size of a potential C sink.

The estimate for total terrestrial C sequestration potential by IPCC (2000) was about 2.5 PgC $y^{-1}$. This estimate includes sinks from the management of agricultural

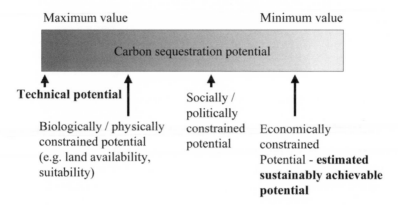

**Figure 28.1.** Relationships between reported carbon sequestration potentials depending on the number and type of constraints considered

land, biomass cropland, grassland, rangeland, and forest; the protection and creation of wetlands and urban forest/grassland; the manipulation of deserts and degraded lands; and the protection of sediments, aquatic systems, tundra, and taiga. Other estimates are as high as 5.65–8.71 PgC $y^{-1}$, around two times the annual increase in atmospheric carbon levels (Metting et al. 1999).

## What Is Meant by Carbon Sequestration Potential?

Estimates of carbon sequestration potentials are often confused by the choice of constraints. Some authors quote biological potentials (Metting et al. 1999), others quote potentials as limited by available land or resources (Smith et al. 2000a), and others also consider economic and social constraints (Cannell 2003; Freibauer et al. 2004; Figure 28.1). An analysis of the estimates presented in Freibauer et al. (2004) and the assumptions used by Cannell (2003) suggest that the sustainable (or conservative) achievable potential of carbon sequestration (taking into account limitations in land use, resources, economics, and social and political factors) may be about 10 percent of the biological potential. Though this value is derived predominantly from expert judgment, it may be useful in assessing how different estimates of carbon sequestration potential can be compared and how they might realistically contribute to $CO_2$ stabilization.

## Methods to Enhance Terrestrial Carbon Sinks

Various management practices have been suggested to manipulate the terrestrial carbon sink (Box 28.1). All of the management options in Box 28.1 have the potential to increase sink strength. Some of these options, however, also have an associated carbon

**Box 28.1.** Management Options to Enhance Terrestrial Carbon Sinks

## Vegetation

- Afforestation/reforestation (planting trees to increase the per area NPP of a unit of land)
- Forest management (fire suppression, thinning, fertilization, other techniques to improve per area NPP of a tree-covered unit of land)
- Revegetation (increasing vegetation carbon stocks using plants other than trees)
- Cropland and grazing land management (increasing the aboveground standing stock of carbon in agricultural systems)

## Soil

### Forestry

- Increase in soil carbon stocks though afforestation, reforestation, improved forest management, or revegetation

### Cropland

- Zero or reduced tillage
- Set-aside or conservation reserve program
- Conversion to permanent crops
- Conversion to deep-rooting crops
- Improved efficiency of animal manure use
- Improved efficiency of crop residue use
- Agricultural use of sewage sludge
- Composting
- Improved rotations
- Fertilization
- Irrigation
- Bioenergy crops
- Extensification or deintensification of farming
- Organic farming
- Conversion of cropland to grassland
- Improved management to reduce wind and water erosion

*(continued on next page)*

## Grazing Land

- Conversion to deep-rooting species
- Improved efficiency of animal manure use
- Improved efficiency of crop residue use
- Improved livestock management to reduce soil disturbance
- Improved livestock management to maximize manure C returns
- Agricultural use of sewage sludge
- Composting
- Fertilization
- Irrigation
- Extensification or deintensification of farming
- Improved management to reduce wind and water erosion

## Revegetation

- Increase of soil carbon stocks by planting vegetation other than trees with higher carbon returns to soil or with litter more resistant to decomposition

## Other Potential Sinks

- Protection and creation of wetlands
- Protection and creation of urban forest and grassland
- Improved management of deserts and degraded lands
- Protection of sediments and aquatic systems
- Protection of tundra and taiga

Compiled from Lal et al. (1998), Metting et al. (1999), Follett et al. (2000), IPCC (2000), Smith et al. (2000a), Freibauer et al. (2004), and Smith (2004).

cost. Schlesinger (1999) has suggested that a number of these potential sink-enhancing activities (e.g., mineral fertilization, irrigation) may be carbon-negative in that more carbon is consumed in producing the resources and implementing the management than is gained in the sink. Others have argued that at least some of these options (e.g., animal manure) are simply a redistribution of the resource produced as a by-product of other activity (Smith and Powlson 2000) and as such have no additional associated carbon cost. In any case, it is important to consider the full C impact of a C sequestration practice (Schlesinger 1999). In fact, the full impact on global warming potential (GWP; including the impact on non-$CO_2$ greenhouse gases) should be assessed (Robertson et al. 2000) because some sequestration options can be severely reduced or negated by increases in non-$CO_2$ greenhouse gas emissions (Smith et al. 2001). Some sequestration practices, such as conversion of cropland to set-aside land, actually derive most of their climate mitigation potential from reductions in non-$CO_2$ greenhouse gases, even though they were originally implemented as carbon sequestration practices (G. P. Robertson, pers. comm.).

Various studies have been conducted to examine the relative potential of each of the measures listed in Box 28.1. Estimates for the sequestration potential of these activities range from about 30 to 80 g C m$^{-2}$ y$^{-1}$, with some estimates lower and others higher (Nabuurs et al. 1999; Watson et al. 2000; Follett et al. 2000; Smith et al. 2000a; West and Post 2002; Lal 2004). Estimates of the total sequestration (or mitigation) potential of these and other mitigation scenarios are discussed in Caldeira et al. (Chapter 5, this volume).

## Time Course of Sink Development and Permanence

The time course of sink development is sink specific. Soil carbon sinks increase most rapidly soon after a carbon-enhancing land management change has been implemented. Vegetation carbon sinks may be most rapid a few years after a land use change, as it may take some time for the new trees and vegetation to become established. Sink strength (i.e., the rate at which C is removed from the atmosphere) in both soil and vegetation becomes smaller as the ecosystems approach new equilibria. At equilibrium, the sink has saturated: the carbon stock may have increased, but the sink strength has decreased to zero (Figure 28.2).

The detailed time course for each sink is, of course, highly variable, as is the time for sink saturation (i.e., new equilibrium) to occur. Soils in a temperate location reach a new equilibrium around 100 years after a land use change (Jenkinson 1988; Smith et al. 1996), but tropical soils may reach equilibrium more quickly. Soils in boreal regions may take centuries to approach a new equilibrium. As a compromise, current IPCC good practice guidelines for greenhouse gas inventories use a figure of 20 years for soil carbon to approach a new equilibrium (IPCC 1997; Paustian et al. 1997). Temperate tree species may reach maturity within 100 years but slower-growing species and those in boreal latitudes may take longer.

Terrestrial sinks are not permanent. Because sink strength decreases as ecosystems approach a new equilibrium (Figure 28.2), the net amount of C removed from the atmosphere decreases. If a land management or land use change is reversed, the carbon accumulated will be lost, usually more rapidly than it was accumulated (Smith et al. 1996). For the greatest potential of terrestrial C sinks to be realized, new C sinks, once established, need to be preserved in perpetuity. Within the Kyoto Protocol (discussed in the next section), suggested mechanisms provide disincentives for sink reversal (i.e., when land is entered into the Kyoto process it must remain in the accounting, and any sink reversal will result in a loss of carbon credits).

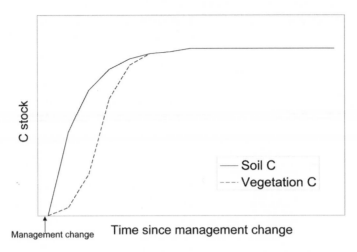

**Figure 28.2.** Illustrative graph showing the possible time course of sink development

In the soil, factors leading to carbon stabilization include chemical, physical, and biological processes, with a varying balance. More research is required to elucidate fully the balance of these processes in determining soil C stabilization under different conditions.

## Sinks in the Political Arena

In the political arena, sinks are perhaps more hotly debated than they are in the purely scientific arena. Terrestrial sinks have received close political scrutiny since their inclusion in the Kyoto Protocol at the 4th Conference of Parties (COP4) to the United Nations Framework Convention on Climate Change (UNFCCC). Under Articles 3.3 and 3.4 of the Kyoto Protocol, biosphere sinks (and sources) of carbon can be included by parties in meeting greenhouse gas emission reduction targets by comparing emissions in the commitment period (the first CP is 2008–2012) with baseline (1990) emissions. The exact nature of the sinks to be allowed and the modalities and methods that could be used were addressed in subsequent meetings of the COP, culminating in agreement at COP7 in 2001, leading to the publication of the Marrakesh Accords.

Under the Marrakesh Accords, Kyoto Article 3.3 activities limited to afforestation, reforestation, and deforestation are included but are subject to a complex series of rules. These rules attempt to factor out any impacts that are not directly human induced. Kyoto Article 3.4 activities can be included under forest management, cropland management, grazing land management, and revegetation. The details of the rules and modalities governing how sinks can be used are too complex to describe in detail but include safeguards to avoid double counting of sinks, to avoid carbon credits being claimed and the land subsequently reverting to low carbon stock levels (e.g., by deforestation), and to ensure that accounting is transparent and verifiable.

An interesting outcome of the Marrakesh Accords is that crop- and grazing land management and revegetation will be reported on a net-net basis, that is, net emissions during the commitment period will be compared with net emissions during the baseline year. Under this form of accounting, a carbon credit will arise if a party reduces a source, even though a sink (i.e., a net removal of C from the atmosphere) has not been created. In this respect, decreasing a source is equivalent to creating a sink.

Scientific issues related to the use of biosphere sinks in the Kyoto Protocol were addressed in the IPCC special report *Land Use, Land-Use Change, and Forestry* (Watson et al. 2000). Though many of these issues no longer apply (because the political negotiations have moved on), this text is still a useful source of information on the potential of biosphere sinks for climate mitigation.

Biological sinks on land will remain in the political spotlight at least until the end of the first commitment period in 2012. With further commitment periods set for negotiation in the near future, sinks are likely to remain politically important for the foreseeable future.

## Vulnerability of Current C Sinks

A number of the biosphere stocks are vulnerable both to future climatic change and to current and future human activity (Gruber et al., Chapter 3, this volume). These impacts will reduce the strength of biosphere C sinks and will militate against activities to increase sink strength in the future. A few examples of potentially threatened C stocks are given here. Highly organic boreal soils contain huge stocks of carbon. Many are in low temperature environments or even under permafrost. Climate change is projected to warm these areas, which will potentially release large amounts of carbon to the atmosphere and dramatically constrain the sink potential in these regions. Another example of a threatened C stock is carbon in tropical and subtropical soils threatened by desertification. Increasing temperatures and more frequent droughts (caused by increased climate variability) threaten to enhance desertification, while an increasing human population and changing dietary habits further exacerbate the problem by overexploitation of these vulnerable soils. Desertification in these regions will also threaten aboveground C stocks. Because degraded soils account for about half of the soil C sequestration potential (Lal 2004), further degradation and desertification will not only increase emissions but will also militate against enhancing the soil C sink through C sequestration.

Among vegetation C stocks, deforestation poses the greatest current threat. The same factors that lead to the overexploitation of vulnerable soils (e.g., population growth and changing dietary habits) also lead to forest clearance. Deforestation reduces the size of the biosphere stock and adds extra C to the atmosphere. It also reduces future sink strength, because over the course of a year, most crops remove less atmospheric C than do trees and cropland loses more soil carbon than forest does. Increases in climate variability also pose a threat to vegetation C stocks. Increasingly frequent severe storms will fell trees, whereas increased frequency of drought may increase the incidence of fire.

In short, increasing pressure on limited land resource, from increasing population and changing dietary habits, is likely to cause conflict between the need to maintain and enhance C sink strength and other competing demands upon the land. Further, a changing climate, including increasing climate variability, is likely to reduce the current terrestrial C sink strength and will militate against efforts to increase future sink strength.

## Costs of Enhancing, Measuring, Monitoring, and Verifying Sinks

The costs of methods to enhance C sinks are likely to vary over a large range. Some technological solutions may be costly (e.g., genetic engineering of new plant varieties), whereas others (e.g., changing animal manure use) could be implemented at very low or no cost. The fact that the parties to the UNFCCC have negotiated for the inclusion of sinks in the Kyoto Protocol and intend to use them to meet greenhouse gas emission reduction targets suggests that at least some methods to enhance C sinks are deemed to be cost-effective. A number of joint implementation (JI) and clean development mechanism (CDM) projects, mainly based on forestry programs (Brown et al. 2000), already exist and have shown that C sink enhancement can be cost-effective.

In addition to the cost of implementing a sink-enhancing measure, other costs are associated with measuring, monitoring, and verifying the enhancement of the sink. These costs are not trivial (Smith 2004). Costs are highest where the change in carbon stock is small, relative to the background C content (as in soils). Smith (2004) suggests that in many cases the value of the carbon sequestered will be less than the cost of demonstrating the increase in the carbon sink. Costs can be reduced in integrated national programs, including benchmark sites and modeling. He also suggests that a low level of verifiability may be achievable by most UNFCCC parties at a small cost but that an intermediate level will be available to only a few countries that have made significant investments in national carbon accounting or inventory systems. Australia, for example, is investing an additional US$5 million annually to upgrade its carbon accounting system (Watson et al. 2000).

In terms of cost-effectiveness, the best options for enhancing sinks occur where there are co-benefits associated with a sink-enhancing activity. For example, a project to create urban woodlands will enhance leisure and amenity opportunities in addition to carbon stocks. Similarly, the creation of woodlands or wetlands for wildlife will not only improve wildlife value and enhance biodiversity, but also enhance carbon stocks. Increasing the organic matter of a soil provides a number of agronomic benefits, including improved water-holding capacity, nutrient status, and workability, while also increasing the soil C sink. It is these "win-win" options that will likely be the most cost-effective mechanisms for increasing carbon sequestration. In agriculture, many low-cost management options could be implemented quickly, would be beneficial agronomically, and would enhance terrestrial C sinks. Such "no regrets" policies could enhance sink strength now and increase the resilience of the sink in the future. Smith and Powlson (2003) developed these ideas for soil sustainability, but they apply equally well to terrestrial C sink enhancement.

## Social Dimensions

Engineered sinks are, by definition, entirely under human influence. As a consequence, the human dimension to terrestrial sink enhancement is strong. Because there will be increasing competition for limited land resources in the coming century, terrestrial sink enhancement cannot be viewed in isolation from other environmental and social needs (Raupach et al., Chapter 6, this volume). Terrestrial C sink enhancement needs to go hand in hand with other aspects of sustainable development. In any scenario, there will be winners and losers. The key to enhancing sinks, as part of wider programs to enhance sustainability, is to maximize the number of winners and minimize the number of losers. Another possibility for improving the social and cultural acceptability of measures to enhance C sinks is to include compensation costs (for losers) when assessing the costs of strategies for implementing C sequestration.

In agriculture and forestry nearly all of the techniques to enhance carbon sinks are well known. Because the implementation of known technology is currently hampered by ignorance at a regional scale (Sanchez 2000), the education of farmers, foresters, land managers, and regional planners would help to enhance soil carbon sinks. Economics, poor infrastructure, and distribution bottlenecks also hamper implementation.

Terrestrial C sink enhancement needs to be tackled hand in hand with other related problems. Global, regional, and local environmental issues such as climate change, loss of biodiversity, desertification, stratospheric ozone depletion, regional acid deposition, and local air quality are inextricably linked (Houghton et al. 2001). Terrestrial C sink enhancement clearly belongs on this list. Recognizing the linkages among environmental issues and their relationship to meeting human needs provides an opportunity to address global environmental issues at the local, national, and regional level in an integrated manner that is cost-effective and meets sustainable development objectives (Houghton et al. 2001). The importance of integrated approaches to sustainable environmental management is becoming ever clearer.

Attempts to solve a raft of environmental problems in an integrated way must also address social and economic problems in the same package. All of the scientific and technical measures outlined in this chapter have the potential to enhance C sinks, but the extent to which these are sustainable also needs to be considered.

Threats to the soil C stock and to large portions of the tropical vegetation forest stock would be reduced by controlling the size of the human population because this would ease the pressure on land for food production. Poverty remains the main driver for landscape degradation (e.g., by slash-and-burn agriculture) and C loss in the poorest parts of the world, where some of the most vulnerable C stocks occur (Barbier 2000). At the global scale, relief of poverty in these regions would probably do more to protect existing C stocks than could be achieved by any of the scientific or technical measures practiced in the developed world.

The political and economic landscape of the future will determine the feasibility of many strategies to enhance C sinks, but there are a number of management practices available that could be implemented to protect and enhance existing C sinks now and in the future (a no-regrets policy). Because these practices are consistent with and may even be encouraged by many current international agreements and conventions, their rapid adoption should be promoted as widely as possible.

## Literature Cited

Barbier, E. B. 2000. The economic linkages between rural poverty and land degradation: Some evidence from Africa. *Agriculture, Ecosystems and Environment* 82:355–370.

Brown, S., M. Burnham, M. Delaney, R. Vaca, M. Powell, and A. Moreno. 2000. Issues and challenges for forest-based carbon-offset projects: A case study of the Noel Kempff Climate Action Project in Bolivia. *Mitigation and Adaptation Strategies for Climate Change* 5:99–121.

Cannell, M. G. R. 2003. Carbon sequestration and biomass energy offset: Theoretical, potential and achievable capacities globally, in Europe and the UK. *Biomass and Bioenergy* 24:97–116.

Cole, C. V., C. Cerri, K. Minami, A. Mosier, N. Rosenberg, D. Sauerbeck, J. Dumanski, J. Duxbury, J. Freney, R. Gupta, O. Heinmeyer, T. Kolchugina, J. Lee, K. Paustian, D. Powlson, N. Sampson, H. Tiessen, M. Van Noordwijk, Q. Zhao, I. P. Abrol, T. O. Barnwell Jr., C. A. Campbell, R. L. Desjardin, C. Feller, P. Garin, M. J. Glendining, E. G. Gregorich, J. Johnson, J. Kimble, R. Lal, C. Monreal, D. Ojima, M. Padgett, W. Post, W. Sombroek, C. Tarnocai, T. Vinson, S. Vogel, and G. Ward. 1996. Agricultural options for mitigation of greenhouse gas emissions. Pp. 745–771 in *Climate change 1995: Impacts, adaptations, and mitigation of climate change: Scientific-technical analyses*, edited by R. T. Watson, M. C. Zinyowera, R. H. Moss, and D. J. Dokken. New York: Cambridge University Press.

DeFries, R. S., C. B. Field, I. Fung, J. Collatz, and L. Bounoua. 1999. Combining satellite data and biogeochemical models to estimate global effects of human-induced land cover change on carbon emissions and primary productivity. *Global Biogeochemical Cycles* 13:803–815.

Follett, R. F., J. M. Kimble, and R. Lal. 2000. The potential of U.S. grazing lands to sequester soil carbon. Pp. 401–430 in *The potential of U.S. grazing lands to sequester carbon and mitigate the greenhouse effect*, edited by R. F. Follett, J. M. Kimble, and R. Lal. Boca Raton, FL: Lewis.

Freibauer, A., M. Rounsevell, P. Smith, and A. Verhagen. 2004. Carbon sequestration in European agricultural soils. *Geoderma* (in press).

Houghton, R. A. 1999. The annual net flux of carbon to the atmosphere from changes in land use 1850 to 1990. *Tellus* 50B:298–313.

Houghton, R. A., J. L. Hackler, and K. T. Lawrence. 1999. The US carbon budget: Contributions from land-use change. *Science* 285:574–578.

Houghton, J. T., Y. Ding, D. J. Griggs, M. Noguer, P. J. van der Linden, X. Dai, K. Maskell, and C. A. Johnson, eds. 2001. *Climate change 2001: The scientific basis (Contribution of Working Group I to the third assessment report of the Intergovernmental Panel on Climate Change).* Cambridge: Cambridge University Press.

IPCC (Intergovernmental Panel on Climate Change). 1997. *Revised 1996 IPCC Guidelines for National Greenhouse Gas Inventories.* Vol. 2, *Workbook.* Paris.

Jenkinson, D. S. 1988. Soil organic matter and its dynamics. Pp. 564–607 in *Russell's soil conditions and plant growth,* edited by A. Wild. 11th ed. London: Longman.

Lal, R. 1999. Soil management and restoration for C sequestration to mitigate the accelerated greenhouse effect. *Progress in Environmental Science* 1:307–326.

———. 2003. Soil erosion and the global carbon budget. *Environment International* (in press).

———. 2004. Soil carbon sequestration to mitigate climate change. *Geoderma* (in press).

Lal, R., J. M. Kimble, R. F. Follett, and C. V. Cole. 1998. *The potential of U.S. cropland to sequester carbon and mitigate the greenhouse effect.* Chelsea, MI: Ann Arbor Press.

Metting, F. B., J. L. Smith, and J. S. Amthor. 1999. Science needs and new technology for soil carbon sequestration. Pp. 1–34 in *Carbon sequestration in soils: Science, monitoring and beyond,* edited by N. J. Rosenberg, R. C. Izaurralde, and E. L. Malone. Columbus, OH: Battelle Press.

Nabuurs, G. J., W. P. Daamen, A. J. Dolman, O. Oenema, E. Verkaik, P. Kabat, A. P. Whitmore, and G. M. J. Mohren. 1999. *Resolving issues on terrestrial biospheric sinks in the Kyoto Protocol.* Report 410 200 030. Amsterdam: Dutch National Programme on Global Air Pollution and Climate Change.

Paustian, K., O. Andrén, H. H. Janzen, R. Lal, P. Smith, G. Tian, H. Tiessen, M. van Noordwijk, and P. L. Woomer. 1997. Agricultural soils as a sink to mitigate $CO_2$ emissions. *Soil Use and Management* 13:229–244.

Robertson, G. P., E. A. Paul, and R. R. Harwood. 2000. Greenhouse gases in intensive agriculture: Contributions of individual gases to the radiative forcing of the atmosphere. *Science* 289 (5486): 1922–1925.

Sanchez, P. A. 2000. Linking climate change research with food security and poverty reduction in the tropics. *Agriculture, Ecosystems and Environment* 82:371–383.

Schimel, D. S. 1995. Terrestrial ecosystems and the carbon-cycle. *Global Change Biology* 1:77–91.

Schlesinger, W. M. 1999. Carbon and agriculture: Carbon sequestration in soils. *Science* 284 (5423): 2095–2095.

Smith, P. 2004. Monitoring and verification of soil carbon changes under Article 3.4 of the Kyoto Protocol. *Soil Use and Management* (in press).

Smith, P., and D. S. Powlson. 2000. Considering manure and carbon sequestration. *Science* 287:428–429.

———. 2003. Sustainability of soil management practices: A global perspective. In *Soil biological fertility: A key to sustainable land use in agriculture,* edited by L. K. Abbott and D. V. Murphy. Amsterdam: Kluwer (in press).

Smith, P., D. S. Powlson, and M. J. Glendining. 1996. Establishing a European soil organic matter network (SOMNET). Pp. 81–98 in *Evaluation of soil organic matter models using existing, long-term datasets,* edited by D. S. Powlson, P. Smith, and J. U. Smith. NATO ASI Series I, Vol. 38. Berlin: Springer-Verlag.

Smith, P., D. S. Powlson, J. U. Smith, P. D. Falloon, and K. Coleman. 2000a. Meeting Europe's climate change commitments: Quantitative estimates of the potential for carbon mitigation by agriculture. *Global Change Biology* 6:525–539.

Smith, P., D. S. Powlson, J. U. Smith, P. D. Falloon, K. Coleman, and K. W. Goulding. 2000b. Agricultural carbon mitigation options in Europe: Improved estimates and the global perspective. *Acta Agronomica Hungarica* 48:209–216.

Smith, P., K. W. Goulding, K. A. Smith, D. S. Powlson, J. U. Smith, P. D. Falloon, and K. Coleman. 2001. Enhancing the carbon sink in European agricultural soils: Including trace gas fluxes in estimates of carbon mitigation potential. *Nutrient Cycling in Agroecosystems* 60:237–252.

Watson, R. T., I. R. Noble, B. Bolin, N. H. Ravindranath, D. J. Verardo, and D. J. Dokken, eds. 2000. *Land use, land-use change, and forestry (Special Report of the Intergovernmental Panel on Climate Change).* Cambridge: Cambridge University Press.

West, T. O., and W. M. Post. 2002. Soil organic carbon sequestration rates by tillage and crop rotation: A global data analysis. *Soil Science Society of America Journal* 66:1930–1946.

# 29

# Abatement of Nitrous Oxide, Methane, and the Other Non-CO$_2$ Greenhouse Gases: The Need for a Systems Approach

G. Philip Robertson

## Contributions to Global Warming

Carbon dioxide currently accounts for about 49 percent of the radiative forcing of the atmosphere that is attributable to greenhouse gases (3.0 watts per square meter [W m$^{-2}$]; Houghton et al. 2001). Tropospheric ozone and black carbon are responsible for another 18 percent, and the well-mixed greenhouse gases (GHGs)—principally methane, nitrous oxide, and various halocarbons—are responsible for the remaining 33 percent (Figure 29.1). Changes in well-mixed GHGs and in black carbon, tropospheric ozone, and ozone precursors (NO$_x$, CO, and NMVOCs), whether engineered or unintentional, could thus have a substantial impact on the radiative forcing of future atmospheres.

All of the well-mixed non-CO$_2$ GHGs are more potent than CO$_2$: 100-year global warming potentials (GWPs) range from 23 for methane to >10,000 for a number of the halocarbons (Table 29.1; Prinn, Chapter 9, this volume). Small changes in the net fluxes of these gases can thus have a proportionately larger effect on radiative forcing than similar changes in CO$_2$ flux. This sensitivity to small changes provides a strong impetus for including the non-CO$_2$ GHGs in the development of effective mitigation strategies: keeping or removing a given quantity of N$_2$O from the atmosphere, for example, can have almost 300 times the impact of removing the same mass of CO$_2$. This impact arises in part from wide differences in atmospheric lifetimes. Short-lived gases such as methane and some halogenated compounds exert most of their radiative influence on decadal or shorter time scales, whereas the influence of N$_2$O and some halogens persist

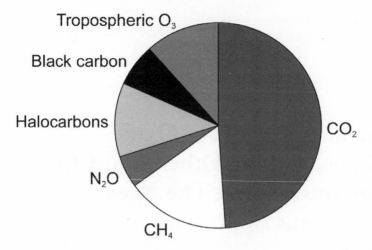

**Figure 29.1.** Estimated climate forcings (globally averaged) between 1750 and 1998 (data from Houghton et al. 2001)

for centuries, and in the case of a few fully fluorinated species such as sulfur hexafluoride ($SF_6$), for tens of millennia (Table 29.1).

The importance of the non-$CO_2$ fluxes are especially apparent when converted to C or $CO_2$-equivalents using 100-year GWPs (Table 29.2). The total anthropic flux of methane, 344 teragrams of methane per year (Tg $CH_4$ $y^{-1}$), for example, is equivalent to 2.2 petagrams of carbon equivalents per year ($PgC_{equiv}$ $y^{-1}$) and that of $N_2O$ is equivalent to 1.0 $PgC_{equiv}$. Together these fluxes are similar to the net annual loading of $CO_2$ to the atmosphere (3.2 PgC; Sabine et al., Chapter 2, this volume).

Until recently, plans for mitigating the buildup of greenhouse gases in the atmosphere have focused primarily on $CO_2$. Strategies include both reducing $CO_2$ emissions and capturing emitted $CO_2$ in biological and other sinks (Caldeira et al., Chapter 5, this volume). More recently the mitigation focus has shifted and grown to include the non-$CO_2$ gases. There is growing recognition that many of the sources of non-$CO_2$ gases are manageable and that small changes in flux can be important and have long-term impact. In one analysis, Hansen et al. (2000) argue that the rapid, observable climate warming of recent decades has been driven mainly by the non-$CO_2$ gases, owing to the offsetting climate forcings of $CO_2$ and aerosols from fossil-fuel burning. They suggest that mitigating the non-$CO_2$ gas fluxes, together with reductions in black carbon emissions (Chameides and Bergin 2002), could lead to a faster decline in the rate of global warming than similar reductions in fossil-fuel burning. Although still controversial (Hansen 2002; Schneider 2002; Wuebbles 2002), the argument that non-$CO_2$ GHGs deserve greater attention is well taken; clearly a combination of mitigation strategies (sensu Hoffert et al. 2002) is warranted.

**Table 29.1.** Global warming potentials (mass basis) for major greenhouse gases over different time periods

| Gas | Atmospheric lifetime (years) | Global warming potential by time horizon | | |
|---|---|---|---|---|
| | | *20 years* | *100 years* | *500 years* |
| $CO_2$ | | 1 | 1 | 1 |
| $CH_4$ | 12 | 62 | 23 | 7 |
| $N_2O$ | 114 | 275 | 296 | 256 |
| HFCs | 1.4–260 | 40–9,400 | 12–12,000 | 4–10,000 |
| PFCs | 2,600–50,000 | 3,900–8,000 | 5,700–11,900 | 8,900–18,000 |
| $SF_6$ | 3,200 | 15,100 | 22,200 | 32,400 |

*Source:* Houghton et al. (2001).

The impact of non-$CO_2$ GHG fluxes on radiative forcing also provides a need for understanding how these fluxes will be affected by $CO_2$ stabilization efforts and by climate change itself. Gases of primarily biological origin—$CH_4$ and $N_2O$—are particularly vulnerable to impact. For example, using nitrogen fertilizer to accelerate carbon capture by crops ignores the potential for the additional fertilizer's stimulating $N_2O$ and $NO_x$ production and thereby nullifying most or all of the benefit of additional $CO_2$ capture and soil C storage (Robertson et al. 2000). The mitigation potential of N fertilizer used to increase productivity is further reduced by the $CO_2$ costs of its manufacture (Schlesinger 1999). Synergistic interactions are also possible, however. For example, planting cover crops in an annual crop rotation or immediately following harvest of an energy crop such as poplar can both stimulate $CO_2$ capture and reduce soil nitrogen levels and consequent $N_2O$ production. Understanding the interactions between $CO_2$ mitigation strategies and fluxes of the non-$CO_2$ greenhouse gases is crucial for effective $CO_2$ stabilization policy.

With one possible exception, there are no significant manageable sinks for the non-$CO_2$ GHGs. The exception, described later in this chapter, is methane oxidation by bacteria in soils abandoned from agriculture. While other sinks are theoretically possible—$N_2O$ reduction to $N_2$ also occurs in soil and aquatic habitats, and $NO_x$ can be captured by plant canopies—none are known to be particularly manageable or significant vis-à-vis their respective sources. Unlike $CO_2$, then, mitigation of the non-$CO_2$ gases depends almost entirely on source attenuation—management intended to reduce the source strength of a given gas. For some gases, notably those with an industrial source such as the halogens, source strength attenuation appears to be working well: the growth rate of the two principal CFCs is near zero and will shortly be negative, owing to Montreal Protocol restrictions on production (Fraser and Prather 1999). For non-$CO_2$ gases of mainly biologic origin, however, attenuation efforts are untested.

**Table 29.2.** Annual fluxes of $CH_4$ and $N_2O$ in carbon-equivalent units (100-year GWP time horizons[a])

| Gas/source | Mass flux $(Tg\ y^{-1})$ | $CO_2$ –equivalent flux $(Pg\ CO_{2-equiv.}y^{-1})$ | C-equivalent flux $(PgC_{equiv}\ y^{-1})$ |
|---|---|---|---|
| *Methane ($CH_4$)* | | | |
| Industry | 162 | 3.73 | 1.02 |
|   Energy sector | 101 | 2.32 | 0.63 |
|   Industrial combustion | 15 | 0.35 | 0.09 |
|   Landfills | 46 | 1.06 | 0.29 |
| Agriculture | 182 | 4.19 | 1.14 |
|   Enteric fermentation | 94 | 2.16 | 0.59 |
|   Rice cultivation | 40 | 0.92 | 0.25 |
|   Biomass burning | 34 | 0.78 | 0.21 |
|   Animal waste treatment | 14 | 0.32 | 0.09 |
| Total $CH_4$ flux | 598 | 13.8 | 3.8 |
|   Anthropic $CH_4$ | 344 | 7.9 | 2.2 |
|   Non-anthropic $CH_4$ | 254 | 5.8 | 1.6 |
| *Nitrous oxide ($N_2O$-N)* | | | |
| Industry | 1.3 | 0.61 | 0.17 |
| Agriculture | 6.8 | 3.16 | 0.86 |
|   Soils | 4.2 | 1.95 | 0.53 |
|   Animal waste treatment | 2.1 | 0.98 | 0.27 |
|   Biomass burning | 0.5 | 0.23 | 0.06 |
| Total $N_2O$-N | 17.7 | 8.2 | 2.2 |
|   Anthropic $N_2O$-N | 8.1 | 3.8 | 1.0 |
|   Non-anthropic $N_2O$-N | 9.6 | 4.5 | 1.2 |

*Sources:* Mosier et al. (1998), Sass et al. (1999), Houghton et al. (2001).

[a] See Table 29.1

## Points of Intersection: Synergies and Liabilities

The intersection between $CO_2$ mitigation and the attenuation of non-$CO_2$ GHGs provides a rich opportunity for exploiting existing synergies. There are $CO_2$ stabilization strategies that promote GHG abatement that are important to leverage. But there are also strategies that promote non-$CO_2$ GHG emissions, and it is equally important to avoid these liabilities.

### Industry and Transportation

Of the non-$CO_2$ GHGs, halocarbons are unique in their exclusively industrial origin. They are, however, little affected by $CO_2$ stabilization technologies. From manufacturing sources—mainly from aluminum smelting, semiconductor production, as replacements for CFCs in refrigerant applications, and from electrical switching equipment (Metz et al. 2001)—there are few direct points of intersection with either fossil-fuel use or engineered sinks for CO$_2$. Rather, many halogen fluxes intersect with efforts to reduce the production of the stratospheric ozone-depleting compounds (chloro- and bromo-carbons) and provide good examples of potential mitigation synergies in general.

Industrial sources of nitrous oxide—mainly adipic and nitric acid production—likewise offer few opportunities for CO$_2$ stabilization synergies and in any case are declining rapidly because of voluntary abatement (Reimer et al. 2000; Metz et al. 2001). Fossil-fuel and, to a lesser extent, vehicle combustion represent industrial sources of nitrous oxide that are reduced by concomitant reductions in fuel use owing to CO$_2$ mitigation—a synergistic effect—but these sources represent <5 percent of total global N$_2$O sources, a small savings.

Of greater importance could be the effects of reduced combustion on NO$_x$ fluxes. These and other ozone precursors (CO and the nonmethane volatile organic compounds [NMVOCs]) exert a greenhouse effect indirectly by affecting concentrations of tropospheric ozone; NO$_x$ is additionally a precursor of nitric acid. Nitric acid is a major constituent of acid rain, which adds nitrate to extensive areas of the Earth's surface at about 25 percent of the global rate of agricultural N fertilizer addition (85 TgN y$^{-1}$). The fate of this deposited N is unknown, but it is likely that some is denitrified by soil bacteria to N$_2$O and thus is a potential contributor to the 6 TgN$_2$O-N y$^{-1}$ now estimated to be emitted by natural forests and grasslands (Nevison et al. 1996).

Industrial methane sources include emissions from landfills, oil and natural gas processing and transport, coal mining, and wastewater treatment. Methane in landfills is produced by bacteria decomposing organic waste anaerobically. Worldwide, landfills contribute around 40 TgCH$_4$ y$^{-1}$ to the troposphere; about twice this amount is contributed by the energy sector (Houghton et al. 2001). Together these sources represent only a modest portion of total global CH$_4$ emissions—about 17 percent of the 600 Tg annual CH$_4$ flux—but both of these sources exceed methane's current atmospheric loading rate of 14 TgCH$_4$ y$^{-1}$. Thus, relatively minor CO$_2$ stabilization strategies, such as biogas energy capture and improvements in pipeline and energy production efficiencies, could significantly affect atmospheric CH$_4$ concentrations. In the United States CH$_4$ emissions from all industrial sources appear to have declined by 7 percent since 1990 (EPA 2003) owing to the economic value of methane recovery and government incentive and regulatory programs.

## Agriculture

Globally, agriculture is responsible for >20 percent of anthropic greenhouse gas emissions; the IPCC (Houghton et al. 2001) estimates that agricultural activities emit 21–25 percent of all anthropic CO$_2$ fluxes, 55–60 percent of total CH$_4$ emissions, and 65–80 percent of total N$_2$O fluxes. Carbon dioxide emissions are from deforestation and fossil-fuel use; $CH_4$ is from enteric fermentation, rice cultivation, biomass burning, and animal wastes; and $N_2O$ is from cultivated soils, animal wastes, and biomass burning. The magnitude of these fluxes and their sensitivity to management makes agriculture an attractive part of many $CO_2$ stabilization schemes, and because most of these fluxes are interdependent, there are numerous opportunities to exploit synergies; at the same time, there are numerous GHG liabilities to avoid.

## $CO_2$ stabilization strategies for agriculture include

- gains in energy efficiency from improvements in the fuel efficiency of farm machinery, irrigation scheduling, and other farm operations that consume fuel;
- carbon sequestration in soil from changes in tillage, changes in crop residue and animal waste management, and changes in the use of cover crops, fallow periods, and other aspects of crop rotation management;
- the production of biofuels and the emergence of bio-based materials technology to offset the use of fossil fuels for energy production and industrial feedstocks; and
- continued gains in the production or yield efficiencies for grain, livestock, and other agricultural products to defray the need to otherwise open new land for agricultural development and consequent carbon loss.

Each of these strategies affects fluxes of $CH_4$ and $N_2O$, sometimes in indirect and unforeseen ways. Examination of the anthropogenic budgets of $CH_4$ and $N_2O$ (Figure 29.2) provides an indication of likely points of intersection; most occur within three types of production systems: animal industry, lowland rice systems, and upland field crops.

### Animal Industry

Animal production is directly responsible for about 31 percent of all anthropic $CH_4$ emissions and about 26 percent of the anthropic $N_2O$ flux. Most of the $CH_4$ flux in the animal industry is from enteric fermentation—methane generated by anaerobic bacteria in cattle and sheep rumen. The rate of $CH_4$ production is nutritional and is thus tightly linked to physiological efficiency: as feed efficiency in developed countries has increased substantially in recent decades, leading to $CO_2$ savings in production and transportation, methane emissions have correspondingly decreased.

Methane production robs the animal of energy that would otherwise be available for the production of meat, milk, or power; there is thus a financial incentive—based on energy and feed costs—to limit methane emissions. Dietary supplements such as lipids, starches, and ionophores (antibiotic feed supplements) are widely used in feedlots and dairies in the United States to improve feed conversion efficiencies, as are hormones such as bST, and all are known to reduce methane emissions by 8–30 percent per kilogram

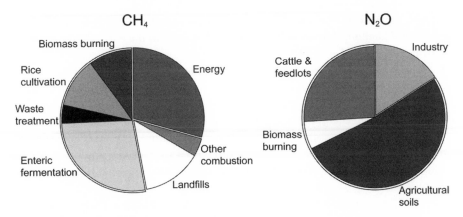

**Figure 29.2.** Estimated global budgets of the anthropic sources of $CH_4$ ($Tg\ y^{-1}$) and $N_2O$ ($Tg\ N\ y^{-1}$). The total annual flux of anthropic $CH_4$ is 344 Tg $CH_4$ or 2.2 PgC equivalent (based on a 100-year time horizon); the total annual flux of anthropic $N_2O$ is 8.1 Tg $N_2O$-N or 1.0 PgC equivalent (from Mosier et al. 1998; Kroeze et al. 1999; Sass et al. 1999; Houghton et al. 2001). Not included are fluxes from deforestation (see text). Agricultural sources are grouped on the left of each circle.

(kg) of feed. Public resistance—including legislation in parts of Europe—may make this approach difficult to implement elsewhere, however.

Methane production by grazed animals appears more difficult to abate. Production efficiencies from intensive grazing management, which can in theory lead to better utilization of deforested landscapes and thus indirectly contribute to $CO_2$ stabilization, has not been shown to reduce $CH_4$ fluxes, although stocking tannin-rich forage has shown promise in New Zealand (Clark et al. 2001). Specific anti-methanogenic vaccines are now under development and may be appropriate for both intensive and extensive grazing systems. Costs, however, will greatly affect their deployment, because they will not necessarily boost feed or forage efficiency. Public resistance may be an additional impediment to adoption.

Animal waste management is responsible for a significant portion of the global $CH_4$ and $N_2O$ flux. Methanogenesis occurs when waste is stored in anaerobic lagoons or in compost mounds, as occurs in most confined animal feeding operations. Nitrous oxide is produced when the nitrogen in waste undergoes nitrification and denitrification. Two strategies related to $CO_2$ stabilization can, however, largely abate both sources of these GHGs (CAST 2004). The first strategy is to store waste in capped lagoons or move it to a centralized storage tank that will allow produced methane to be captured and used as an energy source. Such systems are technically feasible but not widely deployed. By keeping the waste anaerobic, any $N_2O$ that is formed from residual nitrate (new nitrate production is inhibited by anaerobic conditions) will be further reduced to $N_2$. Once digested, the waste can be spread back onto fields as nitrogen-rich compost, contributing to soil C buildup, although there is a risk that some of the nitrogen added will be subsequently emitted as $N_2O$.

The second strategy is to immediately spread the waste onto a cropped field or pasture. By keeping the waste aerobic and applying it only when a crop is present, methane production will be minimal and available nitrogen will offset the need for synthetic fertilizer with its high energy and $CO_2$ cost. Additionally the waste can contribute to soil C accretion. This second strategy is less suitable for intensive centralized confined animal feeding operations, however, owing to the need to transport waste to dispersed fields, and storage will be needed during periods when crop growth is nil, such as wintertime.

## Rice Production

Lowland rice accounts for about 12 percent of the anthropic $CH_4$ budget (Figure 29.2); some $40\ TgCH_4\ y^{-1}$ are emitted by methanogenic bacteria in submerged rice soils (Sass et al. 1999; Houghton et al. 2001). Much of the methane produced is dependent on the rice plant, which provides both substrate (fixed carbon) to the methanogens and a physical conduit to the surface. Like most macrophytes, rice lacks leaf stomates; thus some portion of the methane produced in the rhizosphere is transmitted directly to the atmosphere via aerenchyma in the rice plant. Managing the rice crop thus provides an indirect means to mitigate $CH_4$ emissions, and recent work (Corton et al. 2000; van der Gon et al. 2002) has shown a remarkable inverse relationship between $CH_4$ production and rice yield (Figure 29.3). Well-managed, high-yielding rice crops have substantially lower $CH_4$ emissions owing to photosynthate-partitioning: higher-yielding plants allocate more C to grain and less C to the rhizosphere where it can undergo methanogenesis. This is an important synergy not yet exploited as increasing rice yields also contributes to $CO_2$ stabilization by reducing pressure to expand agricultural areas.

Other means for managing $CH_4$ production in rice fields includes residue management and irrigation scheduling. Incorporating crop litter early in the season significantly diminishes $CH_4$ release, as does pre-fermentation of added compost (Wassmann et al. 1993); in both cases less labile C is available to methanogens after soil flooding. Likewise, mid-season drainage or alternate flooding and drying markedly decreases the seasonal methane flux; in a recent Chinese study (Lu et al. 2000) a five-day, mid-season drainage reduced seasonal $CH_4$ loss by almost 50 percent.

Rice cultivation also emits $N_2O$, but because the soil is anoxic during much of the year and nitrification is consequently low, methane dominates the total greenhouse gas flux. In the few studies of total GHG emission in lowland rice (e.g., Abao et al. 2000), $N_2O$ emissions are a small proportion of the total GWP-corrected flux. Where $CH_4$ emissions are mitigated, however (by adjusting organic matter inputs and the timing and duration of irrigation, for example), $N_2O$ fluxes can dominate and are of a magnitude

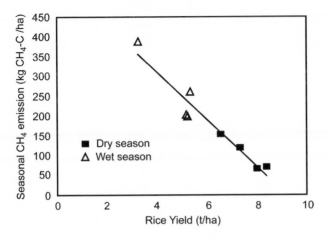

**Figure 29.3.** Seasonal CH$_4$ emission in lowland rice as a function of grain yield (from van der Gon et al. 2002, after Corton et al. 2000)

similar to those in other annual cropping systems (Bronson et al. 1997; Wassmann et al. 2003).

## Row-Crop Agriculture

Nitrous oxide from agricultural soils accounts for >50 percent of the global anthropic N$_2$O flux. N$_2$O is formed during nitrification, the microbial oxidation of NH$_4^+$ to NO$_3^-$, and during denitrification, the microbial reduction of NO$_3^-$ to N$_2$O and then N$_2$. Nitrifiers are especially active in well-aerated soil with available NH$_4^+$, and denitrifiers in poorly drained soils with available C and NO$_3^-$. Denitrifiers are also active in well-aerated soils, particularly following rain events and in anaerobic or partially anaerobic microsites such as the interior of soil aggregates (Robertson 2000).

Nitrous oxide fluxes are highest where inorganic nitrogen is readily available. As a percentage of total nitrogen inputs, N$_2$O flux appears to be relatively low; for example, the IPCC national inventory methodology calls for calculating total flux as 1.25 percent of N inputs from fertilizer, legumes, compost, and crop residue (Metz et al. 2001). Because modern crop yields are heavily dependent on high nitrogen availability, however, on an absolute basis agricultural N$_2$O fluxes are very high even where inputs of synthetic fertilizer are low.

Efforts to abate N$_2$O fluxes in agricultural soils have met with limited success. Specific inhibitors of nitrification (such as nitrapyrin, nitrogen dicyandiamide, and CaC$_2$) are expensive and work inconsistently. The effects of specific types of fertilizers can be equally inconsistent. Probably the most promise for mitigating field crop N$_2$O production comes from increasing the efficiency of crop N use by manipulating the magnitude, placement, and timing of nitrogen fertilizer. Nitrogen taken up by the crop is not available for microbial uptake; ergo, $N_2O$ production will be low. Effecting this synchrony in most annual cropping systems without affecting crop yields is technically feasible but difficult, and there is little economic payoff with today's low fertilizer cost. For irrigated crops and high-value perennial crops, however, chemigation (the addition of nitrogen and other chemical inputs in irrigation water) and other fertilizer technologies are less difficult and can markedly abate $N_2O$ flux (e.g., Matson et al. 1996).

Because carbon and nitrogen cycling are so tightly linked in most soils, $CO_2$ stabilization strategies that affect soil C will likely also affect nitrogen cycling and thus $N_2O$ flux. Cover crops that are planted to stabilize or build soil carbon and crop residues that are left in place rather than taken as secondary harvest will potentially immobilize soil nitrogen otherwise available to nitrifiers and denitrifiers. Thus efforts to stabilize $CO_2$ by sequestering carbon in soil may have synergistic effects on $N_2O$ flux, so long as the soil organic matter in which the nitrogen is stored is not mobilized at a later time, when plant nitrogen uptake is low and $N_2O$ emission potentials high. This potential has not yet been demonstrated experimentally, however, and effects will likely vary by sequestration method. No-till cultivation, for example, does not in theory change nitrogen availability in soil, but the effects of no-till cultivation on $N_2O$ flux are widely variable, and most likely reflect site-specific responses to simultaneous changes in soil aggregate structure, water-filled pore space, and carbon availability (Robertson et al. 2000; Dobbie and Smith 2002). The complexity of this response, coupled with the extreme spatial and temporal variability of soil $N_2O$ fluxes, argues for the development and testing of effective $N_2O$ models (e.g., Li et al. 1997; Del Grosso et al. 2002).

Although theoretically possible, soil methane oxidation is not known to be promoted by any existing agronomic practice. Conversion of forest and savanna soils to agricultural production reduces soil $CH_4$ oxidation by 80–90 percent (Mosier et al. 1991; Smith et al. 2000). To date, the only documented recovery of a lost $CH_4$ oxidation capacity is after decades of secondary succession and reforestation (Robertson et al. 2000).

## Deforestation and Other Land Use Change

Nitrous oxide fluxes increase following deforestation. Where cessation of deforestation is used to mitigate $CO_2$ emissions, $N_2O$ emissions will be similarly abated. High emissions persist for only a few years, however, especially under pasture conversion. And since nitrogen gas fluxes prior to deforestation are likely to be higher than subsequent fluxes in either pasture or secondary succession (Robertson and Tiedje 1988; Keller et al. 1993), over the long term, accelerated fluxes during the several years post-clearing may very well be mitigated by the following decades of depressed fluxes.

Likewise, wetland drainage may substantially reduce $CH_4$ fluxes at the same time

**Table 29.3.** Sources of greenhouse gas flux in agricultural and forest systems at a U.S. Midwest site

| | Flux (g $CO_2$-equivalents $m^{-2} y^{-1}$) | | | | | | |
|---|---|---|---|---|---|---|---|
| | | $CO_2$ | | | | | |
| Ecosystem management | Soil C | N fertilizer | Lime | Fuel | $N_2O$ | $CH_4$ | Total net |
| Annual crop | | | | | | | |
| Conventional tillage | 0 | 27 | 23 | 6 | 2 | 4 | 114 |
| No-till | −110 | 27 | 34 | 12 | 6 | −5 | 14 |
| Organic | −29 | 0 | 0 | 9 | 56 | −5 | 41 |
| Energy crop (poplar) | −117 | 5 | 0 | 2 | 0 | −5 | −105 |
| Native forest | 0 | 0 | 0 | 0 | 1 | −25 | −4 |

*Source:* Robertson et al. (2000)

that it causes the emission of large quantities of $CO_2$ from newly aerobic soil organic matter (and possibly large quantities of $N_2O$). In both of these cases, a long-term, total GHG analysis is needed to evaluate the full effects of land conversion.

## The Need for a Systems Approach

The difficulty of assessing the effects of $CO_2$ stabilization strategies on other parts of a system, and especially on the non-$CO_2$ GHG fluxes, argues for a systems approach to their implementation. This is especially true in ecosystem settings, where indirect effects can cause counterintuitive, unintentional, and possibly unwanted flux changes. A systems approach can also illuminate synergies not otherwise recognized.

In a U.S. Midwest corn-soybean-wheat rotation, for example, soil carbon sequestration following no-till implementation almost completely mitigated the GHG cost of the cropping system (Table 29.3). Systems analysis showed that substantially more mitigation could potentially be achieved were the system managed to reduce $N_2O$ flux and minimize synthetic fertilizer and agricultural lime use. Similar potentials have been suggested for sugar cane (Weier 1998), subtropical wheat (Robertson and Grace 2003), and rice (Wassmann et al. 2003) ecosystems.

Fluxes following land use change can be similarly complex, as noted in the preceding section. Analyses of these systems at a watershed or larger scale could inform landscape management decisions to optimize $CO_2$ stabilization. In many landscapes it is likely that the value of lower non-$CO_2$ GHG fluxes will exceed the value of $CO_2$ mitigation over time, especially for $CO_2$ strategies that employ sinks that will eventually saturate, such as trees and soil (Caldeira et al., Chapter 5, and Smith, Chapter 28, both this volume). Even in the absence of $CO_2$ stabilization per se, including the non-$CO_2$ gases in mitigation efforts will be highly worthwhile. For many non-$CO_2$ GHG sources, abatement technology is currently available, and for a number of other sources, new mitigation technology awaits a modest investment in research and development.

# Literature Cited

Abao, E. B., K. F. Bronson, R. Wassmann, and U. Singh. 2000. Simultaneous records of methane and nitrous oxide emissions in rice-based cropping systems under rainfed conditions. *Nutrient Cycling in Agroecosystems* 58:131–139.

Bronson, K. F., H. U. Neue, and E. B. Abao Jr. 1997. Automated chamber measurements of methane and nitrous oxide flux in a flooded rice soil. I. Residue, nitrogen, and water management. *Soil Science Society of America Journal* 61:981–987.

CAST (Council on Agriculture Science and Technology). 2004. *Agriculture's response to global climate change*. Ames, IA.

Chameides, W. L., and M. Bergin. 2002. Climate change: Soot takes center stage. *Science* 297:2214–2215.

Clark, H., C. de Klein, and P. Newton. 2001. *Potential management practices and technologies to reduce nitrous oxide, methane, and carbon dioxide emissions from New Zealand agriculture*. Wellington, New Zealand: Ministry of Agriculture and Forestry.

Corton, T. M., J. B. Bajita, F. S. Grospe, R. R. Pamplona, C. A. Assis Jr., R. Wassmann, R. S. Lantin, and L. V. Buendia. 2000. Methane emission from irrigated and intensively managed rice fields in Central Luzon (Philippines). *Nutrient Cycling in Agroecosystems* 58:37–53.

Del Grosso, S. J., D. Ojima, W. Parton, A. Mosier, G. Peterson, and D. Schimel. 2002. Simulated effects of dryland cropping intensification on soil organic matter and greenhouse gas exchanges using the DAYCENT ecosystem model. *Environmental Pollution* 116:S75–83.

Dobbie, K. E., and K. A. Smith. 2002. Nitrous oxide emission factors for agricultural soils in Great Britain: The impact of soil water-filled pore space and other controlling variables. *Global Change Biology* 9:1–15.

EPA (Environmental Protection Agency). 2003. *Inventory of U.S. greenhouse gas emissions and sinks: 1990–2001*. Washington, DC: U.S. Environmental Protection Agency.

Fraser, P. J., and M. J. Prather. 1999. Atmospheric chemistry: Uncertain road to ozone recovery. *Nature* 398:663–664.

Hansen, J. E. 2002. A brighter future. *Climatic Change* 52:435–440.

Hansen, J., M. Sato, R. Ruedy, A. Lacis, and V. Oinas. 2000. Global warming in the twenty-first century: An alternative scenario. *Proceedings of the National Academy of Sciences* 97:9875–9880.

Hoffert, M. I., K. Caldeira, and G. Benford. 2002. Advanced technology paths to global climate stability: Energy for a greenhouse planet. *Science* 298:981–987.

Houghton, J. T., Y. Ding, D. J. Griggs, M. Noguer, P. J. van der Linden, X. Dai, K. Maskell, and C. A. Johnson, eds. 2001. *Climate change 2001: The scientific basis (Contribution of Working Group I to the third assessment report of the Intergovernmental Panel on Climate Change)*. Cambridge: Cambridge University Press.

Keller, M., E. Veldkamp, A. M. Weitz, and W. A. Reiners. 1993. Effect of pasture age on soil trace-gas emissions from a deforested area of Costa Rica. *Nature* 365:244–246.

Kroeze, C., A. Mosier, and L. Bouwman. 1999. Closing the global N$_2$O budget: A retrospective analysis 1500–1994. *Global Biogeochemical Cycles* 13:1–8.

Li, C., S. Frolking, G. J. Crocker, P. R. Grace, J. Klir, M. Korchens, and P. R. Poulton. 1997. Simulating trends in soil organic carbon in long-term experiments using the DNDC model. *Geoderma* 81:45–60.

Lu, W. F., W. Chen, and B. W. Duan. 2000. Methane emissions and mitigation options in irrigated rice fields in southeast China. *Nutrient Cycling in Agroecosystems* 58:65–73.

Matson, P. A., C. Billow, S. Hall, and J. Zachariassen. 1996. Fertilization practices and soil variations control nitrogen oxide emissions from tropical sugar cane. *Journal of Geophysical Research* 101:18,533–18,545.

Metz, B., O. Davidson, R. Swart, and J. Pan, eds. 2001. *Climate change 2001: Mitigation (Contribution of Working Group II to the third assessment report of the Intergovernmental Panel on Climate Change).* Cambridge: Cambridge University Press.

Mosier, A., D. Schimel, D. Valentine, K. Bronson, and W. Parton. 1991. Methane and nitrous oxide fluxes in native, fertilized and cultivated grasslands. *Nature* 350:330–332.

Mosier, A., C. Kroeze, C. Nevison, O. Oenema, S. Seitzinger, and O. van Cleemput. 1998. Closing the global N$_2$O budget: Nitrous oxide emissions through the agricultural nitrogen cycle. *Nutrient Cycling in Agroecosystems* 52:225–248.

Nevison, C. D., G. Esser, and E. A. Holland. 1996. A global model of changing N$_2$O emissions from natural and perturbed soils. *Climatic Change* 32:327–378.

Reimer, R. A., C. S. Slaten, M. Sepan, T. A. Koch, and V. G. Triner. 2000. Adipic acid industry: N$_2$O abatement implementation of technologies for abatement of N$_2$O emissions associated with adipic acid manufacture. In *Non-CO$_2$ greenhouse gases: Scientific understanding*, edited by J. van Ham, A. P. M. Baede, L. A. Meyer, and R. Ybema. Dordrecht, the Netherlands: Kluwer.

Robertson, G. P. 2000. Denitrification. Pp. C181–190 in *Handbook of Soil Science*, edited by M. E. Sumner. Boca Raton, FL: CRC Press.

Robertson, G. P., and P. R. Grace. 2004. Greenhouse gas fluxes in tropical and temperate agriculture: The need for a full-cost accounting of global warming potentials. *Environment, Development and Sustainability* 6:51–63.

Robertson, G. P., and J. M. Tiedje. 1988. Denitrification in a humid tropical rainforest. *Nature* 336:756–759.

Robertson, G. P., E. A. Paul, and R. R. Harwood. 2000. Greenhouse gases in intensive agriculture: Contributions of individual gases to the radiative forcing of the atmosphere. *Science* 289:1922–1925.

Sass, R. L., F. M. Fisher, and A. Ding. 1999. Exchange of methane from rice fields: National, regional, and global budgets. *Journal of Geophysical Research-Atmospheres* 104:26943–26951.

Schlesinger, W. H. 1999. Carbon sequestration in soils. *Science* 284:2095–2097.

Schneider, S. H. 2002. Can we estimate the likelihood of climatic changes at 2100? *Climatic Change* 52:441–451.

Smith, K. A., K. E. Dobbie, B. C. Ball, L. R. Bakken, B. K. Situala, S. Hansen, and R. Brumme. 2000. Oxidation of atmospheric methane in Northern European soils, comparison with other ecosystems, and uncertainties in the global terrestrial sink. *Global Change Biology* 6:791–803.

van der Gon, H. A. C. D., M. J. Kropff, N. van Breemen, R. Wassmann, R. S. Lantin, E. Aduna, T. M. Corton, and H. H. van Laar. 2002. Optimizing grain yields reduces $CH_4$ emissions from rice paddy fields. *PNAS* 19:12021–12024.

Wassmann, R., H. Papen, and H. Rennenberg. 1993. Methane emission from rice paddies and possible mitigation strategies. *Chemosphere* 26:201–217.

Wassmann, R., H. U. Neue, J. K. Ladha, and M. S. Aulakh. 2004. Mitigating greenhouse gas emissions from rice-wheat cropping systems in Asia. *Environment, Development and Sustainability* 6:65–90.

Weier, K. L. 1998. Sugarcane fields: sources or sinks for greenhouse gas emissions. *Australian Journal of Agriculture Research* 49:1–9.

Wuebbles, D. J. 2002. Oversimplifying the greenhouse. *Climatic Change* 52:431–434.

# List of Contributors

**Paulo Artaxo**
Universidade de São Paulo
Instituto de Fisica USP
Rua do Matao, Travessa R, 187
São Paulo, Brazil
artaxo@if.usp.br

**Dorothee C. E. Bakker**
School of Environmental Sciences
University of East Anglia
Norwich NR4T7J, United Kingdom
T 44 1603 592648
d.bakker@uea.ac.uk

**Dennis Baldocchi**
Ecosystem Science Division
151 Hilgard Hall, University of
    California–Berkeley
Berkeley, CA 94720, USA
T 1 510 6422874
baldocchi@nature.berkeley.edu

**Peter G. Brewer**
Monterey Bay Aquarium Research
    Institute
7700 Sandholdt
Road Moss Landing
Monterey, CA 95039, USA
T 1 831 7751706
brpe@mbari.org

**Ken Caldeira**
Climate and Carbon Cycle Modeling
    Group (CCCM), Atmospheric Science
    Division
P.O. Box 808, L. 103
Livermore, CA 94551-0808, USA
T 1 925 4234191
kenc@llnl.gov

**Josep G. Canadell**
GCP Project Office
CSIRO Land and Water, Pye Laboratory
P.O. Box 1666
Canberra, ACT 2601, Australia
T 61 2 62465631
Pep.Canadell@csiro.au

**Chen-Tung Arthur Chen**
National Sun Yat-sen University
Institute of Marine Geology and
    Chemistry
Kaohsiung 80424, Taiwan
T 886 7 5255146
ctchen@mail.nsysu.edu.tw

**Philippe Ciais**
Laboratoire des Sciences du Climat
UMR CEA-CNRS Bât. 709 CE
L'Orme des Merisiers
F-91191 Gif-sur-Yvette, France
T 33 169087121
ciais@lsce.saclay.cea.fr

**Jae Edmonds**
Pacific Northwest National Laboratory
Joint Global Change Research Inst. at
  University of Maryland
8400 Baltimore Avenue, Suite 201
College Park, MD 20740-2496, USA
T 1 301 3146749
jae@pnl.gov

**JingYun Fang**
Department of Ecology
College of Environmental Sciences
Peking University
Beijing 1000871, China
T 86 10 62765578
jyfang@urban.pku.edu.cn

**Christopher B. Field**
Department of Global Ecology
Carnegie Institution of Washington
260 Panama Street
Stanford, CA 94305, USA
T 1 650 3251521 x 213
cfield@globalecology.stanford.edu

**Jonathan A. Foley**
CPEP - Institute for Environmental
  Studies
University of Wisconsin
1225 West Dayton Street
Madison, WI 53706, USA
T 1 608 2655144
jfoley@facstaff.wisc.edu

**Pierre Friedlingstein**
LSCE Unité mixte CEA-CNRS
CE-Saclay - Bât. 709
L'Orme des Merisiers
F-91191 Gif-sur-Yvette, France
T 33 1 69088730
Pierre@lsce.saclay.cea.fr

**Manuel Gloor**
Max-Planck-Institute for
  Biogeochemistry
PF 100164, D-07701
Jena, Germany
mgloor@bgc-jena.mpg.de

**Jeffery B. Greenblatt**
AOS Program, Princeton University
Sayre Hall, Forrestal Campus
P.O. Box CN710
Princeton, NJ 08544-0710, USA
T 1 609 258 2462
buygreen@splash.princeton.edu

**Nicolas Gruber**
IGPP and Department of Atmospheric
  Sciences
University of California, Los Angeles
5853 Slichter Hall
Los Angeles, CA 90095-4996, USA
T 1 310 8254772
ngruber@igpp.ucla.edu

**Martin Heimann**
Max-Planck-Institut für Biogeochemie
Postfach 10 01 64
07701 Jena, Germany
T 49 3461 576350
martin.heimann@bgc-jena.mpg.de

**Fortunat Joos**
Climate and Environmental Physics
Sidlerstr. 5
CH-3012 Bern, Switzerland
T 41 31 6314461
joos@climate.unibe.ch

**Louis Lebel**
Unit for Social and Environmental
  Change (USER)
Faculty of Social Sciences
Chiang Mai University
Chiang Mai 50200, Thailand
llebel@loxinfo.co.th

**Corinne Le Quéré**
Max-Planck-Institut für Biogeochemie
Postfach 10 01 64
07701 Jena, Germany
T 49 3641 576217
lequere@bgc-jena.mpg.de

**Alan S. Manne**
Department of Management Science
  and Engineering
Stanford University
Stanford, California 94305, USA
asmanne@attglobal.net

**Jerry M. Melillo**
The Ecosystems Center
Marine Biological Laboratory
7 MBL Street
Woods Hole, MA 02543, USA
T 1 508 2897494
jmelillo@mbl.edu

**Nicolas Metzl**
Laboratoire de Biogeochimie et Chimie
  Marines
CNRS/IPSL, UPMC
Case 134
4 Place Jussieu
75252 Paris Cedex 5, France
metzl@ccr.jussieu.fr

**M. Granger Morgan**
Lord Chair, Professor in Engineering
Carnegie Mellon University
Baker Hall 129, 5000 Forbes Avenue
Pittsburgh, PA 15213, USA
T 1 412 2682672
granger.morgan@andrew.cmu.edu

**Gert-Jan Nabuurs**
ALTERRA
P.O. Box 47
NL-6700 AA Wageningen
The Netherlands
T 31 317 4777897
G.J.Nabuurs@Alterra.wag-ur.nl

**Nebojsa Nakicenovic**
International Institute for Applied
  Systems Analysis (IIASA)
A-2361 Laxenburg, Austria
T 43 2236 807411
naki@iiasa.ac.at

**I. Colin Prentice**
Max Planck Institute for
  Biogeochemistry
Winzerlaer Str. 10
D-07745 Jena, Germany
cprentic@bgc-jena.mpg.de

**Ronald G. Prinn**
Department of Earth Atmospheric and
 Planetary Sciences
Massachusetts Institute of Technology
Building 54-918
Cambridge, MA 02139-4307, USA
T 1 617 2532452
rprinn@mit.edu

**Navin Ramankutty**
Center for Sustainability and the Global
 Environment (SAGE)
Gaylord Nelson Institute for
 Environmental Studies
University of Wisconsin
1710 University Avenue
Madison, WI 53726 USA
nramanku@facstaff.wisc.edu

**Michael R. Raupach**
CSIRO Earth Observation Centre
GPO Box 3023
Canberra, ACT 2601, Australia
T 61 2 62465573
Michael.Raupach@csiro.au

**Richard G. Richels**
Electric Power Research Institute
3412 Hillview Avenue
Palo Alto, CA 94304, USA
rrichels@epri.com

**Jeffrey E. Richey**
School of Oceanography
University of Washington
P.O. Box 355351
Seattle, WA 98195-7540, USA
T 1 206 5437339
jrichey@u.washington.edu

**G. Philip Robertson**
W. K. Kellogg Biological Station
Michigan State University
Hickory Corners, MI 49060, USA
T 1 269 6712267
Robertson@kbs.msu.edu

**Christian Rödenbeck**
Max-Planck-Institute for
 Biogeochemistry
PF 100164
D-07701 Jena, Germany
Christian.Roedenbeck@bgc-jena.mpg.de

**Patricia Romero Lankao**
Universidad Autonoma Metropolitana
 (UAM)
Department of Politics and Culture
Campus Xochimilco
Calzada del Hueso 1100
Villa Quietud, 04960 Mexico City,
 Mexico
Rolp7543@cueyatl.uam.mx

**Christopher L. Sabine**
NOAA/PMEL
7600 Sand Point Way NE
Seattle, WA 98115, USA
T 1 206 5264809
sabine@pmel.noaa.gov

**Maria José Sanz**
Fundacion Centro de Estudios
 Embientales del Mediterraneo
 (CEAM)
Parque Tecnologico
Calle 4, Sector Oeste
46980 Paterna, Valencia, Spain
T 34 96 1318227
mjose@ceam.es

**Jorge L. Sarmiento**
AOS Program, Princeton University
Sayre Hall, Forrestal Campus
P.O. Box CN710
Princeton, NJ 08544-0710, USA
jls@Princeton.EDU

**Jayant A. Sathaye**
International Energies Studies
Lawrence Berkeley National Laboratory
1 Cyclotron Road, MS 90-4000
Berkeley, CA 94720, USA
T 1 510 4866294
jasathaye@lbl.gov

**Ernst-Detlef Schulze**
Max-Planck Institut für Biogeochemie
Winzerlaer Str 10
07745 Jena, Germany or
PO Box 100164
07701 Jena, Germany
Tel: +49 3641 576100
dschulze@bgc-jena.mpg.de

**Géraud Servin**
Global Terrestrial Observing System
Food and Agriculture Organization of
   the United Nations
Viale delle Terme di Caracalla
Rome, 00100 Italy
Geraud.Servin@fao.org

**Pete Smith**
Department of Plant and Soil Science
University of Aberdeen
Cruikshank Building
Aberdeen AB24 3UU, Scotland, United
   Kingdom
T 44 1224 272702
pete.smith@abdn.ac.uk

**John W. B. Stewart**
118 Epron Road
Salt Spring Island, BC V8K 1C7,
   Canada
T& F 1 250 5379250
jwbStew@island.net

**Jeff Tschirley**
Global Terrestrial Observing System
S.D.R.N., Food and Agriculture
   Organization of the United Nations
Via delle Terme di Caracalla
00100 Rome, Italy
T 39 06 52253450
Jeff.tschirley@fao.org

**Riccardo Valentini**
Disafri, University of Tuscia
Department of Forest and Environment
   and Resources
Via San Camillo de Lellis
01100 Viterbo, Italy
T 39 0761 35734
rik@unitus.it

**Reynaldo Victoria**
Centro de Energia na Agricultura CENA
Av. Centenario 303
13400-970 Piracicaba, São Paulo, Brazil
T 55 19 34335060 or 34294611
reyna@cena.usp.br

# SCOPE Series List

SCOPE 1–59 are now out of print. Selected titles from this series can be downloaded free of charge from the SCOPE Web site (http://www.icsu-scope.org).

SCOPE 24: *Noise Pollution*, 1986, 466 pp
SCOPE 25: *Appraisal of Tests to Predict the Environmental Behaviour of Chemicals*,
1985, 400 pp
SCOPE 26: *Methods for Estimating Risks of Chemical Injury: Human and Non-Human
Biota and Ecosystems, SGOMSEC 2*, 1985, 712 pp
SCOPE 27: *Climate Impact Assessment: Studies of the Interaction of Climate and
Society*, 1985, 650 pp
SCOPE 28: *Environmental Consequences of Nuclear War*
*Volume I: Physical and Atmospheric Effects*, 1986, 400 pp
*Volume II: Ecological and Agricultural Effects*, 1985, 563 pp
SCOPE 29: *The Greenhouse Effect, Climatic Change and Ecosystems*, 1986, 574 pp
SCOPE 30: *Methods for Assessing the Effects of Mixtures of Chemicals, SGOMSEC 3*,
1987, 928 pp
SCOPE 31: *Lead, Mercury, Cadmium and Arsenic in the Environment*, 1987, 384 pp
SCOPE 32: *Land Transformation in Agriculture*, 1987, 552 pp
SCOPE 33: *Nitrogen Cycling in Coastal Marine Environments*, 1988, 478 pp
SCOPE 34: *Practitioner's Handbook on the Modelling of Dynamic Change in
Ecosystems*, 1988, 196 pp
SCOPE 35: *Scales and Global Change: Spatial and Temporal Variability in Biospheric
and Geospheric Processes*, 1988, 376 pp
SCOPE 36: *Acidification in Tropical Countries*, 1988, 424 pp
SCOPE 37: *Biological Invasions: a Global Perspective*, 1989, 528 pp
SCOPE 38: *Ecotoxicology and Climate with Special References to Hot and Cold
Climates*, 1989, 432 pp
SCOPE 39: *Evolution of the Global Biogeochemical Sulphur Cycle*, 1989, 224 pp
SCOPE 40: *Methods for Assessing and Reducing Injury from Chemical Accidents*,
*SGOMSEC 6*, 1989, 320 pp
SCOPE 41: *Short-Term Toxicity Tests for Non-genotoxic Effects, SGOMSEC 4*, 1990,
353 pp
SCOPE 42: *Biogeochemistry of Major World Rivers*, 1991, 356 pp
SCOPE 43: *Stable Isotopes: Natural and Anthropogenic Sulphur in the Environment*,
1991, 472 pp
SCOPE 44: *Introduction of Genetically Modified Organisms into the Environment*,
1990, 224 pp
SCOPE 45: *Ecosystem Experiments*, 1991, 296 pp
SCOPE 46: *Methods for Assessing Exposure of Human and Non-human Biota
SGOMSEC 5*, 1991, 448 pp
SCOPE 47: *Long-Term Ecological Research. An International Perspective*, 1991, 312 pp
SCOPE 48: *Sulphur Cycling on the Continents: Wetlands, Terrestrial Ecosystems and
Associated Water Bodies*, 1992, 345 pp

# SCOPE Executive Committee 2001–2004

*President*
Dr. Jerry M. Melillo (USA)

*1st Vice-President*
Prof. Rusong Wang (China–CAST)

*2nd Vice-President*
Prof. Bernard Goldstein (USA)

*Treasurer*
Prof. Ian Douglas (UK)

*Secretary-General*
Prof. Osvaldo Sala (Argentina–IGBP)

*Members*
Prof. Himansu Baijnath (South Africa–IUBS)
Prof. Manuwadi Hungspreugs (Thailand)
Prof. Venugopalan Ittekkot (Germany)
Prof. Holm Tiessen (Canada)
Prof. Reynaldo Victoria (Brazil)

# Index